Monographs of the Palaeontographical Society

The Palaeontographical Society was established in 1847, and is the oldest Society devoted to study of palaeontology worldwide. Its primary role is to promote the description and illustration of the British fossil flora and fauna, via publication of an authoritative monograph series. These monographs cover a wide range of taxonomic groups, from microfossils, trilobites and ammonites through to Coal Measure plants, mammals and reptiles, and from all ages from Cambrian to Pleistocene. They form a benchmark for understanding the past life of the British Isles and many include the original descriptions of numerous key species. The first monograph (on the Crag Mollusca) was published in March 1848 and the Society still continues this work today. Notable authors in the series include Charles Darwin (fossil barnacles) and Richard Owen (dinosaurs and other extinct reptiles). Beginning in 2014, the Cambridge Library Collection and the Society are collaborating to reissue the earlier publications, focusing on monographs completed between 1848 and 1918.

A Monograph on the British Fossil Echinodermata of the Oolitic Formations

Urged by his colleague Edward Forbes, Thomas Wright (1809–84) devoted himself to completing this monograph of the echinoderms ('spiny-skinned animals') of Britain's Oolitic Formations. These would be referred to as Middle Jurassic by the modern geologist. This is a notable contribution, describing as it does the echinoderms following a major stratigraphic gap. In the British Isles, apart from some minor occurrences in the Permian and Lower Jurassic, echinoderms are almost entirely absent from the Lower Carboniferous (Mississippian), a span we now know to represent 150 million years. Although common and diverse elsewhere during this interval, the British Oolitic echinoderms show many changes from those of the Mississippian. Wright's two-volume monograph includes thorough descriptions and locality details, all supported by beautiful plates. Volume 1, originally published in four parts between 1857 and 1861, considers the many and varied echinoids (sea urchins) of the Middle Jurassic.

A Monograph on the British Fossil Echinodermata of the Oolitic Formations

VOLUME 1

THOMAS WRIGHT

CAMBRIDGE
UNIVERSITY PRESS

CAMBRIDGE
UNIVERSITY PRESS

University Printing House, Cambridge, CB2 8BS, United Kingdom

Cambridge University Press is part of the University of Cambridge.

It furthers the University's mission by disseminating knowledge in the pursuit of
education, learning and research at the highest international levels of excellence.

www.cambridge.org
Information on this title: www.cambridge.org/9781108081153

This edition first published 1857–61
This digitally printed version 2015

ISBN 978-1-108-08115-3 Paperback

A MONOGRAPH

ON THE

BRITISH FOSSIL

ECHINODERMATA

OF

THE OOLITIC FORMATIONS.

BY

THOMAS WRIGHT, M.D., F.R.S.E.

PART FIRST,

CONTAINING

THE CIDARIDÆ, HEMICIDARIDÆ, AND DIADEMADÆ.

LONDON:
PRINTED FOR THE PALÆONTOGRAPHICAL SOCIETY.
1855.

J. E. ADLARD, PRINTER, BARTHOLOMEW CLOSE.

PREFACE.

In presenting the First Part of my Monograph on 'British Fossil Echinodermata' to the members of the Palæontographical Society, I deem it necessary to make a few remarks in order to explain—1st, how I came to occupy the position of an author in the magnificent volumes published by your Society; and 2dly, to state the manner in which I have endeavoured to discharge the duties of the task I have undertaken.

After the publication of my 'Memoirs on the Echinodermata of the Oolites,'[*] in the 'Annals of Natural History,' my much lamented friend, the late Professor Edward Forbes, as a member of your Council, asked me to contribute a Monograph on the same subject to the Palæontographical Society. Knowing that he had in preparation a supplementary chapter on the Echinoderms of the Great Oolite, for Messrs. Morris and Lycett's Monograph on the Mollusca of that Formation, I thanked him for the good opinion he had formed of my ability for such a work, but declined, lest, by complying with his request, I might possibly have interfered with any intentions of his own on the subject, knowing how ardently he loved all that related to this class of the Animal Kingdom. As Professor Forbes, however, on another occasion, renewed, in the most pressing manner, his solicitation, I then proposed to join him in a Monograph on the British Fossil Echinodermata of the Secondary Formations, which he at once agreed to, and the proposal for this joint work was submitted to the approval of your Council, and received its sanction.

The numerous and constantly increasing duties of my esteemed colleague at the School of Mines prevented him from taking any share in the collection of materials for the preparation of the Monograph on the Oolitic Echinodermata, and, with his usual candour, he told me that, as he was unable, from want of time, to contribute to this division of our proposed joint work, his name must be withdrawn from its title page; and, if agreeable to me, that I should undertake the Monograph on the Oolitic species, whilst he would devote himself to the description of the Cretaceous forms. After this arrangement, I directed my attention with redoubled energy to the subject of my special studies.

On his appointment to the chair of Natural History in the University of Edinburgh, Professor Forbes took with him the materials for the first part of his promised Monograph

[*] 'Annals and Magazine of Natural History,' new series, vol. viii, 1851.

on the Cretaceous Echinoderms; but, alas! his untimely and much-lamented death prevented him from even commencing that work upon which his mind had been so long and busily engaged, and which was looked forward to with so much interest by all who knew the high qualifications of my friend for his selected task. But the mysterious decrees of Providence disappointed our expectations, and at the same time deprived Natural Science, in this country, of one of its brightest ornaments and warmest advocates. It would be doing violence to my own feelings if I did not, on this occasion, record the high estimation in which I held the opinions of my distinguished colleague on all points relating to the work we had undertaken together, and the uniform deference I paid to his suggestions, as to the best mode of executing the same, which were always dictated by that kindness, frankness, and wisdom so characteristic of the man.

Having been thus thrown entirely on my own resources, before the real difficulties of the work began, I have experienced more than ever the deep responsibility of the task I have undertaken. I trust the circumstances I have narrated will entitle me to the consideration and indulgence of all who know the nature and amount of the difficulties to be grappled with in a work like that in which I am engaged, and the time and labour necessary to overcome them. I can only add, that I have spared neither time, labour, nor research, in order to make this Monograph worthy of the confidence originally reposed in me; but how far I may have succeeded in my efforts, it remains for others to decide.

At the suggestion of my excellent friend, Thomas Davidson, Esq., author of the magnificent Monograph on the Brachiopoda, and several other kind friends interested in the success of this work, it was thought advisable that, at the conclusion of my Monograph on the Oolitic species, I should proceed with the description of the Cretaceous forms, in order that a greater unity in the arrangement and management of the subject might be observed in the two Monographs on the Echinodermata of the Secondary rocks; and a proposal to this effect has been submitted to your Council, and received its sanction.

I have ventured to propose some important alterations in the classification of the Echinoidea, and have grouped the genera into thirteen natural families, many of which are entirely new. My object has been to attain a more natural method, and thereby facilitate the study of the different groups. I have given an analysis of these families at the commencement of the work, and enumerated the most common types of each.

In the description of the species, I have taken them in their stratigraphical order, always commencing with the species found in the oldest rock in which the genus is discovered, thus — a, Lias, Lower, Middle, and Upper; b, Inferior Oolite; c, Great Oolite, including Fullers-earth, Stonesfield Slate, Great Oolite, Bradford Clay, Forest Marble, and Cornbrash; d, Oxford Clay; e, Coralline Oolite, including Calcareous Grit, and Coral Rag; f, Kimmeridge Clay; g, Portland Oolite; h, Purbeck Beds;—so that my work has the double advantage of being stratigraphical and palæontological at the same time, a mode of treating the subject which I hope will prove useful and convenient to geologists.

Many of the readers of this Monograph will probably be surprised to find some old generic names reproduced, which have long been superseded by those of modern writers; but a sense of justice to such authors as Van Phelsum, Breynius, Klein, and Leske, has led me to consult their original works, and restore the genera first described and figured by them, but omitted from the treatises of later authors on the same subject. In the nomenclature of the Echinodermata, had I merely gone back to the time of Linnæus, as suggested by the committee of the British Association in their report made in 1842, I must necessarily have excluded the important work by Breynius,* in which, for the first time, were proposed seven well-described and accurately figured genera of Echinoidea, which, by some strange oversight, were not adopted by his contemporaries, although they have reappeared under new names in the works of later authors. On the principle of priority, therefore, I have restored the original genera so clearly defined by Breynius, even although it may occasion a temporary inconvenience in the names of some well-known forms of urchins.

In every case, where practicable, the name of the author who either first recorded, described, or figured the species, follows the specific name of the object, without the addition of " Sp." adopted by some authors. By this mode justice is done to the original author, and confusion avoided. The modern practice of inventing and changing generic names, and appending to the old specific name that of the individual who has merely changed a name, but discovered nothing, cannot be sufficiently discountenanced, as it greatly increases the confusion arising from an already overloaded synonymy, and thereby retards the real progress of the natural history sciences.

The accurate determination of species, and their distribution in time and space, form problems of the highest importance to the palæontologist, as their true solution are the only certain guides of the geologist in his investigations in the field, and his generalizations in the study: for the classification of strata, the subdivision of rock groups, and the boundary lines between different formations, are all points which are more or less affected by the soundness of his conclusions.

In determining the species of Echinodermata, therefore, the most careful comparison has been made with the true type forms to which they are referred, and the extent of the section in the description of the species, on the affinities and differences exhibited by each with other Foreign and British congeneric forms, will show how much care has been taken to arrive at a correct determination.

The range and stratigraphical position of the species described in this work has occupied much time and attention, as many errors found in previous lists of Oolitic Echinodermata required considerable research to correct; for experience has taught me that, unless the palæontologist can verify for himself the statements of his collectors, he will frequently be led into similar errors. In every instance, with the exception of the Northamptonshire beds, which have been carefully noted by my friend the

* De Echinis et Echinitis, sive Methodica Echinorum distributione, Schediasma. Gedani, 1732.

Rev. A. W. Griesbach, I have visited the different localities given in this work, and with my own hammer ascertained the presence of the species in the rock whence they are stated to be obtained; the most perfect confidence may therefore be placed in the notes on the stratigraphical distribution of the species, as the greatest care has been taken in order to arrive at the truth.

As the Oolitic rocks of Europe were deposited in basins of greater or less extent, it follows that many contemporary species which lived on different shores of these ancient seas will, from time to time, be discovered; and Foreign species, hitherto found only in the Oolitic rocks of the Continent, will doubtless be discovered in strata of the same age in England, and *vice versá*. I have, therefore, at the end of the description of the species of each genus, for the purpose of easy reference in the event of new species being found, appended original notes on Foreign Oolitic species of that genus most nearly allied to our own forms, but which have not as yet been found in the English Oolites. The Foreign species are printed in a different type, and the notes are placed at the end of the section to which they belong. The short diagnosis I have given of each species is drawn from authentic specimens kindly contributed by several distinguished foreign friends, whose names are mentioned in connection with their specimens. A reference is made to the best figures of each species extant; and for the localities in which they are found, and the collections in which the types are contained, I have consulted with much advantage M. Desor's excellent 'Synopsis des Échinides Fossiles,' now in course of publication.

It is now my pleasing duty to return my most sincere thanks, either for the loan of specimens, or permission to inspect their collections in quest of new forms, to Mr. Pickering and Mr. King, Malton; Mr. Charlesworth, York; Mr. Waite and Mr. Duck, Calne; Mr. William Buy, Sutton; Mr. Bean, Scarborough; the Hon. Mr. Marcham; Mr. H. C. Sorby; Mr. W. Cunnington, Devizes; Mr. Walton and Mr. Bush, Bath; Mr. Mackneil, Wotton-under-Edge; the Rev. P. B. Brodie, Rowington Vicarage, near Warwick; Mr. John Lycett, Minchinhampton; Mr. John Jones, Gloucester; Professor Buckman and Mr. Bravender, Cirencester; Professor Morris, Professor Tennant, and Mr. J. S. Bowerbank, London; Mr. W. M. Tartt, Mr. Charles Pierson, Mr. Thomas Bodley, and Mr. Edward Hull, F.G.S., Geological Survey, Cheltenham.

I beg to tender my especial thanks to the Rev. A. W. Griesbach, of Wollaston, for several valuable contributions, consisting of many fine series of different species of Echinoderms from the Great Oolite, Forest Marble, and Cornbrash of Northamptonshire, likewise for the labour he has bestowed in finding some rare species, and ascertaining many valuable facts relative to the distribution of the species found in his county; to Mr. J. Graham Lowe, Kensington Park, for the gift of *Pygaster umbrella*, Lamk., from the Coral Rag; and to Mrs. Lowe for the gift of the rare *Asterostoma excentricum*, Agass.; to Dr. Symes, Bridport, for a fine *Clypeus Agassizii*, from the Inferior Oolite of Chideock; to Mr. Charles Moore, Bath, for the gift of some rare specimens from the

Upper Lias of Ilminster; to Mr. Etheridge, Bristol, for several rare urchins; to Mr. G. E. Gavey, C.E., for the donation of several fine CRINOIDEA and ASTEROIDEA, from the Middle Lias of Chipping-Campden, and for the loan of his finest specimens for figuring in this work; to the Earl of Ducie, for the loan of his unique *Solaster Moretonis*, Forbes, and several fine Cretaceous Cidaris and Star Fishes; to Mr. John Leckenby, Scarborough, for much useful information relative to the distribution of the Yorkshire Oolitic Echinodermata, and for the gift of several specimens; to Dr. Murray, Scarborough, for the donation of several rare Coralline Oolite *Pygasters* and *Pyguri*, collected by him at Ayton; to Mr. Reed, York, for much valuable information regarding the Whitwell beds of Inferior Oolite, and for the gift of type specimens of *Pygaster semisulcatus*, Phil., and *Echinus germinans*, Phil.; to Mr. Wood, Richmond, Yorkshire, for the gift of fine specimens of *Echinobrissus orbicularis*, Phil., *Echinobrissus dimidiatus*, Phil., and *Woodocrinus macrodactylus*, de Koninck, and for kindly placing his beautiful collection of CRINOIDEA at my disposal; to Mr. Charles Fowler, Cheltenham, for the gift of *Cidaris Fowleri;* to Mr. Davidson, of Brighton, for the uniform interest he has taken in the success of this work, for the specimens he has contributed, the manuscript plates he has lent, and the introductions he has given me to several distinguished Continental naturalists, who have kindly supplied much useful information.

I desire to make my warmest acknowledgments to M. Michelin, of Paris, who possesses the finest collection extant of living and fossil Echinodermata, for the magnificent series of type specimens he most generously contributed to my cabinet for comparison with English forms; to M. Bouchard-Chantereaux, of Boulogne, for a series of Echinoderms from the Oolitic rocks of the Boulonnais; to M. Cotteau, of Coulommiers, for the types of the species described by him in his 'Études sur les Échinides Fossiles du département de l'Yonne;' to M. Triger, of le Mans, for a suite of specimens collected by him from the Oolites in the departement de la Sarthe; to Professor Deslongchamps, of Caen, for the specimens collected by him from the Oolites of Calvados, and determined by M. Agassiz; to M. de Lorière, of Paris, for many rare urchins from the département de la Sarthe; to Professor Roemer, for the types of several of his brother's species from the Oolites of Hanover; to Dr. Fraas, of Stuttgart, for the types of many of Count Münster and Professor Goldfuss's species from the Royal Museum of Württemberg; to Professor de Koninck, of Liège, Dr. Oppel, of Stuttgart, and M. Sæmann, of Paris, for good types of many Foreign species.

My warmest thanks are likewise due to my friend Mr. S. P. Woodward, of the British Museum, for kindly acting as my referee in the prosecution of this work, and for the many valuable suggestions he has made during its preparation and progress, as well as for the assistance he has given me in comparing my specimens with Foreign types in the British Museum, and aiding in the determination of dubious forms.

The late Sir Henry de la Beche, Director-General of the Geological Survey of Great Britain, most liberally gave me free access to all the specimens contained in the Geological

Museum in Jermyn Street ; and the same privilege has been most kindly renewed by his distinguished successor, Sir Roderick I. Murchison, to whom I beg to tender my warmest acknowledgments.   I am under many obligations to my friend Mr. Waterhouse, of the British Museum, for his kindness in allowing me to examine all the Echinoderms in the National Collection, and his permission to figure those I have selected for this purpose. Professor Sedgwick, of Cambridge University, at my request, most liberally communicated the types of Professor M'Coy's new species of urchins, described in the 'Annals of Natural History.'   Mr. Rupert Jones has at all times given me free admission to examine the rich cabinets of the Geological Society of London.   Professor Phillips, of Oxford, has afforded me much useful information relative to the species of Echinoderms first figured in his valuable work on the 'Geology of Yorkshire.'   To each of these kind friends I beg to tender my most grateful acknowledgments.

My best thanks are likewise especially due to Messrs. Bone and Baily, for the great care they have bestowed on the beautiful plates that enrich my Monograph, which, for scientific accuracy in details, and artistic effect in execution, are second to no lithographs of similar objects extant.

THOMAS WRIGHT.

EXETER PLACE, CHELTENHAM ;
August, 1856.

A MONOGRAPH

ON THE

FOSSIL ECHINODERMATA

OF THE

OOLITIC FORMATIONS.

SUB-KINGDOM—RADIATA.

THIS great division of the Animal Kingdom includes classes which differ widely from each other in form, organization, and habits. Some have the body circular, globular, or ringed; or vermiform, plant-like, or amorphous. Some are enclosed in a soft arachnoid, transparent membrane, and float like crystal masses through the water, as the *Infusoria* and *Acalephæ*; others hang like living stalactites from the roofs of submarine caverns, like the *Amorphozoa*; or, assuming the forms of the Vegetable World, they develope ramose stems, with numerous branches, of which myriads of zoophytes are at once the builders and inhabitants, as the *Polypifera*. Some are enclosed in exquisite shells, microscopic in size, but unrivalled in symmetry, although the structure that produces them is but a mere film of jelly, as the *Foraminifera*; others have a complicated calcareous skeleton, composed of many thousands of separate elements, which, for beauty and contrivance, is unsurpassed by that of any other class, as the *Echinodermata*. Where the nervous system has been discovered, it consists of a simple gangliated filament, surrounding the entrance to the digestive organs; but in by far the greater number of animals grouped in this division, no distinct nervous system is found, although the creatures themselves possess an exquisite sensibility.

The sub-kingdom RADIATA is formed of classes which are more remarkable for their

1

negative than for their positive characters: hence it wants that unity of composition so well displayed in the sub-kingdoms MOLLUSCA, ARTICULATA, and VERTEBRATA. Some naturalists have proposed to separate the RADIATA into two sections, under the names *Aneura* and *Cyclo-neura*, or *Acrita* and *Nemato-neura*; but, unfortunately, the nervous system of only a very few genera of the Cyclo-neura is known, so that, by generalizing too much upon these isolated facts, we are in danger of reasoning on an error in order to establish a method.

We include in the sub-kingdom RADIATA the six following classes, which may, for the sake of convenience, be subdivided into two sections;—in the one, the form of the body is more or less globular, sometimes it is symmetrical, often it is irregular or amorphous,—these form the GLOBULAR RADIATA. In the second section the body is stellate, and the divisions are arranged in the form of rays around a common centre,—this stellate form can often be shown to consist of a bilateral symmetry. These classes form the STELLATE RADIATA. The following table exhibits the sections and classes, to which are added the names of typical genera as examples of each:

Sub-Kingdom.	Sections.	Classes.	Examples of Genera.
RADIATA.	Globular Radiata	1. AMORPHOZOA	*Halichondria, Spongilla, Spongia.*
		2. FORAMINIFERA	*Orbitoides, Nummulites, Rotalia.*
		3. INFUSORIA	*Plæsconia, Dileptus, Paramecium.*
	Stellate Radiata	4. POLYPIFERA	*Alcyonium, Oculina, Meandrina.*
		5. ACALEPHÆ	*Medusa, Physalus, Cassiopæia.*
		6. ECHINODERMATA	*Encrinus, Asterias, Echinus.*

CLASS—ECHINODERMATA.

The name ECHINODERMATA was given by J. T. Klein, in 1734,[*] to the shells of Sea-urchins, which were called Echini. Bruguière[†] subsequently called that class which comprised the Star-fishes, and the Sea-urchins, ECHINODERMATA. Cuvier[‡] included in his class ECHINODERMES, with *Asterias* and *Echinus*, the *Holothuria*, animals that are destitute of the prickly skin, of the more typical forms, and have many external affinities with some Mollusca; and subsequently, in his ' Règne Animal,'[§] he grouped in this class, *les Echinodermes sans pieds*, forming the order Sipunculoidea, which connect the Radiata with the Annelidous Articulata.

[*] 'Naturalis Dispositio Echinodermatum,' Jacobi Theodori Klein, 1734.
[†] 'Tableau Encyclopédique des trois Règnes de la Nature,' 1791.
[‡] 'Tableau Elémentaire de l'Histoire naturelle des Animaux,' 1798.
[§] 'Règne Animal distribué d'après son Organisation,' 1834.

The Echinoderms are most highly organized animals, and for the most part are covered with a coriaceous integument. In several orders it is strengthened with numerous calcareous pieces, which together form a complicated skeleton. The external surface of the skin, in many families, developes spines of various forms and dimensions, which aid in locomotion, and serve as defensive instruments to the creatures possessing them. By far the largest number of these animals have a complicated system of vessels for the circulation of water through their bodies. These aquiferous canals are intimately connected with the locomotion of the animal; for by means of it, most of the typical groups put in motion those remarkable suckers which protrude in rows from different divisions of the body. In the *Echinoidea* they escape through the holes in the poriferous zones, and in the *Asteroidea* they pass through apertures in the intervals of the small plates which form the middle of the rays, whilst in the *Sipunculoidea* these organs are altogether absent.

No class of the Animal Kingdom more clearly exhibits a gradation of structure than the *Echinodermata;* for, whilst some remain rooted to the sea bottom, and in this sessile condition and other points of structure resemble the *Polypifera*, others exhibit the true rayed forms, clothed in prickly armour, which characterise the central groups of this class. These conduct us, through a series of beautiful gradations, to soft elongated organisms, whose forms mimic the *Ascidian Mollusca;* whilst others have the long cylindrical body and annulose condition of the skin, with the reptatory habits of the *Apodous Annelida*.

With so fertile a field for investigation, it is not surprising that the minute anatomy of the Echinodermata should have engaged the attention of some of the most distinguished Naturalists of our age—Tiedemann, Müller, Van Beneden, Agassiz, Desor, Forbes, and Sharpey,—and have yielded fruits which the physiologist reckons as among the most marvellous contributions to morphological science.

The class Echinodermata is divided into eight orders, which, in descending sequence, may be thus arranged:

1. SIPUNCULOIDEA.
2. HOLOTHUROIDEA.
3. ECHINOIDEA.
4. ASTEROIDEA.
5. OPHIUROIDEA.
6. BLASTOIDEA.
7. CYSTOIDEA.
8. CRINOIDEA.

Order I. SIPUNCULOIDEA—form the apodal Annelidous Echinoderms; they have a long cylindrical body, divided into rings by transverse folds of the integument: they have no tubular suckers, nor calcareous parts in their body, nor is it divided into a quinary arrangement of longitudinal lobes: some have horny bristles, like the feet of many Annelida, which they somewhat resemble: their mouth is sometimes surrounded by tentacula, which are not, however, regulated by a definite number, nor disposed with the same regularity as in the next Order. They are unknown in a fossil state.

Type. *Sipunculus edulis*. Cuv. Pallas.

Order II. HOLOTHUROIDEA.—Body in general elongated; skin usually soft and leathery, in a few genera strengthened by calcareous or horny spines. Five avenues of suckers, which divide the body into as many longitudinal, nearly equal, lobes or segments; mouth surrounded by plumose tentacula, the numbers of which are in general multiples of five; anus at the opposite extremity of the body; digestive organs consist of a long intestine, which makes some coils in passing through the body; respiration performed by internal ramified tubes, like a miniature tree; locomotion effected by contractions and extensions of the body, and by rows of tubular suckers, similar to those in the Star-fishes and Sea-urchins. The softness of their naked integument prevents their preservation in the stratified rocks. We know none in a fossil state.

Type. *Cucumaria frondosa.* Grüner.

Order III. ECHINOIDEA.—Body spheroidal, oval, or depressed, without arms; furnished with a distinct mouth, sometimes armed, which is always below, and an anus which occupies different positions. Body enclosed in a shell or test, composed of twenty columns of calcareous plates, and ten rows of holes for the passage of retractile tubular suckers; the surface of the test is studded with tubercles, which possess, jointed with them, moveable spines, of various sizes and forms in the different families and genera: at the summit of the test is the apical disc, composed of five genital plates, perforated for the passage of the ovarial and seminal tubes, and five ocular plates for lodging the five eyes. The intestine winds round the shell, attached by a mesentery, the surface of which, as well as the membrane lining of the test, is covered with vibratile cilia.

Type. The common Sea-urchin, *Echinus sphæra.* Müller.

The ECHINOIDEA are represented by one family in the Palæozoic rocks, and by numerous families in the Mesozoic and Tertiary rocks, several of which characterise these great periods of geological time. They likewise abound in our present seas.

Order IV. ASTEROIDEA.—Body stelliform, depressed, provided with five or more lobes or hollow arms, which are a continuation of the body, and contain prolongations of the viscera; the mouth, which is always below and central, serves likewise as an anus; rows of retractile tubular suckers occupy the centre of the rays. Skeleton complicated, composed of numerous solid calcareous pieces, variable as to number, size, and disposition; skin coriaceous, studded with calcareous spines of various forms; a madreporiform plate on the upper surface, near the angle between two rays; eyes placed at the extremity of the rays; reptation performed by the tubular suckers.

Type. The common Star-fish, *Uraster rubens.* Linnæus.

This order is represented in the Silurian rocks by two genera. The Oolitic, Cretaceous, and Tertiary rocks contain many extinct forms. The existing species are very abundant in all the present seas.

Order v. OPHIUROIDEA.—Body discoidal, distinct, depressed, provided with long, slender arms, in which there is no excavation for any prolongation of the viscera; they are special organs of locomotion, and independent of the visceral cavity; they have spines developed from their sides; the mouth, surrounded by membranous tentacula, is always below and central, and serves at the same time as the anus. Skeleton complicated, composed of calcareous pieces, of which the size and number vary in different genera. The arms, long and slender, are sustained internally by central vertebral-like pieces, but they are not hollow or grooved underneath, as in the *Asteroidea:* they are special organs of locomotion, independent of, and superadded to, the visceral cavity, and have numerous plates or spines regularly disposed along their sides to assist in reptation.

Type. The common Sand-star, *Ophiura texturata.* Lamarck.

This order is represented by one genus in the Silurian rocks. Several genera are found in the Oolitic, Cretaceous, and Tertiary rocks, as well as in our present seas.

Order vi. BLASTOIDEA.—Body in the form of an oval calyx, composed of solid, calcareous plates, provided with five interambulacra and five ambulacra, the latter united superiorly, striated transversely, and having a deep furrow down the middle; ten ovarial holes, opening into five at the summit, and a central mouth aperture; a short, slender stem; but the body is destitute of arms.

Type. *Pentremites inflatus.* Sowerby. Carboniferous Limestone.

The genera are all extinct, and belong to the Palæozoic rocks. One species appertains to the Upper Silurian, six to the Devonian, and twenty-four are special to the Carboniferous rocks.

Order vii. CYSTOIDEA.—Body more or less spherical, supported on a jointed stem; the bursiform calyx is formed of close-fitting polygonal plates, varying in number in the different genera, and investing the surface like a coat of mail, except above, where there are three openings, one for the mouth, one for the anus, and one with a valve for the reproductive organs; the fourth aperture is below, and is continuous with the canal in the stem. Some have two or four arms, others are armless; certain species possess articulated tentacula, and curious comb-like appendages, or pectinated rhombs, in connection with the plates.

Type. *Pseudocrinites quadrifasciatus.* Pearce. Upper Silurian.

This order is extinct. All the genera are found in the Silurian and Devonian rocks.

Order viii. CRINOIDEA.—Body bursiform, distinct, formed of a calyx composed of a definite number of plates, provided with five solid arms, which are independent of the visceral cavity, and are adapted for prehension; a mouth and anus distinct; no retractile suckers; ovaries at the base of the arms opening into special apertures. Skeleton complicated, calcareous, composed of thick plates closely articulated together, the number and arrangement of which are determinate in the different families, the multiples of five being the numbers which predominate; the central plate of the body is supported on a long, jointed column, which is firmly rooted to the sea-bottom. The mouth is central and prominent; the anus is situated at its side; the arms are mostly ramose and multi-articulate, and when extended form a net-like instrument of considerable dimensions. The mouth is always placed upwards, and retained in that position by the column being jointed to the central plate of the calyx. The normal station of the CRINOIDEA is the reverse, therefore, of the ASTEROIDEA and ECHINOIDEA.

Type. *Pentacrinus Caput-medusæ*. Miller. From the seas of the Antilles.

Extinct families of Crinoids have existed in all seas from the Silurian downwards, and one or two representatives are now living.

From the above analysis of the class ECHINODERMATA, it appears that, as the *Sipunculoidea* and *Holothuroidea* are not found in a fossil state, and the *Blastoidea* and *Cystoidea* are special to the Palæozoic period, our field of investigation in this Monograph is limited to the ECHINOIDEA, ASTEROIDEA, OPHIUROIDEA, and CRINOIDEA, which we now propose to consider seriatim, commencing with the ECHINOIDEA.

Order—ECHINOIDEA.

The body is spheroidal, oval, depressed or discoidal, and is enclosed in a calcareous test or shell, composed of ten columns of large plates, *the inter-ambulacral areas:* and ten columns of small plates, *the ambulacral areas*, which are separated from each other by ten rows of holes, constituting *the poriferous zones*. The external surface of the plates is studded with tubercles of different sizes, in the different families; these are articulated with the spines by a kind of moveable ball-and-socket joint: the spines are of various forms and dimensions, and serve well to characterise the genera and species.

At the summit of the test is the apical disc, composed of five genital plates, perforated for the passage of the ovarial and seminal canals, and five ocular plates, notched or perforated for lodging the eyes. There are two great apertures in the test, one for the mouth, and the other for the anus; the relative position of these oral and anal apertures varies in the different families, and forms an important character for their systematic classification.

The mouth is sometimes armed with a complicated apparatus of jaws and teeth, but sometimes it is edentulous. The internal organs of digestion consist of a pharynx, œsophagus, stomach, and intestine, which winds round the interior of the shell, attached thereto by a delicate mesentery, its surface, as well as that forming the lining membrane of the shell, is covered with vibratile cilia, the play of which causes currents of sea-water to traverse incessantly the interior of the body, and to perform an important part in the function of respiration ; their blood is circulated in arteries and veins, aided by a central pulsating organ or heart. The five ovaries and testicles occupy the ambulacral divisions, and open externally through the holes in the genital plates. Their locomotion is effected by the joint action of the tubular retractile suckers and the spines. Many sea-urchins attach themselves to rocks by these tubular feet, and some bury themselves in limestone, and sandstone or even in granitic rocks, by the abrading action of the spines.*

The nervous system consists, according to M. Van Beneden, of a circular cord, which surrounds the entrance to the digestive organs, and sends branches into the divisions of the body. Professor Agassiz, and the late Professor Edward Forbes, regarded the organs situated in the ocular plates as eyes, but M. Dujardin† denies them even a nervous system. In the absence of a greater amount of direct anatomical evidence on the point, the following observation, related by M. Alcide d'Orbigny,‡ has an important bearing on the question, and supports it affirmatively.

Captain Ferdinand de Candé, who commanded the 'Cléopâtre' in the Chinese seas, told M. d'Orbigny that he had captured, on their coasts, an urchin with long spines, probably a *Diadema*, which he examined in a vessel of water. " I hastened to seize it," he observed, " when it instantly turned all its spines in the direction of my hand, as if to defend itself.

" Surprised at this manœuvre, I made an attempt to seize it on the other side, when immediately the spines were directed towards me.

" I thought from this that the urchin saw me, and that the motion of the spines was intended as an act of self-defence ; but, to prove whether this movement of the animal

---

* M. Eugène Robert exhibited to the Academy a block of old red sandstone, obtained from the shore of the great Bay of Douarnenez, which was perforated with numerous holes, evidently formed by the Echini which were lodged in them. Each rounded cavity is in exact proportion, both as to size and form, with the body of the Echinoderm. M. Lory, Professor at Grenoble, and well known for his numerous and excellent works on geology, has begged me to exhibit several specimens of perforating Echini, which have taken up their abode in the granite of the Bay of Croisic, not far from Piriac. It is the same granite as that from the Pouliguen, and in the same state of alteration. This igneous rock is there perforated by Mollusca and Echinodermata for an extent of several kilometers. Those which M. Lory has just discovered are certainly of the same species as the Echini which burrow in the old red sandstone of the Bay of Douarnenez. They closely resemble the Mediterranean *Echinus*, mentioned by Lamarck under the name of *Echinus lividus*. It is one of the most abundant Echini on the coast, and in the market of Marseilles, whence Lamarck obtained his specimens. I have never heard that these individuals possessed perforating habits; and probably a careful examination of living specimens of the Echinus from the coast of Brittany may show that it belongs to a distinct species, notwithstanding its apparent identity with that of the Mediterranean. In this case it might be called *Echinus terebrans*. 'Observations on Echini perforating the Granite of Brittany,' by M. Valenciennes.—*Comptes Rendus*, Nov. 5, 1855, p. 755.

† Lamarck, 'Animaux sans Vertèbres,' 2d ed., tom. iii, p. 200.

‡ 'Paléontologie Française Terrains Cretacés,' tom. vi, p. 12.

was produced by my approach, or merely by the agitation of the water, I repeated the experiment very slowly, and even over the water with a stick: the urchin, whether in the water or out of it, having always directed its defensive spines towards the object which approached it. From these observations I arrived at the conclusion that these urchins see, and that their spines serve them as defensive instruments."

It is worthy of remark, that Captain Candé, at the time he watched this urchin, was ignorant of the anatomical fact that eyes had been detected in the Echinidæ, and his inference was simply the legitimate conclusion drawn from carefully-made observations.

The calcareous test of the Echinoidea is the only part of the structure of these animals which is preserved in a fossil state. It has hitherto failed to attract that amount of attention from the palæontologist which the importance of its study demands, although in a stratigraphical point of view it is not inferior to the skeletons of any other class of the Animal Kingdom. The fact seems to have been almost entirely overlooked, that most of the generic characters of the different groups of the Echinoidea are more indelibly impressed on the separate pieces of their test than in the skeletons of any other class of the Invertebrata.

Unlike the shells of the Mollusca, the test of the Echinoidea constitutes an internal and integral portion of the animal, being secreted by, and enclosed within, organized membranes, and participating in the life of the organism; portions of the skeleton are likewise intimately connected with the organs of digestion, respiration, and generation, as well as with those of vision and locomotion.

As the analysis of the test of the Echinoidea, with full anatomical details of the structure of the skeleton in the Echinodermata in general, will be given in the Introduction to this Monograph, it is unnecessary to enter at present minutely into the subject; but, as many of our readers are doubtless unacquainted with the terminology employed in the description of the test of the Echinoidea, and the characters on which a diagnosis of the species is made, it is desirable now to preface our description of the species with brief explanations of the same, in this part of the work, illustrating the terminology by a reference to the plates for accurate figures of the different parts of the test, and the magnified details of its anatomical characters.

Terminology, or descriptive analysis of the component elements of the test of the Echinoidea.

The test of the Echinoidea is composed of the following parts :

  a. Five ambulacral areas.
  b. Five inter-ambulacral areas.
  c. Ten poriferous zones.
  d. An anal opening, and anal membrane and plates.
  e. A mouth opening, and buccal membrane and plates.
  f. Five jaws when organs of mastication exist; some are edentulous.
  g. Tubercles of various sizes, developed on the outer surface of the plates.
  h. Spines of various forms and dimensions, which are jointed with the tubercles.

These are the parts essential to be known ; but there are others of secondary importance, which will be described hereafter in their proper place.

The Body of the Echinoidea is divisible into three parts :

1st. The calcareous envelope or skeleton has a globular, circular, oval, pentagonal, hemispherical, conoidal, or discoidal form; it is composed of a framework of pentagonal, hexagonal, and polygonal calcareous plates. This testaceous box is called the test. It is the *form*, the *test* of Agassiz; the *general form*, the *test* of Desmoulins; *la coquille* of d'Orbigny.

2d. The visceral cavity, which contains the organs of digestion, respiration, circulation, and generation, is formed entirely by the calcareous skeleton.

3d. The external surface of the test is covered with spines, which are moveably articulated with the tubercles developed on the surface.

The normal position of the body.—In describing the different parts of the test of the Echinoidea, it is assumed that the urchin, the common purple-heart urchin, *Spatangus purpureus*, Müller, for example, is placed before the observer; or, the common chalk urchin, *Micraster cor-anguinum*, Klein, will answer equally well. The side with the single ambulacrum lodged in the sulcus, and the mouth in that third of the base, is the *anterior region*. The four other ambulacra are disposed in pairs, and correspond to the right and left sides of the observer's body; there is, therefore, a right antero-lateral and a right postero-lateral; a left antero-lateral and left postero-lateral ambulacral area. The side having the single inter-ambulacrum in the middle, and the anal opening in the upper part of the border, is the *posterior region*. The four other inter-ambulacra are likewise disposed in pairs, two of which, with the single ambulacrum, form the anterior part of the test; the other pair, with the pairs of ambulacra, forming the sides, and the single inter-ambulacrum its posterior part.

All the Echinoidea have the mouth situated at the under side of the body. The surface in which this opening is placed is called the *base;* that region of the test opposite to the base is the *upper* or *dorsal surface.*

The most convex part of the margin, border, or sides, between the base and the upper surface, is called the *circumference,*—the *ambitus* of some authors. It is round, flat, convex, angular, or carinated, according to the general form and thickness of the test.

The *length,* or *antero-posterior* diameter, is the distance between the anterior and posterior regions, and corresponds to the middle line of the body.

The *breadth,* or *transverse diameter,* is the distance between the greatest lateral convexity of the circumference, in the direction of a line which will cut the line of length at right angles.

The *height* is the distance between the most convex part of the upper surface and a plane on which the base of the test can rest. The apical disc is generally situated at the vertex, but it is not always so. The height has reference to the highest point of the test, quite irrespective of any other consideration.

The test has invariably two openings, the one for the mouth, the other for the anus.

The *mouth-opening* is always situated at the under surface; to its circumference is attached the buccal membrane, and through its central aperture, when they exist, protrude the five jaws. (Pl. IV, fig. 1 *b;* Pl. XI, fig. 1.) The buccal, like the anal membrane in many families, is clothed with numerous small plates. In *Cidaris, Rabdocidaris, Goniocidaris, Diplocidaris,* and probably in other *Cidaridæ,* the mouth-opening is central, circular, or slightly pentagonal (Pl. I, figs. 2 *a,* 4 *a;* Pl. II, figs. 1 *a,* 2 *b*); but in *Hemicidaris, Diadema, Hemipedina, Pedina, Echinus,* and other ECHINIDÆ, the mouth opening is more or less decagonal, its margin being divided by notches (*entailles*) into ten lobes. (Pl. III, figs. 2 *b,* 6 *b;* Pl. IV, fig. 2 *b, e;* Pl. V, fig. 1 *e;* Pl. VI, figs. 2 *b,* 5 *c*.) The lobes are, in general, unequal in size; those which correspond to the base of the ambulacra are the largest, and are called the *ambulacral lobes;* those corresponding to the base of the inter-ambulacra are the *inter-ambulacral lobes.* The margin of the mouth-opening is called the *peristome;* to it the buccal membrane, which closes the base of the test, is attached.

The *mouth-opening* is central, and is armed with jaws, in the CIDARIDÆ, ECHINIDÆ, SALENIDÆ, GALERITIDÆ, and CLYPEASTERIDÆ. It is more or less excentral and edentulous in the ECHINONIDÆ, COLLYRITIDÆ, ECHINOLAMPIDÆ, ECHINOCORYDÆ, and SPATANGIDÆ. In these families it is round, oval, or pentagonal; sometimes its margin is ray-like, or surrounded by five prominent lobes; in others it is distinctly bilabiate.

The *anal opening* is always in the upper surface, directly opposite to the mouth, in the centre of the genital and ocular plates, and is either central or sub-central in the CIDARIDÆ, ECHINIDÆ, and SALENIDÆ. (Pl. III, figs. 1 *c,* 3 *c;* Pl. IV, figs. 1 *g,* 2 *f.*) In the other families its position varies much; sometimes it opens on the upper surface, as in some GALERITIDÆ and CASSIDULIDÆ; sometimes it opens near the margin, or is *supramarginal,* marginal, or *infra-marginal;* often it opens at the base, between the mouth and the border. During the life of the animal this opening was closed by an anal membrane, and a series of small angular anal plates, the number and disposition of which vary in the different genera. The anal plates are seldom preserved in fossil species; and the term anal opening is given to all that part of the test occupied by them, the anal membrane, and the vent.

The Ambulacral and Inter-ambulacral Areas.

The test is composed—1st, of twenty columns of calcareous plates of different sizes, the *plaquettes*, *Täfelchen*, *Assulæ*, of authors. They are pentagonal in form, and are united by harmonial sutures to form rays, which proceed from the mouth, where they have their greatest breadth, to the apical disc, where they are narrowest. 2d, of a series of hexagonal or polygonal plates, forming a disc, which occupies the upper surface of the test. 3d, of ten rows of small plates, which are notched on their margins, to form holes: these constitute the poriferous zones. 4th, of moveable spines, which are jointed with eminences on the outer surface of the columnar plates.

The *ambulacral plates* form two narrow columns, which are bounded by two poriferous zones. The space thus circumscribed is the *ambulacral area*. There are five of these areas in the test of the Echinoidea; in the CIDARIDÆ the ambulacral areas are very narrow, and support only granules (Pl. II, fig. 1 *a*, *b*); but in the ECHINIDÆ (Pl. VIII, fig. 1) they are much wider, and have large tubercles on their surface. The comparative width of the ambulacra, as compared with the inter-ambulacra, has led some authors[*] to divide the family Cidaridæ, including therein the Echinidæ, into two tribes,—the ANGUSTISTELLÆ, or Cidaridæ with narrow ambulacra, and the LATISTELLÆ, or Cidaridæ with broad ambulacra. These two tribes nearly represent our two families: the CIDARIDÆ are equal to the ANGUSTISTELLÆ, and the ECHINIDÆ are nearly equal to the LATISTELLÆ.

Pl. I, figs. 1 *b*, 2 *c*, 3 *c*, 4 *c*; Pl. II, figs. 1 *c*, 2 *g* represent the structure of the ambulacra in the *Cidaridæ*; and Pl. VI, fig. 2 *d*, and Pl. VII, figs. 3 and 4, show the structure of the ambulacral areas in some *Echinidæ*.

One of the ambulacral areas is single, and always represents the anterior region of the test. In the spheroidal *Echinidæ* and *Salenidæ*, this is detected by its relation to the apical disc, as the right antero-lateral genital plate always carries the madreporiform body; but in the oval, pentagonal, and elongated forms, its position and relation to the mouth render it unmistakeable. The four other ambulacra are disposed in pairs.

The *inter-ambulacral plates* form two broad columns, which compose the inter-ambulacral areas. Of these there are five, which alternate with the ambulacral areas in the architecture of the test: the poriferous zones form the line of demarcation between these

[*] Albin Gras, 'Description des Oursins Fossiles de département de l'Isère,' p. 20.
E. Desor, 'Synopsis des Echinides Fossiles,' p. 1.

two classes of columnar plates. The inter-ambulacral plates are all pentagonal, and many times larger than the ambulacral; they carry on their external surface the large primary tubercles. Of the five inter-ambulacral areas, one is single and posterior; and in all the Echinoidea which have the anal opening external to the apical disc, it is in this single inter-ambulacrum that the vent terminates. The other four inter-ambulacra are disposed in pairs, and form the greater part of the anterior and lateral parts of the test. They are called *anterior pairs* and *posterior pairs* respectively, to distinguish them from the odd area, which is called the *single inter-ambulacrum*.

Pl. I, figs. 1 *b*, 4 *e*; Pl. II, fig. 1 *a, b, c*, show the form and structure of the inter-ambulacra in the *Cidaridæ*; and Pl. VI, fig. 2 *a, b*; Pl. VIII, fig. 1 *a, b*, the structure of these areas in the *Echinidæ*.

In the *Cidaridæ*, and in some *Echinidæ*, the inter-ambulacral areas have only two rows of primary tubercles; but in most of the *Echinidæ* there are four, six, eight, or even ten rows of primary tubercles in these areas.

The Poriferous Zones.

The *poriferous zones* are situated on each side of the ambulacral areas. (Pl. I, fig. 4 *a*; Pl. II, figs. 1 *a, b*, 2 *b, c*.) They are composed of a very great number of small pieces, articulated together in such a manner as to form a series of holes, the corresponding edges of the plates, remaining uncalcified at certain definite intervals, to produce foramina, which are destined for the passage of retractile tubular suckers. The form and structure of the poriferous zones afford a good generic character. As there are two poriferous zones bordering each ambulacral area, it follows that there are ten zones. Some authors give the collective name *ambulacra* to the *zones* and the *area*, but, for obvious reasons, we consider them distinct sections of the test, and treat of them as such. The pores are *round, oblong*, or *elongated;* the pores forming a pair may be *equal* or *unequal*, or, in relation to each other, they may be *transverse* or *oblique;* they may be *contiguous* or *remote*, and when they are united by a transverse sulcus, they are said to be *conjugate*.

The pores are differently arranged in the zones in the different families. When they are disposed in single pairs they are said to be *unigeminal* (Pl. II, fig. 1 *c*); when in double pairs they are *bigeminal* (Pl. VI, fig. 3 *c, d*); when in triple oblique pairs they are *trigeminal* (Pl. VI, fig. 2 *g*; Pl. XII; Pl. XIV; Pl. XV), or when they are grouped in a greater number, as in many living species of the genus Echinus, they are *polygeminal*.

When the zones extend in a straight uninterrupted line from the mouth to the apical disc, they are said to be *simple*, as in the *Cidaridæ, Echinidæ, Salenidæ, Galeritidæ, Echinonidæ;* when the zones, after parting from the apical disc, expand, and again contract, thereby forming a leaf-like figure on the upper surface of the test, they are said to be *petaloidal*, as in the *Clypeasteridæ*; when the petal is not so complete as in the *Cassidulidæ*, it is *sub-petaloidal*. The zones are *complete* when they extend without interruption from the mouth to the disc; they are *interrupted* when they terminate on the upper surface, and reappear again at the base near the mouth; they are *limited* when they form only a star on the dorsal surface. These terms all represent generic and specific characters of greater or less value, which require to be carefully noted in the description of the species.

The Apical or Genital Disc.

The *apical disc* occupies in general the centre of the summit of the test, and is composed, in most of the genera, of ten plates,—namely, five genital plates, and five ocular plates. (Pl. III, figs. 1 *e*, 2 *f*, and Pl. IV, fig. 2 *f*.) In the *Salenidæ* there is one or more additional plates introduced. The *five genital* or *oviductal plates* correspond to the summits of the inter-ambulacral areas (Pl. III, figs. 1 *a*, 3 *a*); two plates form an antero-lateral pair, two a postero-lateral pair, and the single plate is placed behind. On the right antero-lateral genital plate (Pl. IV, figs. 1 *g*, 2 *f*) is placed a spongy prominent mass, called the *madreporiform body*. The plate, supporting this body, was supposed by Agassiz and Desor always to represent the posterior part of the test, but we shall show in the sequel it is invariably placed on the right antero-lateral plate.

The *ocular plates* are placed at the summit of the ambulacral areas. They are small heart-shaped bodies (Pl. IV, fig. 1 *g*), and are wedged into the angles of the genital plates around the circumference of the disc.

The *sur-anal plates* are found only in the *Salenidæ*. They consist of one or many elements placed in the centre of the genital circle, and almost always before the anal opening. (Pl. XVI.)

The *anal plates* are very small bodies, and are variable as to size and number; they clothe the membrane of the anal opening, and are well seen in recent urchins, but are seldom preserved in fossil species.

The Tubercles.

The plates which compose the test of the *Cidaridæ*, *Echinidæ*, and *Salenidæ*, have large tubercles developed on their external surface. They are divided into *primary tubercles*, *semi-tubercles*, *secondary tubercles*, *minute tubercles*, *granules*, and *miliary granulation*. In the other families the tubercles are smaller, more numerous, and less complicated.

The *primary tubercles* form two rows in the inter-ambulacral areas of the *Cidaridæ* (Pls. I and II), and in the genus *Hemicidaris* (Pls. III and IV), and a section of the genus *Diadema* (Pls. VI and VII); and four, six, eight, or ten rows in many *Echinidæ*.

The *semi-tubercles* are found at the base of the ambulacral areas of the genus *Hemicidaris*. (Pl. III, fig. 1 *b*, *c*; Pl. IV, fig. 1 *c*, *d*.) In a section of the genus *Hemipedina*, they likewise are found at the base of the ambulacra (Pl. XI, fig. 2).

The *secondary tubercles* are found in many of the *Echinidæ*, ranged in general on the ambulacral side in the primary tubercles. They are found likewise on the centro-sutural side of the primary rows (Pl. VII, fig. 4, and Pl. VIII, fig. 1) of many *Diadema*, *Pedina*, *Hemipedina*, and *Echinus*.

The *minute tubercles* are the small tubercles which are found on the margins of the ambulacral areas in the genera *Cidaris* and *Hemicidaris*. (Pl. III, fig. 1 *b, c;* Pl. IV, fig. 1 *d.*) They sometimes fill up spaces at the base of the inter-ambulacral areas, as in some *Diademas;* (Pl. VI, fig. 2 *g*) they are always raised on little eminences, and are in general perforated like the secondary and primary tubercles. In this respect they are distinguished from the granules.

The *granules* are small, round, hemispherical elevations, scattered more or less regularly, and distributed over different parts of the plates of the test. In some species of the genus *Cidaris*, they are arranged in rows in the centre of the ambulacral areas (Pl. 1, fig. 4 *e;* Pl. II, figs. 1 *c*, 2 *g;* Pl. III, fig. 2 *d*); or they form circles around the circumference of the areolas of the primary tubercles. (Pl. II, fig. 2 *g.*)

The *miliary granulation* is formed by a number of small granules closely set together in the centre of the ambulacra (Pl. III, fig. 6 *a:* Pl. V, fig. 2 *d*), or on the inter-tubercular surface of the large plates forming the inter-ambulacra. (Pl. II, fig. *b, c;* Pl. VI, fig. 2 *c, d.*)

The primary tubercles of the genera *Cidaris, Hemicidaris, Diadema, Hemipedina, Pedina*, &c., consist of the following parts:

The *hemispherical tubercle* or *mamelon*, which is sometimes perforated in the centre, as in all the *Cidaridæ* (Pl. II, fig. 1 *c, d*): sometimes it is imperforate, as in many *Echinidæ* (Pl. XVI).

The *boss* or *mamillary eminence* is the conical prominence which rises from the surface of the tubercular plate (Pl. II, fig. 1 *d*); its summit supports the tubercle; its margin is sometimes crenulated in *Cidaris, Hemicidaris*, and *Diadema*, and sometimes smooth in *Hemipedina* and *Pedina*. These characters are important for generic distinctions.

The *areola* or *scrobicule* (Pl. II, fig. 1 *c*) is the round, oval, or elliptical, smooth, excavated space which surrounds the base of the boss. This space is sometimes wide (Pl. II, fig. 1 *c*), sometimes narrow (Pl. IV, fig. 1 *d*); sometimes its margin is elevated into a ridge which encircles it completely (Pl. II, fig. 2 *g*); the *scrobicular* or *areolar circle* is then said to be complete; sometimes the upper and under sides of the circle are wanting, when it is *incomplete*, and the areolas in this case are said to be *confluent*. (Pl. I, fig. 1 *b;* Pl. IV, figs. 1 *d*, 2 *c.*)

The row of granules, which encircles the areola, is called the *areolar* or *scrobicular circle*.

The *miliary zone* is the space comprised between the two ranges of primary tubercles. This zone is sometimes destitute of miliary granules, when it is said to be naked; sometimes it is covered with a close-set granulation. We have, therefore, the ambulacral miliary

ECHINODERMATA.

zone, and the inter-ambulacral miliary zone, when they occupy these different regions of the shell.

When the granules are microscopic, and are closely clustered together to form certain narrow, circumscribed bands or *fascioles* which intersect the general tubercular surface, and occupy fixed positions on the test, the microscopic granules composing these zones are called *fasciolar*.

Fascioles are only found in two families, in one genus of the *Echinocorydæ*, and in almost all the genera of the *Spatangidæ*. The form, width, structure, and position of these bands must be carefully noted, as they afford important characters, both positive and negative, which serve in the diagnosis of the genera.

When the fasciole surrounds the circumference of all the petaloidal portions of the ambulacral areas, it is called *peripetalous;* when it encircles the single ambulacrum alone, it is *internal;* when it extends along the flanks, it is *lateral;* when it passes in whole or in part along the circumference, it is *marginal:* and when it surrounds the base of the single inter-ambulacrum, it is *sub-anal:* sometimes there is only *one*, sometimes there are *two* or *three*, of these fascioles in different genera.

Each family has a special arrangement of the tubercles, granules, and fascioles, which will require a detailed notice in the general outline of the structural characters prefixed to the description of each natural group.

The *sutural impressions* are the lines along which the plates are united together. Sometimes they are mere lines (Pl. I, fig. 4 *b;* Pl. II, figs. 1 *b*, 2 *c*); sometimes the impressions amount to excavations out of the borders of the plates, and give rise to cavities therein, as in the genus *Temnopleurus*.

The *angular* or *sutural pores* are small impressions, situated in some genera at the angles of the plates, upon the median line of the miliary zones.

The external appendages of the Test.

The *spines*, or *Radiolii* as they were called by Plott, Langius, and other old authors, are the calcareous appendages which are moveably articulated with the tubercles of the test. They present numerous modifications as to size, form, and sculpture, which are all intimately connected with specific characters. Some are short, elongated, flattened, cylindrical, fusiform, or subulate; others are compressed, spatuliform, or triangular; others, on the contrary, are expanded, pyriform, or claviform. The surface of the spines is smooth, or striated with fine or coarse longitudinal lines; some have verticellate processes at regular intervals; others have asperities, prickles, or granules, disposed with more or less regularity over the surface; the different parts of the spine have received the following names:

The *articular cavity*, or *acetabulum*, is the socket by which the spine articulates with the tubercle.   Its margin is smooth, or crenulated, according as the summit of the boss is smooth or crenulated. (Pl. II, fig. 2 e.)   In all the genera in which the tubercles are perforated, there is a corresponding pit in the socket of the spine for the attachment of a round ligament, which passes from the tubercle to the spine.

The *head* is that part of the spine which contains the articulating cavity, and is united to the stem near the neck. (Pl. I, fig. 1 f; Pl. II, fig. 2 e.)

The *milled ring* surrounds the head.   It is a prominent ridge, more or less deeply crenulated, around which the muscular fibres that move the spine are firmly attached. (Pl. I, fig. 5 b.)

The *neck of the spine* is the smooth space between the line of junction above the milled ring and the rugose body of the spine. (Pl. I, fig. 1 f.)   It is often finely striated with longitudinal lines.   In long slender, tapering spines, (Pl. IV, fig. 1 o,) the neck often cannot be distinguished from the body, into which it passes.

The *stem*, or *body of the spine*, is the part which exhibits the greatest variety of forms. Smooth and muricated varieties are figured in Plates I, II, IV.

The organs of mastication, forming the lantern of Aristotle, are rarely preserved in fossil species.   They consist of five jaws, each carrying a long tooth. (Pl. IV, fig. 1 b, f.)   As these parts form a complicated mechanism, their analyses and description will be given at length in the anatomical part of the Introduction.

In this descriptive terminology of the skeleton, we have limited our observations to the ECHINOIDEA, the iconography and description of the British Oolitic species of which forms the first part of this Monograph: it is our intention to preface the description of the ASTEROIDEA, OPHIUROIDEA, and CRINOIDEA, which follow seriatim, with similar analyses of the specialities of their organization.

On the relative value of the external organs in the classification of the Echinoidea.

The mouth is always basal, central, subcentral, or excentral, but the excentricity is invariably towards the anterior border.   This opening does not, therefore, afford a character of primary importance, although, in connection with others, it is valuable in the definition of families.   The mouth is sometimes armed with jaws, but sometimes it is edentulous.

The position of the anal opening affords a character of primary importance.   In one great section the anus opens *within* the centre of the apical disc, surrounded by the genital and ocular plates.   The relation, therefore, of the digestive organs to those of generation and vision, is an important primary character for the zoologist.   In another section the anal opening is *without* the apical disc, and is more or less external to, and at a greater or less distance from, the genital and ocular plates.   The physiological importance of the external relation of the organs of digestion, generation, and vision, to each other, imparts great value to the position of the anal opening; and hence it forms the basis of the subdivision of this order into two sections, which are thus defined:

Echinoidea endocyclica.

A. Test circular, spheroidal, more or less depressed, rarely oblong; mouth in the centre of the base. Anus in the centre of the upper surface, directly opposite to the mouth, surrounded by the five perforated genital plates, and having external to them the five ocular plates. Mouth always armed with five powerful calcareous jaws, formed of many elements, disposed in a vertical direction.

Echinoidea exocyclica.

B. Test sometimes circular and hemispherical; oftener oblong, pentagonal, depressed, clypeiform or discoidal; mouth central or ex-central. Anus external to the circle of genital and ocular plates, never opposite the mouth, but situated in different positions in relation to that opening: four of the genital plates are generally perforated. The mouth is sometimes armed with five jaws, but it is oftener edentulous. The elements of the lantern are disposed in a more or less horizontal direction.

The structure of the ambulacral areas, and the poriferous zones, which are in relation with the organs of respiration and locomotion, afford good characters of secondary importance, for grouping the genera into natural families, especially when taken in connection with the position of the anal opening, which varies in its relation to that of the mouth opening in different families.

The form, number, and arrangement of the tubercles, and the spines which are jointed with them; the miliary granulation; the bands of microscopic granules forming the fascioles, which have determinate and permanent positions on the test; added to the size and number of the elements of the apical disc, and the position of the anus, afford collectively good characters for defining the genera.

The minute details in the structure of the plates; the form, and size, and number of the tubercles on each of them,—the form and arrangement of the pores in the zones; their proximity or remoteness from each other; the general outline of the test, which has only certain limits of variation; the character of the sculpture on the plates; the form of the areolas; the presence or absence, the size and distribution of the granules forming the areolar or scrobicular circle; the completeness or incompleteness of the same. The length of the spines, their form and sculpture, are all points which afford good specific characters, as they are persistent details which are developed on every considerable fragment of the test and spines of Echinoidea.

3

Taking these general principles for our guidance, we subdivide the Echinoidea into the following natural families :

Order—ECHINOIDEA.	*Echinoidea endocyclica.* Anus within the genital plates, always opposite the mouth.	Cidaridæ. Hemicidaridæ. Diademadæ. Echinidæ. Salenidæ.
	Echinoidea exocyclica. Anus without the genital plates, never opposite the mouth.	Echinoconidæ. Collyritidæ. Echinonidæ. Echinobrissidæ. Echinolampidæ. Clypeasteridæ. Echinocoridæ. Spatangidæ.

Family 1. CIDARIDÆ.—Test thick, spheroidal; inter-ambulacral areas very wide; primary tubercles large, perforated; bosses crenulated or uncrenulated, spines large, thick, and mostly claviform; ambulacral areas very narrow; poriferous zones narrow, pores unigeminal, rarely bigeminal; mouth opening large, inferior, central, circular or pentagonal; peristome destitute of notches, always armed with large and powerful jaws. Anal opening wide, superior, opposite to the mouth, surrounded by five genital plates, perforated for the genital canals, and five ocular plates excavated for lodging the eyes; buccal and anal membranes covered with scales.

Types. *Cidaris florigemma*, Phillips. *Cidaris Smithii*, Wright. (Pl. II.)

Family 2. HEMICIDARIDÆ.—Test thick, spheroidal, or more or less depressed; ambulacral areas narrow or wide, with semi-tubercles at their base only, or extended throughout the area; inter-ambulacral areas with two rows of primary tubercles, rarely more than eight in each row; the tubercles of both areas are perforated, and the bosses are deeply crenulated; the poriferous zones are narrow and undulated; the pores are unigeminal throughout, except near the peristome, where they are bigeminal and trigeminal. The mouth opening is large; the peristome is decagonal, and is divided by more or less deep notches into ten lobes; the jaws are large and powerful; the apical disc opposite the mouth is small, and is composed of five genital plates and five ocular plates. The spines are long, thick, cylindrical, tapering, or claviform, or stout, compressed, and angular; their surface is smooth, or covered with fine longitudinal lines, but, as far as is known, neither prickles nor asperities are developed thereon.

Types. *Hemicidaris intermedia*, Fleming (Pl. IV). *Acrocidaris formosa*, Agassiz.

Family 3. DIADEMADÆ.—Test thin, circular, or pentagonal, more or less depressed;

ambulacral areas wide, with two or four rows of primary tubercles ; inter-ambulacral areas with two, four, six, or more rows of tubercles, nearly of the same size and structure as those of the ambulacra; the tubercles are perforated or imperforated, crenulated or uncrenulated, in different genera ; the apical disc opposite the mouth is small, and composed of five genital and five ocular plates ; the poriferous zones are narrow; the pores are unigeminal or bigeminal. The mouth opening is large and decagonal; the peristome is divided into ten lobes by deep notches; the spines are long, cylindrical, more or less slender, and are either tubular or solid ; sometimes they are encircled by spiral verticillate processes, or their surface is sculptured with fine longitudinal lines.

Types. *Astropyga radiata*, Leske. *Diadema pseudo-diadema*, Lamarck (Pl. VIII, fig. 1). *Cyphosoma Kön9igii*, Mantell. *Hemipedina Marchamensis*, Wright (Pl. XI, fig. 1).

Family 4. ECHINIDÆ.—Test thin, spheroidal; inter-ambulacral areas, with primary tubercles, small and of various sizes, perforate or imperforate ; bosses crenulate or uncrenulate ; ambulacral areas wide, always supporting two or more rows of primary tubercles ; poriferous zones narrow or wide, pores unigeminal, trigeminal, or polygeminal, and disposed in arcs; spines short, mostly subulate. Mouth opening large, inferior, always decagonal; the peristome divided into lobes by notches more or less deep. Anal opening small, superior, opposite the mouth, surrounded by five genital and five ocular plates ; buccal membrane naked.

Types. *Arbacia Forbesii*, Wright (Pl. XIII). *Glypticus hieroglyphicus*, Münster (Pl. XIII). *Echinus bigranularis*, Lamarck (Pl. XIV).

Family 5. SALENIDÆ.—Test thin, spheroidal ; inter-ambulacral areas wide, with few primary tubercles, which are either perforate or imperforate, crenulate or uncrenulate ; ambulacral areas narrow, carrying secondary tubercles. Mouth opening small or large, inferior, decagonal ; peristome more or less notched. Anal opening superior, surrounded by the plates of a large apical disc, composed of more than ten pieces, and which occupies a wide aperture in the superior part of the test. Poriferous zones narrow, pores unigeminal, except near the peristome, where they are trigeminal. Spines long, subulate, circular, or flattened. Species all extinct ; the genera distributed in the Oolitic, Cretaceous, and Tertiary rocks.

Types. *Salenia petalifera*, Defrance. *Acrosalenia hemicidaroides*, Wright (Pl. XVI). *Goniophorus lunulatus*, Agassiz.

Family 6. ECHINOCONIDÆ.—Test thin, circular, elongated or pentagonal, elevated or depressed ; inter-ambulacral areas wide, ambulacral areas narrow ; the external surface of the plates of both covered with numerous small, perforated, and crenulated tubercles ; poriferous zones narrow, pores unigeminal, except near the base, where they are trigeminal. Mouth opening inferior, central, circular, or pentagonal, armed with five jaws ;

peristome notched, dividing the circumference into ten nearly equal lobes. Apical disc central, superior, composed of five genital and five ocular plates; the madreporiform body being large, and extending from the right antero-lateral genital plate into the centre of the disc. Anal opening situated at the upper surface, in the margin, or at the inferior surface of the test. Spines small, short, subulate. The species are all extinct; the genera are distributed in the Oolitic and Cretaceous rocks.

Types. *Echinoconus albo-galerus*, Klein. *Pygaster semi-sulcatus*, Phillips (Pl. XVII). *Holectypus depressus*, Leske.

Family 7. COLLYRITIDÆ.—Test thin, circular, or oval; ambulacral areas meeting at two points on the upper surface, which are more or less apart; poriferous zones narrow, pores unigeminal; tubercles small, numerous, perforated, and crenulated. Mouth opening excentral, small, round, oval; peristome feebly fissured; jaws unknown. Anus round, oval, supra-marginal; elements of the apical disc detached; four genital holes. The species are all extinct, and distributed in the Oolitic and Cretaceous rocks.

Types. *Collyrites ringens*, Desmoulins. *Collyrites ovalis*, Parkinson.

Family 8. ECHINONIDÆ.—Test thin, oval; poriferous zones narrow, meeting at the apical disc; pores unigeminal; tubercles of both areas nearly equal, but neither perforated nor crenulated; spines stout, subulate. Mouth opening nearly central, irregularly pentagonal, and edentulous. Anal opening oblong or pyriform, basal or marginal, closed by anal plates; apical disc nearly central; four genital pores. One group is living in tropical seas; another is found fossil in the Cretaceous rocks.

Types. *Echinoneus cyclostomus*, Leske. *Pyrina Desmoulinsii*, D'Archiac.

Family 9. ECHINOBRISSIDÆ.—Test thin, circular, oblong, sub-pentagonal or clypeiform, covered with small tubercles, surrounded by excavated areolas; ambulacra narrow, enclosed by poriferous zones, more or less petaloidal; pores set at different distances apart, and united by connecting sutures. Mouth opening small, nearly central, pentagonal, and edentulous, in general surrounded by five lobes. Anal opening, lodged in a sulcus, in the upper surface of the single inter-ambulacrum, or in a marginal depression thereof; apical disc small, four genital lobes; madreporiform body extending into the centre of the disc. One species is living, the others are all extinct, and distributed in the Oolitic, Cretaceous, and Tertiary rocks.

Types. *Echinobrissus clunicularis*, Llhwyd. *Clypeus sinuatus*, Leske. *Catopyus carinatus*, Goldfuss. *Pygaulus cylindricus*, Desor.

Family 10. ECHINOLAMPIDÆ.—Test thin, oblong, oval, elevated or sub-discoidal; ambulacra large, petaloid; poriferous zones wide; pores placed apart, and united by a suture; the zones extend near to the margin. Mouth small, surrounded by five lobes, sometimes it is transversely oblong. Anal opening transversely oblong, and infra-marginal; apical disc small, excentral, with four genital holes. Some species are living in warm seas, but the greatest number are extinct, and distributed in the Oolitic, Cretaceous, and Tertiary rocks.

Types. *Echinolampas orientalis*, Gray. *Pygurus depressus*, Agassiz. *Conoclypus Leskii*, Goldfuss.

Family 11. CLYPEASTERIDÆ.—Test thick, elevated or depressed, circular, elliptical, or pentagonal; the surface closely covered with small, nearly equal-sized tubercles, sunk in the plates, and surrounded by ring-like areolas, the tubercles carrying short hair-like spines. The mouth large, central and pentagonal, and armed with five strong jaws, which carry the same number of teeth. The anus posterior, marginal or infra-marginal; the interior of the test is sometimes divided by pillar-like processes formed of the inner layer of the plates. The dorsal portions of the ambulacral areas have a petaloid form, circumscribed by large poriferous zones; the basal portions are narrow, rectilineal, or branched; the five genital plates form a circle around the madreporiform body, and between these are wedged the five ocular plates. This family includes the genera *Clypeaster*, Lamk., *Laganum*, Klein., *Echinarachnius*, Van Phels., *Arachnoides*, Klein., *Scutella*, Lamk., *Dendraster*, Agass., *Lobophora*, Agass., *Encope*, Agass., *Echinodiscus*, Breynius, *Mellita*, Klein., *Runa*, Agass., *Moulinsia*, Agass., *Scutellina*, Agass., *Echinocyamus*, Van Phels., *Fibularia*, Lamk., *Lenita*, Desor.

Types. *Clypeaster rosaceus*, Lamarck. *Scutella subrotunda*, Lamarck. *Echinarachnius placenta*, Gmelin.

Family 12. ECHINOCORIDÆ. — Test thick, oval and elevated, sometimes cordate; ambulacral areas narrow; poriferous zones narrow, pores unigeminal; test covered with small tubercles, which are perforated and crenulated. One genus has a marginal fasciole. Mouth opening small, excentral, transversely oblong. Anal opening nearly of the same size, oblong, marginal, or supra-marginal. Apical disc elongated, nearly central, with four genital pores; the cordate forms have an anterior central depression. The species are all extinct, and are limited to the Cretaceous rocks.

Types. *Echinocorys vulgaris*, Breynius. *Holaster sub-globosus*, Leske. *Cardiaster granulosus*, Goldfuss.

Family 13. SPATANGIDÆ.—Test thin, oval, oblong or cordiform, and satisfactorily exhibiting the bilateral symmetry of the *Echinoidea*. The anal opening is posterior and supra-marginal, and is closed by a complicated series of small plates. The apices of the ambulacral areas are united at the summit of the test. The anterior single ambulacrum has a different structure from the antero- and postero-lateral pairs, and is in general lodged in a depression of the test, which extends to the anterior border, forming the anteal sulcus; the test is extremely thin, and is covered with small tubercles, which support hair-like spines. Besides these, there are some larger crenulated and perforated tubercles, which support large spines. There are two or four genital pores, which are sometimes placed close together, but are in other genera apart. The eye-plates are five in number, and are placed at the apices of the ambulacra, in a pentagonal form, around the genital plates. On the surface of the test of some *Spatangidæ* certain delicate lines called *fascioles* are observed, having a smoother appearance than the tubercular surface of the test : they are furrows, strewed with microscopic tubercles, and destined to carry very delicate spines, which, when seen under the microscope, appear to have the same structure as the *Pedicellariæ*. The fascioles have a different disposition in each genus, and afford a good character in making definitions of the same. When the fasciole surrounds the ambulacral petals like an undulating groove, as in *Hemiaster, Schizaster,* &c., it is said to be *peri-petalous;* when it surrounds the single ambulacrum, as in *Amphidetus*, it is *internal;* when it extends along the sides, as in *Schizaster*, it is *lateral;* when it surrounds the circumference of the test, as in *Pericosmus*, it is *marginal;* when it is limited to the base of the anal opening, it is *sub-anal.* Sometimes, in the same genus, more fascioles than one exist; thus the sub-anal and peri-petal are frequently associated together.

Types. *Spatangus purpureus*, Müller. *Brissus lyrifer*, Forbes. *Brissopsis Ducieii*, Wright.

Family 1—CIDARIDÆ.

Test thick, turban-shaped, more or less depressed at the oral and anal apertures. Mouth opening wide, central; peristome circular or pentagonal, but without notches; aperture closed by a buccal membrane, covered with small spines, metamorphosed into imbricated scales, upon which the pores from the zones are prolonged.

Opening for the apical disc very large; the disc composed of five large equal-sized angular genital plates, and five ocular plates; anus opens in the centre of the disc, directly opposite to the mouth; anal membrane clothed with small angular plates, unequal in size, and variable in number.

Ambulacral areas extremely narrow, composed of a great number of very small plates, which have only minute tubercles, or rows of small granules on their surface, but never supporting tubercles carrying primary spines.

Inter-ambulacral areas very wide, composed of large plates, rarely more than from six to eight in one column; the external surface of each plate carries a large perforated tubercle, raised on a prominent boss, and encircled by an areola, which has either a round or oval form; the areola is surrounded in general by an elevated margin, on which are placed a circle of granules, the scrobicular circle, usually larger than those filling the miliary zone.

The poriferous zones are narrow, and extend without interruption from the margin of the buccal membrane to the apical disc; the pores in general are strictly unigeminal, but in one genus they are bigeminal; the pores are contiguous, or separated by septa more or less thick.

The jaws, five in number, form a very powerful lantern, which is moveably connected with, and supported by, a series of calcareous processes or auricles, arising from the inner surface of the test; the teeth are more simple, and the lantern less complicated, however, than in the *Echinidæ*.

The spines in this family are large, strong, cylindrical, fusiform, prismatic, club-shaped, or flattened; their surface is sometimes covered with longitudinal lines, or with prickles or granules, which in general have a linear arrangement, or they are more or less irregularly disposed; but the form and sculpture of the spine has invariably, as far as is known, a specific value, its dominant characters being always persistent.*

* The form and general character of the spine should, in every case, be examined with scrupulous attention, and, whenever in fossil species the spines are found attached to their test, the facts connected therewith should be noted with the greatest accuracy. The neglect of this caution has been the cause of much confusion, and led to some serious errors, as will appear in the sequel.

The *Cidaridæ* are the most ancient type of the Echinoidea. The remains of different forms of this family are found in the Silurian, Devonian, and Carboniferous rocks, as well as in those of the Secondary and Tertiary periods. In his valuable Synopsis, M. Desor describes six genera in this tribe: these are *Cidaris*, Klein.; *Rabdocidaris*, Desor; *Diplocidaris*, Desor; *Porocidaris*, Desor; *Goniocidaris*, Desor; *Palæocidaris*, Desor. Of this number three are extinct—*Diplocidaris*, *Porocidaris*, and *Palæocidaris;* two contain both extinct and living forms—*Cidaris* and *Rabdocidaris;* and one is only found living—*Goniocidaris.*

Genus—CIDARIS, *Klein.* 1734.

ECHINUS, *Linnæus, Hissenger,* &c.
ECHINITES or HISTRIX, *Bourguet.*
CIDARIS, *Klein, Leske, Defrance, Blainville, Agassiz, Desor.*
CIDARITES, *Lamarck, Goldfuss, Gray, Desmoulins.*

General form circular or turban-shaped; test thick, sub-spheroidal, nearly equally flattened on the under and upper surfaces.

Ambulacral areas very narrow and undulated, supporting only minute granuloid tubercles, or small granules, disposed in two, four, or six close-set rows.

Inter-ambulacral areas at least four times the width of the ambulacral, furnished with two rows of large primary tubercles, from four to six in a row,—rarely there are ten.

The tubercles are perforated; the bosses are large and prominent, and their summits are smooth or crenulated; the areolas are wide, and in general are deeply excavated; they have a round or oval figure, and their elevated margin is in general surrounded by a complete circle of granules (the scrobicular circle).

The miliary zones are concave, and more or less wide in proportion to the size of the primary tubercles; they are filled with numerous rows of small, close-set granules.

The poriferous zones are narrow; the pores are unigeminal and contiguous; they are separated by septa more or less thick.

The spines are robust and massive, cylindrical, fusiform, or claviform; their surface is covered with longitudinal lines, or furnished with prickles, granules, or other asperities, which often assume a linear arrangement.

The mouth opening is circular or pentagonal, without notches; the buccal membrane is covered with imbricated scales, on which the pores from the zones are prolonged.

The apical disc is large, and is composed of pentagonal, nearly equal-sized, genital plates, and triangular ocular plates.

The lantern is powerful, composed of massive pyramids, the branches of which are not united at their summits; the teeth are canaliculated, and formed of a folded plait, without a carina on the inner surface.

Notwithstanding the limitation of the genus *Cidaris* of Klein., by Lamarck, Goldfuss, Agassiz, and Desor, it still forms an extensive group of Urchins, which range from the Palæozoic rocks into our modern seas.

This genus divides itself into two types; in the one, the mammillary bosses have smooth summits; in the other, they are more or less crenulated.

THE FIRST TYPE.—*Tubercles with smooth and uncrenulated bosses, exist in our present seas, and are found fossil in the Carboniferous, Triassic, Cretaceous, and Tertiary rocks.*

4

THE SECOND TYPE.—*Tubercles with the summits of the bosses more or less crenulated, are found in the Triassic and Oolitic rocks.*

There are some exceptions to these rules, for *Cidaris marginata*, Goldfuss, and *Cidaris lævigata*, Desor, both from the Coral Crag, have smooth bosses; and some species from the Neocomian and Cretaceous strata are said to have the summits of these eminences crenulated.

A. *Cidaris from the Lias.*

CIDARIS EDWARDSII, *Wright*. Pl. I, fig. 1 *a, b, c, d, e, f.*

CIDARIS EDWARDSII, *Wright*. Annals and Magazine of Natural History, 2d series, vol. xiii, p. 161, pl. 11, fig. 1 *a—f.*

Test crushed; the true form therefore unknown. Ambulacral areas narrow, gently flexed, and furnished with two rows of small perforated marginal tubercles, and a median row of smaller tubercles irregularly interspersed amongst them; the marginal tubercles being alternately larger and smaller; poriferous zones wide, with large, closely approximated, oblong pores; inter-ambulacral areas about four times the width of the ambulacral, with two rows of large primary tubercles, the areolas of which are confluent throughout; miliary zones wide, and covered with numerous small granules; primary spines long, showing a complex structure; secondary spines short, with blunt apices; the surface of both covered with delicate longitudinal lines; mouth armed with five powerful jaws, each having carinated ridges on their convex surface; upper part of the test unknown.

Description.—The great argillaceous deposits of the Oolitic group—the Lias, the Oxford Clay, and the Kimmeridge Clay—were formed under conditions which appear to have been unfavorable to the development of Urchin life; for, although these rocks have been industriously explored, they have hitherto yielded very few remains belonging to the Echinoidea: this remark does not apply only to the English Oolitic group, but is applicable to the whole series as developed on the continent of Europe. Although a few forms of *Cidaridæ* lived in the Silurian, Devonian, Carboniferous, Permian, and Triassic seas, still it was in the Jurassic ocean that they existed in any considerable numbers; and this fact gives increased interest to the study of the anatomy of these Oolitic representatives of this most beautiful family, one of the oldest of which we have figured in Pl. I, fig. 1. It is much to be regretted that the specimen before us is the only one of the species that has been found in Gloucestershire in anything like a state of preservation, a few fragments of its test having been rarely collected in two or three other localities. At Lyme Regis one specimen of this species, with some of its spines attached, has been discovered; and one was found in the Lias of Yorkshire; but these are the only examples with which we are acquainted. The specimen which forms the subject of our figure was discovered by G. E. Gavey, Esq., C.E., whilst cutting through the shales of the middle Lias, to form the Oxford, Worcester, and Wolverhampton Railway; the zone of Lias in which it was found was that containing *Ammonites maculatus*, Young and Bird; it was associated with three or four other remarkable species of *Asteriadæ*, *Ophiuridæ*, and *Crinoidæ*, in a high state of preservation, and which will be figured and described in the future parts of this Monograph. The conditions under which these Echinoderms were found are curious and deserve notice. *Uraster Gaveyi*, Forbes, was found lying on the upper surface of a slab of sandstone, twelve inches thick, at twenty-five feet below the surface, associated with fragments of *Extracrinus*, *Ammonites*, and other fossils. All the specimens of *Tropidaster pectinatus*, Forbes, and *Cidaris Edwardsii*, Wright, were found imbedded in the under surface of a thick slab of ironstone, at twenty feet below the surface. Almost all the specimens show the under surface uppermost, and most of them had their spines attached to the spinigerous tubercles. *Cidaris Edwardsii*, when first discovered, was so entirely covered with spines, that the inter-ambulacral plates were almost concealed. For further details relating to this section, and the fossils found therein, the reader is referred to Mr. Gavey's paper.*

The specimen is crushed and flattened, and lies with its mouth and jaws uppermost. At our request, Mr. Baily, in his beautiful drawing, has restored the urchin to its globular form; but it must be added, that all the parts of the test, and the details he has given of its structure, are faithful copies of the original. The ambulacral areas are provided with two rows of small perforated tubercles on the margin of the areas, each plate thereof developing alternately a smaller and a larger marginal tubercle (Pl. I, fig. 1 *b*); amongst these a row of small granules are scattered somewhat irregularly throughout the areas. All these tubercles supported short, stout, blunt-pointed spines (fig. 1 *d, e*), the surface of which is delicately sculptured with longitudinal lines: many of these spines are seen *in situ* in the specimen, although many have been removed to expose the surface of the test. The poriferous zones are wide, and follow the gentle flexures of the ambulacral areas; the pores are large and oblong, and the two pores forming a pair are separated by very thin partition walls, a circumstance which forms a good diagnostic character (Pl. I, fig. 1 *b*); there are five pairs of pores opposite each of the large tubercular plates.

The inter-ambulacral areas are four times the width of the ambulacral; they are composed of rather narrow tubercular plates, which support two rows of primary tubercles, set, therefore, rather closely together, in a vertical direction, so that the areolas are all confluent above and below (Pl. I, fig. 1 *b*); the miliary zone is wide, and filled with six rows of minute, close-set, perforated tubercles, each of which is raised on an elevation of the test;

* 'Quarterly Journal of the Geological Society,' vol. ix, p. 29, with sections of the Railway Cuttings.

the semi-circlet surrounding the outer and inner margin of the areolas is composed of granules very little larger than those which fill up the miliary zone. The primary tubercles are of moderate size; the number in each column is uncertain; but, judging from the eight tubercular plates still remaining in an imperfect column, and estimating the number necessary to complete the same, it is probable that the perfect test had from twelve to fourteen tubercles in each row, gradually increasing in size from the peristome upwards; they are proportionately small when compared with the size of the test itself. The areolas are oblong, and the narrowness of the plates causes them to open into each other above and below; the bosses are neither large nor prominent, but the crenulations on their summits (Pl. I, fig. 1 *b*) form a well-marked circlet of granules, which surrounds the base of the tubercles (fig. 1 *c*); these are small, with deep and wide perforations on their summits. The peristome is wide, and, in the only three specimens we know, part of the dental apparatus is preserved *in situ :* the portion remaining in the figured specimen consists of three prominent jaws (fig. 1 *a*); one of these is shown in fig. 1 *g*. It is convex, and the external surface is strengthened by three prominent ridges; the teeth are large, but are all fractured. As the test rests on its upper surface, the ovarial plates and dorsal surface are concealed from observation.

The spines remaining *in situ* on the test are of two kinds (fig. 1 *a*); those articulated with the primary tubercles, the primaries; and those articulated with the smaller tubercles, the secondaries. The primary spines exhibit a peculiar structure (fig. 1 *f*); the head is large, gradually increasing in diameter from the articular cavity to the circular band; the rim of the acetabulum is coarsely and deeply crenulated, and the raised band is narrow, prominent, and finely milled; the neck tapers gradually from the band to the point where it joins the stem, which has the same structure as the head; and the surface is delicately sculptured with fine longitudinal lines (fig. 1 *f*).

The stem is united to the neck by an oblique harmonia suture; in those spines which are denuded of their external layer, it has a horny semi-transparent aspect; but when this external layer is present, its surface is seen, with a pocket lens, to be sculptured with fine longitudinal lines; besides these, there are a number of small processes arranged in rows, the points of which have a direction forwards, but their linear arrangement is by no means regular; the stem is long, slender, and circular, its length in proportion to the diameter of the test has not been ascertained, as all the spines are fractured. The secondary spines are very uniform in size and structure, and are preserved in abundance in the specimen before us (fig. 1 *a*). Each of the minute tubercles of the ambulacral areas, and of the miliary zones, carried one of these spines, which measure about $\frac{6}{20}$ths to $\frac{1}{20}$ths of an inch in length; they are stout and blunt pointed, tapering very little from the milled band to the apex, and having their surface covered with microscopic longitudinal lines (fig. 1 *d, e*).

Affinities and differences.—This Cidaris belongs to the same group as *Cidaris Fowleri,*

Wright, and *Cidarites maximus*, Münster, but it presents many important characters by which it is distinguished from both. From *Cidaris Fowleri* it differs in wanting the regular close-set row of marginal granules, with the four internal rows of minute granules in the ambulacral areas, which form so fine a character in that Inferior Oolite species (Pl. I, fig. 4 *e*) ; the irregular distribution of the small tubercles in these areas in *Cidaris Edwardsii* forms a marked distinction between them : the poriferous zones are occupied by large pores, separated from each other by thin septa; in *Cidaris Fowleri* the pores are small, oblong, and separated by a thick portion of the test (fig. 4 *e*). The tubercular plates in *Cidaris Edwardsii* being narrow, there are consequently a greater number of primary tubercles in the inter-ambulacral areas : the areolas, likewise, are oblong, and always confluent in the vertical direction; whereas in *Cidaris Fowleri* the areolas are nearly circular (fig. 4 *e*), and each is surrounded by a circle of granules. The inter-tubercular space or miliary zone in *Cidaris Edwardsii* is filled with small perforated tubercles, but in *Cidaris Fowleri* the same space is occupied by small imperforate granules, more numerous, but less regular in size and distribution (fig. 4 *e*) : the spines in *Cidaris Fowleri* are stout, and have well-developed rows of forward-directed prickles (fig. 4 *d*) ; whereas in *Cidaris Edwardsii* they are long and slender, with small irregular elevations of the surface, which is likewise sculptured with microscopic longitudinal lines.

Locality and Stratigraphical position.—This fine urchin was discovered in the upper shales of the Lower Lias, or Middle Lias, by G. E. Gavey, Esq., C.E., at Mickleton tunnel, near Chipping Campden, in the zone of that formation containing *Ammonites maculatus*, Young and Bird. The bed which yielded *Cidaris Edwardsii* contained likewise other species of Echinoderms in a fine state of preservation : these were—

Cidaris Edwardsii, Wright	*Tropidaster pectinatus*, Forbes
Hemipedina Bowerbankii, Wright, plates of	*Ophioderma Gaveyi*, Wright
Uraster Gaveyi, Forbes	*Extracrinus robustus*, Wright
	„ Two species undetermined.

Besides these Echinodermata, 40 species of Conchifera, 8 species of Gasteropoda, and 10 species of Cephalopoda, with Serpulæ, and 2 new species of *Pollicipes*, were associated with the same. For more details, the reader may consult Mr. Gavey's paper.[*] Isolated plates of *Cidaris Edwardsii* have been collected from the same zone of the Lias in other localities in Gloucestershire, but the specimen we have figured is the only one which displays the true character of the species. Professor Morris has kindly communicated a Cidaris with spines, collected from the Lias at Lyme Regis, which has proved to be a small specimen of *Cidaris Edwardsii* : this beautiful fossil shows many of the long slender

[*] 'Quarterly Journal of the Geological Society,' vol. ix, p. 29.

spines *in situ* on the slab.   Like the specimen we have figured, it lies crushed upon its
upper surface; the mouth-opening contains some of the jaws and teeth.   A crushed
Cidaris, on a slab of Lower Lias shale, from Brockeridge Common, Gloucestershire,
has been communicated by Professor Buckman.   The areolas in this specimen are
oblong, and some of them are entirely surrounded by a scrobicular circle of small, close-
set granules, whilst a few only are confluent.   The character of the bosses, however, with
their flat, deeply-crenulated summits, and small, perforated tubercles, convince me that
it is *Cidaris Edwardsii.*   Had a doubt existed, it would have been removed by the
discovery of one of the primary spines laid along the concealed portion of the test,
which exhibited the crenulated head and milled neck, with longitudinal lines, so very
characteristic of this Middle Lias urchin.   The entire confluence of the areolas in the
large specimen may prove to be an adult character; as the crowding together of so many
tubercles in each column of tubercular plates, may have necessitated the absorption of
a portion of the scrobicular circles, which existed in juvenile individuals.   On this
interesting slab there are two separate vertebræ of an Ichthyosaurus.   Professor Phillips, in
his 'Geology of Yorkshire,' figures a spine, Pl. XIII, fig. 17, which resembles those of
our urchin, but its characters are not given in detail.   Goldfuss* mentions spines, which
he erroneously referred to *Cidaris Blumenbachii,* that were collected from the " Gryphi-
tenkalke der Lias-Formationen bei Pretzfeld und Theta, und ist der einzige Echinit, der
in Baiern dieser Formation angehort."   One of the " aculeorum variæ formæ fragmenta"
(i, fig. 4, Pl. XXXIX) so very much resembles the spines attached to the test of *Cidaris
Edwardsii,* that it is possible it may be the same.   Quenstedt† has found spines in the
Lias, " am Donau-Mainkanal bei Dörlbach," one of which (Pl. XLVIII, fig. 30) resembles
those of our species.   Another spine, from the Inferior Lias of Dasslingen, Wurtemberg
(figs. 31, 32), has fine longitudinal lines, with small wart-like eminences, not arranged in
rows.   All these Liassic urchins present nearly allied characters, and prove that Cidarites,
belonging to the same natural group, and if not to the same, at least to allied species
of the genus Cidaris, lived over an extensive area during the deposition of the upper
shaly beds of the Lower Lias, or the Middle Lias of Continental geologists.

In the Scarborough Museum I found a beautiful specimen of this species, with many
spines attached, as large as the one figured in this Monograph.   It was collected from
the Middle Lias of the Yorkshire coast.

This species is dedicated to Professor Milne Edwards, of the Museum of Natural
History at the Jardin des Plantes, Paris.

---

* 'Petrefacta Germaniæ,' vol. i, p. 117, tab. 39, fig. 4 *c—k.*
† 'Handbuch der Petrefaktenkunde,' p. 574, pl. 48, figs. 28—32.

CIDARIS ILMINSTERENSIS, *Wright*, n. sp.   Pl. V, fig. 6 *a, b*.

Test large, form unknown; ambulacral areas narrow, with two marginal rows of small granules; poriferous zones narrow, pores nearly round, and closely approximated; inter-ambulacral plates large, as deep as they are wide; areolas large, circular, surrounded by a complete scrobicular circle of small granules; mammary bosses not much elevated; summits deeply crenulated; tubercles of moderate size.

Description.—We only possess, of this species, one tubercular plate, and a portion of a second, with one half of the corresponding part of the ambulacral area, but still this fragment is sufficient to enable me to give its diagnosis, and point out those characters by which it is distinguished from its congeners. The test must have attained a size quite equal to that of *Cidaris Fowleri;* the ambulacral areas are very narrow, and have only two rows of marginal granules, there being one granule on each alternate plate; the poriferous zones are narrow, the holes are contiguous and nearly round, and each pair has a thin septum separating them (Pl. V, fig. 6 *b*); there are fifteen pairs of holes opposite each large inter-ambulacral plate.

The inter-ambulacral areas are composed of large, nearly quadrate, tubercular plates (fig. 6 *b*); the number of the primary tubercles is unknown; the areolas are large and circular, and are surrounded by a scrobicular circle, composed of very small granules (fig. 6 *b*); down the inter-tubercular space there were other four rows of granules of nearly the same size; the scrobicular circle extends so very near to the poriferous zones, that there is no granular space intervening between them (fig. 6 *b*); the mammary boss is not very prominent, but its summit is sculptured with well-defined crenulations; the tubercle is of moderate size, and has a wide perforation in its summit.

Affinities and differences.—This species differs from *Cidaris Edwardsii* in having a complete scrobicular circle of granules around the areolas, in the narrowness of the ambu-lacral areas, with only two marginal rows of granules, and in the scrobicular circle abutting against the poriferous zones without a row of granules separating them; the smallness of the holes in the zones, and their proximity to each other, is another point of difference. Compare Pl. I, fig. 1 *b*, with Pl. V, fig. 6 *b*.

Cidaris Ilminsterensis differs from *Cidaris Fowleri* in having no granules between the scrobicular circle and the zones, the latter having two or three rows (fig. 4 *e*, Pl. I); the breadth of the ambulacra, and its four rows of granules, and the wide poriferous zones, with the holes kept far apart by very thick septa, serve further to distinguish that Inferior Oolite form from this Upper Lias species.

Locality and Stratigraphical position.—It was collected by Mr. Moore in the Upper

Lias of Ilminster, and was found associated with *Ammonites serpentinus*, Schlotheim, *Ammonites annulatus*, Sowerby, and other Upper Lias fossils. The fragment figured was the only one ever found.

B. *Species from the Inferior Oolite.*

CIDARIS FOWLERI, *Wright.*　Pl. I, fig. 4 *a, b, c, d, e.*

CIDARIS CORONATA.		Murchison's Geology of Cheltenham, 2d ed., by Buckman and Strickland, p. 73.
—	—	Morris's Catalogue of British Fossils, Inf. Oolite, Cotteswold Hills, 1st ed., p. 49.
—	FOWLERI.	Wright, Annals and Magazine of Natural History, 2d series, vol. viii, p. 246, pl. 11, fig. 5 *a, b.*
—	—	Desor's Synopsis des Echinides Fossiles, p. 6, tab. 3, fig. 13.
—	—	Morris's Catalogue of British Fossils, 2d ed., p. 74.

Test spheroidal, depressed at both poles; ambulacral areas prominent, flat, narrow, and slightly undulated, with two marginal rows of small, equal-sized, close-set granules, and two rows of central, irregular, and almost microscopic granules, with still smaller granules interspersed between them; poriferous zones broad, each as wide as the ambulacra; pores oblong, set wide apart, with a thick septum between the two pores of each pair; inter-ambulacral areas with two rows of primary tubercles, six to eight in each row, all the areas surrounded by a complete scrobicular circle of granules; miliary zone wide, filled up with small close-set granulations; spines large, covered with irregular rows of forward-directed prickles.

Dimensions.—Height one inch and one tenth, transverse diameter one inch and eight tenths; two crushed specimens, the relative dimensions of which, therefore, cannot be accurately ascertained, measure considerably more.

Description.—This beautiful Cidaris, which, when first discovered, more than fifteen years ago, was supposed to be the *Cidaris coronata*, Goldfuss, and was catalogued as such in the works cited in the synonyms of the species. Apart from the organic characters to be pointed out in the sequel, it is a form which has hitherto only been found in the Inferior Oolite, whilst *Cidaris coronata*, Goldf., as constantly characterises the upper zone of the middle division of the Jurassic group. In the Swiss Jura, for example, it is found in the "terrain à chailles," a local formation, the greatest similarity to which, it appears, is the lower Calcareous Grit of England; in the Coral Rag, or Argovien, of Randen, of Birmansdorf, of the valley of the Birse, and of the same stage in Bavaria and

Wurtemberg. It is found likewise in the Coral Rag of Besançon, in the Corallian stage at Chatel-Censoir, and at Druyes, Yonne, in Ain, Isere, and other departments of France. On the Continent *Cidaris coronata* is therefore a characteristic Coral Rag urchin; but it has never yet been found in this or any other stage in England.

The ambulacral areas in *Cidaris Fowleri* are slightly prominent and undulated; they are nearly uniform in width throughout, and are furnished with two marginal rows of nearly equal-sized, close-set granules (Pl. I, fig. 4 *e*), which become larger near the peristome; within these marginal pairs are two other rows of smaller granules, with other microscopic granulations interspersed among them; these inner rows gradually disappear above and below, and the marginal rows then come into juxtaposition; the poriferous zones form wide ribbon-like depressions, nearly equalling in breadth that of the area (Pl. I, fig. 4 *e*); the pores are oblong, and the holes forming a pair are set widely apart, with a thick septum, equalling the long diameter of the pores, separating them from each other. In the large specimens there are seventeen pairs of pores opposite each of the large tubercular plates.

The inter-ambulacral areas are formed of large plates, nearly as broad as they are long (Pl. I, fig. 4 *b*), their lines of sutures being very distinct in young as well as in old individuals; each column consists of from six to eight tubercular plates, and each plate externally has a large flat, nearly circular areola (Pl. I, fig. 4 *e*), slightly furrowed towards its outer border; in the centre rises abruptly the mammillary boss, the summit of which is sculptured with about fifteen deep crenulations; from the interior of these rises a short cylindrical stem, which terminates in a small hemispherical, deeply perforated, spinigerous tubercle, the diameter of the tubercle slightly exceeding that of its stem; the margins of the areolas are bounded by a circle of fifteen scrobicular granules (fig. 4 *e*), which are arranged with much regularity, each granule having a microscopic circlet of granulets around its base; the scrobicular circles on the upper half are more fully developed than those on the lower half of the test; from the equator upwards, each areola has its own distinct scrobicular circle; but from that line to the peristome, one scrobicular line of granules separates two areolas above and below from each other; the miliary zone is wide, and is thickly covered with very small granules, which diminish in size from the scrobicular circle to the median suture; among the principal granules filling up this interspace are others of still smaller size; the entire granulation does not observe much regularity, although it follows lines which somewhat describe the tract of the median suture: the space between the scrobicular circle and the poriferous zones is not so wide as that between the scrobicular circle and the centro-sutural line; this space is likewise filled up by granulations of the same size and character as those which fill up the median miliary zone.

The mouth opening measures half the diameter of the test at the equator, and the peristome has a pentagonal form. (Pl. I, fig. 4 *a*.) In two of the specimens before us, the lantern is preserved, consisting of five strong pyramidal jaws, armed with conical teeth.

(Pl. I, fig. 4 *a*.)   The plates composing the apical disc are absent in all the specimens hitherto found.

The spines (Pl. I, fig. 4 *d*) are large; the milled band below the neck is close to the cup-like articulating cavity; the neck increases slightly in thickness, and the nearly round stem is covered with longitudinal rows of short, sharp prickles, which have their points directed forwards; the stem of the spine is slightly flattened.   Fortunately one specimen was found by Mr. Gibbs, of the Geological Survey, with the spine *in situ* attached to the test, so that all doubt about the species to which it belongs is removed by this discovery. The specimen with the spine attached is in the Geological Museum in Jermyn Street. The secondary spines are short, and blunt pointed, but only a very few of these have been found.   Those belonging to the scrobicular circle are larger than those which armed the small granules.

Affinities and differences.—This urchin very much resembles an undescribed species collected by M. De Loriere, from the Inferior Oolite of the department of the Sarthe.   There are some slight shades of difference between the French specimens and *Cidaris Fowleri*, but the general resemblance between them is so very great, that we believe them to be identical.   The rock from which the French urchin was collected is referred to the " étage Callovien ;" but it contains several species which hitherto have only been found by us in the Inferior Oolite of England, as *Holectypus gibberulus*, Agassiz, *Pygurus depressus*, Agassiz, and *Clypeus Agassizii*, Wright.   From these facts we are disposed to think that there must be a mistake about the true stratigraphical position of the formation from whence these urchins have been collected.

In the general structure of the test, and in its ornamentation, *Cidaris Fowleri* closely resembles *Cidaris Orbignyana*, Agassiz,* from the Kimmeridge Clay of Havre.   If the palæontologist had to decide the species from the anatomy of the test alone, he would have great difficulty in distinguishing the one form from the other; but fortunately the spines of both are now known, which settles the question as to the specific difference existing between them.   We have the test and spines of *Cidaris Orbignyana* before us, but the reader will find fine figures of both in M. Desor's Synopsis, where it is admirably figured for the first time, and to this work we beg to refer the reader.   Compare, for example, Pl. III, fig. 13, spine of *Cidaris Fowleri*, with Pl. VIII, figs. 7—9, the spine of *Cidaris Orbignyana.*

In the general form and structure of the test, *Cidaris Fowleri* resembles *Cidaris florigemma*, but it differs from that well-known Coral Rag species in the flatness of its ambulacral areas, in the greater breadth of the poriferous zones, its smaller primary tubercles, the depth and character of the crenulations on the bosses, and, above all, in the structure of its spines.   Compare Pl. I, fig. 4 *d*, with Pl. II, fig. 2 *a*.

* A beautiful figure of this fine species is published in M. Desor's ' Synopsis des Echinides Fossiles,' pl. 1, fig. 3 ; pl. 8, figs. 7—9, contains the spines.

It is distinguished from *Cidaris Edwardsii* by the width of the poriferous zones, and the thickness of the septum which separates the pores composing a pair; by the circular form of the areas, and the complete scrobicular circle of granules that surround and limit them in *Cidaris Fowleri*: whereas in *Cidaris Edwardsii* the areolas are oblong and confluent. In the structure of the spines there are likewise important differences. Compare Pl. I, fig. 1 *f*, with fig. 4 *d*.

Cidaris Fowleri is distinguished from *Cidaris Bouchardii* (Pl. I, fig. 2 *a*, *b*, and Pl. VIII, fig. 3 *a*, *b*, *c*), which occurs with it in the same bed, by the latter being in general a rather more depressed form, its ambulacral areas having only two marginal rows of granules; the poriferous zones being much narrower, and the areolas of the primary tubercles more deeply sunk in the plates; the tubercles themselves are larger, the scrobicular circle is more prominent, and the granulations in the miliary zone are fewer and larger.

Cidaris Fowleri differs from *Cidaris Wrightii* (Pl. I, fig. 3), which is likewise found rarely in the Pea Grit, by the latter having a very thick test, large primary tubercles, with very small perforations; the poriferous zones are narrow, and the ambulacral areas have only two rows of marginal granules (fig. 3 *c*). Moreover, it is a much smaller and a more inflated form than *Cidaris Fowleri*.

Only one tubercular plate of *Cidaris Ilminsterensis* from the Upper Lias (Pl. V, fig. 6 *a*, *b*) is known; but so finely are the specific characters defined upon that fragment, that we are enabled to point out the differences between these two species, so nearly related to each other in time. In *Cidaris Ilminsterensis* the scrobicular circle of granules abuts very close upon the poriferous zones, which are narrow, and have the pores placed close together; whereas in *Cidaris Fowleri* there is a considerable granulated space between the scrobicular circle and the poriferous zones, which are wide, and have their pores placed far asunder.

Locality and Stratigraphical position.—*Cidaris Fowleri* was obtained from the Peagrit of the Inferior Oolite at Crickley Hill, Gloucestershire. It has been likewise found in the same bed at Birdlip and Leckhampton Hills. The remarkable rock in which these Cidarites are found seems to have been a tranquil littoral deposit in shallow water, as the most of the Echinoderms are finely preserved, with all the sharpness of their sculpture quite intact. Associated with this urchin are—

Cidaris Bouchardii, Wright	*Hemipedina Bakeri*, Wright
Cidaris Wrightii, Desor	*Hemipedina perforata*, Wright
Diadema depressum, Agassiz	*Hemipedina tetragramma*, Wright
Echinus germinans, Phillips	*Hemipedina Waterhousei*, Wright
Polycyphus Deslongchampsii, Wright	*Hemipedina Bonei*, Wright

BRACHIOPODA.

Terebratula simplex, Buckman. (*Terebratula trigonalis*, Lhwydd.)
Terebratula plicata, Buckman.
Terebratula submaxillata, Davidson.

CEPHALOPODA.

Ammonites corrugatus (*Murchisonæ*), Sowerby.

This species is dedicated to our friend, Charles Fowler, Esq., who generously added to our collection the fine specimen figured in detail.

CIDARIS BOUCHARDII, *Wright.* Pl. I, fig. 2 *a, b, c;* Pl. VIII, fig. 3 *a, b, c.*

CIDARIS ELEGANS. Morris's Catalogue of British Fossils, 1st ed., p. 49, 1843.
— BOUCHARDII. Wright, Annals and Magazine of Natural History, 2d series, vol. xiii, p. 163, pl. 11, fig. 2.

Test circular, much depressed; ambulacral areas narrow and flexuous; poriferous zones deep and narrow; inter-ambulacral areas with two rows of primary tubercles, six to seven in each row; areolas circular, deeply excavated, and entirely surrounded by an elevated scrobicular circle of large granules; a zigzag depression extends through the centre of the inter-ambulacral areas; centro-sutural line strongly marked.

Dimensions.—A large specimen: Height, nine tenths of an inch; transverse diameter, one inch and seven tenths. A moderate-sized specimen: Height, eleven twentieths of an inch; transverse diameter, one inch and two tenths.

Description.—This beautiful urchin was entered in the Catalogue of British Fossils as *Cidaris elegans*, Goldfuss. A comparison, however, of several individuals of *Cidaris Bouchardii*, with typical specimens of Goldfuss's species, kindly sent us by our friends, Prof. Roemer, of Bonn, which he had identified with the original *Cidaris elegans*, Goldf., in the Bonn Museum, and Dr. Fraas of Stuttgard, has enabled us to separate these two forms.

The test of *Cidaris Bouchardii* is circular, and much depressed, from the great flattening of both of the upper and under surfaces; the ambulacral areas are narrow, and much undulated; they have two marginal rows of small granules, with a few more minute ones scattered irregularly between them; the poriferous zones are narrow and slightly sunk; the holes are circular and contiguous, and are separated by a prominent granule rising from the surface of the septum. (Pl. I, fig. 2 *c.*)

The inter-ambulacral areas are five times the width of the ambulacral (Pl. I, fig. 2 *a, b, c;* Pl. VIII, fig. 3 *a, b*); the primary tubercles are small, and from six to seven in each row;

the areolas are circular and deeply sunk (fig. 2 c); around the margin of each there is a complete elevated scrobicular circle of small close-set granules; the mammary boss rises abruptly from the centre, and its summit is sharply crenulated; the tubercle is proportionately small, and deeply perforated with a small hole; the elevation of the scrobicular circle gives an excavated air to the areolas of this species, and produces, at the same time, a marked zigzag depression down the centre of the areas; the miliary zone is wide, and filled with six rows of fine granules, rather smaller in size than those forming the scrobicular circles.

The mouth opening is small, being less than one half the diameter of the test. In a small specimen measuring $\frac{11}{20}$ths of an inch in diameter, that of the peristome was a little more than $\frac{7}{20}$ths of an inch. This opening lies in a slight depression of the under surface; the peristome has a pentagonal form; and the ambulacral areas retain their full diameter to the margin of the same.

The apical disc is absent in all the specimens; but the dimensions of the aperture occupied by these plates is very considerable, being more than half the diameter of the test. In a small individual before us, measuring $\frac{16}{20}$ths of an inch in diameter, the diameter of the opening for the apical disc measures $\frac{9}{20}$ths of an inch. In this small specimen, in fine preservation, which we collected in Dorsetshire, the margin of the discal opening is entire, so that we can state with certainty its comparative measurement, with that of the equatorial diameter of the test itself. It is the only individual that has afforded this information, all the others being more or less crushed, broken, or otherwise distorted.

Affinities and differences.—*Cidaris Bouchardii* has been mistaken for, and registered as *Cidaris elegans*, Goldf., but, although it has some affinities with that species, its differences are numerous and important. The ambulacral areas in both species have two rows of marginal granules, and the poriferous zones in both are of about the same width and depth. In the inter-ambulacral areas the differences are conspicuous: the number of tubercles is greater in *Cidaris Bouchardii* than in *Cidaris elegans;* and the sunk areolas, with their elevated scrobicular circle, form another important distinction between our urchin and the German form. The tubercles, moreover, are larger and much more prominent in *Cidaris elegans:* it likewise belongs to a higher stratigraphical zone, the specimens before us having been found in the Coral Rag of Streitberg and Sigmaringen. Another fact, of a negative character it is true, but not the less valuable in palæontological investigations, is this, that we have never seen spines at all approaching the curious and singular character exhibited by those of *Cidaris elegans* in the Inferior Oolitic rock, from whence our urchins have been collected.

Cidaris Bouchardii resembles *Cidaris coronata*, Goldf., in many points. It has the depressed form, the flexuous ambulacra, and the prominent scrobicular circle of this German species, but is distinguished from it by the following characters : The equatorial portion of the ambulacral areas in *Cidaris coronata* has four rows of granules, which diminish to two rows, above and below ; these areas, therefore, are enlarged in the middle, and taper towards the peristome and towards the apical disc, whereas in *Cidaris Bouchardii* they retain very nearly a uniform width throughout.   There are seldom more than four tubercular plates in each column in *Cidaris coronata;* whereas in *Cidaris Bouchardii*, although a small species, the numbers are from six to seven.   In *Cidaris coronata* the scrobicular circles of the upper tubercular plates are separated by a considerable granulated space, whereas in *Cidaris Bouchardii* the scrobicular circles are contiguous ; the granules of the scrobicular circles are likewise larger and more spaced out in *Cidaris coronata* than in *Cidaris Bouchardii*, and the primary tubercles are proportionately larger in the former. Although both urchins have zigzag depressions down the centre of the inter-ambulacral areas, in *Cidaris coronata* the miliary zone is much wider, and covered with larger granules. The poriferous zones are sunk in very deep depressions in *Cidaris coronata*, arising from the prominence of the granules on the ambulacral and inter-ambulacral areas, but these zonal depressions are not so deep in our species.

Cidaris Bouchardii resembles *Cidaris propinqua*, Münster, only in the depressed form of the test.   In all other points it is sufficiently distinct from that species.   In *Cidaris propinqua* there are four primary tubercles in each column of tubercular plates ; and those on the upper surface are very large in proportion to the size of the test ; their areolas have entire scrobicular circles of large spaced-out granules, which are closely contiguous ; the ambulacral areas are much flexed, and have two rows of close-set marginal granules ; the poriferous zones are slightly sunk, and have very small holes.   The spines of *Cidaris propinqua* have an elliptical-shaped body, covered with longitudinal rows of small tubercles, and united together by connecting calcareous threads.

Cidaris Bouchardii resembles *Cidaris marginata*, Goldf., in the depressed form of the test, and in having, like it, deep sunk areolas, with a wide, depressed, much granulated miliary zone, between the primary tubercles ; but it is distinguished from *Cidaris marginata* in having narrower ambulacral areas, with only two rows of granules ; whilst *Cidaris marginata* has four rows throughout.   The primary tubercles are likewise smaller, and their bosses are sharply crenulated ; whilst in *Cidaris marginata* they are smooth and uncrenulated.   This magnificent Coral Rag German urchin is, moreover, a much larger form than our species from the Inferior Oolite.

Locality and Stratigraphical position.—This is a very rare urchin, and was discovered by us about three years ago in the Pea Grit of the Inferior Oolite of Birdlip and Crickley Hills.   Since Pl. I was finished, we have obtained the finest specimen yet known, which presents many points of importance that are not shown in our earlier found specimens. This urchin is figured in Pl. VIII, fig. 3 *a, b, c.*   We collected, with Dr. Syme, *Cidaris Bouchardii* in the Inferior Oolite near Bridport, Dorsetshire, in beds which are equivalent to the Upper Ragstones of the Inferior Oolite of Gloucestershire.   The Dorsetshire specimens were associated with *Collyrites ringens, Collyrites bicordatus*, and *Clypeus altus*, being all species belonging to that zone of the Inferior Oolite which contains *Ammonites Parkinsoni*, Sow., and *Ammonites subradiatus*, Sow. Some separate plates of an urchin collected with spines from the Bradford Clay, near Tetbury-road Station, Great Western Railway (Pl. V, fig. 7 *a, b*), closely resemble this form; but no entire test of this Cidaris has been found, and it is doubtful whether or not it is distinct from *Cidaris Bouchardii*.

We dedicate this urchin to our friend M. Bouchard Chantereaux, of Boulogne-sur-Mer, to whom we are indebted for some beautiful and rare Jurassic Echinoderms and other fossils, sent by him in the kindest manner to aid us in the production of this Monograph.

CIDARIS WRIGHTII, *Desor*. Pl. I, fig. 3 *a, b, c, d, e, f.*

> CIDARIS PROPINQUA. Wright, Annals and Magazine of Natural History, 2d series, vol. viii,
> p. 250, pl. 11, fig. 6.
> WRIGHTII. Desor's Synopsis des Echinides Fossiles, p. 7.

Test thick, circular, inflated, not much depressed at the poles; ambulacral areas very narrow, sinuous, and furnished with two rows of close-set marginal granules throughout; the inter-ambulacral areas have six tubercles in each column of plates; the tubercles are large, and slightly perforated; the upper mammillary bosses only are crenulated; the areolas are surrounded by complete scrobicular circles of small, prominent, well-spaced-out granules; apical disc unknown; spines never found attached to the test.

Dimensions.—Height, six tenths of an inch; transverse diameter, 1 inch.

Description.—When this urchin was found, five years ago, we were then unable to compare it with a type specimen of *Cidaris propinqua*, Münster. A subsequent comparison with the German species has convinced us that we were mistaken in our determination, and it has long lain marked as a new species in the trays of our cabinet. M. Desor, in his 'Synopsis des Echinides Fossiles,' has likewise detected the difference between this species and *Cidaris propinqua*, and justly observes—" Petite espèce voisine du *Cid. propinqua*, mais plus renflée; les tubercules paraissent aussi être moins gros. Ambulacres très étroits, composés de deux rangées seulement de granules."* This rectification removes another of those apparent exceptions to the law which regulates the distribution of species in time and space, for it is now evident that no species of Echinoderm is common to the Inferior Oolite and Coral Rag of England; but, on the contrary, that both these formations are characterised by distinct and well-defined species of this class.

The test of *Cidaris Wrightii* is very thick for so small a species. It is inflated at the

---

* 'Synopsis des Echinides Fossiles,' p. 7.

sides, and not much depressed at the poles (Pl. I, fig. 3 *a, b*); the ambulacral areas are very narrow and sinuous, having two rows of small, equal-sized, close-set granules arranged on the margins thereof; the poriferous zones are rather deeply sunk, from the prominence of the ambulacral granules, and those of the scrobicular circles, which extend to the zones; the pores are round, and the septa between the pores are about the thickness of the diameter of one of the holes: there are thirteen pairs of pores opposite one inter-ambulacral plate. (Pl. I, fig. 3 *c*.)

The inter-ambulacral areas, at the equator, are five times the width of the ambulacral; each tubercular column consists of six plates, which support very large, prominent tubercles; the areolas are circular, and surrounded by a complete scrobicular circle of prominent, well-spaced-out granules (fig. 3 *c*); the mammillary bosses are small; the summits of the three inferior mammæ are smooth and destitute of crenulations, those of the two or three upper ones are feebly crenulated; the tubercles are disproportionately large to the size of the bosses, and even of the test itself; they are hemispherical eminences terminating a stout stem, and all their perforations are in the form of oblong slits; the size and prominence of the tubercles form an excellent diagnostic character for this species; the scrobicular circle consists of fourteen small, round, prominent granules, raised on little basal eminences, and forming a distinct beaded wreath around each areola (fig. 3 *c*).

The miliary zone, between the two rows of tubercles, is slightly concave; it does not extend throughout to the apical disc, in consequence of the large upper tubercles encroaching on each other; in its widest part it is filled with three rows of granules, much smaller in size than those forming the scrobicular circle.

The mouth opening is circular, about one half the diameter of the test at the equator; the primary tubercles near the peristome are well developed, and very prominent; they increase gradually in size towards the upper surface; the opening for the apical disc is of moderate dimensions.

The spines which are referred to this species (Pl. I, fig. 3 *d, e*) belong to the same group as that to which *Cidaris florigemma* belongs. They are from an inch and a quarter to an inch and a half in length; the stem swells gently outwards towards its inferior third, and then tapers to the extremity; it is covered with longitudinal rows of granules, united together by lines of calcareous threads (fig. 3 *f*), which shows the surface of the spine enlarged several diameters. As these spines have not been found attached to the test, it is only the smoothness of the acetabulum, and the similarity in size, that gives probability to the supposition that they really belonged to the test to which we have provisionally assigned them. As many grave errors have been committed by assigning spines to tests to which they never belonged, we caution the reader, that unless spines have been found on the test, or *in situ* on the tubercles thereof, we must treat the supposition of their belonging to the same individual as a question of doubt.

Affinities and differences.—*Cidaris Wrightii* resembles *Cidaris propinqua* in the disproportionate size of its tubercles to that of the test, and in having some of the bosses smooth, and others feebly crenulated, but it is distinctly separated from that form by having six plates in the tubercular columns, *Cidaris propinqua* having only five; the test is more globular and inflated, the areolas are smaller, the circles of scrobicular granules are not so prominent, and the tubercles are more crowded together; the spines of *Cidaris Wrightii* are larger in proportion to the diameter of the test than in *Cidaris propinqua;* their stems are smaller, and taper more, and the rows of tubercles are not very prominent; whereas in *Cidaris propinqua* the stems of the spines are thick and elliptical, and the tubercles are large and bead-like on their surface.

Cidaris Wrightii is distinguished from *Cidaris Bouchardii*, with which it is associated in the same bed, by its globular and inflated test, the size of the tubercles, and the shallowness of the areolas; *Cidaris Bouchardii* having a depressed test, with small tubercles, and deeply sunk areolas.

From *Cidaris Fowleri* this species is distinguished by the size and prominence of the tubercles, the narrowness of the ambulacral areas and poriferous zones, and the greater proportionate height and inflation of the test; the ambulacral areas in *Cidaris Fowleri* are broad, with four rows of small granules; and the pores in the poriferous zones are wide apart. The spines in these two associated species are likewise very different; those of *Cidaris Fowleri* are oval, and covered with irregular rows of small spines; whilst those of *Cidaris Wrightii* have regular longitudinal rows of small tubercles.

Cidaris Wrightii has many points of affinity with *Cidaris florigemma*, both as regards the general structure of its test and the form and decoration of its spines; but it is distinguished from that Coral Rag form in having larger tubercles and smaller areas, with a much wider granulated miliary zone between the primary tubercles. The spines assigned to *Cidaris Wrightii* belong to the same group as those of *Cidaris florigemma*, but the stem is longer and narrower, and the granules forming the rows are set closer together, and not so prominent as in *Cidaris florigemma*, where they form lines of tubercles very distinct from each other. Although connected by delicate calcareous threads, the apices of the tubercles are likewise directed forwards, and the summit of the stem terminates in a truncated star-like extremity. (Pl. I, fig. 3 *e*.)

Locality and Stratigraphical position.—This is a very rare Cidarite. Three specimens only have been found by me in the Pea Grit of Crickley Hill, which are all more or less imperfect; and I have never seen an example in any other cabinet. An urchin found in the Stonefield Slate at Eyeford, Gloucestershire, is referred to *Cidaris propinqua*. Whether this is identical with *Cidaris Wrightii* I have no means of judging, not having seen the type specimen catalogued under that name.

History.—It was first figured in my 'Memoir on the Cidaridæ of the Oolites,' and subsequently entered in M. Desor's 'Synopsis des Echinides' under the name it now bears.

6

CIDARIS CONFLUENS, *Forbes*, MS.

> CIDARIS CONFLUENS. Morris, Catalogue of British Fossils, 2d edition, p. 74.
> —   —   Woodward, Memoirs of the Geological Survey, Decade V, Note on
>   species of Cidaris.

The specimen consists of four consecutive plates, of nearly equal size, from one of the inter-ambulacral rows. The set measures one inch in length and half an inch in width. The areolas are oblong and excavated, with acute lateral margins, but they are all confluent above and below; the bosses have broad, prominent, deeply crenulated summits; the tubercles are small, and widely perforated; the lateral borders of the plates supported small granulations. The spines imbedded in the same rock were long, cylindrical, and longitudinally striated; their surface was armed, at intervals, with short, stout, forward-directed prickles. The ambulacral areas are absent; and the surface of the plates is so much weathered, that I did not consider it necessary to figure the specimen.

This fragment resembles very much *Cidaris Lorierii*, Wright, from the étage Bajocien, or Inferior Oolite, of the department of the Sarthe (a description of which will be found in the notes of foreign species of the genus Cidaris), but the weathered and fractured condition of the fragment renders it impossible to make a more correct diagnosis until better specimens are found.

Locality and Stratigraphical position.—The specimen belongs to the Museum of Practical Geology, Jermyn Street; and was obtained from the Inferior Oolite, near Frome, Somersetshire.

C. *Species from the Bradford Clay.*

CIDARIS BRADFORDENSIS, *Wright*.   Pl. V, fig. 7 *a, b, c, d.*

Form and size unknown; ambulacral areas narrow, with two rows of marginal granules; inter-ambulacral plates thick, areola circular, boss prominent, summit feebly crenulated, tubercle large; a complete scrobicular circle of fifteen large granules around the areola; miliary zone concave, with six to eight rows of granules; the spines, associated with the plates, have the acetabulum small, the neck long and smooth; the stem elliptical, and covered with waved lines of granules, neither uniform in size nor arrangement.

Description.—The meagre materials at my disposal, illustrative of this urchin, only permit me to give a very imperfect diagnosis of this Oolitic form, which I at one time thought was identical with *Cidaris Bouchardii;* but on making a careful examination of the ambulacral and the inter-ambulacral plates, the only portions of the test I have seen with those of that species, I am inclined to think them distinct. The plates are thick and deep (Pl. V, fig. 7 *a, b*); the areolas are circular, with a complete circle of scrobicular granules around them; the bosses are raised a little way above the margin of the areola, and the summits are feebly crenulated; the tubercle is large, and supported on a short stem; the miliary zone is wide and concave, and there are six or eight rows of coarse granules filling up the same. The ambulacral areas are narrow, with two rows of marginal granules, one granule being opposite each pair of holes; the poriferous zones are narrow, the holes are round, and the septum has a tubercle developed on its surface; there are seventeen pairs of holes opposite one large tubercular plate. The spines, associated with the plates in the same bed, are well preserved (Pl. V, fig. 7 *d, e*); the acetabulum and head are small; the neck is long and smooth; and the stem is covered with granules, which are arranged in lines that are not longitudinal, but slightly waved. Most of the fragments appear to have been drifted; they are more or less covered with a species of small serpula, which seems to have been abundant in the seas of that period.

Affinities and differences.—This species very much resembles *Cidaris Bouchardii,* but its tubercles are proportionately larger, and the areolas are smaller; the scrobicular circle of granules is more defined, the miliary zone contains a greater number of granules, and there is a granular space between some of the scrobicular circles which does not exist in *Cidaris Bouchardii.*

Locality and Stratigraphical position.—The plates and spines of this species have been collected from the Bradford Clay, near the Tetbury-road Station of the Great Western Railway, associated with *Rhynchonella concinna,* Sow., *Terebratula cardium,* Lamarck, *Terebratula digona,* Sow., and other Bradford Clay forms.

D. *Species from the Coralline Oolite.*

CIDARIS FLORIGEMMA, *Phillips.* Pl. II, fig. 2 *a, b, c, d, e, f;* Pl. VIII, fig. 4 *a, b, c, d.*

"LAPIDES JUDAICI OF OXFORDSHIRE." Plott, Natural History of Oxfordshire, pl. 6, figs. 8, 9, spines only, p. 125, 1677.

"RADIOLUS ECHINITÆ MAXIMI LATICLAVII." Lhwydd (Luidius), Lithophylacii Brittanici Ichnographia, editio Altera Oxonii, 1760, t. 12, fig. 1002, p. 49.

DIE NATURGESCHICHTE DER VERSTEINERUNGEN, KNORR II, i, t. E, figs. 4, 5 ; t. E, vi, fig. 9, 1768.

"A GLOBOSE MAMMILLATED ECHINITE FROM OXFORDSHIRE." Parkinson, Organic Remains, pl. 1, fig. 9, and pl. 4, figs. 15 and 17, vol. iii, p. 13, 1811.

CIDARITES BLUMENBACHII. Goldfuss, Petrefacta Germaniæ, p. 117, t. 39, fig. 3 *c, d, e* (aculei non testa), 1820.

CIDARIS FLORIGEMMA. Phillips, Geology of Yorkshire, pl. 3, figs. 12 and 13, p. 127, 1829.

— — De la Beche, Geological Manual, 3d edition, p. 535, 1833.

CIDARITES ELONGATUS. Roemer, Die Versteinerungen des Norddeutschen Oolithen-Giberges, pl. 1, fig. 14, p. 27, 1836.

— FLORIGEMMA. Desmoulins, Etudes sur les Echinides, p. 338, No. 31, 1837.

— — Agassiz, Prodromus Echinoderm., No. 31.

CIDARIS BLUMENBACHII. *Id.* Echinodermes Fossiles Suisse, part ii, t. 20, figs. 5, 6 (non fig. 7), p. 56, 1840.

CIDARITES BLUMENBACHII. *Id.* Catalogus systematicus ectyporum Echinodermatum Fossilum, p. 10, 1840.

CIDARITES BLUMENBACHII. Lamarck, Histoire Naturelle des Animaux sans Vertèbres, 2d edition, tome iii, p. 386, 1840.

CIDARIS BLUMENBACHII. Morris, Catalogue of British Fossils, p. 49, 1st edit., 1843.

— — Wright, Annals and Magazine of Natural History, 2d series, vol. viii, p. 248.

— — Bronn, Lethæa Geognostica, 2d edition, p. 140, 1851.

— — Cotteau, Etudes sur les Echinides Fossiles du département de l'Yonne, pl. 10, figs. 7, 8 (non fig. 6), p. 108, 1852.

— — Desor, Synopsis des Echinides Fossiles, t. 3, fig. 14, p. 5, 1854.

— FLORIGEMMA. Morris, Catalogue of British Fossils, p. 74, 1854.

— — Woodward, Memoirs of the Geological Survey, Decade V, Notes to Cidaris.

Test round, much inflated at the sides, and depressed at both poles ; ambulacral areas narrow, elevated, and sinuous, with two marginal rows of granules set on basal eminences, and in the widest part of the area, two other rows of very small granules, without basal eminences, the middle rows disappear at both ends of the areas : inter-ambulacral areas wide, with two rows of primary tubercles, from six to seven in each column ; areolas rather oblong, and surrounded by a prominent elevated scrobicular circle of well-spaced-out granules, set on shield-like bases ; miliary zone wide, concave, and filled with several rows of small granules; mouth opening small, peristome pentagonal; apical disc opening large and pentagonal; primary spines with large, thick, cylindrical stems, ornamented with longitudinal rows of prominent, forward-directed granules; secondary spines short and spatulate, covered with fine longitudinal lines; tertiary spines small, conical, or oval shaped.

Dimensions.—A. Large specimen: Equatorial diameter, two inches and four tenths of an inch; height, one inch and seven twentieths of an inch.

B. Large specimen: Equatorial diameter, two inches and two tenths of an inch; height, one inch and four tenths of an inch.

C. Usual-sized specimen: Equatorial diameter, one inch and four tenths of an inch; height, one inch and four tenths of an inch.

Description.—This noble urchin was very abundant in the Corallian Seas of Europe, and its test and spines form characteristic fossils of this stage of the Jurassic group. Much confusion regarding this species has been caused by Goldfuss having figured, along with the test of *Cidaris Blumenbachii*, Münster, the spines of three or four other species of urchins, and especially in having erroneously described the spines of *Cidaris florigemma* as belonging to the test of *Cidaris Blumenbachii*. I am indebted to my friend Mr. S. P. Woodward for having called my attention to this subject, as he has always maintained that *Cidaris florigemma*, Phil., was distinct from *Cidaris Blumenbachii*, Goldf. Having been fortunate in finding the slab, figured in Pl. II, fig. 2 *a*, containing a small *Cidaris florigemma* with spines attached, I had direct evidence that the spines figured by Goldfuss as those of *Cidaris Blumenbachii* in reality belonged to the Wiltshire urchin. The next point to be ascertained was, whether the test figured by Goldfuss was different from the test of this species. A critical examination and comparison of good type specimens of *Cidaris Blumenbachii*, one in the collection of the British Museum, and another kindly sent me by Dr. Fraas, of Stuttgard, with the specimen figured in Pl. II, fig. 2 *b*, *c*, has proved that they are very distinct from each other.

Plott, in 1677, figured the spines of this urchin under the name of *Lapides Judaici*. He says—" We find them here (Oxfordshire) of different sizes, from about two inches in length and an inch and a half in circuit, downwards to an inch and less in length and not much above half an inch round. Most of them have a pedicle from which they seem to have had their growth, and are ridged and channelled the whole length of the stone, the ridges being parted with small knots set in quincunx order. As to their texture, I find it to be very curious, made up of lamellæ or little thin plates, not unlike the stone *Selenites;* only these are opaque, and the whole bulk of the stone indeed much different. The plates, as in the Selenites, seem to be made up of strings, which in most of them run three, but in some but two ways : according to the running of these strings the stones will easily cleave, but generally some one way rather than any other, which most commonly is agreeable to the helical running of the ridges of knots or furrows between them, yet always obliquely to the axis of the stone, as is perfectly shown, tab. vi, fig. 9, which represents the stone broken three several ways."*

Lhwydd, in 1690, figured spines of the same species from the Coral Rag of Oxfordshire. Parkinson, in his 'Organic Remains of a Former World,' gave a beautiful figure of the test (Pl. I, fig. 9), which, it is but just to state, has been entirely overlooked, the references to his work having been made only to the figures of the spines (Pl. IV, figs. 15, 17); the first good figure of the test of this species therefore is, in reality, that of Parkinson's. In the 'Petrefacta Germaniæ,' Goldfuss gave good figures of the spines which he stated appertained to the test of *Cidaris Blumenbachii;* and subsequent palæontologists, believing his statement, have nearly all followed his error. In the 'Description des Echinodermes Fossiles de la Suisse,' M. Agassiz has figured the spines of *Cidaris florigemma* with the test of *Cidaris Blumenbachii;* this is shown in the figure by the smallness of the granules of the scrobicular circle, in the smallness of the tubercles, and the depth of the crenulations on the summits of the bosses, a group of characters which belong to *Cidaris Blumenbachii.* In the 'Geology of Yorkshire,' Professor John Phillips figured a test and spine of this species, under the name *Cidaris florigemma,* from a Wiltshire specimen now in the Museum of the Yorkshire Philosophical Institution ; the correctness of the determination then made has been proved by the tedious investigation which was necessary to clear away the existing confusion relative to this species. M. Desor, in his 'Synopsis des Echinides Fossiles,' restricts the name *Cidaris Blumenbachii* to the spines, " aculei non testa," figured by Goldfuss ; but I cannot agree with my friend in this conclusion, because Münster gave the name to *the test*, about which there can be no mistake. Goldfuss's error consisted in describing and figuring spines as belonging to this test which appertained to another species ; therefore I say " testa non aculei" in the synonym : as the test is the body of the animal, and the spines are merely appendages of the same, it follows that the name given by an author to a species must in every case relate to the major, and not to the minor part described. Münster's name, therefore, must still be given to the German form ; which, as far as is at present known, has not yet been found in England. M. Desor observes : " J'ai été fort longtemps dans le doubte sur les limites de cette espèce, par la raison que les radioles et le test que Goldfuss a réunis, n'appartiennent pas au même oursin. C'est tout récemment que la découverte de quelques échantillons avec leurs radioles attachés au test, m'a permis de rectifier l'erreur dans laquelle j'étais tombé avec d'autres paléontologistes. Le test figuré par Goldfuss n'a rien de commun avec les radioles qu'il lui attribue ; il appartient à une autre espèce décrite ulterieurement par M. Agassiz sous le nom de *C. Parandieri.* Or, comme les radioles sont bien plus abondants

---

* Plott's ' Natural History of Oxfordshire,' p. 125.

que les tests et qu'ils sont connus de tous les paléontologistes c'est à eux, et partant au test décrit ci-dessus, que je crois devoir conserver le nom primitif de *Cidaris Blumenbachii.*"*

The test of *Cidaris florigemma* is round, and much inflated at the sides; it is moderately depressed at both poles, but is most so on the upper surface; the ambulacral areas are narrow, and nearly of a uniform breadth throughout; they are very sinuous and prominent, and are furnished with two rows of granules on the margins of the areas; between these, for about the length of two of the large tubercular plates, there are two rows of very small granules internal to the marginal ones (Pl. II, fig. 2 *g*); the marginal granules are raised on small basal eminences, but the internal granules are not; the six or eight marginal pairs nearest the peristome are very much larger than those in the middle and upper parts of the areas: the poriferous zones are of moderate width; the pores forming a pair are separated from each other by a septum equal in width to the diameter of a pore (fig. 2 *g*); there are nineteen or twenty pairs of pores opposite each of the large equatorial tubercular plates; the zones are rather deep, from the prominence of the ambulacral areas in the middle, and that of the scrobicular circles external to them.

The inter-ambulacral areas are nearly five times the width of the ambulacral areas and poriferous zones; the plates of the tubercular columns are deep, there being only from six to seven plates in each column; the areolas are large and circular, especially above, but they incline towards an oval from below; their margins are surrounded by a prominent scrobicular circle of fifteen or sixteen well-spaced-out granules (fig. 2 *g*), each of which is raised on an oval, shield-like base. In consequence of the size of the areolas in the upper part of the areas, the scrobicular circles of the two uppermost pairs closely approximate; but from this point to the peristome there is a considerable inter-tubercular space, which is filled up with miliary granules of different sizes; those nearest the areolas are raised on small basal elevations, which alternate with those of the scrobicular circle, and the rows internal to them diminish in size as they approach the median sutural line, where they become quite miliary: the mammillary bosses rise from a wide base (fig. 2 *g*); the three or four lower pairs have smooth summits, and the two or three upper pairs only are crenulated; the crenulations, however, are by no means either deeply sculptured, or very persistent in different specimens; the tubercles are large, and are raised on a slightly contracted neck; the perforation in the hemispherical head has the form of an oblong slit, which passes through the head, and extends to the summit of the boss.

The mouth opening is large, and the peristome has a pentagonal form (fig. 2 *b*). In specimen B, it measures nine tenths of an inch in diameter, that of the equatorial diameter being two inches and two tenths; the primary tubercles in the vicinity of the peristome are large and well developed, although smaller than those on the sides and upper surface of the test; the minute tubercles at the base of the ambulacral areas are only a little larger than the marginal granules of these areas.

* 'Synopsis des Echinides Fossiles,' p. 5.

The apical disc is absent in all the specimens I have examined. It was of considerable size; the diameter of the opening in specimen B being $\frac{17}{20}$ths of an inch.

The spines are of three kinds, the primary, the secondary, and the tertiary. (Pl. II, fig. 2 d, e, f.) The primary spines or radioles are very elegant bodies, and as they are often preserved in the Corallian stage, when the test to which they belonged is not discovered, a knowledge of them is of stratigraphical importance to the student of the Jurassic rocks. The concave articulating cavity has a deep pit in its centre (fig. 2 e) for the insertion of a ligament, and the rim of this acetabulum is surrounded with a circle of moniliform crenulations; the head is small, and is surrounded by a narrow ring, nearly smooth, and covered only with a microscopic milling of longitudinal lines; the ring is midway between the rim of the acetabulum and the point where the head articulates with the neck (fig. 2 e); the neck is short, and is covered with fine longitudinal lines; it soon expands to form the body of the spine, the thickest part of which is just beyond the neck, from whence the stem gradually tapers to the apex, which is always more or less truncated (fig. 2 d); the stem is covered with small granulations, very uniform in size, and regularly disposed in longitudinal rows, forming from twenty to thirty lines of tubercles on the body of each spine, the number of the rows varying in different spines; the tubercles of the adjoining rows alternate, so as to produce a quincuncial arrangement; and the tubercles of each series are connected together by a calcareous filament which passes from one tubercle to another; the tubercles are all inclined towards the apex of the spine, and many of them terminate in short, prickly processes, which have their points directed forwards; the surface of the spine between the rows of granules is covered with numerous longitudinal lines: at the summit of the spines the granules coalesce, forming so many plates, which expand, and produce a radiated or star-like disc at the truncated extremity thereof. Some of the spines attain the length of two inches and three quarters. The secondary and tertiary spines are short and spatulate (fig. 2 f), and their surface is covered with fine longitudinal lines.

Affinities and differences.—In the form and structure of the test in general, *Cidaris florigemma* resembles *Cidaris Blumenbachii, marginata, coronata,* and *monilifera,* but it is distinguished from all these forms by its greater height, and a consequent increase in the number of plates in the tubercular columns; *Cidaris marginata, coronata,* and *monilifera* having five plates in each column, whilst *Cidaris florigemma* has from six to seven. In the plates on the upper part of the test of these species, the miliary zone is much wider, and the circles of areolar granules are likewise separated by many rows of miliary granules, which are altogether wanting in *Cidaris florigemma.* The ambulacral areas in *Cidaris marginata, coronata,* and *monilifera* have four distinct rows of small, nearly equal-sized granules, whilst in *Cidaris florigemma* the two marginal rows alone are well developed throughout.

Cidaris florigemma is distinguished from *Cidaris Fowleri* by the narrowness of its ambulacral areas, and by the size and prominence of the marginal granules; the poriferous zones are likewise very much narrower, and the pores are smaller and set closer together; the inter-ambulacral areas are wider, whilst the miliary zone is narrower; the scrobicular circles are likewise much more prominent, and have larger granules, raised on distinct eminences; the smooth summits of the mammillary bosses of the lower tubercles, and the faint crenulations on those of the upper ones, form a striking contrast to the broad, deeply crenulated summits of the mammillary bosses in *Cidaris Fowleri*.

The same group of characters which serve to distinguish *Cidaris florigemma* from *Cidaris Fowleri* form the diagnosis between it and *Cidaris Orbignyana*.

Cidaris florigemma resembles *Cidaris Smithii*, Wright, with which it is occasionally associated in the same rock in Wiltshire and Yorkshire, in the general form and height of the test, but it is distinguished from this much rarer species by having fewer tubercles in the inter-ambulacral areas, and much more prominent scrobicular circles around them; *Cidaris Smithii* having ten primary tubercles in each column, and the granules of the scrobicular circles not being much larger than those filling the miliary zone; the ambulacral areas are likewise narrower, and the poriferous avenues are narrower and deeper; the spines, likewise, are longer, narrower, and differently sculptured.

The primary spines of *Cidaris florigemma* most nearly resemble those of *Cidaris coronata*, but their length, and the regularity of their rows of granules, serve to distinguish them from those of that species. The only other species of Cidaris with which it is necessary to compare them, is that of *Cidaris Smithii*, which sometimes occurs with *Cidaris florigemma* in the same rock in England. Although the tests of these two species resemble each other when of the same size, still the spines show that they belong to two distinct groups of urchins, the stems of *Cidaris florigemma* being thick and massive, whilst those of *Cidaris Smithii* are long and slender, tapering very gradually from neck to point; the surface is covered with elevated longitudinal ridges, covered with sharp, forward-directed prickles; these spines are sometimes nearly twice the length of those of *Cidaris florigemma*. (Pl. V, fig. 5.) The test of *Cidaris Blumenbachii* has the areolas deeply excavated; the bosses not very prominent, the summits all sharply crenulated, and the tubercles small; the granules forming the scrobicular circle are not larger than those of the small granulation filling the miliary zone. These characters are so completely diagnostic, that it is impossible to mistake the true German form for *Cidaris florigemma*, when the two urchins are placed side by side for comparison. If to these characters of the test, however, are added those obtained from the spines, we learn how entirely distinct *Cidaris Blumenbachii* is from *Cidaris florigemma*.

Locality and Stratigraphical position.—*Cidaris florigemma* is found in fine preservation in the Coralline Oolite of Wiltshire, Berkshire, Oxfordshire, Yorkshire, and Dorsetshire. The specimens figured in Pl. II were collected near Calne, in Wilts; and I have seen a beautiful specimen found in that formation at Hildenly, near Malton, Yorkshire.

Judging from the spines figured by Goldfuss, Agassiz, Roemer, and Cotteau, this species must have had a wide European area of distribution, as its test or spines have been collected in the Coral Rag of different parts of Germany, Switzerland, and France.

History.—The history of this species has been given with so much detail in the introductory remarks, that it is unnecessary to reproduce the facts under this section.

Cidaris Smithii, *Wright*, nov. sp.　Pl. II, fig. 1 *a, b, c, d, e ;*　Pl. V, fig. 5 *a, b, c, d, e.*

Test large, much inflated at the sides, and depressed at the poles ; ambulacral areas flat, with two marginal rows of small granules, and two inner rows of microscopic granulations ; poriferous zones wide, the pores separated by very thick septa ; inter-ambulacral areas with ten primary tubercles in each row ; areolas oval, and deeply sunk ; scrobicular granules not larger than the miliary granules ; miliary zone wide and concave, and filled with six or eight rows of nearly equal-sized miliary granulations ; mouth opening wide, peristome pentagonal, jaws and teeth large and powerful ; primary spines long, slender, and tapering ; surface of the stem with thick longitudinal ridges, from which stout forward-directed processes proceed.

Dimensions.—Height, one inch and eight tenths of an inch ; transverse diameter, two inches and three fourths of an inch.

Description.—This noble urchin is a much rarer form than the preceding species, with which, however, it is sometimes associated in the same beds of the Coralline Oolite in Wiltshire and Yorkshire. As a distinct species, it has hitherto escaped the notice of palæontologists.

The test is large, much inflated at the sides (Pl. II, fig. 1 *b*), and depressed at both poles ; the ambulacral areas are slightly elevated, of moderate width, and gently undulated ; they are widest at the equator, and diminish above and below ; they have two marginal rows of small granules, which are alternately larger and smaller (fig. 1 *c*) ; sometimes two of the smaller granules are interposed between every two of the larger ones ; within these marginal rows are two rows of close-set, equal-sized, microscopic granules, which disappear at the upper and lower parts of the areas.

The poriferous zones are very superficial, and are nearly as wide as the areas ; the pores are separated by thick septa, and the pairs of pores are directed obliquely upwards ; the holes of the inner row are round, whilst those of the outer row are oblong or pyriform ; the septa are slightly scooped out on their margins above and below, so that each pair of pores, with its intervening partition, somewhat resembles the frame of a pair of spectacles, with a straight bridge (fig. 1 c); the bevelled margins of the septa produce an appearance like a third row of blind holes in the centre of the zone; there are from thirteen to fourteen pairs of pores opposite one of the large tubercular plates.

The inter-ambulacral areas have a greater number of tubercular plates than is found in any other Oolitic species; in each column there are ten plates, the primary tubercles of which are proportionately small; the areolas are oblong, and are rather deeply sunk, with sharp, prominent margins; the granules forming the scrobicular circle are not larger than those filling the miliary zone; the bosses are not very prominent, and are only a little higher than the rim of the areolas; their summits are flat, and sculptured with fourteen deep crenulations (fig. 1 d); the tubercles are small in proportion to the dimensions of the test, and are very uniform in size throughout the area; the areolas on the upper part of the areas are nearly circular, those at the under part are oblong; only the upper five areolas have complete circles of granules round their margins; and the three or four lower areolas are separated by a transverse ridge of the test, on which no granules are developed.

The miliary zone is broad and concave, and is filled with eight or nine rows of small granules, set on eminences, around the bases of which circles of microscopic granules are very regularly disposed; between the poriferous zone and the adjoining row of tubercles, there is a miliary zone filled with three or four rows of granules.

The mouth opening is large (fig. 1 a); the peristome is pentagonal, but its figure cannot be accurately described, as several of the plates are fractured; the lantern is large, and was composed of five powerful jaws, which remain with their teeth *in situ*. In the fine specimen communicated by my friend, the Rev. P. B. Brodie, and figured in fig. 1 a, each jaw consists of a broad central portion and two lateral carinæ; the long, curved, conical, triangular teeth are bevelled into a chisel-like form on the inner surface, and project five twentieths of an inch beyond the alveoli.

The spines are long, slender, and tapering (Pl. II, fig. 1 e, and Pl. V, fig. 5 a, b, c, d, e); the head is short, with a slightly prominent milled ring (fig. 5 c); the acetabulum is sharply crenulated (fig. 5 d); the neck is marked with fine longitudinal lines (fig. 5 c); the stem is round or oval (fig. 5 e), and is covered with eight or nine prominent longitudinal ridges, which have short, stout, oblique, forward-directed prickles developed from their surface (fig. 5 a, c); a fragment of the stem (fig. 5 a) is nearly three inches in length.

Affinities and differences.—This species very much resembles *Cidaris Blumenbachii*, Münster, in the general structure of the test, but it is distinguished from that species by the following characters: In *Cidaris Blumenbachii* the poriferous zones are narrow, and the pores are all round; the inter-ambulacra have only from six to seven tubercles in each row, and the areolas are sunk and circular. In *Cidaris Smithii* the poriferous zones are wide; of each pair of pores, the one is round, and the other is oval; the inter-ambulacra have ten tubercles in each row, and the areolas of the inferior tubercles are oval. Both forms have wide miliary zones, but those of *Cidaris Smithii* are the widest; the granules of the scrobicular circle in both are not larger than those of the miliary zone, and both have deeply sunk areolas.

A comparison of figs. 1 and 2, in Pl. II, will show in what *Cidaris Smithii* differs from *Cidaris florigemma*. The latter has circular areolas, with a well-developed scrobicular circle of large granules (fig. 2 *g*); narrow poriferous zones; large tubercles, with smooth bosses, and seven tubercles in each row. The spines of *Cidaris florigemma* (Pl. II, fig. 2 *d*) have a .thick, clavate stem, with longitudinal rows of moniliform murications; whilst those of *Cidaris Smithii* are long, slender, and tapering, with longitudinal elevations carrying forward-directed processes.

Cidaris Smithii resembles *Cidaris maxima*, Münster, in the general structure of both areas, but the former has more tubercles in each row in the inter-ambulacral areas than the latter species. The spines of *Cidaris Smithii* have a more regular arrangement of the longitudinal elevations and processes thereon; the stem likewise tapers gently from the neck to the apex, and wants the central swelling and irregular forward-directed prickles which characterise the spines of *Cidaris maxima*.

Locality and Stratigraphical position.—This species was collected long ago, from the Coral Rag of Hillmarton, Wiltshire, by the late Dr. William Smith, the father of English palæontology. The original specimens are now in the British Museum. Mr. Lowe collected it from the Coral Rag of Calne, Wilts. Dr. Murray, of Scarborough, found a beautiful specimen in the Coralline Oolite of Ayton, near Scarborough, which is now in his cabinet. The Rev. W. F. Witts collected some plates and spines in the Upper Calcareous Grit near Scarborough Castle. Mr. William Buy has found two or three very perfect specimens in the Coral Rag near Calne; and Mr. Gibbs has collected plates and spines from the same formation in other localities in Wiltshire.

This species, therefore, characterises the Coralline Oolites of England, and belongs to the same horizon as *Cidaris florigemma*, *Diadema pseudo-diadema*, *Diadema versipora*, with which it is associated in the same rock.

History.—This species has, by some, been mistaken for *Cidaris florigemma;* and, by others, has been considered to be the true *Cidaris Blumenbachii*. The zootomical details into which I have entered show that it is very distinct from both.

It is now figured and described for the first time. I dedicate this fine species to the memory of our distinguished countryman, Dr. William Smith, whose accurate observations, large views, and many researches conducted in a true philosophical spirit, led to the discovery of that great stratigraphical law, that each of the different strata of the earth's crust contains its own specific forms of organic life.

E. *Species from the Kimmeridge Clay.*

[*Spines of which the test is unknown.*]

CIDARIS SPINOSA, *Agassiz.* Pl. XII, fig. 4.

CIDARIS SPINOSA. Agassiz, Echinodermes Fossiles de la Suisse, part ii, p. 71, pl. 21, fig. *a.*
— — Desor, Synopsis des Echinides Fossiles, p. 26, t. 3, fig. 2.
— — Morris, Catalogue of British Fossils, 2d edit., p. 75.

The principal character of this species consists in the prominent thorny processes which project from the surface of its stem, which is round and slender, and is likewise finely marked with longitudinal lines; the neck is short. It is exceedingly difficult to determine from a single spine, without comparing it with the type, whether that which is figured is identical with the *Cidaris spinosa*, Agass. It certainly comes nearest to that form with which it is provisionally placed. It was collected from the Kimmeridge Clay of Aylesbury.

CIDARIS BOLONIENSIS, *Wright.* Pl. XII, fig. 5.

Spine long, nearly two inches in length; acetabulum deeply crenulated; ring prominent, acutely carinated; neck long, slender, smooth; stem flattened, covered with finely granulated longitudinal ridges, which have prominent, thorn-like, forward-directed processes, developed at irregular intervals on the stem, which is flat and oar-like at its distal extremity. On this blade-like termination there are sometimes prominent, longitudinal carinæ, formed by the excessive development of the terminal granulated lines.

Affinities and differences.—This spine resembles that of *Cidaris Orbignyana* in its general characters; but it has a longer neck, a more prominent and acutely carinated ring, and the stem is likewise furnished with stouter thorn-like processes; in the flattened condition of the distal extremity it resembles that species. It is altogether a stronger and larger spine than that referred to *Cidaris spinosa*, and its long, smooth neck serves to distinguish it from that species.

Locality and Stratigraphical position.—Collected by the late Hugh Strickland, Esq., from the Kimmeridge Clay, Dorset. The same species* has been found by M. Bouchard Chantereaux, who kindly sent me a series of spines, from the Kimmeridge Clay near Boulogne-sur-Mer, where it is very rare.

* 'Notes on Foreign Jurassic Species,' page 63.

Genus 2—RABDOCIDARIS,* *Desor.*

This genus was formed by M. Desor to include all the large inflated Cidarites, which are often as high as they are in transverse diameter; their poriferous zones are likewise wider than in the genus Cidaris, the pores forming a pair are placed wider apart, and connected by a small horizontal sulcus. The ambulacra are straight, or very little flexed. The tubercles are large, and always strongly crenulated, and, in the fossil species at least, proportionately more numerous than in the true Cidaris. The areolas are large, and often elliptical; the miliary zones are wide. The spines are large, and have a long and much developed stem.

The species at present referred to this genus are Oolitic and Neocomian.—Desor.

A. *Species from the Lias.*

RABDOCIDARIS MORALDINA, *Cotteau.*   Pl. V, fig. 8.

CIDARIS MORALDINA.   Cotteau, Etudes sur les Echinides Fossiles, p. 33, pl. 1, figs. 1—3.
RABDOCIDARIS MORALDINA.   Desor, Synopsis des Echinides Fossiles, p. 42.

Test large, form unknown; inter-ambulacral plates large; areola oval, occupying a considerable portion of the plate; mammillary boss with a broad base, not much elevated, summit deeply crenulated; tubercle proportionately small, with a large perforation.

The only specimen I have seen of this urchin is the one figured in Pl. V. It was collected by Mr. Moore from the Marlstone of Somerset. M. Cotteau, who first described the species, founded it upon the imprint of a fragment. Judging from the size of the single inter-ambulacral plate, it must have been a very large Cidarite; the areolas are wide and oval, and, from the space they occupy, they must have been confluent above and below, and surrounded only by incomplete scrobicular circles; the miliary zone was flat and wide. This species has some affinities with *Cidaris maxima*, Münster; but it is distinguished from it by many marked characters: the inter-ambulacral plate is one half larger; the areola is not excavated out of the substance of the plate, as in *Cidaris maxima*, but is level with the surface, and the mammillary boss forms a much larger eminence thereon; the summit is likewise broader and more deeply crenulated.

* ῥαβδοτὸς, striated, canaliculated.

Locality and Stratigraphical position.—This fragment was collected from the Marlstone near Ilminster; the specimen figured is all that has ever been found in that locality; the bed in which it was discovered is characterised by the presence of *Ammonites margaritatus*, Montfort. It is interesting to find that M. Cotteau's specimen was collected " dans les couches à *Gryphæa Cymbium* de Vassy près Avallon," where it is very rare: this Gryphæa bed corresponds very nearly with the horizon of the Middle Lias in Somersetshire, where the fragment before me was found.

B. *Species from the Great Oolite.*

RABDOCIDARIS MAXIMA, *Münster*.   Pl. XII, fig. 6 ; Pl. A, fig. 16.

CIDARITES MAXIMUS.	Goldfuss, Petrefact. Germaniæ, p. 16, t. 39, fig. 1 *a, b.*
KNOTTED SPINE OF CIDARIS.	Phillips, Geology of Yorkshire, t. 9, fig. 5.
CIDARIS MAXIMA.	Morris, Catalogue of British Fossils, 2d edit., p. 74.
RABDOCIDARIS MAXIMA.	Desor, Synopsis des Echinides Fossiles, p. 39.

The large spine, figured by Professor Phillips, from the Grey Limestone (Great Oolite) of Gristhorpe Bay, Yorkshire, (two of which I collected last summer,) I have identified with a spine kindly sent me by Dr. Oppel, of Stuttgart, from the Inferior Oolite of Neuffen, Württemberg, as typical of Münster's species. A very large and fine specimen of this Great Oolite spine is figured in Pl. A, fig. 16 ; it measures five inches and a quarter in length. The head is large ; the rim of the acetabulum is coarsely and deeply crenulated ; the neck is short and smooth ; the stem is long and sub-fusiform, nearly five inches in length, it gradually swells out in some specimens towards the middle, and tapers very little towards the distal extremity ; the surface is covered with short, stout, thorn-like prickles, with their points slightly directed forwards, which are irregularly developed, at considerable distances apart ; they do not form straight lines, but stud the surface in a somewhat spiral order.

None of the plates belonging to this large spine have yet been found in Yorkshire, nor in the Great Oolite of the West of England.

Genus 3—DIPLOCIDARIS,* *Desor.*

The genus *Diplocidaris* is formed of large urchins, which have all the characters of the true *Cidaris*, but differ from them only in one particular, in having the pores in the zones so arranged as to form double rows, or double oblique pairs, every third pair being vertically above the third pair below them; whereas in the genus *Cidaris* the pores are strictly unigeminal throughout. This important modification in the structure of the poriferous zones has corresponding relations with the number and disposition of the tubular suckers, and probably with the function of respiration. The ambulacral areas are narrow, straight, and flexed, with two rows of small marginal granules; the inter-ambulacral areas are wide; the areolas are round or elliptical, and the bosses have deeply crenulated summits; the tubercles are of moderate size; the miliary zones are very wide, and the surface of the plates is covered with large well-defined granules, set at some distance from each other.

"The spines are short, massive, and cylindrical; their surface is covered with granules, or pustules, instead of spines as in *Rabdocidaris.*"—Desor.

All the species known have been found in the Oolitic rocks.

A. *Species from the Upper Lias?*

DIPLOCIDARIS DESORI, *Wright*, nov. sp. Pl. VIII, fig. 5.

Test large, form unknown; ambulacral areas narrow and slightly flexed, with two rows of small marginal granules; poriferous zones follow the flexures of the area; the pores round, and contiguous, the septum as thick as the diameter of the pore, with a prominent tubercle on its surface; every two pairs of pores are set obliquely to each other, so that the pores of every third pair stand vertically over each other; inter-ambulacral plates large, rhomboidal, as deep as they are broad; areola large, circular; scrobicular circle complete, abutting against the poriferous zones; granules of the circle of the same size as those covering the rest of the plate; areola shallow, boss not prominent, tubercle small; miliary zone wide, with four or six rows of well-spaced-out granules.

The fragment which I have figured was kindly communicated by my friend, Mr. S. P. Woodward; it was collected near Yeovil, from a rock supposed to be Inferior Oolite.

It is very interesting to find this fragment of a *Diplocidaris* so far down in the Oolitic series, as the specimens of this genus hitherto discovered have all been found in the Coral

* From διπλοῦς, double.

Rag of Nattheim, Besançon, Châtel-Censoir, and Druyes (Yonne); the only exception to this being a spine (Pl. I, fig. 5) found by me in the Pea Grit, Inferior Oolite, of Crickley Hill, and placed by M. Desor in this genus.

The inter-ambulacral plate is large, as deep as it is wide, and must have formed part of a very large test; the areola is rather more than two thirds of the width of the plate, it is circular, and is placed so near the poriferous zone that its scrobicular circle abuts against the pores. The areola is smooth, superficial, and not excavated; the boss forms an inconsiderable prominence in the centre, and has a broad summit, which was crenulated, although the remains of the markings are now nearly effaced. The tubercle is small in proportion to the size of the plate; the miliary zone must have been very wide, and filled with four or six rows of large granules, which stand at distinct intervals apart. The scrobicular circle is complete; the granules forming it are set upon an elevated rim of the areola, but they are not larger than the other granules covering the surface of the plates.

The ambulacral areas are very narrow and flexuous, having two rows of small granules on their margins; the poriferous zones follow the windings of the area, which are more flexuous in this species than in the German Coral Rag forms, in consequence of the proximity of the scrobicular circle to the zones themselves; in fact, the absence of flexures in *Diplocidaris gigantea* led M. Desor to state that the ambulacra were straight in this genus; whereas they are both straight and flexed, in proportion as the tubercles occupy the middle or the zonal sides of the plates.

The poriferous zones are narrow; the pores are round and contiguous, separated by septa, about as thick as the diameter of the holes; on the surface of the septa the test is elevated, and forms small, blunt granules; the pores are not, strictly speaking, bigeminal, but form an irregular series, every pair being more or less oblique to the pair above them and below them; they may be described as forming double oblique pairs, thus, ⋮⋮ so that every third pair of holes stand vertically above a third pair below them, a form of arrangement very different from that prevailing in the genus Cidaris, in which the pores are strictly unigeminal throughout the entire zones.

Affinities and differences.—*Diplocidaris Desori* differs from *Diplocidaris gigantea* in having deeper rhomboidal plates; and the areolas and tubercles are placed closer to the poriferous zones; the ambulacral areas are likewise narrower; the pores form double oblique pairs in the zones, instead of making two distinct series as in *Diplocidaris gigantea*.

Locality and Stratigraphical position.—This fragment was discovered near Yeovil, in a rock supposed to be Inferior Oolite, but which probably may be Upper Lias. It is impossible to say anything upon this point, unless the Ammonites with which it was associated were before me, inasmuch as the stratigraphical line between the Upper Lias

8

and Inferior Oolite of Somersetshire has not yet been accurately defined; and indeed it is only by palæontological characters that the one rock can be distinguished from the other. Lithologically, the rock in which *Diplocidaris Desori* is imbedded reminds me more of the Upper Lias of that region than of the Inferior Oolite.

I dedicate this species to my friend M. Desor, who has established the genus to which this urchin is referred.

B. *Species from the Inferior Oolite.*

DIPLOCIDARIS WRIGHTII, *Desor*.   Pl. I, fig. 5 *a, b.*

[*Species of which the test is unknown.*]

DIPLOCIDARIS WRIGHTII.   Desor, Synopsis des Echinides Fossiles, p. 45, pl. 7, fig. 24.

The spine on which M. Desor established this species was figured in my ' Memoir on the Cidaridæ of the Oolites,' and was, by mistake, drawn with the spines since ascertained to belong to *Cidaris Fowleri*. The articulating head is small, the rim of the acetabulum deeply crenulated, and the cup largely perforated; the ring is carinated and finely milled; one half of the neck is covered with longitudinal lines, and the other half is smooth (Pl. I, fig. 5 *b*); the stem is round, and gradually swells out towards the middle, where it is fractured; it is covered with small pustules, which are rather irregularly disposed on the surface, and not arranged in lines. M. Desor says it resembles the spines of *Diplocidaris gigantea*, but the stem is more swollen out in the middle than in that Corallian species.

Locality and Stratigraphical position.—I have fragments of three spines, which were all collected in the Pea Grit, Inferior Oolite, at Crickley Hill, along with the test and spines of *Cidaris Fowleri*.

NOTES

ON SOME FOREIGN JURASSIC SPECIES OF CIDARIDÆ NEARLY ALLIED TO BRITISH FORMS, BUT WHICH HAVE NOT YET BEEN FOUND IN THE ENGLISH OOLITES.

Genus—CIDARIS.

A. *Species from the Lias.*

CIDARIS AMALTHEI. Quenstedt, Petrefactenkunde, p. 574, t. 48, figs. 28—30.

The inter-ambulacral plates only are known; they attain a large size, nearly an inch in breadth; the tubercles are not so large in proportion, but they are deeply crenulated, and have a large perforation in their summit; areolas elliptical, large, distinct; scrobicular granules not larger than those filling the miliary zone. Spines long, slender; stem with small, forward-directed prickles; head and acetabulum large; rim deeply crenulated; ring prominent; neck partially covered with fine longitudinal lines.

Formation.—Lias, Donau-Mainkanal bei Dörlbach.

Collections.—Museums of Bonn, Stuttgart; Professor Quenstedt. Specimens in my Cabinet sent by Dr. Fraas.

B. *Species from the Inferior Oolite.*

CIDARIS LORIERII, *Wright*, nov. sp., 1855.

Test high, one inch and two tenths, transverse diameter two inches, inclining to a pyriform figure, narrow towards the base, and broad and flat at the upper surface; ambulacra flat, with four rows of granules; poriferous zones as wide as the area, pores round, separated by thick septa; inter-ambulacra with seven or eight plates in each column; areolas wide, superficial, with sharply defined margins, surrounded by small granules, of the same size as those filling the miliary zone, which is narrow, and shows the centro-sutural zigzag line well defined; bosses prominent, with broad, deeply crenulated summits; tubercles deeply and largely perforated, gradually increasing in size from the peristome to the coronal plates; disc opening small and circular.

Formation.—Collected by M. De Loriere from the étage Bajocien, département de Sarthe.

Collection.—My Cabinet.

This species comes very near to *Cidaris Fowleri;* but it is distinguished from that species by its pyriform figure, the circle of minute scrobicular granules, the wide areolas, broad and deeply crenulated bossal summits, and wide and deep perforations in the tubercles. These characters closely resemble those of *Cidaris confluens,* with which it has many affinities.

Sent me most kindly by M. De Loriere, to whom I dedicate this fine new species.

C. *Species from the Coral Rag.*

CIDARIS BLUMENBACHII.   Münst., in Goldf., Petrefacten, p. 117, t. 39, fig. 3 *a, b* (testa non aculei).

Agassiz, Echinodermes Suisses, part ii, p. 57, t. 20, figs. 2—6.

Cotteau, Études sur les Échinides Fossiles, p. 108, t. 10, fig. 6 ?

Test inflated, depressed at both poles; ambulacra narrow and flexuous, with two rows of small, very close-set granules; poriferous zones narrow; pores small, round, contiguous, from 20 to 22 pores opposite one of the largest tubercular plates; areolas large, nearly circular, deeply excavated, six to seven in each row, with a prominent margin; boss small, summit broad, deeply crenulated; tubercle small and prominent; scrobicular circles complete, margins touching, granules scarcely larger than those filling the wide miliary zones. Spines not known, those referred to this species I have proved to belong to *Cidaris florigemma.*

Formation.—Coral Rag.

Localities.—Besançon, Châtel-Censoir, et Druyes, Yonne; Saint-Mihiel, Vaches-Noires. Corallien blanc de Hoggerwald, Canton de Soleure. Jura superieur, Jurakalkes bei Thurnau und Muggendorf, Bavaria. Formation ε, *Quenst.*, Sigmaringen.

Collections.—British Museum; very rare in English Collections; in almost all the Foreign Collections; my Cabinet.

CIDARIS PARANDIERI.   Agassiz, Echinodermes Foss. Suisse, ii, p. 58, t. 20, fig. 1.

Test inflated; ambulacra narrow, with two rows of granules at the base, and sometimes two other intermediate rows—four rows in the middle; the areolas, bosses, and tubercles similar to the preceding species, of which it appears to be only a variety.

The spines referred to this species by M. Merian cannot positively be said to belong to it.

Formation.—Terrain à Chailles, Corallien.

Localities.—Fringeli, Canton de Soleure.
Corallien de Besançon, du dèpartement de l'Yonne.
Nattheim, Württemberg.   Formation ε, *Quenst.*, d'Ulm, Württemberg.

Collections.—Museum of Vienna (Dudressier Collection).
British Museum.

CIDARIS DROGIACA.   Cotteau, Études sur les Échinides Foss., p. 111, t. 11, figs. 1, 2.

Test very large, inflated, depressed both on the upper and under surfaces; tubercles large and prominent; bosses large, with deeply crenulated summits; uppermost plates of the inter-ambulacra with rudimentary tubercles, destitute of areolas; scrobicular circle elliptical; granules large and perforated; granules of the miliary zone likewise large, distant, and perforated; ambulacra narrow, with two rows of perforated granules; poriferous zones very narrow; pores very small and contiguous.

Formation.—Calcaire à Chailles de Druyes (Yonne).
Coral Rag Inferieur de Châtel-Censoir.

Localities.—Druyes, Châtel-Censoir, Yonne.

Collection.—M. Cotteau, Écôle des Mines, Paris.
Plaster mould, in my Cabinet, sent by M. Cotteau.

CIDARIS SUEVICA.   Desor, Synopsis des Échinides Foss., p. 7, t. 1, fig. 2.

Test large, inflated, depressed on the upper and under surfaces; areolas circular, deeply excavated; bosses prominent, summits finely and deeply crenulated; tubercles moderate sized; miliary zone covered with a fine, abundant, homogeneous granulation; scrobicular circle with granules of the same size as those filling the zone; ambulacra with two rows of unequal-sized granules.

Formation—Argovien ?

Locality.—Württemberg.

Collections.—Museums of Tübingen, Zurich.

CIDARIS PROPINQUA.   Münst., in Goldfuss, Petrefact., p. 119, t. 40, fig. 1.
                      Agassiz, Echin. Fos. Suisse, ii, p. 62, t. 21, figs. 5, 6, 7, 9, 10.

Test small, depressed; ambulacra much flexed, with two rows of marginal granules; poriferous zones narrow, sunk; pores small, contiguous; inter-ambulacra with four or five plates only; areolas small; bosses small, summits nearly smooth; tubercles very large and prominent; scrobicular circles complete, touching, formed of prominent spaced-out granules. Spines short, elliptical, clavate; stems much swollen out in the middle, and tapering to a point at their distal extremity, covered with longitudinal rows of large, distinct granules.

Formation.—Coral Rag, Jura blanc, (Argovien étage?)

Localities.—Baireuth.
            Argovien des Lægern, du Randen.
            Argovien (formation γ, *Quenstedt*) Sirchingen, Württemberg.
            Environs de Metz, Moselle.

Collections.—Museums of Zurich, Bale, Tübingen, Neuchatel.
             British Museum, my Cabinet.

CIDARIS ELEGANS.   Münst., in Goldfuss, Petrefact., p. 118, t. 39, fig. 5.

Test small, depressed; areolas small, circular; tubercles large and prominent, bosses crenulated; scrobicular circle formed of spaced-out granules; ambulacra with two rows of marginal granules. Spines short, clavate; stem covered with small granules.

Formation.—Coral Rag of Baireuth, Bavaria.
            Coral Rag of Nattheim, Sigmaringen, Württemberg. Formation ε, *Quenst.*

Collections.—Museums of Bonn, Tübingen, Stuttgart, my Cabinet.

CIDARIS CORONATA.—Goldf., Petrefact., p. 119, t. 39, fig. 8.
                  Agassiz, Echinod. Foss. Suisse, ii, p. 59, t. 20, figs. 8—17.
                  Quenstedt, Petrefactenkunde, t. 48, figs. 16—21, with beautiful
                      and accurate details.
                  Desor, Synopsis des Échinides Foss., p. 9, t. 1, fig. 1.

Test moderate sized, much depressed; ambulacra flexuous and prominent, with four rows of small, close-set granules in the middle, diminishing to two rows in the upper and lower parts of the area; poriferous zones sunk; pores small, round, contiguous; inter-ambulacra with from four to five plates in each column; areolas large, circular; rim prominent, surmounted by a scrobicular circle of round, prominent, spaced-out granules; bosses with feebly crenulated summits above, nearly, if not quite, smooth below; tubercles large and prominent; disc opening very large; miliary zone filled with numerous small, round granules; a considerable granulated space between the scrobicular circles. Spines round, with the head large; neck long, smooth; stem swollen out, and then gradually tapering to a blunt point; surface covered with a confluent granulation, arranged in parallel carinated, longitudinal lines.

Formation.—Coral Rag of Sigmaringen, Württemberg, and Bavaria. Formation γ, *Quenstedt.*

Coral Rag Inferieur, et Calcaire à Chailles, Châtel-Censoir et Druyes, Yonne.

Collections.—In all the Foreign Collections.

British Museum, Bristol Museum, Scarborough Museum. It is often mistaken for and ticketed as a British fossil. My Cabinet.

CIDARIS MARGINATA. Goldf., Petrefact., p. 118, t. 39, fig. 7.

Test nearly as large as *Cidaris Blumenbachii,* slightly depressed at both poles; ambulacra prominent, with six rows of small, compressed granules in the middle of the area, diminishing to four rows above and below; five or six pairs of plates in the inter-ambulacra; the areolas large, circular, and much excavated; margin prominent, and surrounded with a circle of scrobicular granules, not much larger than those filling the miliary zone; bosses small, with smooth summits; tubercles large; poriferous zones deeply sunk; pores small, round, and contiguous; miliary zone wide, filled with small, close-set granulations. Spines round; head and acetabulum large, with a smooth rim; neck thick, short, and smooth; stem covered with lines of granules.

Formation.—Coral Rag of Nattheim. Formation ε, *Quenstedt.*

Jura Superieur Heidenheim.

Collections.—Museums of Bonn, Tübingen, Stuttgart, and others.

British Museum, my Cabinet.

D. *Species from the Kimmeridge Clay.*

Cidaris Boloniensis, *Wright,* nov. sp., 1855. Figured in Mr. Davidson's MSS., pl. 1,
figs. 11, 12.

Form unknown, plates and spines only found. Ambulacra very narrow, with two rows
of small granules; poriferous zones as wide as the areas; pores oblong, separated by thick
septa; inter-ambulacra wide, plates twice and a half as broad as they are deep; areolas
elliptical, central, much excavated; bosses small, summits broad and deeply crenulated;
tubercles large, gradually increasing in size from below upwards; scrobicular circle complete in the upper areolas; granules small, a little larger than those in the zone; miliary
zone wide; the granules diminish in size between the scrobicular circle and the centro-
sutural line; a second miliary zone between the scrobicular circle and the poriferous zones.
Spines long, cylindrical, or a little compressed; head very large; rim of the acetabulum
deeply crenulated; ring very prominent, forming a carina milled with microscopic lines;
neck long, lower half covered with fine longitudinal lines, upper half smooth; stem nearly
round, and closely covered with small granular longitudinal lines, compressed at the extremity; arising, at irregular intervals, from among the granular lines, are a number of
short, stout, thorn-like, forward-directed prickles; in some of the more spatulate varieties
of the spines the prickles are not conspicuous, but in the round forms they are very
prominent processes.

M. Bouchard Chantereaux, of Boulogne, who kindly sent me the specimens, at the same
time states, "they are the only *débris* of the species that I have yet found. If we may
judge of the size of this urchin by the dimensions of the tubercular plates, this species
must have been at least four times larger than *Hemicidaris Boloniensis,* which you have
now."

This Cidaris was most accurately figured in detail by my excellent friend Mr.
Davidson, in Pl. I of an 'Atlas of Plates illustrative of the Fossils of the Boulonnais,' and
which plates have been most obligingly placed at my disposal for this work.

Formation.—Kimmeridge Clay, near Boulogne-sur-Mer.

Collections.—M. Bouchard Chantereaux, Mr. Davidson, my Cabinet.

Genus—RABDOCIDARIS.

B. Species from the Inferior Oolite.

RABDOCIDARIS MAXIMA. Syn. *Cidarites maximus.* Münst., in Goldf., Petrefact., p. 116, t. 39, fig. 1.

Test large, very high, having at least seven tubercular plates in each inter-ambulacral column; areolas large, elliptical, confluent; bosses with deeply crenulated summits; tubercles large; scrobicular circles incomplete; granules small, not larger than those filling the miliary zones; ambulacra broad, with two rows of marginal granules, and an interspace between; poriferous zones nearly straight, and broad; pores oblong, wide apart. Spines long, round; head large, neck long and smooth, stem swollen out in the middle; surface armed with irregularly disposed, forward-directed, thorny spines.

Formation.—Inferior Oolite of Baireuth. Formation, Brauner Jura δ, *Quenst.*

Collection.—Museums of Bonn and Stuttgart; plates and spine in my Cabinet.

C. Species from the Coral Crag.

RABDOCIDARIS NOBILIS. Syn. *Cidarites nobilis.* Münst., in Goldf., Petrefact., p. 117, t. 39, fig. 4.
Agassiz, Echinod. Foss. Suisse, ii, p. 65, t. 21 *a*, fig. 21.
Desor, Synopsis Échinides Foss., t. 8, fig. 10.

Test large, thin, spheroidal, depressed at both poles; ambulacra broad, slightly flexed, with six rows of small granules; poriferous zones narrow; pores small, round, separated by thick septa: inter-ambulacra wide; seven large plates in each column; areolas large, circular, superficial; bosses small, not prominent, summits deeply crenulated; tubercles large and prominent; scrobicular circles complete; granules small, a little larger than those filling up the zones. The scrobicular circles of the lower plates touch; those of the upper plates are separated by a miliary space; miliary zone very wide, covered with numerous granules, which diminish in size from the scrobicular circle to the centro-sutural line, which is very distinctly visible. Spines long, cylindrical or sub-prismatic; the head large; neck smooth, concave; stem very long, sometimes twelve inches in length, *Quenstedt;*

surface covered with short, thorn-like, forward-directed prickles, which in prismatic varieties are disposed in lines; base of the stem slightly enlarged; the remainder of the stem is narrower, and of a uniform width throughout.

Formation.—White Jura (Argovien)? Baireuth.
White Jura, Coral Rag, Württemberg. Formation ε, *Quenst.*

Collections.—Museums of Vienna, Tübingen, Stuttgart.
British Museum, my Cabinet.

D. *Species from the Kimmeridge Clay.*

RABDOCIDARIS ORBIGNYANA. Desor, Synopsis des Échinides Fossiles, p. 40, t. 1, fig. 3.

Test large, spheroidal; ambulacra broad, flat, with four rows of granules; poriferous zones as wide as the area; pores small, round, placed far apart by thick septa; inter-ambulacra with seven or eight plates in a column; areola circular, superficial; boss prominent, summit broad and deeply crenulated; tubercles large and widely perforated; scrobicular circles complete; granules small, well spaced out, and raised on a base; miliary zone moderately wide, and filled with small granules, which gradually diminish in size between the scrobicular circle and the centro-sutural line, which is well defined. Spines long, tricarinate or prismatic, from three to four inches in length; sometimes they are compressed and flattened near their distal extremity; the head is small, the rim of the acetabulum is deeply crenulated, the ring is narrow, the neck short and smooth; at the base, and along the edges of the carinæ, there are rows of stout, short, forward-directed, thorn-like prickles; the intermediate surface of the stem is covered with longitudinal lines of small, irregular-sized granules.

Formation.—Kimmeridge Clay, Rochelle, Villersville, Cap la Hève, Havre.

Collections.—MM. Michelin, Cotteau, Thurmann.
British Museum, Jermyn Street Museum, my Cabinet.

Genus—DIPLOCIDARIS.

DIPLOCIDARIS GIGANTEA. Syn. *Cidaris gigantea.* Agass., Echin. Foss. Suisse, ii, p. 66,
t. 21, fig. 22.

,, ,, Quenstedt, Petrefactenkunde, t. 48,
fig. 45.

Diplocidaris gigantea. Desor, Synopsis Échinides Foss.,
t. 1, fig. 5.

Test large; ambulacra straight, with two rows of marginal granules; poriferous zones wide; pores bigeminal throughout; interambulacra wide; areolas superficial, small in proportion to the size of the plates; bossal summits deeply crenulated; tubercles small and perforated; scrobicular circle composed of large, mammillated, spaced-out granules. Miliary zone wide, filled with granules placed wide apart; between the areolas and the poriferous zones, there is a wide miliary zone; the uppermost plates of the inter-ambulacra are destitute of primary tubercles, the miliary granules covering their entire surface. Spines stout, cylindrical; stem covered with numerous equal-sized granules, which are nearly longitudinal in their disposition, but have no connecting calcareous filament; acetabulum large, rim deeply crenulated, ring prominent, neck short and smooth.

Formation.—Coral Rag of Besançon, Salins, Châtel-Censoir, and Druyes, Yonne.
Coral Rag of Nattheim, " Weissen Jura, &c., bei Ulm," *Quenst.* Württemberg.

Collections.—Museum of Vienna (Dudressier Collection), Tübingen, Collections of M. Cotteau and M. Michelin.

Family 2—HEMICIDARIDÆ.

Test thick, spheroidal, more or less depressed; mouth opening large, central; peristome decagonal, and divided by notches, more or less deep, into ten unequal-sized lobes.

Apical disc small, directly opposite the mouth, composed of five genital plates, and five ocular plates; the anterior pair of genital plates are larger than the posterior pair, and the right antero-lateral, which always supports the madreporiform body, is the largest. Anal opening round or oval in the centre of the genital circle.

Ambulacral areas wider than in the CIDARIDÆ, with semi-tubercles at their base only, as in *Hemicidaris*, or extended throughout the area as in *Acrocidaris;* these tubercles are perforated, and provided with crenulated bosses like those occupying the inter-ambulacra; the areas are straight or undulated.

Inter-ambulacral areas wide, composed of large plates, rarely more than eight in each column; the external surface of the plates supports large perforated tubercles, raised on very prominent bosses; the areolas are in general oblong, and confluent above and below; the incomplete scrobicular semicircles form two crescents on the sides, and they alone form the narrow, central, miliary zone. One small group, *Acropeltis*, has the bosses smooth, and the tubercles imperforate.

The poriferous zones are narrow and undulated; the pores are small, contiguous, and unigeminal throughout, except near the peristome, where they are bigeminal and trigeminal.

The jaws are large and powerful, and armed with stout tricarinate teeth. The spines in general are long, cylindrical, and tapering; sometimes they are claviform or stout, compressed, and angular. Their surface is in general covered with fine longitudinal lines; but, as far as we know, neither prickles nor asperities are developed on the stem.

I have grouped in this family the genera *Hemicidaris*, *Acrocidaris*, and *Acropeltis*, which are all extinct, and found in the Oolitic, Cretaceous, and Tertiary Formations.

Genus—HEMICIDARIS, *Agassiz.* 1840.

CIDARITES (pars), *Lamarck.* 1816.
CIDARITES (pars), *Goldfuss.* 1829.
DIADEMA (pars), *Desmoulins.* 1835.

The genus *Hemicidaris* was established by Agassiz,[*] to receive certain urchins which Lamarck and Goldfuss had included in the genus *Cidarites*, and Desmoulins had placed among his *Diademas*. The dismemberment of these forms from the genera with which they had been associated, was an important progressive step in stratigraphical palæontology, as they form extinct types of the Echinoidea, which, up to the present time, have only been found in the Oolitic, Cretaceous, and Nummulitic Formations.

The species of this genus exhibit a group of characters which are easily recognised, even when portions of the test only are preserved.

If the reader will please to compare Plates I and II with Plates III and IV of this work, he will discover at a glance how widely different the general *facies* and structure of the test in *Hemicidaris* is from that of the true *Cidaris*.

The *Hemicidaris*, in general, have the test thick, of a medium size, more or less subglobose, and generally flattened at the base; the altitude being greater in proportion to the latitude than in either the CIDARIDÆ, DIADEMADÆ, or ECHINIDÆ. The distinctive character of the genus lies in the structure of the ambulacral areas, which are narrow, and more or less flexuous. At the enlarged base of each area (Pl. IV, fig. 1 *b, c, d*) there are three or four pairs of mammillated tubercles, which occupy the lower fourth part of the area. These *semi-tubercles* are smaller in size, but identical in structure, with the primary tubercles of the inter-ambulacral areas, and, like them, they gradually increase in magnitude from below upwards. The upper three fourths of the area has two marginal rows of minute perforated tubercles, which contrast strongly with the large semi-tubercles they immediately succeed. Between the minute marginal tubercles there is a narrow miliary zone, containing two or more rows of small, close-set granules.

The poriferous zones are narrow; the pores are unigeminal, and approximated on the sides, but near the peristome they become bigeminal and trigeminal, according as the space in that region is more or less augmented.

The inter-ambulacral areas are wide, with two rows of primary tubercles, from four to eight in each row. They have large, prominent bosses, with deeply crenulated summits, especially those near the equator of the test; the areolas are wide, and mostly confluent above and below, so that the scrobicular circle of granules is generally incomplete. In some species the tubercles increase and diminish gradually in the area; in others,

[*] 'Description des Echinodermes Fossiles de la Suisse,' part ii, p. 42.

the large equatorial tubercles are succeeded by others, which diminish suddenly in size.

The mouth opening is large; the peristome is decagonal, its margin being always divided into ten lobes by deep notches; the five jaws, when preserved (Pl. IV, fig. 1 *b*), are always large and powerful.

The apical disc is small, its elements are very solid, and so well articulated with the areal plates, that it is very often preserved in the fossil state. It forms, in this respect, a remarkable contrast to the apical disc in *Cidaris*, which is large, and almost always absent. The anterior pair of genital plates are larger than the posterior pair, the right antero-lateral plate being the largest, and always supporting a prominent spongy madreporiform body; the single posterior plate is often small and rudimentary, from the encroachment of the vent; the five ocular plates are small, prominent, heart-shaped bodies, with marginal orbits, whereas the oviductal holes of the genital plates are pierced at some distance from the border; the surface of all the discal elements is covered with small granules.

The primary spines are in general long, tapering, and cylindrical, but sometimes they are claviform; their surface is smooth, and sculptured with fine longitudinal lines (Pl. IV, fig. 1 *h*, *n*, *o*); the secondary spines are small, short, blunt processes (fig. 1 *j*, *k*).

The size of the species in this genus varies from half an inch to two inches in diameter, and from two tenths of an inch to an inch and three quarters in altitude.

Hemicidaris is distinguished from *Cidaris* by the breadth and structure of the ambulacra, and the presence of semi-tubercles at the bases of these areas. It is distinguished from *Diadema* by the narrowness of the ambulacra, and the presence of minute marginal tubercles only on the sides of the ambulacra, *Diadema* having primary tubercles as large as those in the inter-ambulacra, ranging throughout these divisions of the test.

Hemicidaris very much resembles *Acrosalenia*, many species of this latter genus having been mistaken for the former. *Hemicidaris* may be readily distinguished by the great difference in size between the uppermost pair of semi-tubercles and the lowest pair of minute marginal tubercles, but, above all, by the size of the discal opening, and the structure of the apical disc itself, which in *Acrosalenia* has one or more sur-anal plates; even when the elements of the disc are absent, there is always a certain amount of excentricity in the opening, one angle thereof intruding farther down the single inter-ambulacrum than into either of the other inter-ambulacral areas.

Hemicidaris is distinguished from *Acrocidaris* by the narrowness of the ambulacra, and by the semi-tubercles being limited to the bases of the areas, whereas in *Acrocidaris* the ambulacra are wide, and furnished with primary tubercles throughout; each of the genital plates likewise is furnished with a large perforated tubercle, raised on a crenulated boss.

The genus *Hemicidaris* may be subdivided into three sections, of each of which we have typical examples in our Oolitic formations:

Section *a.*—Test elevated, tubercles large in the upper parts of the inter-ambulacra.

> Examples: *Hemicidaris intermedia, Hemicidaris crenularis, Hemicidaris Bravenderi.*

Section *b.*—Test large, depressed; tubercles suddenly diminishing in size on the upper part of the inter-ambulacra.

> Examples: *Hemicidaris diademata, Hemicidaris Stokesii, Hemicidaris pustulosa.*

Section *c.*—Test in general small; ambulacra very flexuous; tubercles in the inter-ambulacra few in number, but large in size, and very prominent.

> Examples: *Hemicidaris minor, Hemicidaris Thurmanni.*

A. *Species from the Inferior Oolite.*=10*th Étage, Bajocien*, d'Orbigny.

HEMICIDARIS GRANULOSA, *Wright*, Pl. III, fig. 2 *a—f.*

> HEMICIDARIS GRANULOSA.   Wright, Annals and Magazine of Natural History, 2d series, vol. viii, p. 257, pl. ii, fig. 4 *a, a.*
> —   —   Desor, Synopsis des Échinides Fossiles, p. 55.
> —   —   Morris, Catalogue of British Fossils, 2d edit., p. 82.

Test sub-spheroidal, depressed at both poles; ambulacral areas straight, with two marginal rows of prominent, well-defined, imperforate granules, and three pairs of semi-tubercles at the bases thereof; inter-ambulacral areas, with three pairs of primary tubercles, extending only as far as the equator, the upper tubercular plates being covered with warty granules; apical disc large, and not prominent; base flat; mouth opening large; peristome slightly notched.

Dimensions.—Height, seven tenths of an inch; transverse diameter, one inch and one tenth.

Description.—This remarkable Urchin is closely allied to *Hemicidaris pustulosa*, Agassiz, and replaces that Dorsetshire species in the same zone of the Inferior Oolite at Dundry. The test is hemispherical, flat below, and slightly depressed above (Pl. III, fig. 2 *c*); the ambulacral areas are straight and prominent, they have two marginal rows of from ten to twelve large, prominent, imperforate granules, which are smooth and deformed, and set regularly in alternate rows, the intervening surface of the plates being filled with small, ill-defined, and irregularly arranged miliary granules (Pl. III, fig. 2 *d*); the bases of the areas are wide, to allow of the development of three pairs of moderately sized semi-tubercles (Pl. III, fig. 2 *b*); the poriferous zones are nearly straight, except where they follow the basal expansion of the areas; the pairs of pores (Pl. III, fig. 2 *d*) are placed slightly oblique, and the septa have raised eminences, which form a moniliform division between the pores; there are ten to eleven pores opposite each large tubercular plate, and at the wide basal region of the zones they fall into triple oblique pairs.

The inter-ambulacral areas are twice and a half the width of the ambulacral; they had from six to seven large plates in each column, the three or four inferior plates alone supporting large primary tubercles; the three upper plates are destitute of them, and, in lieu thereof, have clusters of granules, similar in size to those of the ambulacra, developed on their surface, and forming triangular, quadrangular, or pentagonal figures, according to the number of granules in each group (Pl. III, fig. 2 *a*); the large primary tubercles occupy the entire surface of the rhomboidal plate (fig. 2 *d*); the boss is large and prominent (fig. 2 *e*), its summit is deeply crenulated, and the tubercle is likewise large and deeply perforated; the areolas are smooth and gently inclined (fig. 2 *e*), and around them is a complete scrobicular circle, formed of from thirteen to fifteen round granules (fig. 2 *d*); the base is flat (Pl. III, fig. 2 *b*); the mouth opening is very large, thirteen twentieths of an inch, that of the diameter of the test being one inch and two tenths; the peristome is not so deeply notched as in many other congeneric forms.

The apical disc is concealed by hard adherent rock in the larger specimen, but is well exposed in a smaller one (fig. 2 *a*); it consists (fig. 2 *f*) of elongated heptagonal genital plates, the one carrying the madreporiform body is the largest; the oviductal holes are pierced near the apices; the ocular plates are distinctly heart-shaped, with a depression down the centre of each, the orbit being formed by the notch; the disc makes a slight prominence on the upper part of the test, and the surface of the plates is destitute of any sculpture; the anal aperture is circular and central (fig. 2 *f*).

Affinities and differences.—This Urchin very much resembles *Hemicidaris pustulosa*, Agassiz, but is distinguished from it in having narrower ambulacra, with larger, fewer, and more prominent granules thereon; the inter-ambulacra have likewise fewer primary tubercles, and their areolas are surrounded with complete circles of well-spaced-out scrobicular granules (fig. 2 *d*, *e*), whereas in *Hemicidaris pustulosa* the areolas are confluent.

It is distinguished from *Hemicidaris intermedia* by the absence of tubercles from the upper parts of the inter-ambulacra, and by the form and size of the granules covering the ambulacra. It is so entirely distinct from all other congeneric forms, that it is impossible to mistake it for either of them.

Locality and Stratigraphical position.—The two specimens figured in Pl. III, fig. 2, were collected from the Inferior Oolite of Dundry Hill, associated with *Diadema depressum*, *Echinus germinans*, and *Polycyphus Forbesii*. In Mr. Lowe's cabinet there is a fine specimen of this species, showing the base very well, which was collected from the Forest Marble near Corsham, Wilts.

History.—This species was first figured in my 'Memoirs on the Cidaridæ of the Oolites,' published in the 'Annals of Natural History,' October, 1851. It is a very rare Urchin; fine specimens of it are contained in the British Museum, Bristol Museum, and in my collection; the only known localities are those already mentioned.

HEMICIDARIS PUSTULOSA, *Agassiz.* Pl. III, fig. 1 *a, b, c, d, e.*

HEMICIDARIS PUSTULOSA.	Agassiz, Catalogus Systematicus, p. 8.
— —	Agassiz and Desor, Catalogue raisonné Annales des Sciences Naturelles, 3e série Zool. tome vi, p. 338.
— —	Desor, Synopsis des Échinides Fossiles, p. 53.
— —	D'Orbigny, Prodrome de Paléontologie, tome i, p. 320, No. 420.

Test large and sub-conoidal; ambulacral areas wide and straight; semi-tubercles large, wide asunder, upper part covered with homogeneous granules, those on the margins the largest; inter-ambulacral areas narrow, with two rows of prominent primary tubercles, which disappear a little way above the equator; the uppermost four or five tubercular plates have clusters of from six to ten homogeneous granules developed on their surface; apical disc large and very prominent, the elements thereof forming a considerable elevation on the test; mouth opening large. Spines thick, oval, with a ridge on one side.

Dimensions.—A. Agassiz's type specimen, from the Great Oolite, Ranville. Height, one inch and one twentieth of an inch; transverse diameter, one inch and seven tenths of an inch.

B. Height of the specimen figured (Pl. III, fig. 1), seven tenths of an inch; transverse diameter, one inch and three tenths of an inch.

Description.—In the examination of the species of the genus *Hemicidaris*, the palæontologist often experiences much difficulty in finding on the test alone good characters for the separation of forms which, from the study of the entire organism, he knows to be distinct, but which, were he to rely merely upon the shell, he would pronounce to be identical; this circumstance has doubtless led to much confusion among the members of this group; no such difficulty, however, exists in the species now under consideration, as it is marked by characters so prominent and well defined, that when it is once seen it cannot possibly be mistaken for any but one of its congeners.

Hemicidaris pustulosa was first discovered by Professor Deslongchamps, in the Great Oolite of Normandy, and entered by Agassiz in his 'Catalogus Systematicus.' These specimens have been most kindly communicated to me by Professor Deslongchamps. But it has never, until now, been either figured or described.

The test is large, with a broad base; it is inflated about the equator; from this point

10

the shell rises into a sub-conoidal form, with a prominent and elevated apical summit; the ambulacral areas are wide and nearly straight, so much so that I have received specimens from France that were ticketed as *Diadema;* the base of the areas is expanded, and there are from six to seven pairs of semi-tubercles in this region, which are well spaced out and very prominent, with a small granule at the angle of each zigzag interspace; the upper part of the areas has two marginal rows of imperforate granules, not very regular, however, in their arrangement, and between them are several smaller granules, equally irregular in size and disposition; the poriferous zones are rather wavy and of moderate width; the pores have an elevated granule rising from the surface of the septa, and separating them, which produces a moniliform line in the track of the zones; the pairs of pores are nearly horizontal, and unigeminal, from the disc to the equator; the zones at this point bend outwards, to form enlarged spaces for the basal semi-tubercles; as the zones approach the peristome, the pores lie in triple oblique pairs.

The inter-ambulacral areas at the equator are only twice the width of the ambulacral areas (Pl. III, fig. 1 *b*), and, what is very unusual among the Echinidæ, they are more prominent than the ambulacra; the primary tubercles are limited to the inferior and middle parts of the areas, one or two only extending above the equator of the test (Pl. III, fig. 1 *c*); they are surrounded by areolas, which are confluent both above and below, and have semicircles of five or six pustulose granules on their outer and inner margins. On the upper part of the areas the true tubercles disappear (Pl. III, fig. 1 *a*); the surface of the tubercular plates in this region develops only clusters of granules, which are very irregular in their mode of arrangement, but uniform in size and form; the first tubercular plates above those having true primary tubercles have a small imperforate tubercle in the centre, and a cluster of granules around it, but the three or four plates between this and the disc are covered with a homogeneous granulation; the areolas of the primary tubercles are separated from the poriferous zones by semicircles of granules (Pl. III, fig. 1 *c*), and like semicircles of scrobicular granules separate the areolas in the median line from each other; there are from eight to ten plates in each tubercular column.

The apical disc is large and very prominent, and forms a conspicuous elevation at the summit of the sub-conoidal test (Pl. III, fig. 1 *a, e*); the antero-lateral genital plates are the largest; the postero-laterals nearly equal them in size and figure, and the single plate is the smallest; the madreporiform body forms a kind of warty eminence on the surface of the right antero-lateral plate, which is the largest; the ocular plates are heart-shaped, and project like tubercles from between the angles formed by the genital plates; the surface of all the elements of the disc, the genital, and even the ocular plates, is covered with numerous granules. (Pl. III, fig. 1 *e*.)

The spines are only known by a fragment, which is imbedded in the base of a specimen from St. Aubin de Langrune. This spine is thick and oval, with a ridge on one of its sides.

The mouth opening is large, and the peristome is not deeply notched; the anal opening is obliquely oval.

Affinities and differences.—This species closely resembles *Hemicidaris granulosa* in the presence of granules instead of tubercles, on the upper part of the test. It is distinguished from it by the following characters: the form is more conoidal, the ambulacra are wider, the granules on the same are smaller and more numerous; the scrobicular circles are incomplete, and the surface of the plates of the apical disc is covered with granules. *Hemicidaris pustulosa* resembles *Hemicidaris Stokesii* (Pl. III, fig. 3) in the sudden diminution in the size of the tubercles on the upper parts of the inter-ambulacral areas, and in the granulated character of the surface of the apical disc; but it is distinguished from that species in having unequal-sized, irregularly disposed marginal granules on the ambulacral areas. The tubercles of the two upper inter-ambulacral plates in *Hemicidaris Stokesii* are perforated, with distinct scrobicular circles around their areolas (Pl. III, fig. 3 *a*); whereas in *Hemicidaris pustulosa* the rudimentary tubercles and the scrobicular granules form clusters on the plates (Pl. III, fig. 1 *a*). The elements of the apical disc, in both species, are much alike in form, structure, and sculpture. The disc, however, in *Hemicidaris Stokesii* does not rise above the surface; whereas in *Hemicidaris pustulosa* it forms a marked projection. There is no other Oolitic species of *Hemicidaris* at present known for which *Hemicidaris pustulosa* can be mistaken.

Locality and Stratigraphical position.—The only English specimen I have seen was presented to me by my excellent friend Mr. Etheridge. The exact locality in Dorsetshire from whence it was collected, however, is not known; but, judging from the lithological character of the rock in which it is imbedded, it is probable it came from the Inferior Oolite near Bridport.

The French specimens were found in the "Grand Oolite (Bathonien), de Luc, St. Aubin, Langrune, Calvados."—*Deslongchamps.*

History.—This species was first entered in the 'Catalogus Systematicus' of Agassiz, and afterwards in the 'Catalogue raisonné' of Agassiz and Desor, but it is now figured and described for the first time.

B. *Species from the Stonesfield Slate, Great Oolite, Bradford Clay, Forest Marble, and Cornbrash.*=11*th Étage, Bathonien,* d'Orbigny.

HEMICIDARIS STOKESII, *Wright*, nov. sp., Pl. III, fig. 3 *a, b, c.*

> CIDARIS FROM STONESFIELD. Stokes, Transactions of the Geological Society of London, 2d series, vol. ii, pl. 45, fig. 17.

Test circular, depressed; ambulacral areas straight, with two rows of small regular marginal tubercles; inter-ambulacral areas with large primary tubercles at the equator, and disproportionately small tubercles on the upper parts of the area; apical disc large, and composed of nearly equal-sized plates.

Dimensions.—Antero-posterior diameter, one inch and five tenths.   Height unknown.

Description.—This beautiful species has been hitherto overlooked by English palæontologists, notwithstanding the very excellent figure given of it in the Transactions of the Geological Society, 2d series, by the late Mr. Charles Stokes.   I have not been able to discover the original specimen, but through the kindness of Professor John Phillips, of Oxford, I am enabled to figure a much better specimen, recently discovered by him at Stonesfield.   Unfortunately the upper surface of this specimen, like that figured by Mr. Stokes, is alone exposed, the under surface being irremoveably surrounded by the rock.

The ambulacral areas are straight, and rather wide; they are provided with two marginal rows of small tubercles, set on slightly prominent elevations (fig. 3 *b*); a few small miliary granules form incomplete circlets around their base; the under surface and semi-tubercles are concealed; but the apex is not much narrowed, and forms a rather obtuse arch (in the figure the ambulacra is drawn rather too lanceolate), over the summit of which the heart-shaped prominent ocular plates are rather conspicuously placed.

The inter-ambulacral areas are nearly three times the width of the ambulacra; the primary tubercles are large and prominent at the equator of the test, but they suddenly diminish in size at the upper part of the areas, so that the two upper tubercles of each row are disproportionately small when compared with those at the equator.   The bosses of the large tubercles rise prominently from a narrow areola, and are sculptured with about twenty crenulations at their summits; the areola is surrounded with a complete circle of small round scrobicular granules (fig. 3 *b*); the pores in the zones are small, and placed widely apart, the rounded surface of the thick septa forming a moniliform line down the middle of the zone.

The apical disc is large (fig. 3 *c*); the genital plates are of nearly the same size; the madreporiform body is spongy and prominent, and occupies almost the whole of the surface of the right antero-lateral plate, which is the largest; the genital foramina open at the centre of mammillated elevations, near the apices of the plates; there are two small tubercles near the base of the postero-lateral and single plates, and four on the surface of the left antero-lateral, as if the pair belonging to the right antero-lateral plate had been transposed to its fellow of the left side, in consequence of the madreporiform body occupying nearly all the surface of the right plate; the anal opening is central and circular; the ocular plates form heart-shaped elevations around the circumference of the disc; each of the three anterior ocular plates support two small tubercles, but on the posterior pair there is only one on each plate.

The under surface of the test is unfortunately so much embedded in the rock, that its removal therefrom is impossible; we are, therefore, in ignorance about many important points relating to the anatomy of this species.

Affinities and differences.—*Hemicidaris Stokesii* resembles *Hemicidaris pustulosa* and *Hemicidaris diademata,* in having straight ambulacral areas and disproportionately small tubercles on the upper parts of the inter-ambulacra. From *Hemicidaris pustulosa* it is readily distinguished by its regular rows of small marginal tubercles in the ambulacral areas, with miliary granules between them, and in having small, single, primary tubercles only on the upper parts of the inter-ambulacral areas, but no clusters of granules thereon, as in *Hemicidaris pustulosa.* The test likewise is much more depressed, and the apical disc is not so prominent. Professor Agassiz observes* of *Hemicidaris diademata,* that the essential character consists in the rapid diminution of the large tubercles on the upper part of the inter-ambulacral areas; a unique peculiarity in this genus, forming a remarkable contrast to the exuberance of these same tubercles in other species of Hemicidaris. The discovery, however, of *Hemicidaris pustulosa* and *Hemicidaris Stokesii,* show that this character is shared in common with other congeneric forms. From *Hemicidaris diademata* this species is distinguished by having the ambulacral areas straighter, the tubercles on the upper parts of the inter-ambulacral areas larger; and the miliary granules are likewise larger and less numerous than those which cover the plates in *Hemicidaris diademata;* but in other respects, as far as I can make a comparison, there is a very close affinity between these species.

Locality and Statigraphical range.—This species was first collected by the late Mr. Charles Stokes, from the Stonesfield Slate at Stonesfield; and the urchin I figure was collected from the same rock and locality by Professor John Phillips, who has kindly communicated the specimen for my monograph. In the Stonesfield Slate at Eyeford, Gloucestershire, a portion of a Hemicidaris has been occasionally found, which belongs likewise to this species. When we consider the enormous surface of this rock which is annually exposed by the splitting of the same into slates, and the very few specimens that have been found during the many years the quarries have been worked, we must consider *Hemicidaris Stokesii* as one of the rarest species of our Oolitic Urchins.

History.—First figured by Mr. Stokes, in the Transactions of the Geological Society; as the type specimen appears to have been unknown, and as it was not named by its discoverer, it was not entered in our lists of species. It is now described for the first time. The specimen figured belongs to the Oxford Museum; I have prepared plaster moulds from the same for the British Museum, Geological Museum, Jermyn Street, and the Bristol Institution.

* 'Echinodermes Fossiles de la Suisse,' part ii, p. 49.

HEMICIDARIS LUCIENSIS, d'Orbigny, Pl. III, fig. 6 *a, b, c, d, e, f.*

HEMICIDARIS LUCIENSIS.   D'Orbigny, Prodrome de Paléontologie, tome i, p. 320.
—   —   Desor, Synopsis des Échinides Fossiles, p. 52.
—   CONFLUENS.   Forbes, Memoirs of the Geological Survey, Decade V, description
                         of pl. 5. Notes on Hemicidaris.
—

Test sub-spheroidal, flat at the base, and depressed on the upper surface; ambulacral areas straight below, sinuous above, more especially in large adult individuals; four pairs of semi-tubercles, which increase gradually in size from the peristome upwards; two marginal rows of small tubercles perforate and crenulate, with intervening miliary granules below, but imperforate and approximated above; inter-ambulacral areas with from six to seven pairs of primary tubercles; apical disc prominent; base flat; mouth opening large; peristome with nearly equal-sized lobes.

Dimensions.—A. Height, seven tenths of an inch; transverse diameter, one inch and three twentieths of an inch.
B. Height, eleven twentieths of an inch; transverse diameter, one inch.
C. Height, five tenths of an inch; transverse diameter, eight tenths of an inch.

Description.—This Hemicidaris, it appears, was mistaken on the Continent for *Hemicidaris crenularis*, Lamarck, of the Coral Rag, from which it has been with justice separated by M. d'Orbigny; it has, up to the present time, been overlooked as an English urchin, the specimens hitherto found at Minchinhampton having been almost indeterminable. One specimen, however, collected from a band of clay in the Great Oolite, and communicated by my friend Mr. Lycett, has enabled me to compare this form with a good series of type specimens sent from Luc, Ranville, and Langrune, by MM. Michelin and Deslongchamps; so that the English specimens are found to occur in the same geological horizon as the original French types.

The test is thick and sub-spheroidal, with a flat base, and a depressed summit (Pl. III, fig. 6 c); the ambulacral areas are narrow, slightly sinuous in young shells, but much more markedly so in large specimens; their bases are a little expanded, to give space to the four or five pairs of semi-tubercles (fig. 6 b), which gradually increase in size from the peristome to the last pair; the upper portions of the areas are provided with two marginal rows of small tubercles (fig. 6 a), set tolerably wide apart, and alternating with each other; two rows of close-set miliary granules occupy the middle of the area (fig. 6 d), and circlets of the same surround the small marginal tubercles; on the upper part of the areas two rows of marginal granules alone fill up the entire areal space (fig. 6 c); the poriferous zones (fig. 6 *d*) are narrow; the pairs of pores are placed obliquely upwards and outwards; between each pair there is a small elevated granule; these interporous granules form a moniliform undulated line, which marks the course of the zones.

The inter-ambulacral areas are very regularly formed, they are three times the width of the ambulacral areas, and occupied by two rows of primary tubercles, from six to seven in each row, which increase very regularly in size from the peristome to the equator, and diminish in like manner as they approach the apical disc; the areolas are large and circular; those of the five or six lower tubercles are confluent above and below, one row of miliary granules forms a series of crescents, which surround their inner margin, separating them from the poriferous zones, and two rows of the same-sized granules form a narrow zigzag inter-tubercular space down the middle of the areas; the two uppermost plates of the tubercular columns have the small tubercles they develop, alone surrounded by a distinct and continuous circle of scrobicular granules; the three largest areolas are channeled at their circumference; the mammillary bosses have a wide base, and are very prominent; their summits are deeply crenulated; the tubercles are moderately large; they have a short stem, and are deeply perforated.

The apical disc (fig. 6 *f*) is large and prominent; the antero-lateral genital plates are much the largest; the postero-lateral are longer and narrower, and the single posterior plate is the smallest and narrowest; the right antero-lateral plate, as usual, supports a conspicuous madreporiform body; the eye plates are heart-shaped, and nearly all of the same size, and the whole of the elements of the disc are covered with small close-set miliary granules; the anal opening is large, and widest in its transverse diameter.

The base is flat, and the mouth-opening (fig. 6 *b*), especially in the larger specimens, is very wide, being rather more than half the diameter of the test; the lobes of the peristome are nearly equal in size, those of the ambulacral being a little larger than those of the inter-ambulacral areas.

Affinities and differences.—*Hemicidaris Luciensis* very much resembles *Hemicidaris Wrightii*, Desor, but it is distinguished from that species by having less prominent ambulacra, without the rows of granules which fill up the area in that form. Like as in *Hemicidaris Wrightii*, the apical disc is very prominent, and the surface of the plates is covered with numerous granules.

Hemicidaris Luciensis is distinguished from *Hemicidaris Bravenderi* by having less prominent marginal granules in the ambulacra, a smaller and more prominent apical disc; the mouth opening is more unequally lobed; and it is altogether a smaller and more depressed form. Although the critical distinction between these Bathonian species is sufficiently clear, still they have so many affinities in common, that unless the specimens are well preserved, and the determination is carefully made, they may be readily mistaken for each other.

Locality and Stratigraphical position.—This species has been collected from the Great-Oolite of Minchinhampton.   Many small specimens are found in the Oolitic shelly beds, but they are not well preserved.   One or two specimens in good preservation have been found in a Clay seam of the same rock.   In France it has been collected from the Bathonien "Calcaire à polypiers," 11th Étage, *d'Orbigny* (Great Oolite of English authors), at Luc, Langrune, Ranville, Calvados.

History.—M. d'Orbigny, in 1847, separated this urchin from *Hemicidaris crenularis*, with which, he says, he was confounded by M. Agassiz.   He further observes, it is easily distinguished from *Hemicidaris crenularis* by a much greater number of small tubercles between the large inter-ambulacral tubercles ; but this is clearly a mistake, for both species have two rows of close-set granules down the centro-sutural line.   The difference resides more in the structure of the ambulacra, and in the size and prominence of the apical disc, than in the number and arrangement of the inter-ambulacral granules.   It is now figured and described for the first time.

HEMICIDARIS MINOR, *Agassiz.*   Pl. III, fig. 5 *a, b, c, d.*

HEMICIDARIS MINOR.	Agassiz, Catalogus Systematicus, p. 9.
— —	Agassiz and Desor, Catalogue raisonné des Échinides, Annales des Sciences Naturelles, tom. vi, p. 339, 3d series.
ACROSALENIA RARISPINA.	M'Coy, Annals of Natural History, 2d series, vol. ii, p. 411.
HEMICIDARIS MINOR.	Wright, Annals of Natural History, 2d series, vol. xiii, pl. 2, fig. 3 *a—c*, p. 165.
— —	Desor, Synopsis des Échinides Fossiles, p. 56.
— —	Morris, Catalogue Brit. Foss. 2d edit., additional species of Echinodermata.

Test hemispherical above, flat at the base ; ambulacral areas slightly flexuous, not prominent, with six large semi-tubercles at the base, and four rows of small, unequal-sized granules in the middle, diminishing to two rows in the upper part of the areas ; inter-ambulacral areas three times the width of the ambulacral, with three primary tubercles on the upper surface, and three smaller ones at the base ; the wide miliary zones are covered with small, distinct, nearly equal-sized miliary granules, which form complete circles around the margins of the areolas of the primary tubercles ; the apical disc is of moderate size, and its ovarial plates are covered with a delicate granulation ; base flat, mouth opening large, peristome unequally decagonal, pores arranged in the zones in single pairs throughout.

Dimensions.—Height, three tenths of an inch ; transverse diameter, nine twentieths of an inch.

Description.—This beautiful little urchin was first discovered in the Étage Bathonien of Langrune, Calvados, the true equivalent of the Great Oolite of English geologists. It was entered in M. Agassiz's ' Catalogus Systematicus '* as *Hemicidaris minor*, from specimens sent to him by M. Michelin. It afterwards found a place in the ' Catalogue raisonné des Échinides' of Agassiz and Desor, accompanied with this remark: " Se distingue entre tous les Hemicidaris par les tubercules très espacés, dont il n'y a que deux ou trois dans une rangée. Terrain Jurassique de France."—*Michelin.* Professor M'Coy, in his paper ' On some new Mesozoic Radiata,' † afterwards described this urchin under the name *Acrosalenia rarispina,* giving the Great Oolite of Minchinhampton for its locality.

As that gentleman has kindly favoured me with pen-and-ink sketches of the species described as new in his paper, I have no difficulty in deciding on the identity of his specimen. Moreover, I have ascertained the collection from whence it originally came. The error committed by this author in the genus must have arisen from the disc in his urchin having been covered with " adhering siliceous matrix," and from his having overlooked the very remarkable character pointed out by Agassiz, " les tubercules très espacés."

I have been fortunate enough, through the kindness of my friend Prof. Deslongchamps, to receive a typical specimen of the original species from the Great Oolite of Langrune, which I have compared with specimens obtained from the same locality as that whence Professor M'Coy's was collected, and there is not a shadow of a doubt about their perfect identity. This pretty little Hemicidaris is very distinct from all others of the group to which it belongs. The test is nearly hemispherical (fig. 5 *a*), and the few primary tubercules stand prominently, at great distances apart from the surface of the test. The narrow ambulacral areas are slightly flexuous above, and have from four to six semi-tubercles at their base; the sides and upper part of the areas having first four (fig. 5 *c*), and then two rows of small, imperforate granules upon their surface, about equal in size to the granulation which covers other parts of the test.

The poriferous zones are depressed, and the pedal pores disposed in pairs throughout (fig. 5 *c*); there are twelve pairs of pores opposite each of the large tubercular plates. The inter-ambulacral areas depart considerably from the typical structure of this portion of the test in the *Hemicidaridæ* (fig. 5 *a*); they are three times the width of the ambulacra, and have at their base three large primary tubercles, two on one side and one on the other, with a smaller tubercle above the single large one (fig. 5 *b*); on the sides and upper part of the areas there are only three primary tubercles, two on one side and one on the other, making only three pairs of primary tubercles in the inter-ambulacral areas, those of the base being closely set together, and those on the sides at great distances apart (fig. 5 *a, b*); the tubercles are large and hemispherical, and only slightly perforated (fig. 5 *c*); the mammillary eminences which support them are small and ring-like, (fig. 5 *a*),

* ' Catalogus Systematicus Ectyporum Echinoderm. Foss. Mus. Neocomensis,' 1840.

† ' Annals of Natural History,' 2d series, vol. ii, p. 411.

with faintly marked crenulations; the areolas are rather wide, and only slightly grooved, so that the tubercles project prominently and abruptly from the surface of the test. The margin of the areola is encircled by a row of thirteen granules (fig. 5 *c, d*), rather larger than those which cover the rest of the inter-tubercular surface of the plates. The miliary granules are close-set, and disposed without much regularity on the surface of the plates. The apical disc (fig. 5 *b*) is of moderate size, and slightly prominent; the five ovarial plates are large, and of a heptagonal form; the ocular plates are small and heart-shaped, and the surface of both is covered with a close-set, delicate granulation; the anal opening is nearly central, and circular; the base is flat; the mouth opening is large and widely decagonal, from the great size of the ambulacral lobes, and the comparative smallness of the inter-ambulacral. The spines are as yet unknown.

Affinities and differences.—This remarkable little urchin is so entirely different from its congeners, that it is impossible to mistake it for any other of the group to which it belongs. The presence of semi-tubercles at the base of the ambulacral areas only, and of granules on the sides of these spaces, associate it with *Hemicidaris diademata*, but the small number of the primary tubercles on the inter-ambulacral areas, added to the great distance at which they are placed apart, serve to distinguish it from the young of that species; in fact, these characters alone are perfectly diagnostic of *Hemicidaris minor* among all other forms of Hemicidaris.

Locality and Stratigraphical position.—It was first found in the " Grand Oolite " of Langrune, Calvados, whence the beautiful specimen before me was obtained, and kindly sent by Professor Deslongchamps, of Caen. I take the present opportunity of recording my grateful acknowledgements to that eminent naturalist for his kindness and courtesy, not only in contributing specimens to my cabinet for comparison and reference, but likewise for communicating many rare species of Oolitic Echinidæ, which served as the types of several of M. Agassiz's new species, and which specimens have been of much service to me in clearing up doubts as to the identity of some other English forms.

Hemicidaris minor was collected in this country by W. Walton, Esq., from the Great Oolite of Hampton, near Bath. I have never found this species in the Great Oolite of Minchinhampton, nor have I seen it in any collection of fossils from that locality.

History.—First named by M. Agassiz from specimens in Professor Deslongchamps and M. Michelin's collections; afterwards described as *Acrosalenia rarispina* by Professor M'Coy, from specimens in the Cambridge Museum, which came from Mr. Walton's series, collected near Bath. It was figured and described in detail, for the first time, in my contributions to the Palæontology of Gloucestershire, published in the ' Annals and Magazine of Natural History' for 1854.

HEMICIDARIS RAMSAYII, *Wright*, nov. sp.   Pl. VIII, fig. 6 *a, b, c, d, e.*

Test small, circular, much depressed.   Ambulacra expanded below, to enclose six large semi-tubercles; very narrow and flexuous above, with two rows of small imperforate granules placed alternately on the margins thereof, and forming a single row only above. Poriferous zones narrow; pores set obliquely in pairs, with a prominent elevation of the septa.   Inter-ambulacra wide, with two rows of very large tubercles, four or five in each row.   Apical disc large and prominent; the genital plates with a depression near their centre; mouth opening large; peristome decagonal, with unequal-sized lobes.

Dimensions.—Height, one fifth of an inch; transverse diameter, two fifths of an inch.

Description.—This is the smallest, but certainly not one of the least interesting of the genus to which it belongs.   It is remarkable for the disproportionate magnitude of three of the primary inter-ambulacral tubercles to the smallness of the test that supports them; the size and prominence of the elements of the apical disc; and the altitude being only one half the latitude of the test.   These three characters readily distinguish this little gem from all its other congeners.

The ambulacral areas are wide, and expanded below, to enclose from six to eight semi-tubercles (Pl. VIII, fig. 6 *d*), which are nearly as large as the inter-ambulacral tubercles in the same region of the test; they increase gradually in size, from below upwards, the two superior pairs being the largest; above the semi-tubercles the area suddenly contracts (fig. 6 *d*), and becomes flexuous above; on its margins there are two rows of small imperforate granules, which, from the extreme narrowness of the area above, form only a single row as they approach the apical disc; the poriferous zones are narrow; the pores are set obliquely in pairs; and the thick septum forms a prominent granule, which separates the two pores forming a pair; there are seven pairs of pores, opposite one of the large inter-ambulacral plates.

The wide inter-ambulacral areas are almost entirely occupied by the two rows of primary tubercles, which, in this species, are much larger in proportion to the size of the test than in any other urchin.   There are four or five plates in each inter-ambulacral column; almost the entire surface of the plate is occupied by the base of the large prominent boss (fig. 6 *d*), the summit of which is sculptured with fine crenulations; the tubercles are very large, especially the two at the circumference, and one in each area above them (fig. 6 *b*); on the upper part of these there is one small tubercle, near the circumference of the apical disc; a double row of small granules extends down the middle of the area, which sends short branches off at right angles, by which the areas are separated from each other (fig. 6 *d*); but there are no granules between the boss and the poriferous zone (fig. 6 *d*).

The apical disc is large, and beautifully preserved in one of the two specimens before me; it forms a considerable prominence on the upper surface of the test (fig. 6 *b*); the genital plates have a heptagonal form, are very thick, and have a remarkable depression near their centre, at the bottom of which the oviductal tubes appear to open (fig. 6 *e*); the anterior pair of plates are the largest, the posterior pair are rather smaller, and the single plate is the smallest; the right antero-lateral, carrying a small madreporiform body in its depression, is the largest of the series; the ocular plates are small heart-shaped bodies; the vent is slightly pentagonal, excentral, and posterior.

The mouth opening is one half the diameter of the test, it has a decagonal form, and the peristome is slightly notched; the ambulacral lobes are larger than the inter-ambulacral; the base is flat, and the tubercles of both areas on this region of the test are nearly all of the same size (fig. 6 *c*); the different appearance which the test presents when viewed on its under surface (fig. 6 *c*), and on its upper surface (fig. 6 *b*), is very marked indeed.

Affinities and differences.—The only *Hemicidaris* this little form can possibly be mistaken for is *Hemicidaris minor* (Pl. III, fig. 5), with which it is associated in the same bed and locality; but it is readily distinguished from that species by its depressed form, and by the inter-ambulacral areas being crowded with large prominent tubercles, raised on very prominent bosses; the ambulacral areas are likewise much narrower above, and the space for the semi-tubercles is much wider below; in fact, the remoteness of the tubercles from each other in *Hemicidaris minor*, is as good a diagnostic character of that species, as the disproportionate largeness of the tubercles, and their consequent crowding near the circumference of the test, is of *Hemicidaris Ramsayii*.

Locality and Stratigraphical position.—This rare little urchin was collected from the shelly ferruginous beds of the Great Oolite, at Sham Castle, near Bath, associated with *Hemicidaris minor*, *Acrosalenia spinosa*, and other Great Oolite forms. I dedicate this species to my friend, Professor Andrew C. Ramsay, F.R.S., Director of the Geological Survey of Great Britain.

HEMICIDARIS BRAVENDERI, *Wright*, nov. sp. Pl. V, fig. 1 *a, b, c, d, e, f*. Pl. XI, fig. 3 *a, b, c*.

CIDARIS CRENULARIS.   Murchison, Geology of Cheltenham, 2d edit., Buckman and Strickland, pl. 13, p. 73.

Test sub-globular, flattened at the base; ambulacral areas nearly straight, with two marginal rows of minute, well spaced out, perforated tubercles, and six pairs of moderate sized semi-tubercles; inter-ambulacral areas with two rows of moderate sized primary tubercles, eight in each row, gradually decreasing in size from the equator to both poles; apical disc large, anal opening excentral behind; mouth opening large; peristome divided into ten nearly equal sized lobes.

Dimensions.—Height, nine tenths of an inch; transverse diameter, one inch and one fifth of an inch.

Description.—It is exceedingly difficult to detect on the test alone of many allied species of *Hemicidaris*, characters sufficiently well marked to distinguish them from each other. When these forms, however, are found with their spines, the distinction is in general so evident, that the difficulty at once disappears; but when deprived of these appendages, the diagnosis becomes obscure. This is well exemplified in several species, and strikingly so in the one now under consideration. At the first glance *Hemicidaris Bravenderi* would be taken by most persons for *Hemicidaris intermedia;* but the details of its structure afford sufficient evidence of its distinctness from that Corallian type. The test is sub-globular, not much inflated at the base; the ambulacral areas are nearly straight, being only slightly undulated in the upper third; on the margin of the areas there are two rows of small perforated tubercles, from fifteen to sixteen in each row; raised on small bosses, and placed rather widely apart; two rows of microscopic miliary granules extend down the centre of the areas, and lateral branches form circlets around the areolas (fig. 1 c); the semi-tubercles are not large, but are regular as to form, size, and arrangement, the six pairs gradually decreasing in magnitude from the upper or largest pair, to the smallest or most inferior pair, which extend to the margin of the ambulacral lobe (fig. 1 b).

The inter-ambulacral areas (fig. 1 b), are nearly three and a half times the width of the ambulacral areas; their two rows of primary tubercles decrease gradually in size from the equator, where they are largest, to both poles; there are seven tubercles in each row; the bosses (fig. 1 b, c, d) are prominent, and their summits have from fourteen to sixteen crenulations, not, however, deeply marked; the spinigerous tubercle is small, and finely perforated; the areolas are wide and confluent (fig. 1 c) above and below; down the centre of the area (fig. 1 a, b) a zig-zag double row of small tubercles, slightly perforated, and raised on little eminences (fig. 1 c), separates the two rows of primary tubercles from each other; small miliary granules fill up the interspaces at the base of these elevations; a single row of the same sized small, perforated tubercles separates the areolas from the poriferous zones (fig. 1 c); among these, likewise, a few miliary granules are irregularly distributed.

The base is flat (fig. 1 b), and nearly two thirds of the whole is occupied by the mouth opening, which is more than one half the diameter of the shell at the equator; the peristome is deeply notched, and divided into ten nearly equal sized lobes (fig. 1 e), and the margin of the notches is reflexed.

The apical disc (fig. 1 f) is large; the antero-lateral ovarial plates are the largest, and the surface of the right plate is entirely covered by the madreporiform body; the postero-lateral plates are smaller, and the right is smaller than the left plate; the odd genital plate is the smallest, and the surface of all the plates is slightly roughened, with small imperfectly developed miliary granules; the genital holes are situated near the apices of the plates (fig. 1 f); the ocular plates are small, heart-shaped, and convex, and form a very inconsiderable portion of the disc; they likewise are covered with numerous small miliary granules; the anal opening (fig. 1 a, f) is large, circular, and slightly excentral.

The test of *Hemicidaris Bravenderi*, on the slab (Pl. XI, fig. 3 a), measures one inch and one fifth, and the longest spine measures three inches in length; the head is small and conical; the milled ring is narrow, and not very prominent; immediately above the ring there is a smooth, narrow, slightly depressed neck, not broader than the thickness of the milled ring; the stem is slender in proportion to its length, and preserves a very uniform diameter throughout, tapering very gently to the point; the spines of *Hemicidaris Bravenderi* differ from those of *Hemicidaris intermedia*, (Pl. IV, fig. 1) in the following details :—the spine is longer in proportion to the diameter of the test; the head is smaller; the milled ring is narrower, and less prominent; the neck is smooth and slightly contracted, instead of having a thick prominent second ring, as in *Hemicidaris intermedia;* (Pl. IV, fig. 1 n, o) the diameter of the spine is less at the base, and more uniform, tapering less than the spine of *Hemicidaris intermedia*.   In fact, the specific distinction between these closely allied species is admirably shown in the spines alone, when the two specimens, figured in Plate IV and Plate XI, are placed side by side.

Affinities and differences.—It requires a minute and careful comparison of the tests to distinguish the differences between *Hemicidaris Bravenderi* and *Hemicidaris intermedia*. In *Hemicidaris Bravenderi* the ambulacral areas are straighter, the marginal tubercles are smaller and fewer, being situated at a greater distance from each other; the semi-tubercles are likewise smaller, and not so conspicuous; the primary tubercles have lower bosses, less deeply crenulated at the summit, and the spinigerous tubercles are much smaller; the apical disc is proportionately larger, and the inequality of size between the antero-lateral and postero-lateral genital plates occasions a slight excentricity in the anal opening, not observable in *Hemicidaris intermedia* (Pl. IV, fig. 1 a, g); the lower part of the test is likewise less inflated than in *Hemicidaris intermedia* (Pl. IV, fig. 1 c); it is altogether a smaller form, with less prominent primary tubercles, semi-tubercles, and minute tubercles, than those which adorn the shell of *Hemicidaris intermedia*.   From *Hemicidaris Luciensis* (Pl. III, fig. 6) it is distinguished by having the test more globose, with straighter and wider ambulacral areas, and smaller semi-tubercles; the apical disc is not so convex and prominent; the mouth opening is larger, and its peristome is likewise divided into more nearly equal-sized lobes.   The absence of primary tubercles from the upper parts of the inter-ambulacral areas in *Hemicidaris pustulosa* (Pl. III, fig. 1) and *Hemicidaris granulosa* (Pl. III, fig. 2) distinguish at a glance these species from *Hemicidaris Bravenderi.*

In its sub-globose form it much resembles *Hemicidaris Davidsoni* (Pl. IV, fig. 2), but it is distinguished from that Portland species in having wider ambulacral areas, with the minute tubercles thereon both longitudinally and laterally more widely apart; the semi-tubercles are likewise much smaller, and disposed regularly in pairs, whereas they run (in many specimens) into a single row in *Hemicidaris Davidsoni* (Pl. IV, fig. 2 *b, c*). The size of the anal opening in the latter is much greater, so much so, that the right postero-lateral and odd genital plates are greatly reduced in size. *Hemicidaris Bravenderi* resembles *Hemicidaris Purbeckensis* (Pl. V, fig. 4) in the straightness of the ambulacral areas, but these divisions of the shell are wider, and the minute marginal tubercles are set wider apart in the former species; the semi-tubercles, likewise, are more regularly disposed in pairs than in the Purbeck form (Pl. IV, fig. 4 *b*), where they assume a linear arrangement. The sudden diminution in the size of the primary tubercles on the upper parts of the inter-ambulacral areas in *Hemicidaris Stokesii* (Pl. III, fig. 3) at once separates that Stonesfield slate species from *Hemicidaris Bravenderi.*

Locality and Stratigraphical position.—This species belongs to the Bathonian Oolitic zone. It has been collected from the Great Oolite of Kill-Devil Hill, near Cirencester, by Mr. J. Brown of that town, but the specimen, formerly in the collection of that gentleman, now presented to the British Museum, is unfortunately not well conserved. Mr. Bravender, F.G.S., of Cirencester, found a fine specimen of this species, with its spines attached, in the Great Oolite at Stratton, near Cirencester, which was figured by Mr. Buckman in the 'Geology of Cheltenham.' Its zoological characters are very well preserved. This species was collected by Mr. Bristow, of the Geological Survey, from the Cornbrash, in a lane leading from Stourton Caundle to Lower Woodacre. This fine specimen is in the Museum of Practical Geology, Jermyn Street, and has been admirably figured by Mr. Bone for this Monograph. *Hemicidaris Bravenderi* is found, likewise, in the Great Oolite of Langrune, Calvados. I have specimens from that locality kindly sent me by Professor Deslongchamps and M. Tesson, of Caen.

History.—First figured, without description, as *Hemicidaris crenularis* in the 'Geology of Cheltenham,' and noticed by Professor Forbes in his Note on the species of Hemicidaris found in British strata, 'Memoir of the Geological Survey,' Decade III: "This fine species," he observes, "is very distinct from any other British one, resembling most nearly *Hemicidaris intermedia,* but differing in having gradually, not suddenly, increasing ambulacral areas, with the tubercles upon them set well apart, except below, where the larger ones are closely packed. Until the spines shall have been discovered, I hesitate to give a name to this form, since it so closely agrees with the figure of the Swiss species, *Hemicidaris crenularis.*" The detailed diagnosis of the points of difference between

Hemicidaris Bravenderi and *Hemicidaris intermedia* apply with equal truth as between our species and *Hemicidaris crenularis*, which differs from *Hemicidaris intermedia* chiefly in the form of its spines. The test is now figured in detail and described for the first time. I dedicate the species to Mr. Bravender, F.G.S., of Cirencester, who first found this urchin in 1844, at Stratton, near Cirencester, and to whom I am indebted for the loan of the original specimen with spines, figured in Pl. XI, fig. 3.

HEMICIDARIS WRIGHTII, *Desor.*   Pl. V, fig. 2 *a, b, c, d, e.*

HEMICIDARIS ALPINA.	Wright, on the Cidaridæ of the Oolites, Annals and Magazine of Natural History, 2d series, vol. viii, p. 256, pl. 11, fig. 3 *a, b.*
— —	Forbes, Memoirs of the Geological Survey, Decade III, pl. 5. Notes on British Species of Hemicidaris.
— —	Morris, Catalogue of British Fossils, 2d ed., 1854, p. 81.
— WRIGHTII.	Desor, Synopsis des Échinides Fossiles, p. 54.

Test sub-globose, depressed above; ambulacral areas prominent, convex, and slightly flexuous, crowded with four rows of small miliary granules on the sides, and four semi-tubercles at the base; primary tubercles of the inter-ambulacral areas suddenly diminishing in size above; apical disc large, prominent, and convex.

Dimensions.—Height, nearly three fifths of an inch; transverse diameter, nine tenths of an inch.

Description.—When this beautiful urchin was discovered some years ago, I provisionally identified it with *Hemicidaris alpina*, Agassiz, from the peculiar structure of the ambulacral areas. Finding that my friend Mr. S. P. Woodward had made a like determination of a specimen contained in the collection of Mr. Lowe, and obtained from the same locality, I figured and described it as *Hemicidaris alpina*, with this remark: "I consider my urchin, however, merely a variety of the Swiss species, for which I propose the name variety *granularis*." My late lamented friend, Professor Edward Forbes, had formed a similar conclusion from the specimens he examined, for he observes, in his notes on the species of Hemicidaris found in British strata, *Hemicidaris alpina*, Agassiz—"A pretty species, easily distinguished from its congeners by the very small and thickly-set ambulacral tubercles." M. Desor's knowledge of the type of *Hemicidaris alpina*, Agassiz, enabled him to point out the distinctive characters between that species and our urchin. He says—"*Hemicidaris Wrightii*, Desor. Syn. *Hemicidaris alpina*, var. *granularis*, Wright. Les ambulacres sont plus saillants et moins larges que dans le *Hemicidaris alpina;* leur rangées externes de granules sont moins accusées. Sur contre il existe à l'intérieur de ces granules marginales quatre à six rangées de très fines granelures comme dans l'espèce précédente. Les tubercules inter-ambulacraires diminuent sensible-ment de grosseur à la face supérieure."* The *Hemicidaris alpina*, Agassiz, was collected from the upper division of the Jurassic group, " Portlandien moyen (Kimméridgien) de Gesné et des Ormonds (Alpes vaudoises)," whilst *Hemicidaris Wrightii* was found in the lower division of the same group, the Bradford Clay and Forest Marble.

The test of this beautiful species is sub-globose; the ambulacral areas are slightly undulated, and of a medium size, rather depressed above; they are prominent and convex, of an elongated conical form, and are thickly covered with small hemispherical granules, without perforations or other sculpture (Pl. V, fig. 2 *d*); the marginal rows are larger and more regular; between them are from four to six rows of smaller granules, closely set together. At the base of the areas are four or six mammillated and perforated semi-tubercles (Pl. V, fig. 2 *e*), which are limited to this region. The pores are set obliquely in pairs, with a smooth, elevated granule between each pair, which forms a moniliform sinuous line, running between the pores (Pl. V, fig. 2 *d*).

The inter-ambulacral areas are of moderate breadth, with two rows of primary tubercles, five or six on each column. The bosses of the two central tubercles are large and promi-nent (Pl. V, fig. 2 *f*); those towards the anal and oral poles are smaller (fig. 2 *a*), and they are all crenulated at their summits (fig. 2 *f*); the tubercles are deeply perforated, and supported on a short stem, the hemispherical head of the tubercle not exceeding in diameter that of the stem (fig. 2 *f*); the areolas around the basis are slightly channeled, and they are all confluent; those towards the anal pole have a circle of granules encircling the areolas (fig. 2 *a*); the miliary zones are narrow, and covered with two rows of small, smooth granules (fig. 2 *b*), similar in form and size to those occupying the ambulacral areas. The apical disc (fig. 2 *a* and *c*) is very prominent; the ovarial plates are large, convex, and much granulated; the two anterior pair are larger than the posterior pair, but the right anterior plate is the largest (fig. 2 *c*); the genital holes are large, and near the apices; the ocular plates are of a proportionate size; the spines are unknown.

The mouth opening is of moderate size, the peristome between the lobes being deeply notched and reflexed; the pores are small, and separated by thick septa (fig. 2 *d*); they are disposed in simple pairs nearly all the length of the poriferous zones, but are arranged in double files around the border of the oral aperture, in such a manner as to occupy all the free space in the ambulacral areas, resulting from the contraction of the inter-ambulacral areas in the region of the mouth; the surface of the septa developes prominent convex elevations, which form a moniliform line separating the pores of each zone.

Affinities and differences.—Our specimen is smaller in size, but, with the differences already pointed out, it much resembles *Hemicidaris Alpina*, Agassiz, from the Calcaire de Saanen, in having the ambulacral areas closely crowded with small, uniform, and hemi-

* 'Synopsis des Échinides Fossiles,' p. 54.

12

spherical miliary granules; the areas, however, are more prominent and convex, and want the defined rows of marginal granules which characterise the Swiss specimen. It very much resembles *Hemicidaris Luciensis*, d'Orbigny; but the four rows of nearly equal-sized granules filling the ambulacral areas serve to distinguish it from that species.

Locality and Stratigraphical position.—This species was collected from the Bradford Clay of Pickwick, Wilts. A valve of *Terebratula digona* was attached to the test, and the specimen is firmly adherent to the surface of *Rhynchonella concinna*. In Mr. Lowe's cabinet there is a fine specimen of this urchin, which was collected at Pickwick; and in the British Museum there is another good specimen, likewise from Wiltshire. Separate plates of this Hemicidaris have been found in the Bradford Clay at the Tetbury-road Station of the Great Western Railway. It is, however, a rare species.

History.—First figured and described in my ' Memoir on the Cidaridæ of the Oolites,' afterwards included in Professor Forbes's descriptions of Jurassic Hemicidaris, in the ' Memoirs of the Geological Survey of Great Britain,' Decade III; entered in the second edition of Professor Morris's ' Catalogue;' lastly, separated from *Hemicidaris alpina* by M. Desor in his ' Synopsis des Échinides Fossiles.'

HEMICIDARIS ICAUNENSIS, *Cotteau.* Pl. III, fig. 4 *a, b.*

HEMICIDARIS ICAUNENSIS.		Cotteau, Études sur les Échinides Fossiles, p. 56, pl. 3, figs. 1—3.
—	—	Forbes, Memoirs of the Geological Survey, Decade III. Notes on Hemicidaris.
—	—	Wright, on the Cidaridæ of the Oolites, Annals and Magazine of Natural History, 2d series, vol. viii, p. 256, pl. 11, fig. 3 *a, b.*
—	—	Morris, Catalogue of British Fossils, p. 82, 2d edit.
—	—	Desor, Synopsis des Échinides Fossiles, p. 53.

Test hemisperical, inflated, and slightly depressed; ambulacral areas with two rows of small marginal tubercles, and with three or four pairs of semi-tubercles at the base; inter-ambulacral areas with two ranges of primary tubercles; mouth large and decagonal; margin deeply notched.

Dimensions.—Height, four fifths of an inch; transverse diameter, one inch and one fifth.

Description.—This species is hemispherical and inflated at the sides, and its transverse diameter is one half more than its height. The inter-ambulacral areas are furnished with two rows of large primary tubercles; in each range there are from six to seven tubercles, which attain their greatest development at the equator of the test, and diminish in size near the anal and buccal openings. The mammillary eminences supporting the tubercles are large, prominent, and surrounded by confluent areolas. The tubercles are small and perforated; one row of granules separates the large tubercles from the poriferous zones, and a double row occupies the middle of the area. The lateral boundaries of the areolas are surrounded by semicircles of granules, whilst the upper and lower boundaries of the same blend into each other.

The ambulacral areas are narrow, slightly undulated, and furnished through nearly all their extent with a double row of small tubercles, which are not very apparent, but are larger on the sides than at the apex of the area; between the size of these and the three pairs of semi-tubercles at the base, a sensible difference exists. The mouth opening is large, and is one half the diameter of the test; it is of a decagonal form, with the peristome deeply notched. The apical disc is not preserved, and the spines are unknown.

Affinities and differences.—*Hemicidaris icaunensis,* in its general form and characters, closely resembles *Hemicidaris intermedia.* It is distinguished from the latter by having the primary tubercles of the inter-ambulacral areas less prominent, by the ambulacral areas being less waved, and in having the semi-tubercles much smaller. This character assimilates *Hemicidaris icaunensis* with *Hemicidaris Thurmanni,* but it is sufficiently distinguished from that urchin by its greater height, less undulated ambulacra, and the greater number of tubercular plates in the inter-ambulacral areas.

Locality and Stratigraphical position.—This rare species was obtained by Mr. Lycett from the Great Oolite of Minchinhampton. M. Cotteau collected it in France from the superior beds of the Bathonian stage at Châtel-Censoir; and M. Bathier found it in the Forest Marble of Châtel-Gérard, where it is likewise rare.

History.—This species was first figured and described by M. Cotteau,* and was provisionally identified by Professor Forbes.† It is figured in Pl. A, fig. 9, of the 'Monograph of Great Oolite Fossils,' to be published by the Palæontographical Society. The specimen before me, which belongs to Mr. Lycett's collection, is only a cast; the determination, therefore, is doubtful, but as it was made by my lamented friend, after a careful consideration of the facts, I have adopted his views without vouching for their accuracy. The specimen is so imperfect, that I have followed M. Cotteau's description.

* Échinides Foss. du Département de l'Yonne, t. 3, p. 56.
† Memoirs of the Geological Survey; Brit. Organic Remains, Decade III, description of pl. 5.

HEMICIDARIS CONFLUENS, *M'Coy.*

HEMICIDARIS CONFLUENS.	M'Coy, Annals and Magazine of Natural History, vol. ii, new series, p. 411.
— —	Forbes, Memoirs of the Geological Survey, Decade V, Notes on Hemicidaris.
— —	Morris, Catalogue of British Fossils, 2d edition, 1854, p. 82.
— —	Wright, On the Cidaridæ of the Oolites, Annals and Magazine of Natural History, 2d series, vol. viii, p. 258.

Through the kindness of Professor Sedgwick, I have been enabled to examine the type specimen of this urchin, which forms part of the geological collection of the University of Cambridge. It proves to be a bad specimen of an *Acrosalenia,* with the test so much defaced as to be specifically indeterminable. It unquestionably was collected from the shelly beds of the Great Oolite of Minchinhampton. This species must now therefore be omitted from the list of Hemicidaris.

C. *Species from the Coralline Oolite, including the Calcareous Grit.*=14th *Étage, Corallien,* d'Orbigny.

HEMICIDARIS INTERMEDIA, *Fleming.*   Pl. IV, fig. 1 *a, b, c, d, e, f, g, h, o.*

CIDARIS PAPILLATA.	Var. of Parkinson, Organic Remains, vol. iii, p. 14, pl. 1, fig. 6, and pl. 4, fig. 20.
— INTERMEDIA.	Fleming, British Animals (1828), p. 478.
HEMICIDARIS CRENULARIS.	Morris, Catalogue of British Fossils (1843), p. 53.
HEMICIDARIS INTERMEDIA.	Forbes, Memoirs of the Geological Survey, Decade III, pl. 4.
— —	Wright, On the Cidaridæ of the Oolites, Annals and Magazine of Natural History, 2d series, vol. viii, p. 252.
— —	Desor, Synopsis des Échinides Fossiles, p. 52.
— —	Morris, Catalogue of British Fossils, 2d edition, 1854, p. 82.
CIDARIS INTERMEDIA.	Phillips, Geology of Yorkshire, p. 127.
— PAPILLATA.	Young and Bird, A Geological Survey of the Yorkshire Coast, pl. 4, fig. 1, p. 211.

Test sub-globose or sub-conoidal; ambulacral areas slightly undulated above, with a double row of minute, perforated, marginal tubercles on the sides, and six pairs of semi-tubercles at the base; inter-ambulacral areas with eight pairs of primary tubercles, on large prominent bosses, having deeply crenulated summits; apical disc not prominent; anus nearly central; mouth opening large, peristome deeply notched and divided into ten unequal-sized lobes. Spines long, round, and tapering to a blunt point; surface with fine longitudinal lines; base tumid, with a prominent milled ring below, and a second smooth ring above. Jaws large and powerful.

Dimensions.—1 *c.* Height, one inch and one tenth of an inch; transverse diameter, one inch and three fifths of an inch.

1 *i.* Height, one inch and three fifths of an inch; transverse diameter, one inch and three fifths of an inch.

Description.—This is one of the most common, and, at the same time, one of the most beautiful and typical examples of the genus *Hemicidaris* in our English rocks. Like its associate, *Cidaris florigemma*, it has long been mistaken for a foreign species (*Hemicidaris crenularis*) which has not yet been found in our Coralline Oolites.

Had the determination of this species rested on the anatomy of the shell alone, it would have been almost impossible to distinguish between these two species; but, fortunately, M. Agassiz has given a fine figure and detailed description* of *Hemicidaris crenularis* of Switzerland, said to be the type of Lamarck's species. This magnificent specimen, with its spines attached, was found in the Corallien étage of Besançon, and formed part of the collection of M. le Comte Dudressier, which Herr Suess informs me is in the Imperial Museum of Vienna. Now as Lamarck's *Cidaris crenularis* came from Switzerland ("Habite, Fossile de la Suisse"†), we cannot doubt the identity of the specimen figured by Agassiz with the Lamarckian type. These points having been settled, it is easy to show that this English *Hemicidaris* is very distinct from the Swiss one, for although the tests can only critically be distinguished from each other, still the spines of the Swiss urchin are so very different from the English form, that there cannot be a doubt about the distinctness of the species. In *Hemicidaris crenularis* the spines form large clubs, which gradually increase in diameter from the head to the extremity, whereas in *Hemicidaris intermedia* they gradually taper from the head to the point.

The first good figure of *Hemicidaris intermedia* was given by Parkinson in his 'Organic Remains,' and described as a mammillated Echinite from Wiltshire, which "should perhaps be considered a variation of *Cidaris papillata.*"‡ Mantell gave a reduced copy of Agassiz's figure in his 'Medals of Creation,' vol. i, p. 340, observing, however,—"This species (*Hemicidaris crenularis*) of mammillated Echinus is common in the Oolite of this country, and is considered to be characteristic of the Upper Jura limestone; it is said to be the same as that figured by Mr. Parkinson, under the name of *Cidaris mammillata*, from Calne, in Wiltshire; but I have never observed spines like those of Agassiz's figure in the English Oolite. These spines are not homogeneous throughout, but their central part appears to have been of a softer texture than the external crust, as may be seen in the

---

* 'Echinodermes Fossiles de la Suisse,' seconde partie, pl. 18, figs. 23 and 24, p. 44.

† 'Histoire Naturelle des Animaux sans Vertèbres,' 2d édit. tome iii, p. 384.

‡ Parkinson, 'Organic Remains,' vol. iii, p. 14.

figure where the spines are imperfect."* The figure given by Young and Bird of *Hemicidaris intermedia* found in the Coralline Oolite of Yorkshire is so bad, that, had I not seen the specimens said to be the type of this figure, and which were undeniable specimens, I should have hesitated before including it in the list of synonyms. The description of the species, however, by Mr. Young,† is very accurate, and is, at the same time, the first given of this species. Dr. Fleming‡ described it under the name *Cidaris intermedia*, giving, as its specific character—"Lesser compartments half the width of the larger ones; tubercles crenulated at the base:" which was rather a generic than a specific diagnosis.

Hemicidaris intermedia is sub-globose, varying from a depressed spheroid to a conoidal form (Pl. III, fig. 1 *c, i*); the upper surface is slightly depressed and the base is flat; the ambulacral areas are narrow, and gently undulated towards the upper part; the semi-tubercles occupy the basal third of the area; of these there are six pairs, which gradually increase in size from the peristome upwards (Pl. III, fig. 1 *b*); the two uppermost pairs are rather prominent (fig. 1 *d*); the sides of the areas support two marginal rows of minute, perforated tubercles (fig. 1 *d*), raised on small elevations—a few microscopic miliary granules separate the rows, and form semi-circlets around the base of the eminences; the poriferous zones are much undulated, especially in the upper part; the pores are in pairs throughout the zones as far down as the third pair of semi-tubercles, where they become irregular, and fall into oblique rows, having three pairs in each; five or six such oblique rows extend to the peristome, and fill up the space left by the smallness of the lower semi-tubercles. There are eleven pairs of pores opposite one of the large plates at the equator.

The inter-ambulacral areas are more than three times the width of the ambulacral, and furnished with eight pairs of large, prominent, primary tubercles (fig. 1 *c*); the mammillary bosses form prominent projecting cones (fig. 1 *e*), the areolas of which touch those of the adjoining ones in the same range; the areolated spaces of the plates in each row are therefore confluent (fig. 1 *d*); down the middle of the areas, and following the zigzag centro-sutural line, there is a conspicuous double row of small, well-marked, perforated tubercles (fig. 1 *d*), amongst which some minute miliary granules are scattered; there are seven of these small tubercles round the centro-sutural edges of each of the larger tubercular plates (fig. 1 *d*); a single row of the same sized small tubercles separates the external or zonal border of the areas from the poriferous zones, there being about six small tubercles around the edge of each large plate (fig. 1 *d*). By this arrangement the upper and lower boundaries of the areolas surrounding the bosses are confluent, whilst the outer and inner boundaries thereof are surrounded by the small tubercles described (fig. 1 *d*). The summits of the bosses are deeply sculptured with fourteen well-marked crenulations (fig. 1 *d, e*); the tubercles are large and deeply crenulated (fig. 1 *e*), and maintain their proportional size on the upper part of the areas (fig. 1 *a*).

* 'Medals of Creation,' vol. i, p. 344.   † 'British Animals,' p. 478.
‡ 'A Geological Survey of the Yorkshire Coast,' p. 211.

The apical disc is about one fourth the diameter of the test (fig. 1 *a*); the antero-lateral pairs of genital plates are the largest, the postero-lateral are smaller, and the single plate is the smallest (fig. 1 *g*); the madreporiform body occupies all the surface of the right antero-lateral plate, which is the largest; the genital holes are situated near the apices of the plates; the ocular plates are heart-shaped. They form very inconspicuous elements of the disc in this species. The eyeholes are very minute, and marginal, and situated opposite the truncated apices of the ambulacra; (fig. 1 *a*) the surface of both the genital and ocular plates is covered with numerous minute miliary granules, very irregularly arranged.

The mouth opening is wide, occupying three fifths of the under surface (fig. 1 *b*); the peristome is deeply notched, and the edge thereof is reflexed at the junction between each inter-ambulacral series and the wide part of the poriferous zones; the lobes of the peristome are of unequal size, those of the ambulacral division being nearly one third wider (fig. 1 *b*) than those of the inter-ambulacral.

The five jaws are large and powerful, and are preserved *in situ* in several specimens in my cabinet (fig. 1 *b*); each jaw has a broad, external, convex surface (fig. 1 *f*), and two lateral ridges, with intervening furrows. The jaw consists of two halves, and the symphysis extends through the middle of the central ridge.

In many specimens of this urchin the spines are admirably preserved in connection with the test (fig. 1 *h*). The primary spines are long, tapering, and nearly cylindrical. They grow in length to nearly twice the diameter of the test to which they belong, some of them measuring three inches and three quarters (fig. 1 *n*); they are closely and minutely striated in the longitudinal direction, but the striæ are rather broader than the raised interstices. The base of each spine has a narrow elevation, or second smooth ring, just above the milled ring, which is prominent, and deeply crenulated (fig. 1 *o*); the acetabulum is likewise crenulated round its rim. The secondary spines, which are attached to the minute marginal tubercles of the ambulacra, and to the tubercles of the same size in the inter-ambulacra (fig. 1 *h, i*), are small, compressed, and spatulate-shaped; their surface is likewise covered with longitudinal lines (fig. 1 *k*), the neck is encircled with a small milled ring, and a second smooth ring. Some of them measure one fifth of an inch in length. On some well-preserved specimens from the Clay Beds, I have occasionally observed on the surface of the test, when cleaning these specimens, small bodies resembling pedicellariæ.

Affinities and differences.—The tests of *Hemicidaris intermedia* and *Hemicidaris crenularis* resemble each other so much, that it is difficult to point out the distinction. In a fine specimen of *Hemicidaris crenularis*, from Lure Doubs, kindly sent me by M. Michelin, the base is less tumid, the minute marginal tubercles on the ambulacral areas are fewer in number and smaller, the primary tubercles of the inter-ambulacra are larger, and their areolas are wider; the mouth opening is likewise larger, and the shell is in general higher than in *Hemicidaris intermedia*. But, as I have already observed, it is in the form and structure of the spines that the true specific distinction is found.

The difference between this species and *Hemicidaris Bravenderi* has been already pointed out in the article on that urchin, and its affinities with *Hemicidaris Purbeckensis* and *Hemicidaris Davidsoni* will hereafter be discussed in the description of those species.

Locality and Stratigraphical position.—This species is sometimes very common, and found in fine preservation in the Coralline Oolite at Calne, Wilts. It is collected in the Coral Rag at Weymouth, and in the same rock near Faringdon, in Berkshire; from the Coralline Oolite of Malton, in Yorkshire, it is likewise rarely obtained. I have before me specimens from all these localities. From Hildenley, near Malton, some very fine shells have been collected, but the Yorkshire specimens in general are not well preserved.

History.—This, in all probability, is the urchin which was figured by Martin Lister* and by Plott,† and has been always a much admired and abundant Echinite. Until separated by Dr. Fleming under the name *Hemicidaris intermedia*, it was confounded with *Hemicidaris crenularis* of continental authors; and I have good reasons for believing that it is often mistaken for that species by foreign palæontologists, as I have received specimens ticketed *Hemicidaris crenularis* which undoubtedly belong to the English species. In *Hemicidaris intermedia* the base is always much more inflated than the same region of the test in *Hemicidaris crenularis*, which tapers more, and gives the shell a more globular and elegant form. Compare, for example, our figures with the excellent drawings of *Hemicidaris crenularis*, given in Goldfuss's 'Petrefacta Germaniæ.'

D. *Species from the Portland Oolite.*=16*th Etage Portlandien*, d'Orbigny.

HEMICIDARIS DAVIDSONI, *Wright*, nov. species.   Pl. IV, fig. 2 *a, b, c, d.*

Test subglobose; ambulacral areas slightly undulated, upper half with two rows of minute marginal tubercles, lower half occupied by an irregular row of large semi-tubercles; primary tubercles of the inter-ambulacral areas well developed, and gradually diminishing in size from the equator to both poles; apical disc large, genital plates of unequal size; anal opening large and excentral; mouth opening large; peristome notched into nearly equal-sized lobes.

Dimensions.—Height, nearly one inch; transverse diameter, one inch and three tenths of an inch.

Description.—This is the only *Hemicidaris* known from the Portland Beds, it is identical

---

* 'Historia Animalium Angliæ Lap. Turb.,' p. 221, pl. 7, fig. 21, 1578.
† 'Natural History of Oxfordshire.'

with one collected many years ago from the Portland Sandstone of the Boulonnais by MM. Bouchard, Chantereaux, and Davidson, and beautifully drawn by the latter gentleman in the first plate of his 'Memoir on the Fossils of the Boulonnais,' the publication of which has been delayed until the completion of his great work on British Fossil Brachiopoda for the Palæontographical Society. Mr. Davidson has kindly sent me his beautiful drawings, to make of them whatever use I may think fit; and M. Bouchard has most generously forwarded his best specimens of this species to enable me to complete its description. The organic characters of *Hemicidaris Davidsoni* are intermediate, like its stratigraphic position, between *Hemicidaris intermedia* and *Hemicidaris Purbeckensis*. The shell has a sub-globose form, slightly flattened at the poles; it is rather more tumid towards the base than above; the ambulacral areas are narrow, and slightly undulated; the base is occupied by large semi-tubercles (Pl. IV, fig. 2 *b*), which are not disposed regularly in pairs, as in *Hemicidaris intermedia* (Pl. IV, fig. 1 *c*, *d*); but in consequence of the narrowness of the areas in this region, the four upper semi-tubercles extend one above another, forming an irregular continuous line, ascending nearly half-way up the area (fig. 2 *b*); the upper half of which has two rows of minute perforated tubercles, raised on small eminences (fig. 2 *c*); these tubercles, about twelve in each row, alternate with each other, and are separated by a zigzag line of miliary microscopic granules. The poriferous zones form a series of crescentic waves round the ambulacral sides of the tubercular plates (fig. 2 *c*); the pores are in pairs, with a slightly elevated portion of the septum between them, and there are nine pairs of pores opposite each of the large inter-ambulacral plates; only at the base of the areas, and close to the peristome, do the pores fall into double pairs; and of these there are only two or three rows.

The inter-ambulacral areas are more than three and a half times the width of the ambulacral; there are eight primary tubercles on each of the two rows that fill this division of the test; from the equator, where they are largest, to both poles, they gradually diminish in size; the tubercles are small, and not deeply perforated; the bosses are large and prominent (fig. 2 *c*), with coarsely crenulated summits; they are surrounded by wide, smooth areolas; each tubercular plate (fig. 2 *c*) has on its zonal border, between the areola and the pores, a series of six or seven minute tubercles, raised on small eminences; and on its centro-sutural border a series of six or seven like minute tubercles, which form two lateral crescents around the base of the areola (fig. 2 *d*); the four uppermost plates have a horizontal line of smaller granules extended across their apical border (fig. 2 *a*, *b*), so that the primary tubercles of these upper plates have their areolas encircled by an uninterrupted series of granules; but in all the lower plates they are more or less confluent above and below (fig. 2 *c*).

The apical disc is large (fig. 2 *f*); the antero-lateral genital plates are more than twice the size of the others; the right plate is the largest, and carries a prominent convex madreporiform body; the postero-lateral and single plates are much reduced in size, in consequence of the width of the anal opening (fig. 2 *af*); the oviductal holes are marginal, the ocular

13

plates are heart-shaped; the eyeholes are small and marginal; the posterior pair extend into the anal circle in consequence of the diminution in size of the postero-lateral and single genital plates (fig. 2 *f*).

The base is flat, the mouth opening is large, and occupies three fifths of this region; the peristome is deeply notched, the lobes are nearly of equal size, and those of the ambulacral areas are the largest (fig. 2 *e*).

The primary spines are not long, rarely exceeding in length one and a half times the diameter of the test; they do not taper suddenly; the head is short; the acetabulum is small, crenulated on the rim, and surmounted by a prominent milled ring, beyond which is a second ring-like elevation; the surface of the stem is covered with very fine longitudinal lines, which are almost always effaced; the secondary spines, articulated with the minute tubercles of the ambulacra, and those surrounding the margins of the inter-ambulacral plates, are short, thorn-like processes.   This description of the spines is made from French specimens.

Affinities and differences.—This urchin resembles very much *Hemicidaris Purbeckensis*, both in its general configuration and in many points of its organization; but differs from it in the manner the semi-tubercles are arranged at the base of the ambulacra, in the greater size of the apical disc and anal opening, and, above all, in the absence of the broad, crenulated band which surrounds the body of the primary spines above the milled ring.

Locality and Stratigraphical position.—The specimen figured was collected from the Portland Sand.   Those found in the Boulonnais were collected from the Portland Beds of the Falaise d'Alprecht, and at Ningle, by MM. Bouchard and Davidson, where it is associated with a small elongated urchin, *Echinobrissus Haimii*, Wright, nov. sp.

E. *Species from the Marine Purbeck Beds.*

HEMICIDARIS PURBECKENSIS, *Forbes.*   Pl. V, fig. 4 *a, b, c, d.*

HEMICIDARIS PURBECKENSIS.	Forbes, Memoirs of the Geological Survey, Decade III, pl. 5.
— —	Morris, British Fossils, 1854, 2 ed., p. 82.
— —	Desor, Synopsis des Échinides Fossiles, p. 5, pl. 11, fig. 5.

Test sub-globose; ambulacral areas narrow and nearly straight, with five pairs of small semi-tubercles; inter-ambulacral areas with eight small tubercles set on moderate-sized bosses; apical disc composed of unequal-sized genital plates; primary spines sub-compressed, with a broad band of fine longitudinal lines at the base, above the prominent milled ring.

Dimensions.—Specimens all more or less distorted. Height, about three quarters of an inch; transverse diameter, about one inch and three tenths.

Description.—The discovery of Echinoderms in the Cinder Bed of the Purbecks was one of the rewards of the careful examination of these strata made by the officers of the Geological Survey of Great Britain. "For several days," observed Professor Forbes,[*] who found the first specimen at Swanage, in Dorsetshire, "I had found spines of an urchin with which I was unacquainted among the marine fossils which occur in a zone on the summit of the well-known 'Cinder Bed,' composed chiefly of *Ostrea distorta*, Sow., and constituting a conspicuous stratum in the middle division of the Purbecks. A careful search, during which I was rewarded by the discovery of several new forms of marine Purbeck Mollusca, resulted in the finding of a very perfect specimen of the body of the *Hemicidaris*, now first described, accompanied by its spines, identical in structure with those previously observed." It is this same urchin, with others since found, that my figures represent.

The body is sub-globose, but was apparently rather depressed above; the specimens being all more or less distorted, it is impossible to describe its form with accuracy.

The ambulacral areas are narrow, and only slightly undulated; there are two rows of minute perforated tubercles, on miniature bosses, on the margins, about sixteen in each row; a zigzag line of very small granules runs down the middle of the area, and sends small branches of granules to encircle each minute tubercle (fig. 4 *b*); at the base of the area there are five pairs of small semi-tubercles, which, in a small specimen before me, have a very regular arrangement but in the larger specimens are more diffusely disposed, so as to alternate with considerable interspaces. The poriferous zones are slightly undulated, the pores are small, the pairs are a little oblique, and there is a slight elevation of the test corresponding to the septa; there are about nine or ten pairs of pores opposite one of the large inter-ambulacral plates.

The inter-ambulacral areas at the equator are hardly three times the width of the ambulacral; the two rows of primary tubercles occupy the centre of the plates (fig. 1 *a*); the tubercles are small, set upon a smooth, slightly elevated boss, with a deeply crenulated summit; around the base is a smooth, well-defined, and grooved areola; the ambulacral and centro-sutural sides of the large plates are bordered by small, rounded granules (fig. 4 *b*), some of which extend between adjacent areolas in the upper part of the areas, but they are absent from the plates below when the areolas are confluent. The areola is wide in comparison with the size of the boss (fig. 4 *b*). There are about eight primary tubercles in each row, gradually increasing in size as they approach the equator.

The apical disc is moderately large (fig. 4 *a*); the antero-lateral genital plates are much the largest; the madreporiform body occupies nearly all the surface of the

---

[*] 'Memoirs of the Geological Survey,' Decade III, description of pl. v, p. 3.

right plate; the postero-lateral and single plates are nearly alike in size, but the single one is the smallest; all the genital holes are marginal; as the anal opening is transversely oval, and enlarged at their expense, that aperture is excentral. The ocular plates are small, heart-shaped bodies, wedged in between the truncated apices of the ambulacra and the genital plates; the eyeholes are quite marginal.

The spines (fig. 4 c, d) are sub-cylindrical, and slightly compressed on the sides; the diameter of the largest spine is about one tenth of an inch; they are all so much weathered, that the longitudinal lines on their surface cannot be seen; above the prominent milled ring (fig. 4 c) there is a broad, well-defined space, marked with longitudinal lines, which forms an important diagnostic mark between this species and its congeners. The articulating head is small, and the acetabulum diminutive.

In none of the specimens is the mouth opening exposed.

Affinities and differences.—*Hemicidaris Purbeckensis* differs from *Hemicidaris intermedia* in having smaller tubercles and less prominent bosses. In this respect it is allied to the *Hemicidaris Bravenderi* (Pl. V, fig. 1; Pl. XI, fig. 3); but the character of the spines, with the striated space above the milled ring, distinguishes it from both. Its closest affinity is with *Hemicidaris Davidsoni;* but from that Portland species it is distinguished by having narrower inter-ambulacral areas, smaller bosses, more confluent areolas, and above all by the comparative regularity of the semi-tubercles, which, on the contrary, extend singly up nearly half the ambulacral areas in *Hemicidaris Davidsoni* (Pl. IV, fig. 2 b, c). The apical disc is likewise much smaller, and the vent not so wide.

HEMICIDARIS STRAMONIUM, *Agassiz.*

HEMICIDARIS STRAMONIUM.	M'Coy, New Mesozoic Radiata, Annals and Magazine of Natural History, vol. ii, new series, p. 420.
— —	Forbes, Memoirs of the Geological Survey, Decade III, notes on Hemicidaris.
— —	Morris, British Fossils, 1854, 2 ed., p. 82.

From the doubtful manner in which this species was quoted by Professor Forbes, in his 3d Decade of the Memoirs of the Geological Survey, I requested Professor Sedgwick to permit me to examine the type specimens belonging to the Geological Collection of the University of Cambridge, a favour which the learned professor most kindly and readily granted. These urchins had been labelled by Professor M'Coy *Hemicidaris stramonium*, Agassiz, and were catalogued as such in the Addenda to his Paper on some new Mesozoic Radiata, published in the 'Annals of Natural History.' On examination, these urchins proved to be two small imperfect specimens of *Hemicidaris intermedia*, from the Coral Rag, Calne. On placing these specimens side by side with a true *Hemicidaris stramonium*, Agass., kindly sent me by M. Michelin, the difference between them is seen to be **very great**. The size and prominence of the inter-ambulacral tubercles in *Hemicidaris stramonium*, the magnitude and spaced-out arrangement of the semi-tubercles at the base of the ambulacra, and the rudimentary condition of the marginal granules on the upper half of these areas, form a group of characters by which this species is readily distinguished from its congeners. It belongs, moreover, to beds newer than the Kimmeridge Clay.

This species must, therefore, be omitted from our list of British *Hemicidaris*.

———

102

NOTES

ON FOREIGN JURASSIC SPECIES OF THE GENUS HEMICIDARIS NEARLY ALLIED TO BRITISH FORMS, BUT WHICH HAVE NOT YET BEEN FOUND IN THE ENGLISH OOLITES.

HEMICIDARIS CRENULARIS.	Lamarck's (sp.), Goldf., Petrefact., p. 122, t. 40, fig. 6.
— —	Agassiz, Echinod. Foss. Suisse, II, p. 44, t. 18, figs. 23, 24 ; t. 19, figs. 10—12.
— —	Cotteau, Études Échinides Foss., p. 122, t. 13, figs. 1—9.

Test globular, flattened at the base, and on the upper surface. Ambulacral areas narrow, and flexuous above; six pairs of close-set semi-tubercles at the base, and two rows of minute tubercles on the sides, running into one row in the upper part of the areas; inter-ambulacra with seven or eight large tubercles in each of the two rows. Mouth opening very large, one half the diameter of the test; peristome deeply incised into ten nearly equal sized lobes. Apical disc small.

Spines large, thick, claviform, gradually increasing in thickness from the head to the distal extremity; surface covered with longitudinal lines. The largest spines are once and two thirds the length of the diameter of the test; milled ring small, stem without any apparent neck.

The greater size of the mouth opening, the depth of the notches in the peristome, the greater equality in the size of the lobes, with less tumidity at the base of the test, added to the claviform character of the spines, serve to distinguish this species from *Hemicidaris intermedia*, which in other respects it most closely resembles.

Dimensions.—Height, one inch and one fifth ; breadth, one inch and a half.

Formation.—Corallien of Switzerland, and of France.
Coral Rag, Nattheim.

Collections.—In all Foreign Collections of Jurassic Fossils.
British Museum, my Cabinet.

HEMICIDARIS DIADEMATA. Agassiz, Echinoderm. Foss. Suisse, II, p. 49, t. 19, figs. 15—17.

Cotteau, Études Échinides Foss., p. 128, t. 14, figs. 1—5.

Test hemispherical, sub-inflated below, and depressed at the upper surface. Ambulacral areas straight, wide and expanded below to enclose six pairs of large semi-tubercles; narrow above, with four rows of small granules; more or less irregularly disposed. Inter-ambulacral areas with two rows of tubercles, which are large and prominent at the equator, but are very small, and become suddenly dwarfed, on the upper surface. Apical disc large. Mouth opening large; peristome deeply notched, divided into ten nearly equal-sized lobes.

Dimensions.—Height, seventeen twentieths of an inch; breadth, one inch and four tenths.

Formations.—" Portlandien inférieur ? Astartien de la Vallée de la Birse, de Porrentruy, du Jura Soleurois, rare."—*Desor.*

" Hobel, Soleure."—*Michelin.* " Corallien étage Drùges, Yonne."—*Cott.*

Collections.—Museum Neuchâtel. Coll. Michelin, Cotteau, Wright.

HEMICIDARIS STRAMONIUM. Agassiz, Echinoderm. Foss. Suisse, II, p. 47, t. 19, figs. 13, 14.

Test small, thick, sub-spheroidal. Ambulacral areas narrow and flexuous, with six or seven large, prominent semi-tubercles, extending in a zigzag line to near the circumference: upper part of the area with two marginal rows of small granules. Inter-ambulacral areas with six tubercles in each row; the bosses are large and prominent and the tubercles are large, especially on the upper part of the test. " Spines cylindrical, with a prominent milled ring, without a distinct neck; surface covered with small sporadic granules."—*Desor.*

Dimensions.—Height, three fifths of an inch; breadth, nine tenths of an inch.

Formations.—Portlandien inférieur (Astartien) de Delémont, Rædersdorf, Pfeffingen, de Chablis (Yonne).

Collections.—Museums Neuchâtel, Zurich, Bale. Coll. MM. Michelin, Cotteau.

The specimen in my Cabinet, sent by M. Michelin, is from Le Loile près la chaux de fonds Neuchâtel.

HEMICIDARIS ALPINA.   Agassiz, Echinoderm. Foss. Suisse, II, p. 52, t. 18, figs. 19—22.

Test large, sub-conoidal.   Ambulacra wide, with six pairs of large, close-set semi-tubercles at the base; two marginal rows of round granules above, and between these four or five rows of a microscopic miliary granulation.   Inter-ambulacra with eight tubercles in each column.   Mouth opening large, peristome deeply notched, lobes nearly equal. Apical disc of moderate size, elements nearly equal.

Dimensions.—Height, nine tenths of an inch ; breadth, one inch and one quarter.

Formation.—" Portlandien moyen (Kimméridgian) de Gesné et des Ormonds (Alpes Vaudoises), très rare."—*Desor.*

Collections.—Museums of Berne and Zurich

HEMICIDARIS THURMANNI.   Agassiz, Echinoderm. Foss. Suisse, II, p. 50, t. 19, figs. 1—3.

Test small, depressed.   Ambulacral areas narrow and very flexuous, with small semi-tubercles at the base, and minute marginal tubercles above.   Inter-ambulacral areas wide, with three or four tubercles only in each row, which have very prominent bosses ; the areolas are each surrounded by a complete circle of scrobicular granules.   Mouth opening large, peristome divided by notches into ten nearly equal-sized lobes.

Dimensions.—Height, eleven twentieths of an inch ; breadth, one inch and one tenth.

Formations.—" Portlandien moyen (Marnes Ptérocériennes) du Banné près Porrentruy, de Delémont, du Jura Vaudois, des environs de Salins."—*Desor.*

Collections.—Museums of Neuchâtel and Bale.   Coll. Michelin, Thurmann ; common.

HEMICIDARIS HOFMANNI. Syn. *Cidarites Hofmanni.*   Roemer, Oolitic Gebirges, p. 25, t. 1, fig. 18.

Test small, sub-spheroidal, flat at the summit.   Ambulacral areas narrow, convex, and slightly flexuous, with five or six pairs of semi-tubercles at the base, and two rows of small marginal granules above. Inter-ambulacral areas with seven tubercles in each row; areolas confluent. Apical disc prominent, the anterior pair of genital plates the largest. Mouth opening of moderate width, peristome nearly equal lobed.

Dimensions.—Height, two fifths of an inch; breadth, nineteen twentieths of an inch.

Formations.—" Portlandien moyen, Kimméridgien, du Spielberg (Hanover), de Fritzow en Poméranie."—*Desor*.

Collections.—Museum Neuchâtel. Coll. Roemer, my Cabinet. Sent me by Professor Roemer.

HEMICIDARIS MITRA. Agassiz, Echinoderm. Foss. Suisse, II, p. 48, t. 17, figs. 7—9.

Test hemispherical, inflated towards the base. Ambulacra prominent and nearly straight, semi-tubercles small. Inter-ambulacral tubercles more crowded together than in *Hemicidaris intermedia;* areolas all confluent. Peristome deeply notched, and nearly equal lobed.

Dimensions.—Height, seven tenths of an inch; breadth, one inch.

Formations.—" Portlandien moyen (Calcaire à Tortues) de Saint Nicolas près Soleure Ptérocérien de Pierre-Percée (Jura) assez rare."—*Desor*.

Collections.—Museum Neuchâtel. Coll. Thurmann.

Family 3—Diademadæ.

This family includes large and small urchins, with the test thin, circular, pentagonal, or sub-pentagonal; more or less depressed; mouth opening large, central; peristome decagonal, and divided into ten lobes by deep notches.

Apical disc small, directly opposite to the mouth, and composed of five genital, and five ocular plates; the anterior pair of genital plates are, in general, a little larger than the posterior pair; and the right antero-lateral plate, with the small, spongy, madreporiform body on its surface, is the largest. The vent is round or oblong, and is generally in the centre of the genital circle.

The ambulacral areas are more or less wide; sometimes they are one half the width of the inter-ambulacral areas, and furnished with two or four rows of primary tubercles, often as large and as numerous as those of the inter-ambulacral areas.

The poriferous zones are narrow, and almost always straight; the pores are unigeminal, bigeminal, or trigeminal in their disposition in different genera.

The inter-ambulacral areas are, in general, twice the width of the ambulacral; there are from eight to fourteen plates and upwards in each column; the areas are occupied by two, four, six, or eight rows of primary tubercles, nearly all of the same size; or there are rows of primary tubercles, with two or four rows of secondary tubercles, much smaller in size, filling up the interspaces of the area. The bosses of the tubercles are small, and their summits are in general crenulated, but sometimes they are uncrenulated; the spinigerous tubercles are small; in general they are perforated, rarely are they imperforate; the presence or absence of these characters, added to the structure of the poriferous zones, afford, when taken collectively, good generic characters.

In general the inter-ambulacral tubercles are larger than the ambulacral, but sometimes they are of equal sizes in both areas, a character which distinguishes the genera of this family from those of the Cidaridæ and Hemicidaridæ.

The spines in the living genera are long, slender, and tubular; sometimes they are thrice the length of the diameter of the shell.* In the fossil genera they rarely attain the length of the diameter of the shell, and are stout and solid; the slender, tubular spines have their surface ornamented with oblique annulations of fringe-like scales; whilst the surface of the solid fossil spines is covered with fine, longitudinal lines; but neither prickles nor other asperities are developed on their stems.

Lamarck† subdivided Klein's genus Cidaris into two sections, the " *Turbans* " and the

---

* Peters, Über die Gruppe der Diademen, p. 7.
† 'Histoire des Animaux sans Vertèbres,' tom. iii, 1st edit., pp. 54—58.

"*Diadems;*" these Dr. **Gray*** afterwards converted into two genera, separating from the Diadems *Cidaris radiata*, Leske,† as the type of his new genus *Astropyga*. The genus Cidarites of Lamarck was considered as a natural family, composed of the genera *Cidaris*, *Diadema*, and *Astropyga*, which he thus characterised:

Family 1. CIDARIDÆ. *Cidarites*, Lamarck.
> Body with two-sized spines; larger ones club-shaped, or very long; spine-bearing tubercles perforated at the apex.

Genus 1. *Cidaris*, Klein, Lamarck; *Turbans*.
> *Body depressed, spheroidal,; ambulacra waved; small spines compressed, two edged, two rowed, covering the ambulacra, and surrounding the base of the larger spines.*
> This genus may be divided according to the form of the larger spines; the extra-ambulacral beads have only two rows of spines.
> *Cidaris imperialis*, Lamk., Klein, t. 7, fig. A.

Genus 2. *Diadema*. *Diadems*.
> *Body orbicular, rather depressed; ambulacra straight, spines often fistulous.*
> *Echinometra setosa*, Rumph., Leske, Klein, t. 37, figs. 1, 2.
> *Echinus diadema*, Linn., Syst. Nat., by Turton, vol. iv, p. 139.
> *Echinus calamaria*, Pallas, Spicil. Zool., t. 2, figs. 4—8.

Genus 3. *Astropyga*.
> *Body orbicular, very depressed; ambulacra straight; ovarian scales very long, lanceolate; beads with several series of spines.*
> *Cidaris radiata*, Leske, Klein, t. 44, fig. 1.

The very meager characteristics by which this author has defined the last two genera only show that a difference exists, but his description is insufficient for a correct diagnosis between them; hence various opinions exist regarding the characters and limits of the genus *Diadema*; and only one of the species enumerated as types, the *Diadema setosa*, Rumph., is admitted to be a true Diadem. The valuable memoir of Herr W. Peters‡ has removed some of the difficulties that surrounded this subject, and his grouping of the living Diadems makes an important step towards a natural classification of one section of

* 'Annals of Philosophy,' new series, vol. x, p. 426, 1825. "An attempt to divide the Echinidæ or Sea Eggs into Natural Families."

† Leske apud Klein, 'Naturalis Dispositio Echinodermatum,' t. 44, fig. 1, p. 116.

‡ Über die Gruppe der Diademen, 'Gelesen in der Königl. Akademie der Wissenschaften,' Berlin, Aug. 1853.

this family.   Although the present state of our scientific knowledge of the *Diademadæ* may be considered as transitional rather than positive, still we know enough to justify the separation of the fossil Diadems from the existing genera, as proposed by M. Desor.*

The Diademadæ, in fact, appear to consist of two types : one of these, with a few rare exceptions, appertains to the present epoch; the other existed during the deposition of the secondary and tertiary rocks.   The living forms are in general large, depressed urchins, with a thin shell, having the tubercles and pores variously arranged in the different genera. They have, in general, very long, slender, *tubular* spines, the surface of which is covered with oblique annulations of small imbricated scales.   The fossil species, on the contrary, are smaller urchins, having a thicker test; the tubercles and pores are variously disposed in the different genera; the spines rarely attain the length of the diameter of the test; they are *solid*, in general cylindrical, sometimes flattened or awl-shaped, and their surface is covered with fine longitudinal lines.

I propose to include the following genera in this natural family :

A Table showing the classification of the Diademadæ.

DIADEMADÆ.	**Section A.** Spines very long, slender, tubular, covered with oblique annulations of imbricated scales. Living in tropical seas. A few annulated tubular spines are found in the Upper Chalk and in the Coralline Crag.	DIADEMA, *Gray.* SAVIGNYA, *Desor.* ASTROPYGA, *Gray.* ECHINOTHRIX, *Peters.*
	Section B. Spines short, slender, solid; surface covered with fine longitudinal lines. Extinct in the Oolitic, Cretaceous, and Tertiary rocks.	PSEUDODIADEMA, *Desor.* CYPHOSOMA, *Agassiz.* HEMIPEDINA, *Wright.* PEDINA, *Agassiz.* ECHINOPSIS, *Agassiz.*

PSEUDODIADEMA,† *Desor.* 1854.

This genus is composed of small urchins, with a moderately thick test, which rarely attains two inches in diameter; the ambulacral areas are in general one third, or even one half the width of the inter-ambulacra; the primary tubercles of both areas are all perforated, and nearly of the same size; their bosses are small, with sharply crenulated summits.

The ambulacral areas have two rows of tubercles; the inter-ambulacral areas have two rows only, or two rows of primary, and two or four shorter rows of small secondary

---

* 'Synopsis des Échinides Fossiles.'
† The specific name given by Lamarck to the fossil urchin described as *Cidarites pseudodiadema.*

tubercles; or they have four, or even six rows of nearly equal-sized primary tubercles at the equator.

The poriferous zones are narrow and straight; the pores in one section are unigeminal throughout; but in another section they are bigeminal in the upper part of the zones.

The apical disc is seldom preserved; it is, in general, small; and the anterior pair of genital plates are larger than the posterior pair.

The mouth opening is large, the peristome is deeply notched, and the oral lobes are not very unequal sized.

The spines rarely attain the length of the diameter of the shell; they are in general much shorter, and are cylindrical or needle-shaped, and have a prominent, milled ring near the articulating head; the rim of the acetabulum is crenulated, and the socket perforated; the surface of the stem is sculptured with delicate longitudinal lines.

The *Pseudodiadema* are all extinct, and are found in the Liassic, Oolitic, Cretaceous, and Tertiary rocks.

Pseudodiadema differs from *Diadema* in having solid spines, with a smooth surface, the sculpture, in most cases, consisting of microscopic, longitudinal lines; whilst the spines of *Diadema* are tubular, and have oblique annulations of scaly fringes on their surface. It differs from *Cyphosoma*, which is a Cretaceous genus, in having the tubercles always perforated, those of *Cyphosoma* being imperforate.

It differs from *Hemipedina* in having a small apical disc, and tubercles with crenulated bosses, those of *Hemipedina* being smooth; and from *Pedina* in having the pores unigeminal or bigeminal, those of *Pedina* being arranged in triple, oblique pairs.

The *Pseudodiademas* may be divided into two sections, from the different manner the pores are arranged in the zones. In one group the pairs of pores are not so numerous, and they are disposed in a single file throughout; in another group the pores are more numerous, and crowded together in the upper part of the zones. Professor M'Coy* has proposed the genus *Diplopodia* for this group. *Cæterus paribus*, the crowding together of a greater number of pores in a zone is, at most, a sectional, and can never form a stable generic character, inasmuch as it is subject to great variation in the diplopodous species themselves, and is, moreover, often only an adult development.

---

* 'Annals and Magazine of Natural History,' 2d series, vol. ii, p. 412.

The following Table exhibits these Sections, and shows at the same time the arrangement of the tubercles in the unigeminal and bigeminal groups.

Section A.—*Pores unigeminal in the upper part of the zones:*

Examples.

a. Two rows of primary tubercles only in the inter-ambulacral areas } *Pseudodiadema depressum.*

b. Two or four rows of primary tubercles, and two or four rows of smaller secondary tubercles in the inter-ambulacral areas } *Pseudodiadema hemisphæricum.*

Section B.—*Pores bigeminal in the upper part of the zones:*

a. Two rows of primary tubercles, and two rows of small secondary tubercles, in the inter-ambulacral areas } *Pseudodiadema versipora.*

b. Four rows of primary tubercles in the inter-ambulacral areas } *Pseudodiadema variolare.*

A. *Species from the Lias.*

PSEUDODIADEMA MOOREII, *Wright.* Pl. VI, fig. 1 *a, b, c, d.*

DIADEMA MOOREII.   Wright, Annals and Magazine of Natural History, 1854, 2d series, vol. xiii, p. 171, pl. 12, fig. 3.

—   —   Morris, British Fossils 2d edit., 1854. Note on additional species of Echinodermata.

DIADEMOPSIS MOOREII.   Desor, Synopsis des Échinides Fossiles, p. 81.

Test circular, depressed; ambulacral tubercles smaller than those of the inter-ambulacral areas; plates of the test covered with a small, wide-set, prominent granulation; mouth large and decagonal; anal opening large; apical disc of moderate size.

Dimensions.—Height, one fourth of an inch; transverse diameter, six tenths of an inch.

Description.—There is much difficulty in distinguishing some of the smaller Diademas from each other, inasmuch as the young condition of many of the larger species so closely resembles the adult state of others, that it is only after obtaining a number of individuals of different species in the various phases of their growth, that the naturalist feels himself upon sure ground when endeavouring to distinguish the affinities and differences existing among them. After a diligent search for urchins in the Lias of Gloucestershire, I have succeeded in collecting from these rocks only a very few examples of this group. In addition to those found near Cheltenham, my friend Mr. Moore, of Bath, kindly presented me with a few specimens which he collected from the Upper Lias near Ilminster, and from these collective materials the species under consideration was discovered.

Pseudodiadema Mooreii has a circular outline, slightly inclining to a pentagonal contour; it is much depressed at the upper surface, and is flattened at the base (fig. 1 *a*). The ambulacral areas are very narrow, being less than one third the width of the inter-ambulacral (fig. 1 *b*); their margins are occupied by two rows of tubercles, about eight in each row (fig. 1 *d*), which, at the base, and up to the equator, are nearly as large as those of the inter-ambulacra, but from that region to the apex of the area they rapidly diminish in size, and are here very disproportionate in magnitude to them; a zigzag line of single granulation separates the two rows of tubercles from each other (fig. 1 *d*). The inter-ambulacral areas are wide and well developed (fig. 1 *b*, *c*), from eight to nine in each row, which occupy the centre of the plates (fig. 1 *a*); the areolas of the tubercles on the upper surface are surrounded with a circle of granules which separates them from each other (fig. 1 *d*), but those of the base are confluent above and below (fig. 1 *c*). The miliary zone at the base of the test has a number of granules scattered over it (fig. 1 *e*); whilst on the upper surface the plates are destitute of any other ornament beyond the faint circles that surround the tubercles (fig. 1 *b*).

The poriferous zones are narrow, the pores are strictly unigeminal (fig. 1 *b*, *d*); the avenues are, however, rather flexuous below; the basal tubercles of both areas are nearly alike in size, but on the dorsal surface those of the ambulacra dwindle into large granules; whilst those of the inter-ambulacra maintain their size up to the last pair, which are small near the margin of the disc. The mouth opening is large (fig. 1 *e*), and the peristome is divided into ten nearly equal-sized lobes.

The apical disc is partly preserved in the specimen here figured (fig. 1 *b*). It consists of five large ovarial plates, of a heptagonal form; two of the sides unite with the inter-ambulacral plates, two with the ocular, two with the adjoining ovarials, and the single surface contributes to form the boundary of the anal opening, which is of moderate size; the five ocular plates are small and heart-shaped; their apex is directed towards the anal opening, and their base to the area; the madreporiform tubercle is slightly elevated on the right antero-lateral ovarial plate; the surface of all the discal elements is almost destitute of sculpture or granulation.

Affinities and differences.—*Pseudodiadema Mooreii* resembles *Pseudodiadema depressum*, Agassiz, in the depression of its upper surface and the flatness of its base, likewise in having the tubercles of both areas nearly of a uniform size below; but it is readily distinguished from *Pseudodiadema depressum* by the number and greater development of the tubercles of the ambulacra, which maintain their size throughout, whilst in *Pseudodiadema Mooreii* the ambulacral tubercles are fewer in number, and more rudimentary in size, in all the upper part of the areas. The contour of the test, moreover, does not assume the pentagonal outline of *Pseudodiadema depressum*, nor has the upper surface of the inter-ambulacral areas the median depression seen on the test of the latter. The mouth opening is larger, and the decagonal lobes are more equal in size in *Pseudodiadema Mooreii* than in *Pseudodiadema depressum*.

Locality and Stratigraphical position.—I have collected *Pseudodiadema Mooreii* in the Upper Lias of Gloucestershire. Mr. Moore found it in the same stratum near Ilminster, with *Ammonites communis*, Sow., and *Ammonites serpentinus*, Schloth. Prof. Deslongchamps has communicated a specimen of this urchin, which he found in the Lias supérieur de May, Calvados, associated with *Leptæna Davidsonii*, Deslong., *Thecidium Bouchardii*, Davidson, and several other species of Upper Lias Mollusca.

I dedicate this species to Mr. Charles Moore, of Bath, whose assiduous practical researches in palæontology have brought to light so many interesting forms from the Middle and Upper Lias beds of Somersetshire.

B. *Species from the Inferior Oolite.*

PSEUDODIADEMA DEPRESSUM, *Agassiz.*   Pl. VI, fig. 2 *a, b, c, d, e, f, g, h.*

DIADEMA DEPRESSUM.	Agassiz and Desor, Catalogue raisonné des Échinides, Annales des Sciences Naturelles, tom. vi, p. 349, 3ᵐᵉ série.
— —	Cotteau, Études sur Échinides Fossiles, pl. 2, p. 43.
— —	D'Orbigny, Prodrome de Paléontologie, tome i, p. 290, No. 512.
— —	Morris, British Fossils, 2d edit., 1854, p. 76.
— —	Wright, Annals and Magazine of Natural History, 2d ser., vol. viii, p. 258, pl. 12, fig. 2 *a, b, c, d.*
PSEUDODIADEMA DEPRESSUM.	Desor, Synopsis des Échinides Fossiles, p. 65.

Test pentagonal, depressed; ambulacral areas convex and prominent; inter-ambulacral areas flattened; two rows of nearly equal-sized primary tubercles in both areas; no secondary tubercles; mouth large and decagonal; peristome nearly equally lobed.

Dimensions.—Height, eleven twentieths of an inch; breadth, one inch and three tenths.

Description.—The ambulacral areas of this urchin are rather more than one half the breadth of the inter-ambulacral areas, and have from ten to twelve pairs of well-developed primary tubercles, separated by a zigzag line of small granulations (fig. 2 *d*). The inter-ambulacral areas are nearly of a uniform breadth throughout; there are about ten pairs of tubercles in each area. In consequence of these segments of the test being double the width of the ambulacra, the tubercles stand more apart (fig. 2 *d*). The tubercles of both areas are nearly uniform in size; those of the ambulacra are the smallest; they have a smooth base, with a finely crenulated summit, and are perforated (fig. 2 *h*); there are no secondary tubercles, but the miliary zones are covered with small granulations, which are closely set together on the surface of the plates; three or four of these at the base are perforated (fig. 2 *g*). The mammillary eminences of both areas are surrounded by smooth areolas, which are nearly all confluent (fig. 2 *d*). The ambulacral areas become rapidly contracted towards the vertex (fig. 2 *a*), whilst the inter-ambulacral maintain their breadth, so that the space between the rows of primary tubercles is very uniform in width throughout (fig. 2 *a*, *e*). The miliary zones, with the exception of the internal border of the four superior inter-ambulacral plates, are covered with small, close-set granulations of different sizes (fig. 2 *d*), which form semicircles around the areolas, and zigzag lines down the centre of the areas. The pores consist of from thirty-six to forty pairs in each avenue, super-imposed in a single file; about four pairs of pores are opposite each large inter-ambulacral plate (fig. 2 *d*); in the wide space of the avenues around the mouth, they form rows of triple oblique pairs (fig. 2 *g*). The mouth is large and decagonal (fig. 2 *b*); the notches of the peristome divide the opening into ten nearly equal-sized lobes, the borders of the notches are reflexed at the angles; the apical disc is unknown; the spines are small, subulate, and delicately striated longitudinally (fig. 2 *e*, *f*, *i*).

Affinities and differences.—This urchin resembles *Pseudodiadema æquale*, Agassiz, but differs from it in the absence of secondary tubercles in the inter-ambulacral areas; by its pentagonal form it resembles *Pseudodiadema versipora* Phillips, but is distinguished from that species in having the pores arranged in a single file; whereas in *Pseudodiadema versipora*, from the equator to the apical disc, the pores fall into double rows. The tubercles are likewise smaller, and more deeply perforated; it belongs, moreover, to a lower zone of the Oolitic group, *Pseudodiadema versipora* being a characteristic urchin of the Coral Rag of England and the "Terrains à Chailles" of Switzerland.* Like *Pseudodiadema versipora*, *Pseudodiadema depressum* possesses a pentagonal form, a peculiarity depending on the prominence of the ambulacral areas, and common to several species of this genus.

Locality and Stratigraphical position.—This urchin is common in the lower ferruginous beds of the Inferior Oolite, the Pea Grit, at Crickley, Leckhampton, and Dundry Hills. I have collected it from the Great Oolite at Minchinhampton, and from the Bradford Clay at Tetbury-road Station. The Inferior Oolite specimens are in general much crushed;

---

* Agassiz, 'Echinodermes Fossiles de la Suisse.'

the apical disc is always broken, but the spines are sometimes adherent to the test (fig. 2 *i*). It has been collected by M. d'Orbigny in the Inferior Oolite of Saint Honorine, Ranville, where it is abundant. I have before me a specimen collected from the " Oolite Ferrugineuse de Bayeux " so closely resembling some of those from Crickley Hill, that it might readily be taken for an urchin from that locality. My friend M. Deslongchamps, who has kindly lent me this specimen for comparison, observes, " seul exemplaire que j'ai trouvé ;" so that it must be rare in Normandy. It has been obtained by M. Cotteau from the Ferruginous Oolite, of " Tour-du-Prè, près Avallon, département de l'Yonne," which bed lies upon the Calcaire à Entroques, the true equivalent of the Dundry, Cotteswold, and Dorsetshire beds of the Inferior Oolite.

History.—*Pseudodiadema depressum* was first mentioned in the 'Catalogue raisonné des Échinides' by Agassiz and Desor, but was neither figured nor described by them. This, however, has been done by M. Cotteau, in his 'Études sur les Échinides Fossiles.' In both countries it appears to characterise beds belonging to the same geological horizon.

C. *Species from the Stonesfield Slate.*

PSEUDODIADEMA PARKINSONI, *Desor.*   Pl. VI, fig. 4.

> PSEUDODIADEMA PARKINSONI.   Desor, Synopsis des Échinides Fossiles, p. 66.
> Parkinson's Organic Remains, vol. iii, pl. 1, fig. 8, not named in this work.

Test small, circular, depressed ; ambulacral areas wide ; tubercles of the inter-ambulacral areas a little larger than those of the ambulacral; no secondary tubercles; spines thick, subulate, and a little bent, nearly as long as the diameter of the test, surface covered with longitudinal lines.

Dimensions.—Transverse diameter, three fifths of an inch ; height unknown.

Description.—This pretty little Diadem has been overlooked by all English authors except Parkinson, who figured it in his ' Organic Remains,' and only observes (vol. iii, p. 10), " This very uncommonly perfect specimen from Stunsfield, Oxfordshire (Pl. I, fig. 8), in which a considerable number of spines are still adherent to the shell, appears to be of the same species with the last fossil " (a Wiltshire *Pseudodiadema*). M. Desor, in his ' Synopsis,' names this species *Parkinsoni*, and adds the following note : " Je ne connais cette espèce que par la jolie figure qu'en a donnée Parkinson. Il est surprenant qu'aucun des auteurs Anglais ne l'ait encore mentionnée, ni ne lui donné un nom. Parkinson s'est borné à la figurer sans lui imposer un nom."

When collecting materials for my 'Memoir on the Cidaridæ,' this species engaged my attention, and I endeavoured, in vain, to find the original of Parkinson's figure, as I was under the impression that it was a juvenile form of *Pseudodiadema versipora*, from the Coral Rag of Wiltshire, very much resembling a specimen in the cabinet of Mr. Mackniel, of Trowbridge; for this reason it was not included in my 'Memoir.' As M. Desor has now, however, named the specimen, and entered it in his 'Synopsis,' I have considered it desirable to have an accurate copy of Parkinson's figure transferred to my work.

The ambulacral areas are moderately wide, and carry two rows of tubercles nearly as large as those of the inter-ambulacral areas; the pores are arranged in a single file throughout; the inter-ambulacral areas are about once and a half as wide as the ambulacral, and have two rows of primary tubercles; nor is there any evidence of the existence of a secondary range; the spines are thick and subulate, slightly bent in the middle, and marked with longitudinal lines; they are nearly as long as the diameter of the shell and are very robust in proportion to its dimensions.

Affinities and differences.—This species resembles *Pseudodiadema depressum* in the absence of secondary tubercles; but that species is distinguished from it by the test being more circular, and its spines being thicker and bent. It may, however, be a juvenile form of *Pseudodiadema versipora*, from the Coral Rag, the artist having overlooked the crowding together and doubling of the pairs of pores in the upper part of the poriferous zones; not having the advantage of examining the specimen, I make these remarks with much hesitation, well knowing from experience how readily we may be deceived by figures, where no details are added, to indicate the specialities of the organism.

Locality and Stratigraphical position.—Mr. Parkinson says it was collected from the Stonesfield Slate at Stonesfield, Oxon, where it must be exceedingly rare. I have collected fragments of a small *Pseudodiadema* from the Yellow Clays of the Stonesfield Slate at Sevenhampton, Gloucestershire, which may belong to this species. My specimens are mislaid, but they were not in a good state of conservation.

D. *Species from the Great Oolite.*

PSEUDODIADEMA PENTAGONUM, *M'Coy.* Pl. VI, fig. 3 *a, b, c, d.*

DIPLOPODIA PENTAGONA.	M'Coy, Annals and Magazine of Natural History, 2d series, vol. ii, p. 412.
DIADEMA PENTAGONUM.	Morris, Catalogue of British Fossils, 2d edition, 1854, p. 77.
— —	Woodward, Memoirs of the Geological Survey, Decade V. Notes on Brit. Foss. Diademas.

Test small, pentagonal, depressed; ambulacral areas with two marginal rows of primary tubercles nearly as large as those of the inter-ambulacra; poriferous zones wide; pores unigeminal at the equator, bigeminal on the upper and under parts of the zones; inter-ambulacral areas thrice the width of the ambulacral, with two rows of primary tubercles, and a short row of secondary tubercles, one third the size of the primaries, between the peristome and equator, situated between the zones and the primary row.

Dimensions.—Height, three tenths of an inch; transverse diameter, seventeen twentieths of an inch.

Description.—This pretty little urchin is said by Professor M'Coy, who first described it, to be not uncommon in the Great Oolite of Minchinhampton, but experience has taught me that it is a very rare form in the rich shelly beds of that remarkable locality.

The test is moderately thick, and much depressed (Pl. VI, fig. 3 *a, b*); it is distinctly pentagonal, from the prominence of the ambulacra, and in its general *facies* very much resembles *Pseudodiadema depressum;* but the structure of the poriferous zones, and the crowding together of the pores in the zones, shows it to be very distinct from that Inferior Oolite form. It belongs, in fact, to that section of the genus *Pseudodiadema* which has the pores bigeminal in the upper part of the zones (Pl. VI, fig. 3 *c*), a structure which Professor M'Coy thought of generic value, and proposed the genus *Diplopodia** for the reception of these bigeminal Diadems, describing the species now under consideration as a type of the same.

The ambulacral areas are narrow and prominent, with two rows of from twelve to fourteen primary tubercles, nearly as large as those of the inter-ambulacral areas (fig. 3 *d*); a few miliary granules separate the tubercles in the middle of the areas, but they disappear in their upper and lower parts, in consequence of the close approximation of the tubercles in these portions of the area (fig. 3 *b, c*).

The poriferous zones are wide, especially above and below; near the equator there are five or six pairs of pores, which are unigeminal (fig. 3 *b*); but above these the pairs of pores are bigeminal, each pair being separated by a diagonal line, formed by an elevation of the surface of the zonal plates (fig. 3 *c*); between the unigeminal portion of the zones and the peristome the pores are rather irregularly bigeminal, they then fall into triple, oblique pairs near the mouth; the upper part of the zones consists of small pyriform plates (Pl. VI, fig. 3 *c, d*), which are neatly dovetailed together, the line of union being marked by diagonal lines of elevation (fig. 3 *d*).

The inter-ambulacral areas are three times the width of the ambulacral; they have two rows of primary tubercles, from ten to twelve in each row, situated on the centre of the plates; they are about the same size as the ambulacral tubercles at the equator, but are much larger in proportion in the upper part of the area (fig. 3 *d*); the bosses are small,

* 'Annals and Magazine of Natural History,' 2d series, vol. ii, p. 412.

having their summits feebly crenulated, and the tubercles not deeply perforated. The areolas are superficial, and confluent above and below (fig. 3 *d*); five or six granules form semicircles on each side of the areola. The miliary zone, which is of moderate width, contains few granules besides those forming the scrobicular semicircles, in the upper part of the test this space is almost naked; between the peristome and the equator there are five or six secondary tubercles, about one third the size of the primaries, situated between them and the poriferous zones, forming a short, irregular row, which scarcely reaches the middle of the test (fig. 3 *b*).

The mouth opening in width is nearly one half the diameter of the test; the peristome is unequally decagonal, the ambulacral being larger than the inter-ambulacral lobes.

The disc opening is large, and nearly circular; but the disc is absent in all the specimens I have examined, and in none of them are the spines preserved.

Affinities and differences.—*Pseudodiadema pentagonum* very much resembles *Pseudodiadema depressum* in its pentagonal and depressed form, in having the primary tubercles of both areas at the equator nearly of equal size, and in having the upper part of the miliary zones almost naked; but it is distinguished from that species in having a short row of secondary tubercles at the base, between the inter-ambulacral primary tubercles and the poriferous zones; in having the poriferous zones wide, and the pores bigeminal in all the upper portion of the zones, a character which is at once evident and distinctive between these two nearly allied forms. *Pseudodiadema pentagonum* resembles *Pseudodiadema versipora* in its depressed and pentagonal form, and in the bigeminal character of the pores in the upper parts of the zones; but it is distinguished from that beautiful Coral Rag urchin in having much smaller primary tubercles, and fewer and larger granules in the miliary zones; the central and lateral zones of the inter-ambulacra in *Pseudodiadema versipora* being much wider, and filled with several rows of small, close-set granules; whereas in *Pseudodiadema pentagonum* there are few granules besides those forming the scrobicular semicircles. The diplopodous character of this species serves to distinguish it from all other British Oolitic Diadems.

Locality and Stratigraphical position.—This beautiful urchin has been found in the shelly beds of the Great Oolite at Minchinhampton, where it is very rare. Associated with it are the following Echinoderms: *Acrosalenia hemicidaroides*, Wright; *Polycyphus nodulosus*, Münster; *Pygaster semisulcatus*, Phillips; *Hyboclypus agariciformis*, Forbes; *Echinobrissus clunicularis*, Llhwyd.

History.—First described by Professor M'Coy in the 'Annals of Natural History,'* in his paper "On some new Mesozoic Radiata," under the name *Diplopodia pentagona*, and regarded by him as the type of a new genus of DIADEMADÆ, which he supposed was

* 'Annals and Magazine of Natural History,' 2d series, vol. ii, p. 412.

limited to the Oolites; but it has been ascertained that the same diplopodous structure of the poriferous zones is found to prevail likewise in several Cretaceous species, as well as in those cited by this author. Neither Prof. Agassiz nor the late Prof. Edward Forbes considered the diplopodous character of the poriferous zones alone of generic importance; but Professor M'Coy and M. Desor think otherwise, the latter author having retained the genus *Diplopodia* in his 'Synopsis des Échinides Fossiles.' I am indebted to Professor Sedgwick for the loan of the type specimen belonging to the Cambridge Museum, which is now figured (Pl. VI, fig. 3 *a, b*) for the first time.

PSEUDODIADEMA HOMOSTIGMA, *Agassiz.*   Pl. VI, fig. 5 *a, b, c, d, e, f.*

DIADEMA HOMOSTIGMA.	Echinodermes Fossiles de la Suisse, part ii, t. 17, figs. 1—5, p. 24.
— —	Agassiz and Desor, Catalogue raisonné Échinides, Annales des Sciences Naturelles, 3ᵉ série, tome vi, p. 347.
— ÆQUALE.	Quenstedt, Petrefactenkunde, t. 49, fig. 29, p. 579?
PSEUDODIADEMA HOMOSTIGMA.	Desor, Synopsis des Echinides Fossiles, p. 65.
DIADEMA HOMOSTIGMA.	Woodward, Memoirs of the Geological Survey, Decade V, Notes on British Fossil Diademas.

Test small, depressed, nearly circular; ambulacral areas nearly as wide as the inter-ambulacral; primary tubercles in both, nearly of the same size; no secondary tubercles; poriferous zones unigeminal; mouth opening large, peristome unequally decagonal.

Dimensions.—Height, one fifth of an inch; transverse diameter, half an inch.

Description.—This beautiful little species was first discovered by M. Nicolet in the Inferior Oolite of Chaux-de-Fonds, and figured and described by M. Agassiz in his 'Echinodermes Fossiles de la Suisse,' who, in describing it, observes, "the species which I have figured under this name offers none of those prominent traits which we recognise in so many species. It also presents a uniformity almost hopeless for description. Up to the present (1840) I only know the specimen I have figured, which is all the more valuable on account of the rock (Inferior Oolite) from whence it was collected, for we know in what a bad state of preservation the most of the fossils of this terrain are found in." It is probable that the small *Diadema æquale*, figured by Quenstedt from the Brown Jura δ of Spaichingen, is identical with *Pseudodiadema homostigma*, as it has been collected from the same geological horizon, and wants the rows of secondary tubercles, which characterise that Coral Rag form.

The ambulacral areas are nearly as wide as the inter-ambulacral areas, and support two rows of primary tubercles, eight to nine tubercles in each row; those at the equator are nearly as large as the tubercles in the inter-ambulacra in the same region of the test (fig. 5 *d*); the summits of the bosses are sharply crenulated, and the tubercles are small and perforated; a zigzag row of small granules extends down the middle of the area (fig. 5 *e*), and small transverse branches shoot out from the sides, and extend to the poriferous zones, separating the bosses from each other, and forming imperfect areolar circles around them.

The poriferous zones are narrow, and the pores are unigeminal throughout (fig. 5 *e*); the zones approach each other very closely in the upper part of the test (fig. 5 *b*), which gives to the inter-ambulacra the uniform width they have (fig. 5 *b, d*).

The inter-ambulacral areas are a very little wider at the equator than the ambulacral (fig. 5 *c*); from the equator upwards they maintain a very uniform width (fig. 5 *b*); there are nine tubercles in each of the two rows, which are set on small bosses, with sharply crenulated summits; the tubercles are prominent and perforated (fig. 5 *f*); the areolas are confluent above and below (fig. 5 *b*); between the tubercles and the poriferous zones a row of unequal-sized granules extends (fig. 5 *e*), which forms one side of the scrobicular circle; the miliary zone is narrow, consisting, from the peristome to the equator, of two rows of granules, from the equator upwards a few more granules are interspersed with them, these form the other portion of the scrobicular circle (fig. 5 *e*); the upper part of the zone is naked in the region of the central suture (fig. 5 *b*).

The mouth opening is large, the peristome is unequally decagonal (fig. 5 *c*), the ambulacral lobes being the largest. The base of the test (fig. 5 *c*) shows how uniform the structure of this *Pseudodiadema* is; and the uniformity in the size of the tubercles in this region gives value to its specific name, *homostigma*.

The discal opening is small, and nearly circular; but the plates of the disc are absent.

Affinities and differences.—This species is distinguished from the smaller forms of *Pseudodiadema depressum* in being nearly circular, in having wider ambulacra, a smaller discal opening, and more equal-sized tubercles. It has many affinities with *Pseudodiadema Bailyi* (Pl. VI, fig. 1); but is distinguished from it in having proportionately wider ambulacra, the primary tubercles likewise are of a more uniform size throughout the area, those of *Pseudodiadema Bailyi* diminishing more rapidly in the upper part of this region.

Locality and Stratigraphical position.—I have collected this species in the Bradford Clay, near the Tetbury-road Station of the Great Western Railway, and Mr. William Buy has collected it in the Cornbrash near Sutton-Benger, Wilts. It is rather a rare form in these beds. My excellent friend, the Rev. W. Griesbach, has kindly sent me a specimen which he found in the marly beds of the Great Oolite at Wollaston, Northamptonshire, associated with several Great Oolite and Cornbrash species of Echinodermata.

E. *Species from the Cornbrash.*

PSEUDODIADEMA BAILYI, *Wright,* nov. sp.   Pl. VII, fig. 1 *a, b, c, d, e, f, g.*

Test sub-pentagonal, depressed; ambulacral areas narrow, a little more than half the width of the inter-ambulacral, with two rows of primary tubercles, smaller than those of the inter-ambulacra; inter-ambulacral areas with two rows of primary tubercles, which suddenly diminish in size above the equator; poriferous zones narrow, and unigeminal throughout; mouth opening large, peristome nearly equally decagonal.

Dimensions.—Height, three tenths of an inch; transverse diameter, seven tenths of an inch.

Description.—This beautiful little Diadem, like the preceding species, presents such a uniformity of structure, and such a total absence of prominent characters, that it is extremely difficult to form an accurate diagnosis of its specific form.

The test is small, depressed, and sub-pentagonal (fig. 1 *a, b*); the ambulacral areas are prominent, and rather more than half the width of the inter-ambulacral; they support two rows of primary tubercles, rather smaller than the inter-ambulacral tubercles, of which there are eight in each row; a central, zigzag line of granules separates the two rows of tubercles from each other (fig. 1 *f*), and short, transverse, granular branches separate the areolas of the larger equatorial tubercles; on the upper part of the areas the tubercles become very small; the poriferous zones are straight and narrow (fig. 1 *e*), and the pores are strictly unigeminal throughout (fig. 1 *f*); there are five pairs of pores opposite one of the larger inter-ambulacral plates, and from three to four pairs opposite the ambulacral plates (fig. 1 *f*).

The inter-ambulacral areas are nearly twice the width of the ambulacral; they are occupied by two rows of primary tubercles, there being eight tubercles in each row; from the peristome to the equator the tubercles gradually increase in size (fig. 1 *d*), but above the equator they diminish rather suddenly in magnitude (fig. 1 *e*), so that the tubercles on the upper parts of the areas are small in proportion to those at the equator and base of the test (fig. 1 *d*); the areolas are wide; the bosses are large, conical, and prominent (fig. 1 *d, f, g*); they are confluent above and below, but laterally they are bounded by semicircles of granules (fig. 1 *g*), five or six granules, with a few other microscopic ones, forming incomplete scrobicular semicircles around their sides; the miliary zone is narrow, and formed only by the granules just described; on the upper part of the area the granules almost entirely disappear, and the surface of the plates is naked near the point where they approach the disc (fig. 1 *e*); the summit of the conical bosses is crenulated (fig. 1 *g*), and the tubercle is small and perforated (fig. 1 *f*). There are no secondary tubercles at the base of this species.

The base is flat (fig. 1 *b*); the mouth opening is large (fig. 1 *b d*); the peristome is deeply notched, and the surface of the test sharply everted (fig. 1 *d*); the ambulacral are rather larger than the inter-ambulacral lobes; the margins of the latter have two pointed processes at the angles of the notches.

The disc opening is large (fig. 1 *a, e*); but as the margin is fractured, its true diameter cannot be accurately ascertained.

Affinities and differences.—This species has so many affinities with *Pseudodiadema homostigma*, that at one time I regarded it only as a large pentagonal variety of that species; the narrowness of the ambulacra, however (these areas being a little more than half the width of the inter-ambulacra, whilst those of *Pseudodiadema homostigma* are nearly as wide as the inter-ambulacra), added to the sudden diminution in the size of the tubercles in the upper part of the areas, and the greater prominence of all the bosses, induced me to separate it from *Pseudodiadema homostigma*.

Locality and Stratigraphical position.—This species was collected by Mr. William Buy, from the Cornbrash near Sutton, Wilts, associated with *Acrosalenia hemicidaroides*, Wright; *Acrosalenia spinosa*, Agassiz; and *Echinobrissus clunicularis*, Llhwyd. I have dedicated this species to Mr. W. H. Baily, whose crayon has so accurately delineated many of the forms figured in this Monograph.

PSEUDODIADEMA BAKERIÆ, *Woodward.* Pl. VII, fig. 2 *a, b, c.*

DIADEMA BAKERIÆ. Woodward, Memoirs of the Geological Survey, Decade V. Notes on British Fossil Diademas.

Test sub-pentagonal, depressed; ambulacral areas narrow, with two rows of primary tubercles, about thirteen in each row; poriferous zones narrow, pores unigeminal, near the mouth trigeminal; inter-ambulacral areas wide, with two rows of primary tubercles, larger than the ambulacral, about thirteen in each row; bosses not prominent, summits slightly crenulated; tubercles small, flat, and finely perforated; miliary zone wide, and sparingly granulated.

Dimensions.—Height, eleven tenths of an inch; transverse diameter, one inch and four tenths of an inch.

Description.—The specimen, unfortunately, is not in good preservation, having been much weathered and bouldered (fig. 2 *a, b*). The ambulacral areas are narrow and

16

straight, with two rows of small primary tubercles (fig. 2 *c*), about thirteen in each row; their bosses are small, and set closely together; a zig-zag central line of granules descends between the rows, and a few scattered granules separate the areolas above and below from each other.

The poriferous zones are straight and narrow, the pores are unigeminal and contiguous, and the pairs of pores are obliquely inclined (fig. 2 *c*); three pairs of holes are opposite one ambulacral plate, and five pairs are opposite one inter-ambulacral plate (fig. 2 *c*); near the peristome the pores fall into triple oblique pairs.

The inter-ambulacral areas are rather more than twice and a half as wide as the ambulacral; they have two rows of primary tubercles, with thirteen in each row; they are considerably larger than those of the ambulacral areas; their bosses have a broad base, but their summits are not deeply crenulated (fig. 2 *c*); the spinigerous tubercles are small, flat, and minutely perforated; the central miliary zone is wide, but sparingly granulated in the middle, and almost naked above; on each plate (fig. 2 *c*) two rows of granules form a lateral zone, separating the tubercles from the poriferous avenues, and two rows of granules form one half of the central zone; in the upper part of the area the areolas are separated by transverse rows of granules, which, with the lateral rows, form three fourths of a scrobicular circle. The weathered condition of the specimen renders a more minute description of the test impossible.

The mouth opening measures eleven twentieths of an inch, that of the diameter of the test being one inch and four tenths; the peristome is deeply notched, and unequally lobed, the inter-ambulacral being one third less than the ambulacral lobes.

The disc opening is four tenths of an inch in diameter, but the plates are absent.

Affinities and differences.—This species very much resembles the larger specimens of *Pseudodiadema depressum*, but the ambulacral tubercles appear to be proportionately smaller, and the ambulacral areas narrower than in that species; the bosses, likewise, are not so prominent, and the spinigerous tubercles are smaller; but from the weathered condition of the specimen, the only one known, it is impossible to make a more accurate comparison between these two nearly allied forms.

Locality and Stratigraphical position.—This urchin was collected from the Cornbrash at Caistor, Northamptonshire, by the late Miss Baker, and belongs to the British Museum Collection. It has been kindly communicated for this work by Mr. S. P. Woodward, who first described it in his excellent Notes on British Fossil Diademas, inserted in the 5th Decade of the ' Memoirs of the Geological Survey.'

PSEUDODIADEMA VAGANS, *Phillips*.

CIDARIS VAGANS.	Phillips, Geology of Yorkshire, pl. 7, fig. 1, 1829.
DIADEMA VAGANS.	Desmoulins, Études sur les Échinides, p. 316, 1835.
— —	Morris, Catalogue of British Fossils, 1843, p. 51.
— —	Morris, Catalogue of British Fossils, 2d edit., 1855, p. 72.
— —	Woodward, Memoirs of the Geological Survey, Decade V, Notes on Fossil Diademas.
PSEUDODIADEMA VAGANS.	Desor, Synopsis des Échinides Fossiles, p. 65.

The type specimen of this species, with many other valuable fossils, was lost in 1835 ; there is, therefore, now some difficulty in making out the true form figured as *Cidaris vagans* in the ' Geology of Yorkshire.' Professor Phillips has, however, most kindly communicated, *ex memoriá*, all the facts he recollects relating to this species, and has sent to me for comparison a Diadem which he collected at Whitwell, Yorkshire, from the Bath Oolite. In speaking of this urchin my friend observes—" It seems, except that it is larger, to revive my recollection of *Cidaris vagans*."

Having carefully compared this specimen with *Pseudodiadema depressum*, which it most nearly resembles, 1 find it agrees so well with that urchin in all its details which admit of comparison, that I believe it is referable to the same species ; the under part of the test being imperfect, the base and mouth opening cannot be made out. The examination of the upper surface, and the poriferous zones, which are well preserved, has convinced me, that if it is regarded as a genuine example of Philips's species, *Cidaris vagans* must be considered a synonym of *Pseudodiadema depressum*, Agassiz. My friend, Mr. John Leckenby of Scarborough, has kindly communicated two Diadems from the Great Oolite of the Nab rock ; these urchins are nearly indentical with Prof. Phillip's Whitwell specimen ; the pores in the upper part of the zones, in the Nab rock specimens, however, manifest a disposition to become slightly bigeminal, an irregularity which I have often seen in true unigeminal forms, and regard only as an occasional variety of the species. In all probability these urchins may represent *C. vagans*, as most of the Great Oolite Echiniderms are likewise found in the Cornbrash.

F. *Species from the Coralline Oolite.*

[a. *Pores bigeminal in the upper part of the zones.*]

PSEUDODIADEMA VERSIPORA, *Phillips.*   Pl. VII, fig. 4 *a, b, c, d, e, f, g.*

DIADEMA VERSIPORA.	Woodward MS., Morris, Catalogue of British Fossils, 1st ed., 1843, p. 50.
— SUBANGULARE.	Agassiz, Description des Echinodermes Fossiles de la Suisse, pl. 17, figs. 21—25, p. 19.
— —	Wright, Annals and Magazine of Natural History, 2d series, vol. viii, p. 270.
DIPLOPODIA SUBANGULARIS.	M'Coy, Annals and Magazine of Natural History, 2d series, vol. ii, p. 412.
— —	Desor, Synopsis des Échinides Fossiles, p. 75.
DIADEMA VERSIPORA.	Woodward, Memoirs of the Geological Survey, Decade V. Notes on Fossil Diademas.

Test circular, depressed; ambulacral areas with two rows of primary tubercles, fourteen in each row; poriferous zones undulated below; above the equator the pairs of pores form a double series; inter-ambulacral areas with primary and secondary tubercles; spines short, stout, pointed, covered with longitudinal lines.

Dimensions.—Height, nine twentieths of an inch; transverse diameter, one inch and a quarter.

Description.—This beautiful Diadem, which is so characteristic a fossil of the Coral Rag of England, was formerly mistaken for the *Cidarites subangularis,* Goldfuss. On comparing, however, a fine series of our urchins with a type specimen from the Coral Rag of Nattheim, sent by Professor Roemer, and a fine large specimen from the White Jura of Sigmaringen, kindly sent by Dr. Fraas, of Stuttgart, the difference is very evident. The Swiss *Diadema* figured by M. Agassiz, and considered to be identical with Goldfuss's species, agrees with our urchins in so many points of structure that I believe them to be identical. *P. versipora* has from twelve to fourteen primary tubercles in the interambulacral areas, whereas *P. subangulare* has only from six to eight; the test, moreover, is higher and more circular, and there are more tubercles in each row in the ambulacral areas. I have, therefore, restored the manuscript name first given to this species by Professor Phillips, and afterwards adopted by Mr. Woodward.

Pseudodiadema versipora has the test depressed; in general it is circular; sometimes, however, it inclines to a pentagonal form (fig. 4 a), in consequence of the prominence of the ambulacral areas, which are finely and regularly formed (fig. 4 a, f); they have two rows of primary tubercles on the margin of the areas, and in each row there are from twelve to fourteen tubercles, which are nearly as large as those in the inter-ambulacra; their summits are not deeply perforated, and they are raised on bosses with finely crenulated summits; in the middle third of the area there is a narrow miliary zone (fig. 4 f).

The poriferous zones, from the peristome to the equator, are slightly undulated (fig. 4 b, e); near the mouth they are arranged in triple oblique pairs (fig. 4 e), but above that point they are in single pairs (fig. 4 f); just above the equator, however, they form double pairs in the rest of the upper part of the zone (fig. 4 a, d); this diplopodous character is very constant in this species, and may be detected even in very young specimens.

The inter-ambulacral areas are nearly twice as wide as the ambulacral (fig. 4 a, c); they have two rows of primary tubercles, from twelve to thirteen in each row, and two rows of secondary tubercles on the ambulacral side of them (fig. 4 c, e). The primary tubercles are very uniform in size, and arranged with great regularity, those on the upper part being only a little less than those in the middle of the area (fig. 4 d); the tubercle is large (fig. 4 g), and only slightly perforated, so much so, that a little friction effaces the aperture; the bosses are small, and their summits are finely crenulated (fig. 4 g); the areolas are circular, or slightly inclined to an oval form; they are all confluent above and below (fig. 4 f), and the miliary zones form crescents of minute granules on their sides, so that the scrobicular circles are incomplete throughout. The miliary zone down the middle of the inter-ambulacra is broad in the middle, contracted below, and preserves its width above : it is composed of numerous minute granules, closely set together, among which an irregular row of larger granules extends up the middle thereof (fig. 4 d, f); about the upper third of the area the granules are absent from the middle of the zone, which is consequently naked (fig. 4 d); between the primary tubercles and the poriferous zones there is another miliary zone, about half as wide as that which occupies the centro-sutural region; it is composed of like minute granules, among which a row of secondary tubercles are disposed in the basal region; five or six of these tubercles are about half the size of the primary tubercles in the same region, but from the equator upwards they are very small, and dwindle into granules (fig. 4 e, f).

The apical disc is absent in all the specimens I have examined; the opening is large and pentagonal, and the terminations of the areas where the disc was inserted are in many shells curved outwards (fig. 4 a). This peculiarity in the structure of the terminal plates may have rendered the union of the discal elements with them less firm than in other *Pseudodiademas.* "There appear to be two varieties of the species; one with the upper surface evenly inclined all round ; the other tumid at the angles, and depressed in the centre above."—*Woodward;* a remark in which I entirely concur.

The mouth opening is nearly one half the diameter of the base (fig. 4 b); the peristome is decagonal and unequally lobed, the ambulacral lobes being one third larger than those of the inter-ambulacral; the notches are deep, and their margins are much reflexed (fig. 4 *c*); the base is slightly concave in many specimens, in others it is almost flat.

The spines are well preserved *in situ* in a specimen in the British Museum, from which our figure (Pl. XII, fig. 7) is drawn; the primary spines in that urchin (fig. 7 *a, b*) are eleven twentieths of an inch in length; the stem is marked with longitudinal lines, the milled ring is not very prominent, and the spine may be said to be stout in proportion to its length; the secondary spines are one fifth of an inch in length, and are miniature forms of the primary spines. In some specimens, as one from Steeple-Ashton, kindly presented to me by Mr. Mackneil, of Wootton-under-Edge, the primary spines are swollen out, and slightly bent in the middle of the stem; the head and acetabulum are large, corresponding with the magnitude of the tubercles.

Affinities and differences.—This urchin was formerly identified with *Cidarites subangularis*, Goldfuss, but on comparing the German type of that species with ours, there can be no doubt of their distinctness. Goldfuss describes his species as having six or eight large tubercles in each row, with crenulated bosses in all the areas; the ambulacra are prominent, and the test is therefore pentagonal; each tubercle is surrounded with a circle of small granules. Now *Pseudodiadema versipora* has twelve primary tubercles in each row in the inter-ambulacral areas: the tubercles are consequently set close together, and all the areolas are confluent, so that each tubercle is not surrounded by a circle of small granules. It resembles *Cidarites subangularis* in possessing a double row of pores in the upper part of the poriferous zones, and in having a row of secondary tubercles between the primary tubercles and the pores. *Pseudodiadema versipora* very much resembles *Pseudodiadema subangulare*, Agassiz, which appears to be quite distinct from the *Cidarites subangularis*, Goldf.; in fact, the beautiful figures which Agassiz has given of the Swiss urchin leads me to think that it may be identical with our *Pseudodiadema versipora*. It has a greater number of tubercles in each row than the German species, in this respect resembling the English form; but, without the specimens were placed side by side, it would be impossible to speak positively on the point of their entire identity.

Pseudodiadema versipora is distinguished from *Pseudodiadema mamillanum*, which occurs with it in the same bed, by the double row of pores in the upper part of the zones, and by having secondary tubercles in the inter-ambulacral areas; the bosses are likewise smaller, and the tubercles are larger; the miliary zones are much wider, and the component granules are smaller.

Pseudodiadema versipora resembles *Pseudodiadema pentagona*, M'Coy, in having the pores in double pairs in the upper part of the zones; but it differs form that Great Oolite Diadem, which is a smaller, more depressed, and pentagonal form, in having larger primary and secondary tubercles, with a wider, and more granular miliary zone, in the centro-sutural region.

Pseudodiadema versipora resembles *Pseudodiadema depressum* in size and form, but is distinguished from it by having the pores in double pairs in the upper parts of the zones, and by having secondary tubercles in the inter-ambulacra. From *Pseudodiadema hemisphæricum* it is distinguished by the primary tubercles in both areas being nearly of the same size, and by there being fewer in each row; by the secondary tubercles being limited to the poriferous side of the primary tubercles, and by the pores being arranged in double pairs in the upper part of the zones.

Locality and Stratigraphical position.—This species is found in the Coralline Oolite of Wiltshire, Oxfordshire, Dorsetshire, and Yorkshire. The finest specimens, however, are those of Wilts, where they are sometimes admirably preserved. In Switzerland the urchins which resemble our specimens have been collected from the "Terrain à Chailles," of the valley of the Birse, of Blochmont, and of Weissënstein.

[b. *Pores unigeminal in the upper part of the zones.*]

PSEUDODIADEMA HEMISPHÆRICUM,* *Agassiz.* Pl. VIII, fig. 1 *a, b, c, d, e, f.*

CIDARITES PSEUDODIADEMA.	Lamarck, Animaux sans Vertèbres, tom. iii, p. 59, No. 17, 1st edit., 1816.
— —	Deslongchamps, Encyclopédie Methodique, tom. ii, p. 197, No. 17, 1824.
DIADEMA HEMISPHÆRICUM.	Agassiz, Prodromus Mém. Soc. des Sciences Nat. Neuchâtel, p. 22, 1836.
— TRANSVERSUM.	Agassiz, Prodromus Mém. Soc. des Sciences Nat. Neuchâtel, p. 22, 1836.
— PSEUDODIADEMA.	Agassiz, Echinodermes Fossiles de la Suisse, 2ᵉ partie, t. 17, figs. 49—53, p. 11, 1840.
— LAMARCKII.	Desmoulins, Études sur les Échinides, p. 316, No. 18, 1837.
DIADEMA PSEUDODIADEMA.	Agassiz and Desor, Catalogue raisonné des Échinides Annales des Sciences Naturalles, 3ᵐᵉ serie, tom. vi, p. 349, 1846.
— —	Bronn, Index Palæontologicus, p. 193, 1849.
— —	Alcide d'Orbigny, Prodrome de Paléontologie Stratigraphique universelle, tome ix, p. 27 14ᵉ étage, No. 423, 1850.

* M. Desor having selected the specific name "*pseudodiadema*," originally given by Lamarck to this urchin, I have most reluctantly been obliged to adopt the name *hemisphæricum*, afterwards given by M. Agassiz to the same species, in order to avoid the double use of a specific name thus, *Pseudodiadema pseudodiadema*, Lamarck. This is one of the inconveniences arising from the modern practice of converting *specific* into *generic* names, an innovation in nomenclature which cannot be sufficiently discountenanced, for it leads, as in this case, to an injustice to the *original describer* of the species, whose name should always be associated with the form he first described, and introduces, moreover, a further confusion into the synonyms of species, a growing evil which every true naturalist should strive to the utmost to avoid.

Cidaris diadema.	Young and Bird, Geological Survey of the Yorkshire Coast, pl. 6, fig. 3, p. 212, 1828.
— monilipora.	Phillips, Geology of Yorkshire, p. 127, 1829.
Diadema pseudodiadema.	Wright, Annals and Magazine of Natural History, 2d series, vol. viii, pl. 12, fig. 1 *a, b, c*, p. 271, 1851.
— —	Cotteau, Études sur les Échinides Fossiles du département de l'Yonne, pl. 17, fig. 1, p. 142, 1852.
— —	Salter, Memoirs of the Geological Survey, Decade V, pl. 2.
Pseudodiadema hemisphæricum.	Desor, Synopsis des Échinides Fossiles, t. 13, fig. 4, p. 68, 1854.
Diadema pseudodiadema.	Morris, British Fossils, 2d ed., p. 77, 1854.

Test hemispherical, depressed on the upper surface, flat below; ambulacral areas with two marginal rows of primary tubercles, eighteen to twenty in each row; poriferous zones narrow, straight, with a moniliform line between the pores; inter-ambulacral areas with two rows of primary tubercles, fifteen in each row, and four rows of secondary tubercles, one row flanking each side of the primary rows; apical disc small, anal opening obliquely oblong. Mouth opening large, peristome decagonal, unequally lobed, and deeply notched; spines long and needle-shaped, finely sculptured with longitudinal lines.

Dimensions.—Height, one inch and one quarter of an inch; transverse diameter, two inches and four tenths of an inch.

Description.—Lamarck's description of his original *Cidarites pseudodiadema* was so meagre and ill defined, that great doubts were entertained by Desmoulins as to the urchin meant by the illustrious author of the 'Histoire Naturelle des Animaux sans Vertèbres;' and, unfortunately, M. Dujardin has added nothing in the new edition of that great work to the original text of this species. It would have been impossible for me to have identified it, had not my excellent friend M. Michelin sent me a small type specimen of Lamarck's species from the Coralline Oolite of Commercy Meuse, which is identical with our English forms of the same size.

Pseudodiadema hemisphæricum may justly be regarded as the type of the genus, exhibiting, as it does, the characters of the group in a most satisfactory manner. It appears, moreover, to have had a wide geographical range, as it is found in the same stage of the Coralline Oolites in Switzerland, in different departments of France, and in England.

When M. Agassiz published his 'Prodromus,' he was unacquainted with the original of Lamarck's species, a subsequent comparison of specimens, however, convinced him that the urchin he had collected in the Swiss Jura, and named *Diadema hemisphæricum*, was identical with *Diadema pseudodiadema*, Lamarck, and that his *Diadema transversum* was merely a distorted form of the same urchin.* It is probable that *Diadema*

* 'Description des Echinodermes Fossiles de la Suisse,' partie 2ᵉ, p. 12.

ambiguum, Desmoulins, from the Terrain Jurassique Ardennes, is another name for the same form, but nothing can be stated positively on this point, as Desmoulins's note on this species is evidently written with much hesitation. M. Desor, in taking the name *Pseudodiadema* to designate his genus, has retained the specific name *hemisphæricum* for Lamarck's species; whilst M. Cotteau has figured and described a Corallian *Pseudodiadema* under that name, which he considers to be distinct from it.

This fine large urchin has a hemispherical form; it is depressed above and flat below, almost entirely circular at the circumference, and rarely having a pentagonal form (Pl. VIII, fig. 1 *a*, *b*). The ambulacral areas have two rows of marginal tubercles, which are smaller, and placed more closely together, than those of the inter-ambulacral areas; they are largest at the equator, and diminish rapidly in size in the upper parts of the areas, so that, when viewed at the equator, the primary tubercles of both areas are nearly of equal size (fig. 1 *c*), whilst above, those of the ambulacra are disproportionately small when compared with those of the inter-ambulacra (fig. 1 *a*); there are twenty tubercles in each row, with a zigzag line of small tubercles, extending two thirds up the areas (fig. 1 *d*); the poriferous zones are straight (fig. 1 *a*, *c*, *d*); the pores are small and unigeminal throughout, except at the base, where a few are bigeminal (fig. 1 *b*), and near the peristome, where some are trigeminal (fig. 1 *f*); the septa between the two holes forming a pair are elevated, and these elevations form a moniliform line, which extends down the centre of the zone, separating the pores from each, and defining the boundaries of the areas (fig. 1 *d*); there are five pairs of holes opposite each of the large inter-ambulacral plates (fig. 1 *d*).

The inter-ambulacral areas are more than twice the width of the ambulacral (fig. 1 *a*, *b*, *c*, *d*); they are furnished with two rows of primary tubercles, having fifteen or sixteen in each row, which have their greatest development at the equator, and gradually diminish as they approach the apical disc (fig. 1 *a*), and the peristome (fig. 1 *b*); the primary are flanked on each side by one or more rows of secondary tubercles down the centre of the area (fig. 1 *c*, *d*); the miliary zone consists of two rows of secondary tubercles, which separate the primary rows from each other (fig. 1 *d*), and two rows of secondary tubercles separate the primaries from the poriferous zones (fig. 1 *d*); the secondary tubercles are very irregular in their size and arrangement; they are best developed at the base and equator of the test (fig. 1 *b*, *c*); besides the primary and secondary tubercles just described, the surface of the plates is covered with numerous granulations, which form circles, more or less complete, about the base of the tubercles (fig. 1 *d*).

The tubercles of both areas, at and near the equator, are raised on prominent bosses, the summits of which are all finely crenulated; the base is surrounded by a narrow areola, around which small granules and minute tubercles form more or less complete scrobicular circles (fig. 1 *d*); the areolas at the equator are all confluent above and below, whilst those in the upper part of the areas are nearly, if not quite, complete (fig. 1 *a*).

It is remarkable how often the apical disc of this species is preserved *in situ* (fig. 1 *a*);

the genital plates are of a pentagonal form (fig. 1 *e*) ; the anterior pair are the largest, and the right antero-lateral, supporting the madreporiform body, is much larger than the left antero-lateral; the postero-lateral and single plates are about the same size; the genital holes are small, and pierced near the external prominent angle of the plates; the ocular plates are small, heart-shaped bodies, placed at the summits of the ambulacra, which they terminate; the eye hole is marginal, and proportionally large; the anal opening is oblong, its longest diameter extending obliquely across the test; the surface of all the elements of the apical disc is covered with small, nearly equal-sized granules.

The base is flat (fig. 1 *c*), the mouth opening is large, being more than one half the diameter of the base; the peristome is decagonal, and divided into ten lobes by deep notches, the ambulacral being much larger than the inter-ambulacral lobes (fig. 1 *b*).

The spines are not preserved in any English specimens I have seen. Those figured *in situ*, by Agassiz, are long and needle-shaped, and their surface is covered with fine longitudinal lines.

Affinities and differences.—*Pseudodiadema hemisphæricum* differs so much from all our other English species of this genus, that it is impossible to mistake it for either of them. Some juvenile specimens, however, have an air of resemblance to *Hemicidaris pustulosa,* arising from the disproportionate size between the tubercles at the base and equator of the ambulacra, and those occupying the upper portion of these regions; but the presence of secondary tubercles in the inter-ambulacra, flanking both sides of the primary tubercles, serves as a diagnostic character, no *Hemicidaris* having any secondary tubercles in the inter-ambulacral regions; moreover, the number of the primary tubercles in each row is considerably more in *Pseudodiadema hemisphæricum* than in *Hemicidaris pustulosa.*

Lamarck's species is distinguished, according to M. Cotteau, from the urchin figured by him under the name *Diadema hemisphæricum*, "d'une maniere positive et constante, par ses tubercles secondaire plus developpés, plus nombreaux, et plus regu-lièrement disposès au milieu des aires inter-ambulacraires; il s'en distingue également par les entailles plus profondes de son ouverture buccale.''*

Pseudodiadema hemisphæricum, Agass., resembles *Pseudodiadema Orbignyanum*, Cotteau, but the latter is readily distinguished from it by the greater number, regularity, and uniformity of its secondary tubercles, and the smaller size of the primary tubercles. These three species, which are all characteristic of the Coral Rag, resemble each other very much, and are only distinguished by the number and disposition of their secondary tubercles, which are rare in *Pseudodiadema hemisphæricum*, Cotteau, more numerous in *Pseudodiadema hemisphæricum*, Agass., and very abundant in *Pseudodiadema Orbignyanum*, Cotteau.

Locality and Stratigraphical position.—This species is very rare in the Coralline

* 'Études sur les Échinides Fossiles,' p. 144.

Oolite of Wilts. The specimen figured in Pl. VIII, fig. 1, was collected in the neighbourhood of Calne. It is more abundant in the Coralline Oolite of Yorkshire. Many fine specimens having been obtained from Malton and its vicinity. Its foreign distribution occupies a wide area in France. It is found in the "Corallien étage of Besançon, Saint-Mihiel, La Rochelle, Druyes (Yonne), and Commercy (Meuse);" and in Switzerland in the neighbourhood of Soleure. It is everywhere a very characteristic fossil of the Coral Rag.

PSEUDODIADEMA RADIATA, *Wright*, nov. sp.　Pl. VII, fig. 3 *a, b, c, d, e, f.*

Test circular, depressed, inflated at the sides; ambulacral areas narrow, with two rows of primary tubercles, twelve in each row, rather smaller than those of the inter-ambulacra; poriferous zones straight, pores unigeminal throughout; inter-ambulacral areas with two rows of primary tubercles, twelve in each row, separated by a wide miliary zone filled with several rows of granules; base concave, mouth opening small, peristome unequally decagonal.

Dimensions.—Height, rather more than seven twentieths of an inch; transverse diameter, nine tenths of an inch.

Description.—The test of this species is circular, and nearly equally depressed on the upper and under surface (fig. 3 *c*); the sides are rather tumid, and the ambulacral tubercles are only a little smaller than those of the inter-ambulacral. The ambulacral areas form no prominence on the sides as in so many other *Pseudodiademas* (fig. 3 *a*); they are more than half the width of the inter-ambulacral; and are occupied by two rows of primary tubercles (fig. 3 *d, e*), of which there are twelve in each row; their bosses are small, and set closely together, and their summits have well-marked crenulations (fig. 3 *f*); down the centre of the area there is a single zigzag line of granules, which becomes double about the middle of the area; transverse branches of granules separate the areolas; the tubercles are of a very uniform width throughout, and diminish gradually in size towards both ends of the area.

The poriferous zones are narrow, and nearly straight, and the pores are unigeminal throughout; the pores forming a pair are contiguous, with thin septa; there are three pairs of pores opposite the ambulacral plates, and four pairs of pores opposite one of the inter-ambulacral plates (fig. 3 *f*).

The inter-ambulacral areas are not quite twice the width of the ambulacral; they have two rows of primary tubercles, of which there are twelve in each row (fig. 3 *d*); each plate (fig. 3 *f*) is occupied by a wide areola, the boss is not prominent, its summit is marked with twelve crenulations, and the spinigerous tubercle is small, flat, and finely perforated (fig. 3 *f*); between the areola and the pedal pores there is a narrow lateral miliary zone, formed of two rows of small granules, among which two or three small secondary tubercles are interspersed (fig. 3 *f*); the central miliary zone is wide; on each plate there are two rows of small, close-set granules (fig. 3 *f*), making four rows of granules at the equator; a little way above that point they bifurcate, and on the upper part of the area a few scattered granules only are observed, the surface of the plates being almost naked (fig. 3 *d*). Like the ambulacral tubercles, those of the inter-ambulacra diminish very gradually in size towards both ends of the area.

The base is slightly concave, from the tumidity of the sides of the test; the mouth opening is small (fig. 3 *b, e*), four tenths of an inch in diameter; the peristome is slightly notched, the ambulacral lobes being larger than the inter-ambulacral (fig. 3 *e*).

The disc opening is nearly round, but the plates are unfortunately absent.

Affinities and differences.—This species resembles *Pseudodiadema mamillanum*, Roemer, in having a few small secondary tubercles near the base, between the poriferous zones and the primary tubercles, and in having the pores unigeminal; but it is readily distinguished from that fine species by having smaller tubercles, much less prominent bosses, and a wider miliary zone; the sides of the test are likewise much more tumid, the mouth opening is smaller, and the peristome is more equally lobed.

From *Pseudodiadema versipora*, which occurs with it in the same rock, it is readily distinguished by the unigeminal character of its poriferous zones, and the absence of large secondary tubercles from the inter-ambulacral plates (Pl. VII, fig. 4 *f*).

Locality and Stratigraphical position.—This is a very rare species; it was collected from the Coral Rag, near Steeple Ashton, Wilts, associated with *Pseudodiadema versipora*.

PSEUDODIADEMA MAMILLANUM, *Roemer.*   Pl. VIII, fig. 2 *a, b, c, d;* Pl. XII, fig. 9.

CIDARITES MAMILLANUM.	Roemer, die Versteinerungen des Norddeutschen Oolithen-Gebirges, t. 2, fig. 1 *a, b, c,* p. 26.
DIADEMA MAMILLANUM.	Agassiz and Desor, Catalogue raisonné Annales des Sciences Naturelles, 3me série, Zool., tom. vi, p. 347.
— SPINOSOSUM.	Agassiz, Catalogus Systematicus, p. 8.
— DAVIDSONII.	Wright, Annals and Magazine of Natural History, 2d series, vol. xiii, p. 170, pl. 13, fig. 2 *a—e.*
— MAMMILLATUM.	D'Orbigny, Prodrome de Paléontologie, tome ii, p. 27, 14e étage Corallien.
PSEUDODIADEMA MAMILLANUM.	Desor, Synopsis des Échinides Fossiles, p. 64.
DIADEMA MAMILLANUM.	Woodward, Memoirs of the Geological Survey, Decade V. Notes on British Fossil Diademas.

Test circular, depressed; ambulacral areas wide, with two rows of primary tubercles nearly as large as the inter-ambulacral; poriferous zones narrow; pores unigeminal on the sides and upper part of the zones, trigeminal near the mouth; inter-ambulacral areas with two rows of primary tubercles, and two short rows of secondary tubercles on the zonal sides of the primaries; the bosses of both areas nearly of the same size, large, prominent, and conical; spinigerous tubercles of both areas of the same size, large, and widely perforated. Mouth opening large, peristome decagonal, ambulacral lobes twice the width of the inter-ambulacral; spines long, round, stout, and covered with longitudinal lines.

Dimensions.—Height, nine twentieths of an inch; transverse diameter, one inch and three tenths.

Description.—This Diadem was first discovered in the Coral Rag of Hildesheim, and figured and described by M. Roemer in his work on the 'Oolitic Fossils of Northern Germany;' it has subsequently been found in the Coralline Oolites of France. When I found this species, some years ago, in the Coral Rag of Wiltshire, I was then unable to identify it with any published species, and figured and described it the 'Annals of Natural History' under the name *Diadema Davidsonii*, M. Roemer's figure being very indistinct, and without details, did not furnish sufficient data for identification. Since the publication of that paper, M. Desor has compared my figures and descriptions with the moulds in the Museum of Neuchâtel, attributed to *Diadema mamillanum*, and he says, " n'ai pas trouvé de raison suffisante pour les distinguer.* Mr. Woodward† states that " the British forms of this species agree perfectly with specimens received from Dr. Roemer, labelled U. Coral Rag, Hildesheim, Hanover." There is, therefore, no doubt that *Diadema Davidsonii* is identical with Roemer's species.

This beautiful urchin has a regular cylindrical test, not at all inclined to the pentagonal form of many of its congeners (fig. 2 *a*, *b*); the ambulacral areas are about three fourths the width of the inter-ambulacral, and nearly of a uniform breadth throughout, tapering gracefully inwards towards their superior third (fig. 2 *a*); the contraction assumes the form of a gentle curve, slightly inclined towards the centre. The double row of tubercles, which are nearly as large as those of the inter-ambulacra, gradually increase in size from the peristome to the equator, where three pairs are about the same size; from this point upwards they gradually decrease, and terminate in two pairs of minute rudimentary tubercles near the apical disc. A single row of granules, arranged in a zigzag form, separates the two rows of primary tubercles from each other, a larger granule marking each re-entrant angle (fig. 2 *d*); the areolas abut against the poriferous zones (fig. 2 *d*), without the intervention of any granules between these eminences and the pores.

---

* Desor, 'Synopsis des Échinides Fossiles,' p. 64.

† 'Memoirs of the Geological Survey,' Decade V. Notes on British Fossil Diademas.

The poriferous zones are very narrow; the pores are contiguous, and unigeminal throughout (fig. 2 *d*), except at the wide inter-tubercular spaces around the mouth, where they fall into triple oblique pairs (fig. 2 *b*); in the upper third of their course, the zones curve slightly inwards, and the pores are set very closely together in this region. There are five pairs of pores opposite each inter-ambulacral plate, and nearly a like number opposite the ambulacral plates (fig. 2 *d*).

The inter-ambulacral areas are four tenths of an inch in width, and only one fourth wider than the ambulacral areas. They retain their breadth very uniformly throughout, and are occupied by two rows of primary tubercles, of which there are ten in each row; the bosses are large and prominent; those at the equator occupying nearly the entire surface of the plates (fig. 2 *d*); their summits are sharply crenulated, and the spinigerous tubercle is widely perforated; from the peristome to the equator (fig. 2 *b, d*), or for about two thirds the length of the area, there is a row of small secondary tubercles between the zones and the primaries, which are raised on mammillated bosses, with crenulated summits (fig. 2 *d*); the miliary zone is narrow, and composed only of two rows of granules, which descend down the tract of the zigzag central suture (fig. 2 *d*), there being five granules on each plate (fig. 2 *d*).

The tubercles of both areas stand prominently out from the surface of the test, in consequence of the size of the mammillary bosses, which are unusually conical in this species; from the mouth to the equator the tubercles of both areas are nearly alike in size and number (fig. 2 *b, c*), but on the upper part of the test those of the ambulacral areas become much smaller and more numerous (fig. 2 *a*) than those of the inter-ambulacral areas.

The mouth opening is wide, about eleven twentieths of an inch (fig. 2 *b*); the peristome is decagonal, and unequally lobed, the ambulacral being twice as large as the inter-ambulacral lobes. Since Pl. VIII was printed, I have obtained a slab with the test and spines *in situ* thereon, and which is figured in Pl. XII, fig. 8. The primary spines (fig. 8 *a, b*) are nine tenths of an inch in length; the head is well defined by a prominent milled ring; the stem, which is cylindrical, tapers uniformly from the ring to the point, which is rather blunt, and the surface is covered with fine longitudinal lines; the secondary spines (fig. 8 *c*) are nearly three twentieths of an inch in length, and are miniature representatives of the primaries.

The disc opening is wide (Pl. VIII, fig. 2 *d*); its form cannot be described, as the plates around its margin are more or less fractured. The discal elements are lost in all the specimens I have yet seen.

Affinities and differences.—The great width of the ambulacral areas, and the uniformity in the size and number of the primary tubercles at the base and equator of the test (fig. 2 *b, c, d*), render the diagnosis between this species and its congeners easy and decisive.

From *Pseudodiadema versipora* it is distinguished by the unigeminal character of its poriferous zones, the greater prominence of its bosses, the smaller size of its tubercles, the greater width of the ambulacral areas, and the narrowness of its miliary zones. From *Pseudodiadema radiata*, which has likewise unigeminal pores, by the size of the bosses and the narrowness of the miliary zones, and the absence of the lateral tumidity which characterises that species. From *Pseudodiadema hemisphæricum* it is distinguished by the uniformity in size between the ambulacral and inter-ambulacral tubercles, which are very unequal in that large species; and by the smallness of the secondary tubercles, and their limitation to the zonal sides of the area; whereas the secondary tubercles are developed on both sides of the primaries in *Pseudodiadema hemisphæricum*, a character which forms a good diagnosis between these two Coralline Oolite forms.

Locality and Stratigraphical position.—I have collected this species in the Clays and Limestones of the Coralline Oolite, at Calne, Wilts, associated with *Hemicidaris intermedia*, *Acrosalenia decorata*, *Echinus gyratus*, and *Echinobrissus scutatus*. On many slabs all these species are sometimes found clustered together in a more or less fragmentary condition. It has been collected from the Coral Rag, at Redcliff, near Weymouth; specimens from this locality are in the British Museum, and Geological Museum, Jermyn Street.

On the Continent, it has been collected by M. Roemer from the Upper Coral Rag of Hildesheim, Hanover, and in the " Corallien étage de la Rochelle, de Verdun, France."—*Desor*. It is everywhere a rare species.

History.—First described and figured by M. Roemer under the name *Cidarites mamillanum;* afterwards figured and described as *Diadema Davidsonii*, in the ' Annals of Natural History.' But a comparison of the English with the Foreign types has proved them to be identical.

NOTES

ON FOREIGN JURASSIC SPECIES OF THE GENUS PSEUDODIADEMA NEARLY ALLIED TO
BRITISH FORMS, BUT WHICH HAVE NOT YET BEEN FOUND IN THE ENGLISH OOLITES.

A. *Pores unigeminal in the upper part of the zones ; no secondary tubercles.*

PSEUDODIADEMA SUBCOMPLANATUM, *d'Orbigny.*   Prodrome de Paléontologie, p. 319.

Test small, depressed, pentagonal; tubercles of both areas at the base, and equator of
the test proportionately large, prominent, and equal sized; ambulacra prominent, with
eight or nine tubercles in each row, nearly as large, but more crowded together at the
base and equator; diminishing more rapidly in size in the upper part of the area than
those of the inter-ambulacra.   Inter-ambulacral areas with eight rows of large prominent
primary tubercles, and four small secondary ones on their zonal sides at the base; miliary
zone narrow, with two rows of granules; nearly related to *P. complanatum*, Agass., and
to *P. depressum*, Agass.   It differs from the former in having a much larger mouth-
opening, and from the latter in having larger tubercles, a narrower miliary zone, and four
small secondary tubercles at the base.

Dimensions.—Height, three tenths of an inch; transverse diameter, four fifths of an
inch.

Formation.—Grande Oolite, 11ᵉ étage Bathonien, de Ranville, Luc (Calvados).

Collections.—Deslongchamps, Michelin, d'Orbigny, my Cabinet.

PSEUDODIADEMA COMPLANATUM, *Agassiz.*   Syn. *Diadema complanatum*, Agass., Echinod.
Suisse, part ii, tab. 17, fig. 31—35.

Test small, very much compressed; the ambulacral are smaller than the inter-
ambulacral tubercles, and they diminish more rapidly in size at the upper surface, but in

other parts those of each area very much resemble one another; there are fewer tubercles in a row in the ambulacra than in the inter-ambulacra; the poriferous zones are narrow, and the pores are unigeminal on the upper surface; the mouth opening is small, and the decagonal peristome is divided into nearly equal-sized lobes; there are no secondary tubercles at the base of the inter-ambulacra.

Dimensions.—Height, three twentieths of an inch ; transverse diameter, two fifths of an inch.

Formation.—" Portlandien inférieure (Astartien) de Rædersdorf (Haut Rhin), du Jura Neuchâtelois, des bords du Doubs, de Laufon.

Collections.—MM. Gresly, Thurmann. Desor. Rare."

PSEUDODIADEMA INÆQUALE, *Agassiz.* Syn. *Diadema inæquale*, Agassiz, Cat. syst., p. 8.

Test small, subpentagonal, and depressed ; tubercles small, uniform, and numerous ; inter-ambulacral tubercles arranged near the poriferous zones ; miliary zone wide, and naked from the equator to the apical disc ; mouth opening of moderate size ; peristome divided into equal-sized lobes ; opening for the disc large, and oblong ; the approximation of the tubercles into groups of four rows, separated by the wide naked miliary zone in the upper part of the areas, serves to distinguish this species from allied forms. It differs from *P. superbum* in having the tubercles set closer together, and from *P. textum* in having wider ambulacral areas.

Dimensions.—Height, two fifths of an inch ; transverse diameter, four fifths of an inch.

Formation.—Kellovien de Marolles, de Chaufour, et Montbizot (Sarthe), Lifol (Vosges). Common.

Collections.—MM. Michelin, d'Archiac, d'Orbigny. The specimens in my Cabinet were kindly sent me by M. Triger, of Le Mans.

PSEUDODIADEMA SUPERBUM, *Agassiz.* Syn. *Diadema superbum*, Agass., Echinod. Foss. de la Suisse, tab. 17, fig. 6—10, p. 23, part ii.

Test small, subpentagonal, and sometimes inflated at the sides ; ambulacral tubercles

18

smaller than the inter-ambulacral, diminishing rapidly in size on the upper part of the areas; inter-ambulacra with two rows of primary tubercles, but no secondary ones; areolas surrounded by complete circles of small granules; miliary zone wide, with fine granulations at the equator, but naked at the upper part; mouth-opening small, situated in a concavity; discal opening large and oblong. This species very much resembles *P. inæquale*, and it may be only a small variety of that Kellovian form.

Dimensions.—Height, one quarter of an inch; transverse diameter, half an inch.

Formation.—Oxfordien marnes, des Vaches-noires, Calvados; de Mont Vohayes, Jura Bernois, de Belfort.

Collections.—Museum Paris, Coll. MM. Gresly, Michelin, my Cabinet.

PSEUDODIADEMA MAGNAGRAMMA, *Wright*, nov. sp. Davidson's MSS., pl. 3 bis, figs. 4—6.

Test circular depressed, with two rows of tubercles in the inter-ambulacra; ambulacra two thirds the width of the inter-ambulacra; tubercles of both areas set on bosses, with a very large base; those of the ambulacra are a little smaller than those of the inter-ambulacra; the ranges of tubercles in both areas are separated by a single row of small granules; the discal opening is small, and the mouth opening is concealed; there are no secondary tubercles, and the pores are strictly unigeminal.

Dimensions.—Height, nine twentieths of an inch; transverse diameter one inch and one twentieth of an inch.

Formation. — Collected from the Portland beds, between Portel and Equihen, Boulonnais.

Collections.—Mr. Davidson, M. Bouchard Chantereaux, my Cabinet.

B. *Pores unigeminal in the upper part of the zones, with two or four rows of secondary tubercles in the inter-ambulacra.*

PSEUDODIADEMA PLACENTA, *Agassiz.* Syn. *Diadema placenta*, Agassiz, Echinod. Foss· Suisse, tab. 17, figs. 16—20.

Test circular, much depressed on the upper and under surfaces; tubercles nearly the same size in both areas; areolas surrounded by a circle of granules; four rows of small secondary tubercles between the peristome and the equator; mouth opening of moderate width; peristome not deeply notched.

Dimensions.—Height, eleven twentieths of an inch; transverse diameter, one inch and nine twentieths.

Formation.—" Corallien inférieure (Terrain à chailles) du Fringeli (Canton de Soleure), de Nantua, des environs de Salins, de Druyes (Yonne)." Desor.

Collections.—Museum Neuchâtel, Coll. MM. Gressly, Marcou, Cotteau. Common.

PSEUDODIADEMA ORBIGNYANUM, *Cotteau.* Syn. *Diadema Orbignyanum*, Cotteau, Échinides Foss., tab. 17, fig. 2—6.

Test circular hemispherical, flattened at the base, and depressed on the upper surface; ambulacra narrow, with two rows of primary tubercles; inter-ambulacra with two rows of primary tubercles, and six or eight rows of small equal-sized secondary tubercles; the primary tubercles of both areas are about the same size; those of the inter-ambulacra are closely surrounded by numerous secondary tubercles; apical disc small, genital plates narrow; poriferous zones narrow; pores falling into triple oblique pairs near the mouth, but strictly unigeminal above; mouth opening large; peristome deeply notched; lobes very unequal.

Dimensions.—Height, nine twentieths of an inch; transverse diameter, one inch and seven tenths of an inch.

Formation.—Corallien de Coulanges sur Yonne, Druyes, et Chatel-Censoir, (Yonne), Profont (Ain), Hobel (Canton de Soleure).

Collections. — Museum Bâle, Coll. MM. Cotteau, Renevier. Plaster mould, my Cabinet.

PSEUDODIADEMA TETRAGRAMMA, *Agassiz.* Syn. *Diadema tetragramma*, Agass., Échinod. Foss. Suisse, tab. 17, figs. 39—43.

Test circular, hemispherical, depressed; inter-ambulacra with four rows of secondary tubercles, which at the base, and up to the equator, nearly equal in size the two rows of primary tubercles; this imparts a tetragrammous appearance to this species, which is lost, however, in the upper part of the area, where there are only two rows of tubercles; ambulacra moderately wide, their tubercles nearly as large as those of the inter-ambulacra; pores unigeminal throughout; apical disc small and ring-like, all the plates covered with a fine granulation; rows of small granules separate the areolas from each other; mouth opening large; peristome not deeply notched.

Dimensions.—Height, nine twentieths of an inch; transverse diameter, one inch and one twentieth of an inch.

Formation.—Corallien de Bensançon, de Chatel-Censoir (Yonne).

Collections.—Museum of Vienna (Collection Dudressier), Coll. M. Cotteau. Very rare.

PSEUDODIADEMA ÆQUALE, *Agassiz.* Syn. *Diadema æquale*, Agass., Echinod. Foss. Suisse, part ii, tab. 17, figs. 36—38.

Test circular hemispherical, depressed above; ambulacral rather smaller than the inter-ambulacral tubercles; areolas confluent; four rows of small secondary tubercles, which only extend to the equator; upper part of the miliary zone naked; mouth opening moderately large; peristome not deeply notched; pores strictly unigeminal, falling into triple oblique pairs near the mouth. This species very much resembles *P. hemisphæricum*, but the smallness of the secondary tubercles, the shortness of the rows, and the nakedness of the upper part of the miliary zone, distinguish it from that allied form.

Dimensions.—Height, eleven twentieths of an inch; transverse diameter, one inch and two fifths.

Formation.—Argovien des environs d'Arau. Kellovien des environs de Quingey.

Collections.—Museum Bâle, Coll. Zschokke, Marcou. Very rare.

PSEUDODIADEMA AFFINE, *Agassiz.* Syn. *Diadema affine*, Agass., Echinod. Foss. Suisse, part ii, tab. 17, figs. 54—58.

Test small, circular, depressed; the secondary tubercles at the zonal sides of the primaries nearly equal the latter in size; those near the centro-suture are much smaller; miliary zone covered with small granulations above; pores strictly unigeminal; areolas confluent at the equator, separated by small granules above; mouth opening small; peristome scarcely notched.

Dimensions. — Height, three tenths of an inch; transverse diameter, seventeen twentieths of an inch.

Formation.—" Jura supérieur du Dèpt. du Doubs. Très rare." Desor.

Collection.—Museum Neuchâtel, Coll. Renaud-Comte (Musée Besançon).

PSEUDODIADEMA PLANISSIMUM, *Agassiz.* Syn. *Tetragramma planissimum*, Agass., Echinod. Foss. Suisse, part ii, tab. 14, figs. 1—3.

Test small, circular, very much depressed; inter-ambulacra with four rows of secondary tubercles, which at the base and equator equal the two rows of primary tubercles in size; these divisions of the test therefore support six rows of equal-sized tubercles; in the upper part of the area the secondary tubercles become much smaller; ambulacra with two rows of tubercles as large as those in the inter-ambulacra; miliary zone very narrow; mouth opening small; peristome scarcely notched.

Dimensions.— Height, one fifth of an inch; transverse diameter, four fifths of an inch.

Formation.—" Portlandien moyen (Calcaire à Tortues) de Soleure. Marnes Strombiennes de Porrentruy." Desor.

Collection.—Museum Neuchâtel, Coll. Gressly, Thurmann. Very rare.

C. *Pores bigeminal in the upper part of the zones ; inter-ambulacra with two rows of secondary tubercles.*

PSEUDODIADEMA SUBANGULARE, *Goldfuss.* Syn. *Cidarites subangularis*, Goldfuss, Petrefacta, Germaniæ, tab. 40, fig. 8.

Test pentagonal, depressed; ambulacra one half the width of the inter-ambulacra, with two rows of tubercles 10 or 12 in each row; inter-ambulacra with two rows of primary tubercles 10 or 12 in each row, and two rows of small secondary tubercles on the zonal sides of the primaries, extending three fourths of the length of the area; bosses small and sharply crenulated; primary tubercles of both areas nearly of the same size, large, round, and prominent; areolas narrow, and encircled with small, round, spaced-out granules; poriferous zones narrow; pores bigeminal in the upper part of the zones; miliary zone of moderate width, with a few granules at the equator, but naked in the upper part of the area; mouth opening less than one half the diameter of the test; peristome decagonal, divided by notches into nearly equal-sized lobes.

Dimensions.—Height, three fifths of an inch; transverse diameter, one inch and a half.

Formation.—Coral Rag of Nattheim and Sigmaringen, Württemberg; of Thurnau and Muggendorf, Bavaria; of Galgenberg, near Hildesheim, and Lindenberg, near Hanover; of Chatel-Censoir and Druyes, (Yonne); Argovien? of Randen; "Oxfordien à *Belemnites hastatus* du Jura Neuchâtelois." Desor.

Collections.—In all the Foreign Collections of Jurassic Fossils. The specimen whence the diagnosis given above has been drawn, was sent me by Dr. Fraas of Stuttgart, as a type of Goldfuss's species. It was collected from the White Jura δ and ε of Sigmaringen.

In this specimen the pores are unigeminal in the upper part of the zones, as in the one figured by Goldfuss; but as several plates belonging to the upper part of the areas are absent, and the pores at the same time are much concealed by the matrix, the bigeminal arrangement may nevertheless have existed in this specimen. In another small Diadem from the Coral Rag of Nattheim, sent me by Professor Roemer, from the Bonn Museum, as typical of Goldfuss's species, the pores are distinctly bigeminal; in a third specimen, from the Coral Rag of Druyes, sent by M. Cotteau, the pores are likewise bigeminal. I have not cited the specimens so beautifully figured as *Diadema subangulare*, by M. Agassiz,[*] because they certainly are not identical with Goldfuss's type; they more probably belong to *Pseudodiadema versipora* of the English Coral Rag. Mr. Woodward[†] says, "We cannot agree with M. Agassiz in considering either of these forms referable to the ' *Cidarites subangularis*' of Goldfuss; German specimens agreeing with Goldfuss's figure and description, in the presence of only a single series of pores, are in the British Museum." Whether the diplopodous forms are only a variety, or a distinct species, it is impossible to decide, without a better series of specimens for comparison than I possess at present.

---

[*] 'Echinodermes Fossiles de la Suisse,' part ii, pl. 17, figs. 21—25.

[†] Notes on British Fossil Diademas. 'Mem. of the Geological Survey,' Decade V.

Genus—HEMIPEDINA, *Wright.* 1855.

This rare genus is composed of neat and highly ornamented urchins, in general much depressed on their upper surface, and with a flat or slightly concave base. The ambulacral areas are narrow and straight; the pores in the poriferous zones are arranged in single pairs; the inter-ambulacral areas are, in general, more than double the width of the ambulacral, with two, four, six, or eight rows of tubercles, arranged in general abreast on the same tubercular plate.

The tubercles are perforated, and set on bosses with smooth, uncrenulated summits; one row of tubercles extends from the peristome to the disc; the other rows, when there are four and six rows in the area, stop short at the equator, or between the equator and the disc; the upper part of the miliary zone is therefore in general wide, and covered with a fine granulation.

The apical disc is large; the genital plates are expanded and foliated; the eye holes are perforated in the centre of the ocular plates, which are large.

The mouth opening is of moderate dimensions, and the peristome is divided into ten nearly equal-sized lobes.

The spines are long, slender, and needle-shaped; those that are known equal in length the diameter of the test, and their surface is sculptured with delicate, longitudinal lines.

Hemipedina is related to *Pseudodiadema* in having the pores unigeminal, and the tubercles perforated; but is distinguished from *Pseudodiadema* by the absence of crenulations from the summits of the bosses.

It is related to *Pedina* in possessing perforated and uncrenulated tubercles; but is distinguished from that genus in having the pores unigeminal (*Pedina* having the pores trigeminal like *Echinus*); the elements of the apical disc are likewise more largely developed.

Hemipedina is related to *Echinopsis* in possessing uncrenulated and perforated tubercles, with unigeminal pores; but is distinguished from *Echinopsis* by the narrowness of the ambulacral areas, the general depressed form of the test, the shape of the mouth opening, and the deep decagonal lobes of the peristome (that of *Echinopsis* being almost deprived of incisions), together with the greater size and development of the elements of the apical disc.

Hemipedina, as far as we at present know, is composed of Oolitic species which commenced in the Lower Lias, and extended into the upper division of the Oolites, each stage possessing its own specific forms.

A. *Species from the Lias.*

HEMIPEDINA BECHEI, *Broderip.*   Pl. IX, fig. 1 *a, b.*

CIDARIS BECHEI.	Broderip, Geological Proceedings, vol. ii, p. 202.
DIADEMA BECHEI.	Morris, Catalogue of British Fossils, 1st edition, p. 51.
— —	Morris, Catalogue of British Fossils, 2d edit., p. 76.
ECHINOPSIS BECHEI.	Woodward, Memoirs of the Geological Survey, Decade V. Notes on Fossil Diademas.
HEMIPEDINA BECHEI.	Wright, Annals and Magazine of Natural History, 2d series, vol. xvi, p. 96.

Test small, much crushed, and covered over with spines; ambulacra with two rows of tubercles; inter-ambulacra with four or six rows of tubercles; spines long, slender, and needle-shaped, four fifths of an inch in length, with longitudinal lines on their surface.

Description.—The test of this specimen, and the one in the collection of the Geological Society, which appears to belong to the same species, is so much covered up with spines and adherent matrix, that it cannot be sufficiently developed to show the details of its structure. The ambulacral areas have two rows of small tubercles placed on the margins of the plates; and the inter-ambulacral areas have four or six rows of tubercles abreast.   The spines are long, slender, and needle-shaped; they measure four fifths of an inch in length, have a small conical head, a prominent, narrow, milled ring, and a long, slender stem, which tapers gradually from the ring to the point; their surface is finely marked with longitudinal lines.

Affinities and differences.—This species might be mistaken for *Hemipedina Bowerbankii,* but its test is smaller, and its spines are longer and more slender; the head is more conical, the milled ring is narrower and more prominent; I believe it is identical with Mr. Broderip's species.

Locality and Stratigraphical position.—This specimen has been kindly communicated, together with the following species, by my friend Mr. Bowerbank, who collected it from the Lower Lias at Lyme Regis, from one of the layers of marl which inter-stratifies the beds of limestone.

Hemipedina Bowerbankii, *Wright.*   Pl. IX, fig. 2 *a, b, c.*

Hemipedina Bowerbankii.   Wright, Annals and Magazine of Natural History, 2d series, vol. xvi, p. 96.

Ambulacral areas narrow, with two rows of marginal tubercles, rather smaller than those of the inter-ambulacra, one tubercle on every third or alternate plate; inter-ambulacral areas wide, with six rows of tubercles abreast, each surrounded by a delicate areolar circle; spines long and needle-shaped, deeply sculptured with longitudinal lines.   Form unknown, as the test is crushed.

Dimensions.—One inch and one twentieth of an inch in diameter; height unknown.

Description.—The only specimen of this species I have seen is the one collected by my friend Mr. Bowerbank, and figured in Pl. IX, fig. 2.   The test is unfortunately much crushed, but a sufficient number of plates are preserved to enable me to make a short description of this beautiful Liassic form.

The ambulacral areas are more than one third the width of the inter-ambulacral; they are composed of narrow, hexagonal plates (Pl. IX, fig. 2 *b*); on every third plate a small tubercle is developed, about midway between the zones and the central suture, but nearer to the latter; on the intervening plates there are only two or three small granules on each; the poriferous zones are narrow and straight, the pairs of pores having an oblique direction upwards and inwards; there are four ambulacral plates opposite each inter-ambulacral one; the number of plates in a column cannot be ascertained, as all of them are more or less incomplete.

The inter-ambulacral areas are more than double the width of the ambulacral; they are composed of narrow, pentagonal plates (fig. 2 *b*); on the centre of each is a primary tubercle, and on each side a smaller secondary tubercle; these three tubercles are disposed nearly on the same line, and two smaller tubercles are developed near the lower angles of the plate; the central tubercle is surrounded by an areola, and an incomplete circle of small, spaced-out granules encircles it; granules of the same size form circles around the secondary tubercles, but in them the areola is small or altogether absent.

At the equator each inter-ambulacra is provided therefore with six rows of tubercles, which are arranged nearly all abreast, and four rows of smaller tubercles placed near the lower angles of the plates; but in the upper part of the area, the rows of secondary tubercles dwindle into granules, and finally disappear (fig. 2 *a*), there are four pairs of pores opposite each large inter-ambulacral plate (fig. 2 *b*).

The apical disc is not preserved; and as the base is imbedded in the clay, the structure of the under surface is unknown.

19

The spines, many of which are preserved *in situ*, are long and needle-shaped ; they are shorter and stouter than those of *Hemipedina Bechei* (fig. 1 *b*), although the diameter of the test of *H. Bowerbankii* greatly exceeds that of *H. Bechei ;* the spines of *H. Bower-bankii* are three quarters of an inch in length ; the head does not taper much towards the acetabulum (fig. 2 *c*) ; the milled ring is thick, but not so prominent as in *H. Bechei ;* the stem likewise is thicker at the base, and tapers regularly to the point ; the surface is deeply sculptured with fine longitudinal lines.

Affinities and differences.—This species very much resembles *Hemipedina Bechei,* but the imperfect condition of the test of that species renders a comparison of this part of the body with that of *H. Bowerbankii* impossible ; the spines, however, of the two species, which are drawn side by side on the plate (figs. 1 *b*, 2 *c*), show that these two nearly allied forms from the same bed are, however, specifically distinct, the spine of *H. Bechei* being longer and more slender than that of *H. Bowerbankii,* although the test to which it was attached is smaller ; the milled ring in *H. Bechei* is narrower and more prominent, and the head more conical, than the homologous parts in *H. Bowerbankii.* It resembles *Diadema seriale,* Leymerie (Pl. IX, fig. 3 *a, b*), from the Inferior Lias of France, which I have copied * for the purpose of comparison, and likewise to show the probable form of our Lias specimen. In this very rare urchin, the primary tubercles are larger (fig. 3 *b*), and there are only two, nearly of the same size, abreast on each plate ; the scrobicular circles are more complete (fig. 3 *b*), and the granulations fewer but more numerous ; the pores are much closer set together in the zones, there being twelve pairs opposite each inter-ambulacral plate.

Locality and Stratigraphical position.—*Hemipedina Bowerbankii* was collected from the marly beds of the Lower Lias, at Lyme Regis, by Mr. Bowerbank, to whose magnificent collection it belongs. I have not ascertained the precise Ammonital zone in which this urchin was found. I have great pleasure in dedicating this species to our amiable, excellent, and indefatigable Secretary, to whose great and continued exertions, through a series of years, the Palæontographical Society owes much of its present success.

HEMIPEDINA JARDINII, *Wright.* Pl. IX, fig. 4 *a, b, c, d, e, f.*

> HEMIPEDINA JARDINII. Wright, Annals and Magazine of Natural History, 2d series, vol. xvi, p. 96.

Test small, much depressed ; ambulacral areas wide, with two rows of marginal

* 'Mémoires de la Société Géologique de France,' t. ii, p. 330, pl. 24, fig. 1, 1839.

tubercles, which extend from the peristome to the disc; inter-ambulacral areas with two rows of tubercles set near the poriferous zones, eleven to twelve in each row; a delicate circle of small granules around each, and a naked space in the centre of the miliary zone; mouth opening small, peristome decagonal; base exhibiting very regular radii of tubercular rows.

Dimensions.—Height, one fifth of an inch; transverse diameter, half an inch.

Description.—The test of this beautiful little Diadem is small, circular, and depressed on the upper and under surfaces; the ambulacral areas are narrow, with two rows of small tubercles (fig. 4 *d*) set alternately on the sides of the area, close to the poriferous zones (fig. 4 *g*); a line of granules occupies the course of the central suture, and additional granules fill up the space between that line and the pores (fig. 4 *g*); there are from twelve to fourteen tubercles in each row; about one third the size of those of the inter-ambulacra, but much smaller in proportion in the upper part of the area (fig. 4 *d*); the zones are narrow and straight throughout; the septa forming slight papillæ on their surface; the pores are very small, and there are four pairs of holes opposite each of the large plates.

The inter-ambulacral areas are nearly three times the width of the ambulacral; they have two rows of primary tubercles, about ten or eleven in each row; the tubercles occupy the centre of the plates (fig 4 *g*); the boss is prominent, the tubercle small, and the areola not well defined; on each side of the tubercle there are two rows of granules, the inner of which forms a circle around the boss (fig. 4 *g*); the miliary zone is wide (fig. 4 *f*), and filled with the granules just described; interspersed among them are several which pass into the condition of minute tubercles, being raised on miniature bosses, and having their surface perforated. On the upper part of the area the granules disappear from the centre, leaving a portion of the plates naked (fig. 4 *d*); as the granules are disposed with much regularity, the test of this species has a highly ornamented appearance (fig. 4 *d, e, f*).

The base is concave, and the mouth opening, which is less than half the diameter of the shell, is situated in a depression; the peristome is decagonal, the ambulacral being larger than the inter-ambulacral lobes: as the primary tubercles of the inter-ambulacra are set close together at the base, they form prominent tuberculated radii (fig. 4 *e*) in this region.

The discal opening is small, but the plates are absent.

Affinities and differences.—This species resembles *H. Etheridgii*, but it is a smaller and more depressed form; the tubercles of the ambulacra are larger in size and fewer in number; and those of the inter-ambulacra want the well-defined areolas around them, seen in the primary tubercles of *H. Etheridgii* (fig. 5 *g*). The discal opening is likewise a much smaller aperture in *H. Jardinii* than in *H. Etheridgii*.

Locality and Stratigraphical position.—This beautiful little Diadem has been found in the Marlstone of Bredon, Alderton, and Dumbleton Hills, Gloucestershire, and in the same rock at Ilminster, Somersetshire. It is everywhere a very rare form. I dedicate this species to my friend Sir William Jardine, Bart., of Jardine Hall, Dumfrieshire, author of a magnificent work on the 'Ichnology of Annandale,' and other valuable treatises on different branches of Natural History.

HEMIPEDINA ETHERIDGII, *Wright.* Pl. IX, fig. 5 *a, b, c, d, e, f, g.*

PEDINA ETHERIDGII.	Wright, Annals and Magazine of Natural History, 2d series, vol. xiii, p. 315, pl. 11, fig. 5 *a—c.*
— —	Morris, British Fossils, 2d edition. Additional species.
HYPODIADEMA ETHERIDGII.	Desor, Synopsis des Échinides Fossiles, p. 61.

Test circular, depressed; ambulacral areas with from six to eight small perforate tubercles at their base, and a double row of small granules on their upper surface; the inter-ambulacral areas with primary tubercles only, the areolas of which are surrounded by regular circles of granules; pedal pores not numerous, arranged in nearly a single file, with a slight elevation between the two pores of each pair; apical disc large; ovarial plates leaf-like; mouth opening small.

Dimensions.—Height, one quarter of an inch; transverse diameter, half an inch.

Description.—This pretty little urchin has a circular outline in the young state, which in larger specimens inclines towards a pentagonal form; the base is flattened, and the upper surface is much depressed (fig. 5 *c*). The ambulacral areas are straight and narrow (fig. 5 *d, e, f*); they have from six to eight small perforated tubercles at their base (fig. 5 *e*), and a double row of from twelve to fourteen minute, imperforate granules in each row on their upper surface (fig. 5 *d*), which, in form and size, resemble those covering the other parts of the test (fig. 5 *g*); between the pedal pores of each pair is a small elevation (fig. 5 *g*); these collectively form a prominent moniliform line, which extends from the margin of the disc to the mouth opening; the pores are unigeminal, and disposed in nearly a single file throughout (fig. 5 *g*). The inter-ambulacral areas are about twice the width of the ambulacral; the rows of primary tubercles occupying the centre of the plates have seven or eight tubercles in each row; they are small in size, and rendered prominent from being raised upon uncrenulated mammillary eminences, the bases of which are sharply defined, and surrounded by complete circles of small scrobicular granules (fig. 5 *g*); the regular disposition of these granulations gives an air of decoration to this beautiful little species, and the entire absence of secondary tubercles from the areas renders the ornamentation even more complete (fig. 5 *d, e, f*). The apical disc is large (fig. 5 *d*); the ovarial plates are widely rhomboidal (fig. 5 *d*); the oculars are small and heart-shaped; and the surface of both is covered with minute granules, nearly as large as those which adorn other parts of the test. The madreporiform tubercle makes a distinct elevation on the surface of the right antero-lateral plate (fig. 5 *d*); the anal aperture is transversely oblong; the base is flat, the mouth opening small, and the peristome divided into ten nearly equal-sized lobes; the spines are unknown.

Affinities and differences.—In its general outline, depressed upper surface, and unigeminal pores, this little urchin resembles a *Pseudodiadema*. From that group, however, it is distinguished by the rudimentary condition of the ambulacral tubercles, and the absence of crenulations from the summits of the bosses. It is distinguished from *Hemipedina Bakeri* by having small primary tubercles set more closely together, and in having a greater number in each row. From *Pedina rotata* it is known by having unigeminal pores, the upper surface more depressed, the pedal pores separated by a moniliform line of granules, and in the absence of secondary tubercles. At a first glance, it has a strong resemblance to *Pseudodiadema Mooreii*, but an examination with the lens at once discloses the points of difference, which are these: the ambulacral areas in *Hemipedina Etheridgii* have imperforate granules on their upper parts, whilst in *Pseudodiadema Mooreii* they are perforated tubercles; the moniliform line between the pedal pores in *Hemipedina Etheridgii* is absent in *Pseudodiadema Mooreii*; the mouth opening is likewise much smaller in *Hemipedina Etheridgii*.

Locality and Stratigraphical position.—This species has been found in the Upper Lias of Gloucestershire, and in the Upper Lias near Ilminster, Somersetshire, associated with *Ammonites (Walcotii) bifrons*, Brug.; *Ammonites serpentinus*, Schloth.; *Ammonites annulatus*, Sow.; and other Upper Liassic forms of Mollusca and Radiata. It is a very rare urchin, and is seldom found in a good state of preservation. Professor Deslongchamps has collected the same species in "le Lias supérieur de May et Fontaine—Etoupefour, Calvados." I dedicate this species to my friend Robert Etheridge, Esq., F.R.S.E., of the Bristol Museum, who has kindly assisted me in comparing my specimens with the fine series of Echinoderms under his care, and who has likewise otherwise aided me in the most friendly manner, in working out the subjects of this Monograph.

B. *Species from the Inferior Oolite.*

HEMIPEDINA BAKERI, *Wright.* Pl. X, fig. 1 *a, b, c, d, e, f.*

PEDINA BAKERI.   Wright, Annals and Magazine of Natural History, 2d series, vol. xiii, p. 312, pl. 11, fig. 4.
HEMIDIADEMA BAKERI. Desor, Synopsis des Échinides Fossiles, p. 58.

Test circular, depressed; ambulacral areas narrow, with one row of small tubercles, disposed in a slightly zigzag line, down the centre of the areas; inter-ambulacral areas broad, with two rows of primary tubercles in the centre of the plates, raised on prominent bosses; margins of the areolas surrounded by circles of small granules; no secondary tubercles.

Dimensions.—Height, seven twentieths of an inch; transverse diameter, three fourths of an inch.

Description.—This rare urchin has the test circular and depressed (fig. 1 *b, c*); the ambulacral areas are narrow, about one third the width of the inter-ambulacral; the usual double row of tubercles in this region is reduced to one row, the tubercles of which are disposed alternately on the right and left sides of the areas, thereby forming a single zigzag line down the centre thereof (fig. 1 *b, c, d*); the tubercles at the equator, and on the upper surface, are small (fig. 1 *b*), but there are two or three of a larger size at the base of the areas (fig. 1 *d*); a few granules form imperfect scrobicular crescents round the narrow areolas (fig. 1 *d*). The inter-ambulacral areas are nearly three times the width of the ambulacral (fig. 1 *b*); they are adorned with five pairs of primary tubercles, nearly of a uniform size throughout (fig. 1 *b, c, d*), raised on prominent bosses, the summits of which are smooth, ring-like, and without crenulations (fig. 1 *d, e*); circles of small granules bound the areolar spaces; there are no secondary tubercles, nor any other sculpture upon the inter-tubercular surface of the plates, so that down the centre of the areas there is a smooth valley between the primary tubercles (fig. 1 *d, c, b*). The apical disc is well preserved (fig. 1 *b, f*); the ovarial plates are of moderate size, and have an irregular, heptagonal form; they are covered with a few granules, scattered irregularly over their surface; the ocular plates are rhomboidal, and have large eye-holes perforated near the centre (fig. 1 *f*).

The base is flat (fig. 1 *c*), and the mouth opening small (fig 1 *a*); the peristome is divided into ten lobes by shallow notches.

Affinities and differences.—This species differs so much from all its congeners, that it cannot be mistaken for either of them. Its diagnostic characters consist in the size and number of the primary tubercles (fig. 1 *b*), the absence of secondary tubercles, the naked valley in the centre of the miliary zone (fig. 1 *c, d*), the narrowness of the ambulacra, with the single row of tubercles therein (fig. 1 *d*).

Locality and Stratigraphical position.—I have collected only one specimen of this singular form in the Pea Grit at Crickley Hill, and have seen fragments only on the surface of other fossils, as one or two plates suffice for the determination of this species. I dedicate this fine urchin to my friend T. Barwick L. Baker, Esq., of Hardwicke Court, Gloucestershire, the President of the Cotteswold Naturalists' Club.

HEMIPEDINA PERFORATA, *Wright*.   Pl. X, fig. 2 *a, b, c, d, e, f, g*.

GONIOPYGUS PERFORATUS.   Wright, Annals and Magazine of Natural History, 2d series,
                         vol. viii, p. 267, pl. 13, fig. 5 *a, b*.
HEMIPEDINA PERFORATA.   Wright, Annals and Magazine of Natural History, 2d series,
                         vol. xvi, p. 98.

Test small, circular, depressed; ambulacral areas with two rows of small tubercles, which extend from the peristome to the disc; inter-ambulacral areas with two rows of tubercles, seven to eight in each row, and three or four secondary tubercles near the base; mouth opening large; peristome decagonal, divided into deep, nearly equal-sized lobes; apical disc large, plates much developed.

Dimensions.—Height, one quarter of an inch; transverse diameter, three quarters of an inch.

Description.—This little Diadem was first figured and described as *Goniopygus perforatus*, at a time when I was ignorant of the true organic characters of the group to which it belongs. "I have placed this urchin," I observed, " provisionally in the genus *Goniopygus*, as it comes nearer to the character of that form than any other. The absence of crenulations on the mammæ, the nearly uniform size of the tubercles, the distinctness with which they stand out from the test, and a fragment of the angular apical disc *in situ*, seem to justify the supposition of its being a *Goniopygus*; but the perforation of the tubercles makes the exception, and suggests the query whether the absence of perforations is a generic or only a sectional character."*   At the time this passage was written I had only found three imperfect specimens of this species, and the search after more perfect urchins of the same natural group led to the discovery of several new congeneric forms, and finally to the establishment of the genus *Hemipedina* for their reception.

The ambulacral areas are narrow, and carry two rows of small tubercles, which, from the peristome to the equator, are regularly developed (fig. 2 *c*); but above the equator they rapidly diminish in size, and dwindle into small granules in the upper part of the area (fig. 2 *a, b, g*).

The poriferous zones are narrow and straight, the septa moderately thick, and at the base of the area the pores fall into triple, oblique pairs (fig. 2 *f*), but in the upper part they are strictly unigeminal.

The inter-ambulacral areas are twice and a half the width of the ambulacral, and furnished with two rows of tubercles, from seven to eight in each row (fig. 2 *b, c*); the bosses are prominent, and surrounded by a narrow areola (fig. 2 *g*, the letter omitted);

* 'Annals and Magazine of Natural History,' 2d series, vol. viii, p. 267.

the spinigerous tubercles are small, and not deeply perforated; they stand out, however, in a well-defined manner, from the surface of the test, and are very uniform in size throughout; a complete circle of granules surrounds each areola (fig. 2 *g*), and one row of granules separates the areolas from the zones; the miliary zone is wide (fig. 2 *b*), and filled with four rows of round granules, which are very uniform in size, and closely crowded together (fig. 2 *g*), thus imparting a granular aspect to the surface; the upper part of the centro-sutural line is naked for the space of the two uppermost plates (fig. 2 *b*).

The apical disc is large (fig. 2 *b*); the five genital plates have a heptagonal form, and are nearly all of the same size; they are perforated at some distance from the apex, and in the centre of each plate is a cluster of small granules (fig. 2 *e*); the ocular plates are pentagonal, and large in proportion to the size of the urchin; they have granules on their surface, and the eye-hole is perforated at some distance from the margin; the anal opening is round and central, and the madreporiform body is very small (fig. 2 *e*).

The base is concave, and the mouth opening, which is situated in a depression, is one half the diameter of the test; the peristome is decagonal, and divided into nearly equal-sized lobes.

Affinities and differences.—This species is nearly allied to *H. Jardinii* and *H. Etheridgii*, out is distinguished from both by the structure of the ambulacra, in which the marginal tubercles are large and well developed at the base of the area (fig. 2 *f*), but suddenly diminished in size, or rather reduced to granules, in the upper part of the same (fig. 2 *g*); whereas in these other two allied forms they are very regularly developed throughout the area (Pl. IX, fig. 4 *d*, and fig. 5 *d*, which compare with Pl. X, fig. 2 *b*, *g*).

Locality and Stratigraphical position.—I collected this species from that remarkable rock, the Pea Grit, at Crickley Hill. The species must have been tolerably abundant, although very few specimens are well preserved. Out of a considerable number I have only obtained two or three in which the structure of the test can be satisfactorily made out.

HEMIPEDINA TETRAGRAMMA, *Wright.*   Pl. X, fig. 3 *a, b, c, d.*

HEMIPEDINA TETRAGRAMMA.   Wright, Annals and Magazine of Natural History, 2d series,
vol. xvi, p. 98.

Test circular and depressed; ambulacral areas narrow, with two rows of small, nearly equal-sized tubercles on the margin of the area, extending from the peristome to the disc; inter-ambulacral areas with two rows of primary tubercles, about fourteen in each row, and two rows of secondary tubercles, ten in each row, extending from the peristome nearly to the upper surface; mouth opening small, situated in a depression; peristome decagonal and unequally lobed.

Dimensions.—Height, two fifths of an inch; transverse diameter, nine tenths of an inch.

Description.—This beautiful Hemipedina is a very rare form, having only found one specimen of the species during the many years I have collected from the locality where it was discovered.

The tubercles of the ambulacra are nearly as large as those of the inter-ambulacra, which imparts a very regular appearance to the test (fig. 3 *b*, *c*).

The ambulacral areas are narrow, about one third the width of the inter-ambulacral; on the margins are two rows of tubercles, about twenty in each row; they are very uniform in size, and arranged in a straight line throughout the area (fig. 3 *b*, *c*, *d*); the bosses abut against the poriferous zones, but they are separated from each other by a delicate, double line of granules, which extends on each side of the central suture (fig. 3 *d*), every third granule between the tubercles being larger. The poriferous zones are narrow, and slightly undulated, in consequence of the ambulacral tubercles being placed so close to the avenues, without the intervention of any granular space between to prevent the encroachment of the tubercles on them (fig. 3 *b*, *c*, *d*).

The inter-ambulacral areas are four times as wide as the ambulacral; at the equator, each of the large plates of these areas has one central primary tubercle, and at its centro-sutural side, one secondary tubercle, and between the central tubercle and the zones two rows of granules (fig. 3 *d*); from the peristome to the equator there are therefore four rows of tubercles (fig. 4 *d*, *c*), but as the secondary tubercles diminish in size above the equator, and disappear on the upper surface, there are only two rows of tubercles near the circumference of the apical disc (fig. 3 *b*); the bosses are not very prominent; the tubercles small, the minute granules forming circles around the rudimentary areolas (fig. 3 *d*); there are fourteen primary tubercles and ten secondary tubercles in each row, and from four to five pairs of pores in each of the large inter-ambulacral plates.

The base is very concave, and the small mouth opening lies in rather a deep depression; unfortunately it is filled with a hard mass of pisolite, which cannot be dislodged without risk to the shell (fig. 3 *c*).

The discal opening is small, but all the plates of the disc are absent.

Affinities and differences.- -The regularity of the ambulacral tubercles, and the nearly uniform size they have in both areas, distinguishes this species from its congeners; the fact that the inter-ambulacra have four tubercles abreast at the equator, forms a good diagonistic character, and serves to distinguish it from other allied species found with it in

20

the same rock.   *Hemipedina perforata* and *Hemipedina Waterhousei* having two rows, and *Hemipedina Bonei* six rows of tubercles in this region.

Locality and Stratigraphical position.—The only specimen I have seen of this species I collected from the Pea Grit at Crickley Hill, along with *Hemipedina perforata, Acrosalenia Lycettii, Polycyphus Deslongchampsii,* and *Pseudodiadema depressum.*

HEMIPEDINA WATERHOUSEI, *Wright.*   Pl. X, fig. 4. *a, b, c, d, e.*

HEMIPEDINA WATERHOUSEI.   Wright, Annals and Magazine of Natural History, 2d series, vol. xvi, p. 98.

Test small, pentagonal, rather inflated at the sides; ambulacral areas with two rows of small tubercles extending from the peristome to the disc; inter-ambulacral areas with two rows of tubercles, eight in a row; scrobicular circles neatly defined; mouth opening small; apical disc narrow and prominent.

Dimensions.—Height, seven twentieths of an inch; transverse diameter, half an inch.

Description.—This is a small, pentagonal, inflated species, having the ambulacral areas with two rows of tubercles, nearly as large as those of the inter-ambulacral; there are from twelve to thirteen tubercles in each row, which are placed further apart than those of *H. tetragramma,* each tubercle being surrounded by small, scattered granules (fig. 4 *c, e*); the poriferous zones are narrow and straight (fig. 4 *e*), and the septa are slightly elevated on the surface; there are four pairs of pores opposite each of the large plates.

The inter-ambulacral areas are not quite twice the width of the ambulacral; they are occupied by two rows of primary tubercles, about nine in each row, which have slightly elevated bosses, surmounted by small tubercles; the base of the boss is closely encircled by small granules (fig. 4 *e*), which form complete scrobicular circles around them; each plate has one row of granules between the tubercles and the zones, and two rows of granules between the tubercles and centro-sutural line (fig. 4 *e*).

The apical disc is well preserved in both my specimens (fig. 4 *b, d*); the genital plates have a heptagonal form, the largest side being placed towards the anal opening; they are perforated at a short distance from the apex, and two or three granules are developed on the surface; the right antero-lateral plate is a little larger than the others, and supports, as usual, a small, spongy, madreporiform body; the ocular plates are small and pentagonal; the holes are perforated near the centre of each; the disc forms a slight, ring-like prominence at the vertex of the test, and the anal opening is circular.

The base is flattened, but the mouth opening is unfortunately concealed in the only two specimens I have found.

Affinities and differences.—I regarded this urchin, at first sight, as a young form of *H. tetragramma*, but the spaced-out arrangement of the ambulacral tubercles, the inflation of the sides of the test, and the presence of only two rows of tubercles in the inter-ambulacra, show that it is quite distinct from that species. It is so entirely different from *H. perforatus* and *H. Bakeri*, that it cannot be mistaken for either of them.

Locality and Stratigraphical position.—I collected this urchin in the Pea Grit at Crickley Hill, with the former. It must be rare, as I only know two examples of the species, which I dedicate to my friend G. R. Waterhouse, Esq., F.Z.S., of the British Museum, well known for his valuable contributions to zoological literature, and by the kindness and urbanity of his manner to all who seek information in that department of the great national collection committed to his care.

HEMIPEDINA BONEI, *Wright.* Pl. X, fig. 5 *a, b, c, d.*

> HEMIPEDINA BONEI. Wright, Annals and Magazine of Natural History, 2d series, vol. xvi, p. 98.
> — — Woodward, Memoirs of the Geological Survey, Decade v, " notes on Echinopsis."

Test small, pentagonal, depressed; ambulacral areas with two marginal rows of close-set tubercles; inter-ambulacral areas with two entire central rows, and four shorter lateral rows of tubercles, which extend only to the equator; the tubercles of both areas are small and nearly of the same size; base concave; mouth opening of moderate width; peristome nearly equally decagonal.

Dimensions.—Height, seven twentieths of an inch; transverse diameter, eight tenths of an inch.

Description.—This small pentagonal depressed urchin has the minute tubercles of both areas nearly of the same size (fig. 5 *b, c*); the ambulacral areas are about one third the width of the inter-ambulacral; they have two rows of tubercles, from eighteen to twenty in each row, set on the margin of the areas (fig. 5 *c, d*), with incomplete circlets of small granules surrounding them (fig. 5 *d*), except on their zonal side, where they are absent; the poriferous zones are narrow, straight, and distinctly unigeminal

21

throughout, and there are three pairs of pores opposite each of the inter-ambulacral plates (fig. 5 *d*).

The inter-ambulacral areas are three times the width of the ambulacral; at the equator there are six rows of tubercles (fig. 5 *d*), nearly of the same size, but only the central row of each column extends from the peristome to the disc, at the base this row is large and very conspicuous (fig. 5 *c*), at the equator the tubercles are a little larger than the lateral rows, and from the equator upwards (fig. 5 *b*), they alone occupy the upper surface of the shell, of the two lateral rows, that on the zonal side is the shortest; there are from sixteen to eighteen tubercles in each of the central rows, and from eight to ten in the lateral rows; the areolas are surrounded by small granules, which form a delicate network on the surface of the plates, (fig. 5 *d*).

The apical disc is absent in all the specimens I have seen, it has an oblong form, and extends further into the single inter-ambulacrum than the other areas (fig. 5 *c*).

The base is slightly concave (fig. 5 *c*), and the tubercles are much larger in this region than on the upper surface (fig. 5 *b*); the mouth opening is large, and the peristome is divided into nearly equal sized lobes.

Affinities and differences.—The pentagonal form of this species, with its elongated opening for the apical disc, reminds me of some young *Pygasters*, which occur with it in the same rock; but the size and arrangement of the tubercles, its depressed form, and small discal opening, serve to distinguish it from them; how far a greater number of specimens, especially if the apical disc were preserved, would lead me to modify my opinion as to its generic position, it is impossible to say; I have, therefore, provisionally placed this beautiful little urchin with the *Hemipedinas*, the characters just enumerated being sufficient to show in what it differs from its congeners.

Locality and Stratigraphical position.—I have collected this urchin from the **Pea grit, Inferior Oolite**, of Crickley Hill, associated with the preceding species from the same bed, where it is rare. I dedicate this species to Mr. C. R. Bone, whose beautiful figures of Echinoderms, have given such a lasting value to the plates which accompany this Monograph.

C. *Species from the Great Oolite and Cornbrash.*

HEMIPEDINA DAVIDSONI, *Wright*.   Pl. XII, fig. 6 *a, b, c, d.*

HEMIPEDINA DAVIDSONI.   Wright, Annals and Magazine of Natural History, 2d series, vol. xvi, p. 99.
Woodward, Memoirs of the Geological Survey, Decade v, " Notes on Echinopsis."

Test circular, much depressed, ambulacral areas with two marginal rows of tubercles very regularly arranged throughout the area; inter-ambulacral areas wide, with two central rows of primary tubercles, and four lateral rows of secondary tubercles; poriferous zones narrow, and straight; under surface of the test crowded with tubercles, upper surface deficient in tubercles, base concave, mouth opening small, peristome unequally decagonal.

Dimensions.—Height, seven twentieths of an inch; transverse diameter, one inch.

Description.—I found this urchin about twelve years ago in the sandy beds of the Great Oolite, at Minchinhampton; and although I have searched diligently since in the same locality for other specimens, I have failed. The test is not in very good preservation, but it is still sufficiently so to enable me to describe its characters; the shell is thin, circular, and much depressed; the ambulacral areas are straight and narrow (fig. 6 *a*), being less than one third the width of the inter-ambulacral, on their margins are two rows of tubercles (fig. 6 *d*) in each row, which are very uniform in size throughout, corresponding with a greater uniformity in the width of the area than in some other species; the poriferous zones are narrow; the pores unigeminal throughout (fig. 6 *a*), except at the base, where three additional pairs are crowded in; there are four pairs of pores opposite each inter-ambulacral plate.

The inter-ambulacral areas are three times the width of the ambulacral; each of the elongated pentagonal plates composing the two columns has one central primary tubercle (fig. 6 *d*); these tubercles form an uninterrupted row, which extends from the peristome to the disc; besides the central row, the plates between the peristome and equator have two lateral rows of smaller secondary tubercles (fig. 6 *d*), disposed on each side of the primary one, so that at the equator the inter-ambulacra have six rows of tubercles abreast; above that line, however, the secondary series disappear, and on the upper surface of the test the two primary rows alone exist (fig. 6 *b*); the primary tubercles are very uniform in size throughout, but the secondaries vary much in magnitude (fig. 6 *d*).

The base is flat, the mouth opening is small, being about one third the diameter of the shell, the peristome is decagonal and nearly equal-lobed (fig. 6 *c*).

Affinities and differences.—This species resembles *Hemipedina tetragramma*, but the ambulacral tubercles are larger and less numerous; the poriferous zones are narrower, and the pores more strictly unigeminal; the inter-ambulacral areas are wider, and less ornamented with granules and tubercles; the secondary tubercles between the poriferous zones and the primary row (fig. 6 *d*), is absent in *Hemipedina tetragramma*. Pl. X, fig. 3 *d*, the aperture for the apical disc is likewise much larger in *Hemipedina Davidsoni*.

Locality and Stratigraphical position.—I collected this urchin from the sandy beds of the Great Oolite, at Minchinhampton Common, where it is extremely rare ; the specimen figured is the only one I know. I dedicate this species to my esteemed friend, Thomas Davidson, Esq., author of the 'Monographs on the Fossil Brachiopoda of Great Britain.'

HEMIPEDINA WOODWARDI, *Wright.*   Pl. XII, fig. 7 *a, b, c, d.*

> HEMIPEDINA WOODWARDI.   Wright, Annals and Magazine of Natural History, 2nd series, vol. xvi, p. 99.
> —   —   Woodward, Memoirs of the Geological Survey, Decade v, " Notes on Echinopsis."

Test small, circular, and much depressed ; ambulacral areas narrow, with two rows of small tubercles which extend from the base to the equator, and diminish to small granules in the upper part of the areas ; inter-ambulacral areas with two rows of primary tubercles, eight in each row, and two rows of secondary tubercles, three to four in each row, which scarcely reach the equator ; the miliary zone is wide on the upper surface and filled with numerous close-set granulations ; apical disc large, genital plates much developed ; mouth opening small, peristome decagonal and nearly equal lobed.

Dimensions.—Small specimen, height seven twentieths of an inch ; transverse diameter thirteen twentieths of an inch.

Description.—This urchin belongs to the type of *Hemipedinas*, which have two rows of primary tubercles, with a wide granulated miliary zone between them on the upper surface of the inter-ambulacral areas ; although undescribed, this species has been long known, as there is a fragment of it, collected by Dr. William Smith, from the Cornbrash, which forms part of his collection now in the British Museum.

The ambulacral areas are straight and narrow (fig. 7 *a*), on their lower half, or from the peristome to above the equator (fig. 7 *c*), there are from nine to eleven pairs of tubercles according to the size of the specimen ; these tubercles are perforated, and raised on bosses with semicircles of minute granules around them (fig. 7 *d*) ; from about the equator to the apical disc the tubercles degenerate into granules (fig. 7 *b*), so that the lower half of the area possesses tubercles, whilst the upper half is occupied by granules ; the poriferous zones are narrow, and straight ; the pores, which are strictly unigeminal, are small (fig. 7 *d*), and there are five pairs of pores opposite each of the inter-ambulacral plates ; the septa have slightly raised eminences on the surface.

The inter-ambulacral areas are four times the width of the ambulacral, they have two entire rows of primary tubercles, and two short rows of secondary tubercles ; each plate, between the peristome and the equator, has one large tubercle on its zonal side, and one secondary tubercle on its centro-sutural side (fig. 7 *d*), so that from the base to the equator there are two rows of primary and two rows of secondary tubercles (fig. 7 *a, c*), between the equator and the apical disc the secondary tubercles are absent, and the wide miliary zone is occupied by several rows of small close-set granules (fig. 7 *a, b*). According to the age of the test, there are from eight to ten primary tubercles and from five to six secondary tubercles in each row.

The apical disc is large (fig. 7 *b*), the heptagonal genital plates are much expanded and perforated near their centre, the ocular plates are proportionately large, and form crescentic arches over the apices of the ambulacra; the surface of both the genital and ocular plates is covered with small granules, the anal opening is central, and the rim of the aperture forms a slight prominence at the vertex.

The mouth opening is small (fig. 7 *c*), the peristome is decagonal, and notched into nearly equal sized lobes; from the development of the tubercles at the base of the test, this region presents a much more tuberculated appearance than the upper surface, where the granules of the ambulacra, and the wide miliary zone of the inter-ambulacra (fig. 7 *b*), form a striking contrast to the tubercles which adorn the lower part of the same divisions of the test at the base (fig. 7 *c* and *d*).

Affinities and differences.—In the semi-tuberculous character of the ambulacra this species resembles *Hemipedina perforata*, Pl. X, fig. 2, and *Hemipedina tuberculosa*, Pl. XI, fig. 2, but it is distinguished from both by having secondary tubercles in the inter-ambulacra, which are absent in these species. It resembles *Hemipedina Davidsoni*, Pl. XII, fig. 6, and *Hemipedina tetragramma*, Pl. X, fig. 3, in possessing secondary tubercles in the inter-ambulacra, but the disappearance of the marginal tubercles from the upper parts of the area shows how distinct it is from these congeneric forms.

Locality and Stratigraphical position.—This species was collected many years ago by Dr. William Smith, and is probably the specimen from the Cornbrash of Melbury, referred to in his " Strata identified by organized fossils."

My small specimen came from the Cornbrash near Trowbridge, Wilts; the same species has been collected in the Cornbrash near Boulogne-sur-Mer.

I dedicate this species to my excellent friend S. P. Woodward, Esq., of the British Museum, as a mark of respect and gratitude for the trouble and interest he has taken in the success of this work.

HEMIPEDINA MICROGRAMMA, *Wright*, nov. sp.   Pl. XII, fig. 4 *a, b, c.*

Test subpentagonal, depressed; ambulacral areas wide, with four rows of tubercles; poriferous zones narrow and straight; inter-ambulacral areas with eight rows of tubercles at the equator, diminishing above to six, four, and two rows; tubercles of both areas very small and nearly of the same size; sides rounded, upper and under surfaces much depressed.

Dimensions.—Height, eleven twentieths of an inch; transverse diameter, one inch and seven tenths.

Description.—The only specimen I know of this species is the one now figured; unfortunately the surface of the test has been rubbed smooth in some places, and distorted and broken in others, so that only an imperfect description of this interesting form can be given.

The ambulacral areas are wide and straight, and retain a very uniform width throughout; at the equator they have four rows of small tubercles (fig. 4 *c*), the two marginal rows are the most regular, as regards size and arrangement, there being about twenty-five tubercles in each row, the two inner rows commence below the equator and extend two thirds of the distance between the equator and the apical disc; they are smaller in size and not so regularly arranged as the outer rows; each row contains from ten to twelve tubercles; between these four rows of tubercles small granules are very regularly interspersed. The poriferous zones are perfectly straight and very narrow, and the pores are small and unigeminal throughout; there are five pairs of pores opposite each of the inter-ambulacral plates. The inter-ambulacral areas are about twice and a half as wide as the ambulacral; each plate at the equator supports four tubercles (fig. 4 *c*), which are so disposed on the consecutive plates of the columns that they form a series of oblique rows (fig. 4 *b*); there are sixteen plates in each column, and in the upper part of the areas the number of tubercles gradually diminishes to three, four, and two on each (fig. 4 *a*). The tubercles of both areas are very small, and nearly of the same size; their bosses are flat, and the narrow areolas are surrounded by a circle of small scorbicular granules, which imparts an ornamental appearance to the test.

The space for the apical disc is large, but as the margin of the aperture is fractured no certain estimate can be made of the probable size of the disc.

The base is concave, and the mouth opening so much concealed by adhering matrix that its form cannot be accurately made out; enough, however, is exposed, to show that the peristome is decagonal and formed of unequal sized lobes.

Affinities and Differences.—This urchin belongs to the type of *Hemipedinas* with several rows of equal sized tubercles at the equator; in the number and smallness of its tubercles it is so unlike all its present known congeners that it cannot be mistaken for either of them. In the width of its ambulacra, and in the mode of distribution of the tubercles in oblique rows, it resembles the *Pygasters*.

Locality and Stratigraphical position.—It appears to have been collected from the Cornbrash of Northamptonshire; but as the history of the specimen is unknown, this is only conjecture. It forms part of the late Miss Baker's (Northampton) Collection in the British Museum.

D. *Species from the Coral Rag.*

HEMIPEDINA MARCHAMENSIS. *Wright.* Pl. XI, fig. 1 *a, b.*

HEMIPEDINA MARCHAMENSIS.	Wright, Annals and Magazine of Natural History, 2nd series, vol. xvi, p. 197.
— —	Woodward, Memoirs of the Geological Survey, Decade v, " Notes on Echinopsis."

Test large, circular, and depressed; ambulacral areas narrow, with two rows of marginal tubercles, nearly as large as those of the inter-ambulacral; extending regularly, without interruption, from the peristome to the apical disc, and separated by a zig-zag line of small granules, the areas retaining a nearly uniform width throughout; poriferous zones narrow, forming a slightly waved line, every three pairs of pores being set obliquely in the line of the zones; inter-ambulacral areas four times the width of the ambulacral, with eight rows of tubercles at the equator, each tubercular plate supporting four nearly equal sized tubercles abreast; mouth opening large, peristome unequally decagonal.

Dimensions.—Transverse diameter, two inches and nine tenths; height, one inch and three tenths.

Description.—In the present state of our knowledge, it would be premature to propose subgenera of the remarkable group of urchins now under consideration, but it is evident that we have at least two sections of the genus *Hemipedina,* among its forms now known. 1st, those with two rows of primary tubercles, a wide miliary zone, and sometimes rows of secondary tubercles in the inter-ambulacral areas. 2d, those with four, six, eight, or even ten rows of nearly equal sized tubercles abreast, in the same region of the test; *Hemipedina Woodwardi,* and *Hemipedina tuberculosa,* are types of the first section; *Hemipedina Marchamensis,* and *Hemipedina Bouchardii,* are types of the second. At first sight, it seems difficult to believe that the two urchins, figured in Pl. XI, fig. 1 and 2, belong to the same genus, but a careful analysis of their structure does not afford any permanent characters for generic separation.

A parallel case is afforded by the genus *Pseudodiadema,* where one section has two rows of primary tubercles in the inter-ambulacra, and another has four rows of primary tubercles in the same region of the test.

The section of which *Hemipedina Marchamensis* is the type, appears to belong to the middle and upper division of the Oolites, as the only species exhibiting these hexagrammous and octogrammous characters, are found in the Calcareous Grit, Coral Rag, and Kimmeridge Clay.

The test of *Hemipedina Marchamensis* is large, nearly perfectly circular at the equator, and depressed on the upper surface; it appears to have had a hemispherical form, but unfortunately the upper portion of the body is fractured, and the outline is traced with difficulty, in consequence of adhering matrix. The ambulacral areas are narrow (fig. 1 *a*), with two marginal rows of tubercles, closely placed together, along the sutural line, two rows of granules separate the tubercles (fig. 1 *b*), and transverse branches from these bound the areolas; the poriferous zones, which closely embrace the bosses, are slightly undulated (fig. 1 *b*), from the pores being grouped in threes, the septa are thin and elevated on the surface, and there are five pairs of pores opposite each large plate (fig. 1 *b*).

The inter-ambulacral areas are four times as wide as the ambulacral (fig. 1 *a*); they are crowded throughout with numerous rows of nearly equal sized tubercles, which, added to the regular rows in the ambulacra, impart a most remarkable tuberculous character to this urchin; the inter-ambulacral plates near the equator (fig. 1 *b*), have three large tubercles placed on the same line, and one or two smaller tubercles on their zonal sides, so that at the equator there are eight rows of tubercles; the bosses are large and prominent, and the tubercles are small and deeply perforated; estimating the number of plates in each column at about twenty, each area would contain $20 \times 2 = 40 \times 4 = 160$ tubercles, which multiplied by five areas, gives 800 tubercles in the inter-ambulacral areas of this species; the areolas are narrow and superficial, and surrounded by a complete circle of small granules (fig. 1 *b*), the tubercles of the inter-ambulacra are a little larger than those of the ambulacra.

The base of the test, which is well preserved, is slightly convex at the sides, and depressed at the centre (fig. 1 *a*); the regular distribution of the numerous tubercles gives this region of the shell a highly ornamented appearance, which is admirably represented in (fig. 1 *a*), the mouth opening is not quite one third the diameter of the body, the peristome is unequally decagonal, the ambulacral lobes being the largest; the jaws are large and powerful, and project from the mouth with the teeth " in situ."

The upper surface of the test is unfortunately absent, the specimen having been detached from the rock, without due care having been taken to preserve the part to which it adhered, and which evidently contained the other portion of the shell.

Affinities and differences.—This splendid urchin resembles *Hemipedina Corallina*, from the Coral Rag, in the number of the tubercles developed on the inter-ambulacral plates, but is distinguished from it by the greater size they attain in *Hemipedina Marchamensis*. It resembles, likewise, *Hemipedina Bouchardii*, from the Kimmeridge Clay of Boulogne-sur-Mer, but as that species has ten small tubercles on the inter-ambulacral areas, our species, which has only eight, is distinct from it, whilst they both belong to that section of the genus with numerous tubercles ranged on the same line on the plates.

Locality and Stratigraphical position.—This species was collected from the lower Calcareous Grit of Marcham, Berks, and belongs to the collection of the Hon. R. Marsham, who kindly communicated it for publication.

HEMIPEDINA CORALLINA, *Wright*, nov. sp. Pl. XII, fig. 1 *a, b, c, d*.

Test large, form unknown; ambulacral areas narrow, with two marginal rows of tubercles; inter-ambulacral areas wide, equatorial plates with five or six tubercles on each; spines long and slender, stem nearly a uniform diameter and covered with fine, close-set longitudinal lines, tubercles of both areas small and nearly of the same size.

Description.—The diagnosis of this species is most imperfect, as I only know the fragments of the test I have figured, which doubtless formed portions of large urchins; fig. 1 *a* was collected by Mr. Bean, from the Coralline Oolite of Malton; the ambulacral areas of this specimen have two marginal rows of small tubercles placed rather wide apart.

The inter-ambulacral areas are three times the width of the ambulacral. On the large equatorial plates (fig. 1 *b, c*), there are four rows of primary tubercles abreast, with two smaller secondary tubercles at the zonal side of the plates (fig. 1 *c*), which would give eight rows of tubercles in the inter-ambulacral areas; the areolas are wide and smooth, and surrounded by circles of small granules (fig. 1 *b, c*).

The mouth opening is large, and the jaws, which are preserved in situ, are powerful, and armed with strong teeth (fig. 1 *a*).

The spines are long and slender; the head is short and stout (fig. 1 *d*); the stem tapers very little; the milled ring is prominent, and on it and on the surface of the stem are numerous fine close-set longitudinal lines.

Affinities and differences.—In the number of its tubercles in the inter-ambulacra, this species resembles *Hemipedina Marchamensis* (Pl. XI, fig. 1), but they are much smaller in size, have much smaller bosses, and wider and smoother areolas than in that species; unfortunately, the fragments admit of no further comparison.

Locality and Stratigraphical position.—The fragment (fig. 1 *a*) was the only portion

22

Mr. Bean ever saw after collecting many years from the Coralline Oolite of Malton, York-shire, and fragment (fig. 1 *b*), is the only portion showing the external surface of the plates I have obtained from the Coral Rag of Calne, Wilts ; from this locality, however, I have seen the interior of a large *Hemipedina* which was upwards of three inches in diameter, having the jaws "in situ," but as the external surface of all the plates was con-cealed, the species was indeterminable, it most probably however belonged to this species, as the figure of the plates was the same as those in (fig. 1 *b*, *c*).

HEMIPEDINA TUBERCULOSA, *Wright.*   Pl. XI, fig. 2 *a*, *b*, *c*, *d*, *e*, *f*.

HEMIPEDINA TUBERCULOSA.	Wright, Annals and Magazine of Natural History, 2nd series, vol. xvi, p. 99.
— —	Woodward, Memoirs of the Geological Survey, Decade v, " Notes on Echinopsis."
— —	Desor, Synopsis des Échinides Fossiles, p. 60.

Test hemispherical, depressed ; ambulacral areas narrow, with two rows of semi-tubercles at the base, which extend as high as the equator ; upper part of the areas with two rows of very small marginal granules ; interambulacral areas with two rows of primary tubercles set on prominent bosses, the areolas surrounded by circles of coarse scrobicular granules ; between the peristome and the equator there are two rows of small secondary tubercles on the zonal side of the plates, and several at the base of the rows ; the miliary zone is wide, and covered with coarse granules, the sutural lines are naked above ; mouth opening large, the peristome very unequally lobed, the ambulacral being double the size of the inter-ambulacral lobes ; anal opening large ; spines moderately stout and covered with longitudinal lines.

Dimensions.—Height, seven tenths of an inch ; transverse diameter, one inch and one fifth of an inch.

Description.—This new and beautiful urchin was only discovered a few months ago in the Coral Rag of Wiltshire. The original type specimen, belonging to the British Museum, has so much of the matrix adhering to its upper surface that its true form is concealed ; the discovery, however, of the fine specimen I have figured, now enables me to give very full details of its structure, including even the spines.

The ambulacral areas are narrow ; their basal half is enlarged, and filled with two rows of semi-tubercles gradually increasing in size from the peristome to the uppermost tubercle (fig. 2 *b*, *c*, *d*), like the semi-tubercles of the same region in the genus *Hemicidaris ;* there are eight tubercles in each row (fig. 2 *c*) which alternate with only a solitary granule at the angles between them (fig. 2 *d*); beyond the semi-tubercles the area becomes contracted, and the upper half of the space is filled with two marginal rows of small granules from sixteen to eighteen in each row (fig. 2 *c*, *d*), the very great difference between the structure of the inferior and superior parts of the ambulacral areas forms a remarkable feature in the diagnosis of this urchin.

The poriferous zones are slightly undulated (fig. 2 *c*, *d*); from the apical disc to the fourth pair of semi-tubercles the pores are unigeminal; but from this point to the peristome the area contracts and the zones expand to fill up the increase of space at the base, in this region (fig. 2 *b*), the pores fall into triple oblique pairs; between the pores, forming a pair, the septa form an elevated moniliform line (fig. 2 *d*); and there are five pairs of pores opposite each inter-ambulacral plate.

The inter-ambulacral areas are twice and a half the width of the ambulacral (fig. 2 *a*, *c*); they are filled with two rows of primary tubercles, eight tubercles in each row, which gradually increase in size from the peristome to the equator (fig. 2 *c*), and then as gradually diminish as they approach the apical disc (fig. 2 *a*); the mammillary bosses of these tubercles, especially those about the centre of the rows, are large and prominent (fig. 2 *d*); their summits are broad and smooth (fig. 2 *c*, *e*), and the areolas are not wide (fig. 2 *e*), there are two rows of secondary tubercles on the zonal side of the primaries (fig. 2 *b*), which extend from the peristome to near the equator, and a few others, about six, at the basal portion of the centro-sutural region (fig. 2 *b*, *d*), the miliary zone is broad (fig. 2 *a*), and filled with from four to six rows of coarse, unequal-sized granules, some of them even assuming the form of minute tubercles (fig. 2 *d*); the areolas are surrounded by these granules, which form complete scrobicular circles in the three uppermost tubercles (fig. 2 *a*), and incomplete circles in the lower ones (fig. 2 *d*), the sutural lines in the centro-sutural region are very distinctly marked.

The mouth opening is large, being one half the diameter of the test at the equator, the peristome is very unequally lobed, the ambulacral lobes being double the size of the inter-ambulacral.

The apical disc is absent, but as the opening is very large it indicates a great development of the plates composing it, which, in the only four specimens I at present know, are entirely absent.

Two of the spines are fortunately preserved in our specimen (fig. 2 *a*, *f*); they are cylindrical, with a prominent milled ring, and a small head (fig. 2 *f*), the surface of the stem is covered with longitudinal lines; the stem is unfortunately broken, so that its length is not known.

Affinities and differences.—This urchin might easily be mistaken for a *Hemicidaris*, like that genus it possesses two rows of semi-tubercles at the base of the ambulacra, and two rows of primary tubercles in the inter-ambulacra; but the absence of crenulations from the summits of the bosses, the presence of secondary tubercles in the inter-ambulacra, and the great size of the discal opening, are characters which serve to distinguish it from *Hemicidaris*.

It forms the best type of that section of the genus, which has two rows of tubercles in the inter-ambulacra, with a wide miliary zone; the contrast between the two sections of the group is strikingly exemplified by comparing figures 1 and 2 of Plate XI, where *Hemipedina Marchamensis*, and *Hemipedina tuberculosa* are drawn side by side.

In the structure of the ambulacral areas, with tubercles below, and granules above, and with two rows of tubercles in the inter-ambulacra, this species resembles *Hemipedina Woodwardi*, but the tubercles in *Hemipedina tuberculosa* are much larger, the granules are more developed, and it is a higher and more inflated form than the Cornbrash species.

Our urchin very much resembles the mould in plaster named *Hemicidaris depressa*, Agassiz, "X 55, R 44. Cat. Syst. p. 8. Espèce plate, subconique, à ambulacres non flexueux" from the Forest Marble, of Ranville,* but it is impossible to make a critical comparison of that form with *Hemipedina tuberculosa*, without a typical specimen; at present I know of no other form for which this beautiful urchin could be mistaken.

Locality and Stratigraphical position.—This species has been found only in the Coral Rag of Wiltshire, by Mr. William Buy; the fine specimen I have figured, was discovered by him near Lyneham; it is a very rare form, as that able and accurate collector has only found three or four specimens of the species.

E. *Species from the Kimmeridge Clay.*

HEMIPEDINA MORRISII, Pl. XII, fig. 2 *a*, *b*, *c*, *d*, *e*.

> HEMIPEDINA MORRISII.   Wright, Annals and Magazine of Natural History, 2d series,
>   vol. xvi, p. 198.
> —   —   Woodward, Memoirs of the Geological Survey, Decade v, "Notes
>   on Echinopsis."

Form and size unknown; test small; ambulacral areas with two rows of prominent marginal tubercles; poriferous zones slightly waved; inter-ambulacral areas with four rows of prominent tubercles at the equator, nearly on the same line, surrounded by incomplete circlets of granules.

Description.—The only portion of this urchin I have seen, is that now figured;

---

* Annales des Sciences Naturelles, 3ème série, tome vi, p. 338, Agassiz and Desor. Catalogue raisonné des Échinides.

the ambulacral areas are moderately wide, they have two rows of regular, prominent, marginal tubercles, which gradually diminish in size from the base to the summit of the areas, and are separated by a zig-zag line of small granules occupying the line of the suture (fig. 2 *b*) ; the poriferous zones are slightly waved, the pores are separated by thick septa, the external surface of which form slight eminences, and there are five pairs of pores opposite each large plate.

The inter-ambulacral areas are three times the width of the ambulacral, each plate supports two primary equal-sized tubercles (fig. 2 *b*), which have prominent bosses with well-defined areolas, surrounded by incomplete circlets of small granules (fig. 2 *b*).

The long, round, slender spines referred to this species (fig. 2 *c*) have their surface sculptured with fine longitudinal lines; the articulation is small, with a smooth rim ; the head is short and stout, with a thin, finely milled, prominent ring ; the stem is much smaller in diameter than the head.

Affinities and differences.—The fragment (fig. 2 *a*) is all that I have seen of the test of this species; it belongs to the section with several rows of tubercles in the inter-ambulacra, but the specimen is too imperfect for comparison with other forms.

Locality and Stratigraphical position.—This species was collected by Z. Hunt, Esq., from the Kimmeridge Clay, at Hartwell, Bucks, and was kindly communicated by Professor Morris, to whose collection it belongs.

HEMIPEDINA CUNNINGTONI, *Wright.* Pl. XII, fig. 3 *a, b.*

HEMIPEDINA CUNNINGTONI. Wright, Annals and Magazine of Natural History, 2d series, vol. xvi, p. 198.
— — Woodward, Memoirs of the Geological Survey, Decade v, " Notes on Echinopsis."

Form unknown ; ambulacral areas with two marginal rows of very small tubercles ; poriferous zones straight, pores obliquely disposed ; inter-ambulacral areas with two rows of tubercles on the zonal sides of the plates, miliary zone wide, filled with from eight to ten rows of small granules ; bosses large and prominent ; areolas surrounded by complete circles of granules.

Dimensions.—Transverse diameter, upwards of one inch ; height unknown.

Description.—The fragment figured 3 *a, b,* is all that is known of this urchin, which shows that it belongs to that group of the *Hemipedinas* having two rows of tubercles on the poriferous side of the plates, and a wide miliary zone between them. It is much to be desired that a more diligent search should be made for Echinoderms, in Kimmeridge Clay districts, as the little at present known of this class consists only of fragments of tests and detached spines.

The ambulacral areas are narrow, with two marginal rows of small tubercles, rather irregular in the mode of their arrangement (fig. 3 *b*); the poriferous zones are straight, the pores are placed obliquely, and there are four pair of pores opposite each large plate.

The inter-ambulacral areas are three times the width of the ambulacral; they have only one row of tubercles on the zonal sides of the plates (fig. 3 *b*), which leaves a wide space between the areolas and the central suture; this is filled with four or five rows of small granules, which, with those on the adjoining plate, form a miliary zone with from eight to ten rows of granules; the bosses are large and prominent, and the tubercles are of proportionate magnitude, the areolas are complete, and surrounded by circles of granules of the same size as those which fill the zone.

Affinities and differences.—The fragment (fig. 3 *a*) formed part of a species belonging to the first section of the genus, with two rows of tubercles, and a wide miliary zone in the inter-ambulacra, but, like the preceeding species, it is too imperfect for comparison.

Locality and Stratigraphical position. — This fragment was collected from the Kimmeridge Clay, near Aylesbury, by Z. Hunt, Esq. and was kindly communicated by Professor Morris.

NOTES.

Of Foreign Jurassic species of the genus HEMIPEDINA nearly allied to British forms, but which have not yet been found in the English Oolites.

Hemipedina seriale, *Leymerie*. Mém. de la Société Géologique de France, 1839, tome ii, Pl. 24, fig. 1.
Wright. Monogr. Brit. Ool. Echinodermata. Pl. IX, fig. 3 *a, b.*

Test hemispherical, subglobose above, flat below; ambulacral areas, with two rows of tubercles, nearly as large as those of the inter-ambulacra; inter-ambulacral areas, with six rows of tubercles abreast at the equator, diminishing to four, and two rows above; a few secondary tubercles are unequally distributed amongst them; mouth opening small; peristome slightly decagonal.

Formation.—Lower Lias, France.

Collection.—M. Michelin, Paris.

Hemipedina Sæmanni, *Wright.* Nov. sp.

Test small, hemispherical; ambulacral areas with two rows of tubercles; inter-ambulacral areas with one row of primary, and two rows of secondary tubercles, the primary alternating with the secondary tubercles, but not disposed on the same line, as in most other species; tubercles of both areas nearly the same size.

Formation.—Coral Rag, Commercey, Meuse.

Collection.—My cabinet, kindly sent by M. Sæmann, of Paris.

HEMIPEDINA NATTHEIMENSE, *Quenstedt.*

ECHINOPSIS NATTHEIMENSIS.   Quenstedt, Handbuch der Petrefactenkunde, pl. 49, fig. 37.

Test small, sub-pentagonal, depressed; ambulacra straight, one half the width of row inter-ambulacra, with two rows of tubercles, nine to ten in each; a row of granules extends along the tract of the suture, and transverse branches pass in lateral directions from the main line; poriferous zones narrow, and strictly unigeminal; inter-ambulacra twice as wide as the ambulacra, with two rows of primary tubercles, nine in each row, placed near the centre of the plates; the bosses surrounded by distinct areolas, complete circles of small granules encircle them; near the base, some of the granules attain the size of secondary tubercles, with perforated summits; the tubercles of both areas small and nearly of the same size, those of the inter-ambulacra a little larger than those of the ambulacra; apical disc small; the ocular plates extend beyond the line of the genitals, and are perforated near their centre; vent round, surface of the disc covered with small granulations; mouth opening large and decagonal, lobes unequal.

Dimensions.—Transverse diameter, eleven twentieths of an inch.

Formation.—White Jura, ε. Nattheim.

Collections.—Professor Quenstedt, Tübingen.
British Museum.

HEMIPEDINA BOUCHARDII, *Wright,* nov. sp. Mr. Davidson's MS., Plate III bis., figs. 1, 2, 3.

Test large, depressed; ambulacral areas with two rows of regular marginal tubercles, extending without interruption from the peristome to the apical disc; and separated by a median row of granules, with transverse branches; poriferous zones narrow, straight, and strictly unigeminal; inter-ambulacral areas three times the width of the ambulacral, with ten rows of tubercles at the equator, each inter-ambulacral plate in this region having five tubercles arranged on the same line; areolas narrow, and surrounded by circles of granules; tubercles of both areas small and nearly of the same size; spines long, slender; stem covered with well-marked longitudinal lines.

Dimensions.—Transverse diameter, two inches and one quarter; height unknown, a the specimen is crushed.

Formation.—Kimmeridge Clay, collected by M. Bouchard Chantereaux, from a cliff near Boulogne-sur-Mer.   Very rare.

Collection.—M. Bouchard-Chantereaux, at Boulogne-sur-Mer.

Genus—PEDINA, *Agassiz*, 1840.

The urchins grouped in this genus are sometimes large, but in general they are of moderate size; their test is much inflated at the sides, and nearly equally depressed on the upper and under surfaces. The shell is extremely thin, and the plates have numerous small tubercles developed on their surface.

The ambulacral areas are narrow, from one third to one fourth the width of the inter-ambulacral; they have two rows of marginal tubercles, often as large as those of the other areas.

The inter-ambulacral areas are wide, with two rows of primary, and two or four rows of secondary tubercles, which extend only from the peristome to the circumference. The tubercles are perforated, the bosses have smooth uncrenulated summits, and the areolas are narrow and superficial.

The poriferous zones are wider than in the other DIADEMADÆ, and the pores are arranged in oblique ranks, with three pairs in each file. In this respect the *Pedinas* resemble the ECHINIDÆ, and form a connecting link between these two natural families.

The mouth opening is small, in general from one third to one fourth the diameter of the test; the peristome is divided by narrow superficial notches into ten nearly equal-sized lobes.

The apical disc is small; the genital plates are nearly equal, but the right antoro-lateral, with the madreporiform body, is a little larger than the others; the ocular plates are small, and both genitals and oculars are perforated near their outer third.

The spines, unfortunately, are unknown.

The *Pedinas* were first described by Professor Agassiz as having perforated tubercles, with crenulated bosses. " Enfin un dernier caractère de ce genre consiste dans la petitesse de ses tubercles, lesquels cependant sont perforés et mamelonnés comme ceux des Diadèmes."*

In their ' Catalogue raisonné des Échinides,'† MM. Agassiz and Desor defined this genus—" Oursins comprimés, à test mince, à bouche petite, peu entaillée. Trois paires de pores obliques. Tubercles perforés et crénulés comme chez les Diadèmes. Toutes les espèces sont fossiles; des terrains Oolitiques et Crétacés."

The same definition was adopted by M. Cotteau;‡ and as the specimen I first described was not well preserved, I repeated§ the statement on the authority of these writers.

The discovery, however, of a number of good specimens of *Pedina rotata* enabled me

* 'Échinodermes Fossiles de la Suisse,' 2de partie, p. 33.
† 'Annales des Sciences Naturelles,' 3me serie, tome vi (1846), p. 370.
‡ 'Études des Échinides Fossiles,' p. 191.
§ 'Annals and Magazine of Natural History,' 2d series, vol. viii, p. 272.

to prove that the bosses are smooth, and without any trace of crenulation, in this genus ;* the accuracy of this observation has been confirmed by M. Cotteau,† who has added the following note to his description of *Pedina aspera*: " M. Wright est le premier qui a constaté contrairement à l'opinion de M. Agassiz, que les tubercles des Pédines sont certainement depourvus de crénelures. Sur tous les échantillons de Pédine, que nous avons examinés depuis, nous avons été à même de reconnaître l'exactitude de cette observation."

When I first pointed out the true structure of the bosses in the *Pedinas* to my late colleague, Professor Edward Forbes, he considered the fact of so much importance, that he proposed to suppress the genus *Pedina*, and merge its species into *Echinopsis*, as one of the characters given of *Echinopsis* was that " the tubercles were perforated but not crenulated." For this reason, in the class Echinodermata in Morris's ' Catalogue of British Fossils,' from the pen of Professor Forbes, and likewise in lettering Plate 3, Decade V, of the ' Memoirs of the Geological Survey,' this view was carried out.

The genus *Pedina* was well described and figured by Agassiz in his ' Échinodermes Fossiles de la Suisse,' and its limits accurately indicated, before *Echinopsis* was proposed. Assuming, therefore, that the diagnosis of both genera was equally well defined, which is not the case, still *Pedina* has the priority.

In the absence of good type specimens, it is impossible to make a correct comparison between *Echinopsis* and *Pedina* from the definitions in the books; and if M. Desor, who has lately carefully examined the types of *Echinopsis*, declares the genus to be unsatisfactory, this is another reason why we should retain *Pedina*. In his note on this genus, M. Desor‡ says : " Mais même tel qu'il est ici défini, le genre *Echinopsis* est encore moins précis qu'on ne pourrait le désirer, car il renferme des espèces à pores simples et d'autres à pores dédoublés. Mais comme ces deux types sont si voisins sous tous les autres rapports, particulièrement par leur forme renflée, la petitesse de leur péristome et la structure de leur plaques, je n'ai pas cru devoir les séparer génériquement. Je me suis borné à en faire deux groupes."

The genera *Pedina* and *Echinopsis* are the only ones in which we find perforated tubercles combined with trigeminal pores; but in *Echinopsis* the ambulacral areas are nearly as wide as the inter-ambulacral, which is not the case in *Pedina*.

The *Pedinas* have perforated tubercles, with smooth bosses, combined with trigeminal pores, by which they are distinguished from *Pseudodiadema*. They closely resemble *Hemipedina*, from which they are chiefly distinguished by the smallness of their tubercles, and the trigeminal arrangement of their pores, those in *Hemipedina* being unigeminal; the apical disc is likewise much larger in *Hemipedina*.

The perforation of the tubercles distinguishes *Pedina* from the true ECHINIDÆ, which it otherwise resembles in having a thin test, with trigeminal pores.

---

* ' Annals and Magazine of Natural History,' 2d series, vol. xiii, p. 173.

† ' Études sur les Échinides Fossiles,' pp. 313, 314.

‡ ' Synopsis Échinides Fossiles,' p. 99.

The *Pedinas* are all extinct, and appear to be limited to the Oolitic rocks. One species is catalogued with doubt as coming from the Cretaceous formation.

PEDINA ROTATA, *Wright* (non Agassiz). Pl. XIII, fig. 1 *a, b, c, d, e.*

ECHINUS LINEATUS.	Murchison, Geology of Cheltenham, 2d edit., p. 73 (1845).
PEDINA ROTATA.	M'Coy, Annals and Magazine of Natural History, 2d series, vol. ii, p. 20 (1848).
— —	Wright, Annals and Magazine of Natural History, 2d series, vol. viii, p. 273 (1851).
ECHINOPSIS ROTATA.	Forbes, in Morris's Catalogue of British Fossils, 2d edit., p. 78 (1854).
— —	Salter, Memoirs of the Geological Survey, Decade V, pl. 3 (1856).
PEDINA ROTATA.	Cotteau, Études sur les Échinides Fossiles, p. 315 (1856).

Test circular, or sub-pentagonal, with tumid sides, more or less depressed; ambulacral areas narrow, furnished with two marginal rows of small, numerous (from twenty-five to thirty in each row,) close-set, equal-sized tubercles, arranged with great regularity throughout; and two inner rows of minute tubercles, which disappear above and below; interambulacral areas wide, with two rows of primary, which extend without interruption from the mouth to the disc, and four rows of secondary tubercles on their outer side, which disappear at the equator; mouth opening small, peristome decagonal, with deep notches and unequal-sized lobes, apical disc of moderate size, genital plates nearly equal, poriferous zones wide, trigeminal ranks oblique, with two granules between each rank.

Dimensions.—Height, seven tenths of an inch; transverse diameter, one inch and four tenths.

Description.—There is much difficulty in distinguishing by good characters the different species of *Pedina* figured by M. Agassiz in his ' Échinodermes Fossiles de la Suisse,' arising in a great measure from the thinness of the test, the delicacy of its sculpture, and the great similarity which prevails among the different species of this group; the absence of good details of structure in the plates, showing the specific characters of each form, and of an accurate diagnosis in their description, tends to increase the difficulty; any attempt, therefore, to clear up the synonymy of these species is hopeless, without an attentive examination of the types themselves; fortunately, this has been done by a most competent and learned observer, M. Cotteau, whose analysis of the species will be given when treating of the affinities of the urchin now under consideration.

In my Memoir on the Cidaridæ of the Oolites, I erroneously identified this species with *Pedina rotata*, Agass., which is now considered by MM. Agassiz and Desor to be a variety of *Pedina sublevis ;* that name having thus become obsolete, so far as it relates to the Swiss urchin, and our species having been beautifully figured, in the ' Memoirs of the Geological Survey,' under the specific name I first gave it, I have retained it in this work.

The test of this urchin is in general circular, but in some specimens it has a sub-pentagonal form (fig. 1 *b*). Its sides are always more or less tumid, and it is nearly equally depressed on the upper and under surfaces (fig. 1 *e*).

The narrow ambulacral areas are furnished with two rows of small, numerous, equal-sized tubercles, from twenty-five to thirty in each row, which are closely set together on the margins of the area, and arranged with great regularity throughout ; in the middle third there are two rows of minute tubercles within the marginal rows, which disappear above and below ; a double line of small granules descends in a zigzag form between the tubercles, and sends small lateral branches, to encircle the marginal rows (fig. 1 *a*, *b*, *e*).

The inter-ambulacral areas are four times as wide as the ambulacral (fig. 1 *c*) ; in the specimen figured there are fifteen plates in each column ; each plate supports one primary tubercle, situated near the zonal border of the plate (fig. 1 *e*), and between the peristome and the circumference, two secondary tubercles, on the sutural side of the primaries (fig. 1 *e*) ; on the upper surface the secondary tubercles gradually disappear, so that the under surface of the test (fig. 1 *a*) is much more tuberculous than its upper surface (fig. 1 *b*) ; the areolas are very narrow, but not at all excavated, and around them circles of small granules are regularly disposed ; the primary tubercles form ten conspicuous rows, which are nearly equidistant from each other, whilst the secondary tubercles are not so regular in their arrangement.

The poriferous zones are wide, in which the holes are closely arranged in triple oblique pairs ; the obliquity, however, is greater on the upper than on the under surface, where the holes are so disposed, that the undermost pair of each trigeminal rank forms an oblique line with the uppermost pair of the rank immediately below it, leaving the middle pair by themselves, thus— Each pair is surrounded by a slight oval rim, which is only seen, however, on the best-preserved specimens ; there are three oblique pairs of holes opposite each large plate (fig. 1 *c*) ; and between each trigeminal rank there are two small tubercles (fig. 1 *b*).

The base is flat, the mouth opening small, being about two sevenths the diameter of the test ; the peristome is decagonal (fig. 1 *a*), and divided by deep notches into ten nearly equal-sized lobes, those corresponding to the ambulacra are the largest.

The apical disc is moderately large, being two sevenths the diameter of the test (fig. 1 *b*) ; it is often well preserved, as in fig. 1 *b* ; the genital plates are nearly all of the same size (fig. 1 *d*) ; the right antero-lateral, supporting the fine spongy madreporiform body, is a little larger than the others (fig. 1 *d*) ; they have all a heptagonal shape, and their apices form angles re-entering into the inter-ambulacral areas (fig. 1 *b*); the ocular plates are small pentagonal pieces, firmly wedged between the genitals; the eye-holes are very minute, and the surface of the discal elements is covered with numerous small granules; the oviductal and eye-holes are perforated about the junction of the outer with the middle third of the plates. The anal aperture is central and circular (fig. 1 *d*), and in diameter is about the length of one of the genital plates.

Affinities and differences.— This species very much resembles *Pedina Gervillii*, Desmoulins, in fact, it requires a close and critical comparison of good specimens to discover the differences between them; *Pedina rotata* has more tumid sides, the ambulacral areas possess a greater number of small, equal-sized tubercles, more closely set together, and arranged with greater regularity than in *Pedina Gervillii;* the poriferous zones are likewise wider, and the pores lie more oblique, with two granules between each rank; the primary tubercles are larger and more prominent; and there is, consequently, a greater disproportion between the tubercles of the ambulacral and those of the inter-ambulacral areas than in *Pedina Gervillii.*

It differs from *Pedina sublevis* in having larger tubercles in the inter-ambulacra, and in the number, approximation, and regularity of the arrangement of the ambulacral tubercles.

M. Cotteau* has examined the types of M. Agassiz' species contained in the magnificent collection of M. Michelin, and has given the following analysis of the affinities and differences which exist between the four species he describes, and which I have translated for this section.

"1st. The *Pedina sublevis*, Agassiz, which is characterised by its great height, its pores disposed by very oblique triple pairs, its primary tubercles slightly developed, spaced out, and confounded, so to speak, especially at the base, with the secondary tubercles which accompany them. We consider the *Pedina ornata*, Ag., as a variety of this species.

"2d. The *Pedina Gervillii*, Desmoulins; remarkable by its depressed form, its principal ambulacral and inter-ambulacral tubercles few in number and spaced out, its secondary tubercles very small, its pores largely open, ranged in triple oblique pairs, with little obliquity, and, consequently, enclosed in narrow poriferous zones.

"3d. The *Pedina rotata*, Wright; assuredly very near to *Pedina Gervillii*, but distinguished from it by its more tumid form, its pores disposed more obliquely, its ambulacral tubercles smaller, closer set together, more numerous, and forming on the borders of the poriferous zones perfectly regular ranges. This last character suffices to distinguish *Pedina rotata* from young individuals of *Pedina sublevis*. We retain to this species the name *rotata*, with this observation, that in the figures of the 'Échinodermes de la Suisse' (pl. xv, figs. 4—6), the mouth is relatively smaller, and the ambulacral tubercles are more spaced out.

* 'Études sur les Échinides Fossiles,' p. 315.

" 4th. The *Pedina aspera*, Agassiz, which is distinguished from the preceding species by its more granular aspect, by its principal tubercles being larger and more prominent, by its more numerous secondary tubercles, and by its poriferous zones being still more narrow than those of *Pedina Gervillii*."

M. Cotteau adds, that stratigraphical geology completely justifies these distinctions; *Pedina rotata* comes from the Inferior Oolite of England, *Pedina Gervillii* is met with in the Callovien stage of the Sarthe, *Pedina sublevis* characterises the inferior layers of the Coral Rag, and *Pedina aspera* appears to be special to the Kimmeridge Clay.

Locality and Stratigraphical position.—This species was first found in the upper ragstones of the Inferior Oolite, in a thin marly vein, which, in some places, rests on the Trigonia grit, in the same bed with *Ammonites Parkinsoni*, Sow. I have collected it from this stratum at Shurdington, Cold Comfort, and Hampen, in Gloucestershire; in the latter locality it is associated with *Holectypus depressus*, Leske, *Holectypus hemisphæricus*, Desor, *Echinobrissus Hugi*, Agass., *Echinobrissus clunicularis*, Llhwyd, *Clypeus sinuatus*, Leske, and *Stomechinus intermedius*, Agass. The Rev. A. W. Griesbach has discovered two fine large specimens in the Cornbrash of Rushden, Northamptonshire. Professor M'Coy gives the Great Oolite of Minchinhampton as the locality for the specimen in the Cambridge Museum. The specimen, figured in the 'Memoirs of the Geological Survey' was found in the upper beds of the Inferior Oolite, at Hampen, in the same bed from whence the specimens figured in Pl. XIII were obtained.

PEDINA SMITHII, *Forbes.* Pl. XIII, fig. 2 *a, b, c.*

CIDARIS, sp. 2 of William Smith's Stratigraphical System of Organized Fossils, p. 109.
ECHINOPSIS SMITHII. Forbes, Memoirs of the Geological Survey, Decade V, pl. 3.
Notes on British Echinopsis.
— — Morris, Catalogue of British Fossils, 2d edit., p. 78.

Test pentagonal, much depressed; ambulacral areas narrow, and extremely prominent, with two rows of tubercles, which closely alternate in the lower half of the area, but abruptly cease on the upper half; inter-ambulacral areas wide, with two rows of primary tubercles, which occupy the zonal sides of the plate, and form a continuous series from the mouth to the disc, and two short secondary rows, which extend from the peristome to the circumference, where they abruptly cease; the miliary zone is wide, and covered with numerous granules, among these several small perforated tubercles occur; mouth opening large, peristome with deep incisions, and unequal-sized lobes.

Dimensions.—Height, about one inch; transverse diameter, nearly two inches.

Description.—I only know two fragments of this remarkable *Pedina*, the one (fig. 2 *a*) was collected by the late Dr. William Smith from the Inferior Oolite at Tucking Mill, the other (fig. 2 *c*) I found in the Inferior Oolite, near Birdlip. The description given by Dr. William Smith of this species was the following :

"Pentangular, depressed, with projecting and rather distant small mamillæ ; two contiguous rows in each areola, and four converging rows in each area, the two middle rows short, and only on the side or widest part of the area ; rough, with small points encircling the mamillæ ; rays obliquely triporous.

"The areolæ form the angles of the pentagon. The two larger rows of mamillæ in each area are parallel to the rays, and converge to the aperture, and the space between them on the side is occupied by two shorter converging rows.

"*Locality.*—Tucking Mill."

Dr. Smith's specimen is a fragment (fig. 2 *a*) which exhibits a part of two inter-ambulacral and an ambulacral area, the upper surface is concealed, and only two of the notches of the peristome are exposed ; the other specimen (fig. 2 *c*) is smaller, but shows more of the form and structure of the test. The ambulacral areas are narrow and straight ; their two rows of tubercles, which are nearly as large as those of the inter-ambulacra, closely alternate (fig. 2 *b*) between the peristome and the circumference, but as suddenly disappear from the upper part of the area (fig. 2 *c*).

The poriferous zones are wide, extremely so below (fig. 2 *a*), where the trigeminal ranks lie at angles of from 15° to 45° (fig. 2 *b*), above the ranks are more oblique, and wider apart, but throughout, the pores are always arranged in triple oblique pairs.

The inter-ambulacral areas are upwards of three times the width of the ambulacral ; from the peristome to the equator each of the inter-ambulacral plates has two tubercles developed on its surface (fig. 2 *b*), the primaries occupy the zonal sides of the plates, and the secondaries their outer margin ; the primary tubercles, nine or ten in each row, are raised on very prominent bosses (fig. 2 *b*), encircled by smooth well-defined areolas, around which circles of small granules are disposed (fig. 2 *b*) ; as the tubercles are placed very near the ambulacra, and the secondaries are absent on the upper surface, there is an unusually wide miliary zone in this region of the test, which is filled with numerous small granules, among which some minute perforated tubercles are interspersed (fig. 2 *b*).

The upper surface of the specimen (fig. 2 *a*) is covered with an extremely hard rock, that of fig. 2 *c* shows the opening for the apical disc, which is of moderate width. Only two of the notches of one angle of the mouth (fig. 2 *a*) are exposed ; they are deep, and have reflected edges ; the peristome is unequally lobed, and those of the ambulacra are much the largest.

Affinities and differences.—The only urchin which this species resembles is *Hemipedina Bakeri*, Pl. X, fig. 1 ; the mode in which the tubercles closely alternate in the ambulacra, and the proximity of the primary tubercles to the poriferous zones, show the near affinity which exists between them; it is not improbable that a series of specimens might prove *H. Bakeri* to be the young condition of *Pedina Smithii*, although, in the present state of our knowledge of these forms, I should not be justified in stating such to be the fact.

Locality and Stratigraphical position.—The large fragment (fig. 2 *a*) was collected by the late Dr. William Smith, at Tucking Mill, in Moreton Combe, south-east of Bath, from a rock which I take to be Inferior Oolite. The specimen (fig. 2 *c*) I collected from the Inferior Oolite at Birdlip, near Cheltenham; it is the only specimen of the species I have found.

History.—As this is one of the Echinodermata belonging to Dr. Smith's original geological collection, deposited in the British Museum, its history is more than usually interesting. My friend, Mr. Woodward, first called Professor Forbes's attention to the specimen, who named it in honour of the father of English geology. In the description of the species, however (Notes on British *Echinopsis*, in the 'Memoirs of the Geological Survey,' Decade V), it is erroneously stated to have been collected from the Coral Rag, instead of from the Inferior Oolite. This remarkable form is now figured for the first time.

NOTES

On Foreign Jurassic species of the genus PEDINA nearly allied to British forms, but which have not yet been found in the English Oolites.

Pedina arenata, *Agassiz.* Échinoderm. Foss. Suisse, II, tab. xv, figs. 1—3.

Test small, hemispherical, flat at the base, convex on the upper surface; ambulacra with two rows of marginal tubercles, placed apart and separated by fine granules; poriferous zones very narrow, trigeminal ranks nearly upright; inter-ambulacra with two rows of small primary tubercles, secondary tubercles very small and scarcely apparent, except on the under surface; miliary granulation more abundant and distinct than in the other species; centro-sutural line naked and conspicuous; mouth opening proportionately large, two fifths the diameter of the test, peristome decagonal, and only slightly notched; the size of the mouth opening distinguishes it from other *Pedinas.*

Dimensions.—Height, six tenths of an inch; transverse diameter, one inch.

Formation.—Collected from the Inferior Oolite (Bajocien) of Goldenthal (Jura Soleurois).

Collection.—M. Gressly, very rare.

Pedina Gervillii, *Desmoulins.* Tableaux synonymiques des Échinides, p. 316, No. 19.

Test sub-pentagonal, depressed; ambulacra with two marginal rows of granules, placed wide apart, a larger and smaller tubercle alternating on the same row; in the structure of the ambulacra the difference between this species and *Pedina rotata* is chiefly observed; inter-ambulacra with two rows of primary and four rows of small secondary tubercles; poriferous zones narrow, pores largely open, trigeminal ranks slightly oblique, mouth opening small, peristome decagonal and deeply notched; apical disc moderate in size, genital plates nearly equal.

Dimensions.—Height, thirteen twentieths of an inch; transverse diameter, one inch and one quarter.

Formation.—Abundant in the " Callovien étage de Chauffour (Sarthe)." Triger.

Collections.—MM. Michelin, Desmoulins, Cotteau, de Lorière, Triger.   My cabinet.

Pedina sublevis, *Agassiz.*   Échinoderm. Foss. Suisse, II, tab. xv, figs. 8—13.
—   —   *Cotteau.*   Études sur les Échinides Fossiles, pl. 26, figs. 1—6.
—   —   *Bronn.*   Lethæa Geognostica, dritte Auflage, tab. xvii, fig. 10.

Test circular, shell extremely thin, equally depressed at both surfaces ; ambulacra with two rows of small tubercles, spaced widely apart, and accompanied by secondary tubercles ; inter-ambulacra three times as wide as the ambulacra, with two rows of primary, and four rows of secondary tubercles nearly as large as the primaries ; tubercles of both areas nearly of the same size, and placed wide apart ; poriferous zones rather wide, trigeminal ranks very oblique below, nearly vertical above ; mouth opening small, peristome decagonal, and deeply notched.

Dimensions.—Height, seven tenths of an inch ; transverse diameter, two inches.

Formation.—Corallien (Terr. à Chailles) des Ravières près de Locle, Val-de-Travers (Jura Neuchâtelois).   *Desor.*—Dans les couches calcareo-siliceuses de Châtel-Censoir et de Druyes. *Cotteau.*   This species is the most abundant of the genus, and is widely distributed through the Corallian stage of France, Germany, and Switzerland.

Collections.—MM. Michelin, Cotteau, and in most Continental museums.   British Museum.

Pedina Michelini, *Cotteau.*   Études sur les Échinides Foss., pl. 23, figs. 2—4.

Mould much inflated, and sub-pentagonal, slightly depressed at both poles ; inter-ambulacra very wide, with a median depression ; ambulacra very narrow ; poriferous zones very narrow, trigeminal ranks nearly straight ; test unknown.   This species is found always as a siliceous mould.

Dimensions.—Height, one inch and seven twentieths ; transverse diameter, two inches.

Formation.—" Dans les couches calcareo-siliceuses de l'étage Corallien inférieur, à Châtel-Censoir et à Druyes." *Cotteau.*

Collections.—Abundant.   My cabinet.

PEDINA CHARMASSEI, *Cotteau.* Études sur les Échinides Foss., pl. 24, figs. 1—3 ; pl. 25, figs. 1—3.

Test large, extremely thin, sub-circular; hemispherical and inflated on the upper surface, flattened on the under surface; ambulacra with two rows of tubercles, as large as those in the inter-ambulacra, set closely and regularly together, on the extreme margins of the area; the intermediate miliary zone covered with fine granulations, poriferous zones wide, trigeminal ranks very oblique; inter-ambulacra four to five times the width of the ambulacra, with four rows of primary tubercles, which extend very regularly from the mouth to the summit; tubercles small, slightly developed, well spaced out, and accompanied with secondary tubercles almost as large, and which form, at the inferior surface, and towards the circumference of the test, some irregular rows, extending to the upper surface; mouth opening small, peristome divided by wide notches into ten unequal lobes.

Dimensions.—Height, two and a half inches; transverse diameter, four inches and a quarter.

Formation.—" Dans les couches calcareo-siliceuses subordonnées au Coral-rag inférieur, avec *Pedina Michelini* et *sublevis* presque toujours à l'état de moule intérieur siliceux." *Cotteau.*

Collection.—M. Cotteau, very rare; plaster mould in my cabinet.

PEDINA ASPERA, *Cotteau.* Études sur les Échinides Foss. pl. 44, figs. 7—12.

Test circular, thin, nearly equally depressed at both poles; ambulacra with two rows of marginal, well-spaced-out tubercles; inter-ambulacra thrice as wide as the ambulacra, with two rows of primary tubercles, larger and more developed than in other *Pedinas*, and surrounded by areolas; several secondary tubercles scattered over the surface of the plates; poriferous zones narrow, trigeminal ranks not very oblique; mouth opening small, peristome not deeply notched.

Dimensions.—Height, nine twentieths of an inch; transverse diameter, one inch and one tenth.

Formation.—" Le *Pedina aspera* caractérise les couches inférieures du Kimmeridge ; nous l'avons rencontré dans les marnes de Baroville et des Riceys (Aube), où il est assez rare.   M. Royer nous en a communiqué deux échantillons fort beaux recueillis par lui dans le Kimmeridge inférieur de Marbeville (Haute Marne)."   *Cotteau.*

Collections.—MM. Cotteau, Royer.

Family 4—ECHINIDÆ.

This extensive natural family comprehends many genera of living and fossil urchins : some of which are large and globular ; others are of moderate size, or small, hemispherical, or depressed : in general the test is thin, and each column is composed of a considerable number of plates.

The ambulacral areas are about one third the width of the inter-ambulacral ; they have two, four, or more rows of tubercles developed on their surface, which are often nearly as large as those of the inter-ambulacral areas.

The poriferous zones present considerable diversity in the number and arrangement of the pores : in one section they are in single pairs ; in a second they form double rows ; in a third they are in triple oblique pairs ; and in a fourth the wide poriferous zones have the pores disposed in three vertical rows.

The inter-ambulacral areas are more or less wide, and their large pentagonal plates are four times as long as they are broad ; they are sometimes perforated at the angles, as in *Mespilia* and *Microcyphus ;* or they have depressions in the line of the sutures, as in *Temnechinus* and *Opechinus ;* the surface is sometimes sculptured with irregular figures in relief, as in *Glypticus*, or finely and microscopically plaited as in *Codiopsis ;* for the most part the plates have numerous tubercles developed on their surface.

The tubercles are in general small, and nearly of the same size in the ambulacral and inter-ambulacral areas ; their bosses have smooth summits, and they are always imperforate ; there are often several rows on the same horizontal line.

The spines are always short and subulate, and their surface is sculptured with fine longitudinal lines.

The mouth opening is sometimes small, and sometimes very large ; the peristome is often pentagonal, and feebly indented ; or it is deeply incised, and divided by notches into unequal-sized lobes.

The apical disc is small, and composed of five genital and five ocular plates ; the spongy madreporiform body is always prominent on the right antero-lateral genital plate.

The large and powerful jaws are composed of the same pieces as in the *Cidaridæ ;* but the pyramids are excavated in their upper part, and the two branches are united by an arc at the summit : the teeth are long and tricarinated.

In the following table I have endeavoured to classify the genera, and to show at one view the most striking characters of the different groups included in this family :

A Table showing the Classification of the ECHINIDÆ.

FAMILY.	SECTIONS.	DIAGNOSIS.	GENERA.
ECHINIDÆ.	SECTION A. Poriferous zones very narrow. Pores in single pairs.	Inter-ambulacra with irregular sculptured plates on the upper surface of the test.	GLYPTICUS, *Agassiz.*
		Upper surface of both areas finely plaited; large tubercles at the base only; mouth small.	CODIOPSIS, *Agassiz.*
		Plates of both areas with pyriform depressions in a portion of the line of the sutures; ambulacra wide.	TEMNECHINUS, *Forbes.*
		Plates with deep depressions along the whole line of the sutures; ambulacra narrow; mouth small.	OPECHINUS, *Desor.*
		Numerous small, equal-sized tubercles in both areas; two rows in the ambulacra, and four to ten in the inter-ambulacra.	ECHINOCIDARIS, *Desmoulins.*
		Surface granular, with numerous small, close-set tubercles; mouth small; peristome feebly indented.	COTTALDIA, *Desor.*
		Inter-ambulacra lobed, with numerous small, close-set tubercles; mouth large; peristome pentagonal.	MAGNOTIA, *Michelin.*
	SECTION B. Poriferous zones narrow. Pores in double rows.	Both areas naked in the middle, tubercles on the sides, with sutural pores, and a tuberculous base.	MESPILIA, *Desor.*
		Tubercles numerous, equal-sized, sporadic; inter-ambulacra with naked horizontal spaces in the line of the sutures; angular pores.	MICROCYPHUS, *Agassiz.*
	SECTION C. Poriferous zones wide. Pores in triple oblique pairs.	Tubercles small, limited to the sides of the areas; test globular, with a very small pentagonal mouth.	CODECHINUS, *Desor.*
		Tubercles small, numerous, equal-sized, in vertical and horizontal rows; larger at the base; mouth large, pentagonal.	POLYCYPHUS, *Agassiz.*
		Tubercles numerous, in many unequal, vertical rows; mouth moderate; peristome feebly notched.	PSAMMECHINUS, *Agassiz.*
		Tubercles numerous, same size in both areas; mouth small; peristome with deep, narrow slits.	ECHINUS, *Linnæus.*
		Tubercles unequal, two or more rows in each area, with secondaries; mouth very large, pentagonal, with wide, deep notches.	STOMECHINUS, *Desor.*
	SECTION D. Poriferous zones very wide. Pores in three vertical rows.	Tubercles small, irregular, sporadic; middle of the ambulacra naked; pores at the angles of the plates.	AMBLYPNEUSTES, *Agassiz.*
		Tubercles small, numerous; the inner row of pores separated from the two outer rows by a line of tubercles.	BOLETIA, *Desor.*
		Tubercles not prominent; the outer and inner rows of pores rectilinear and regular, the middle row irregular.	TRIPNEUSTES, *Agassiz.*

Genus—GLYPTICUS, *Agassiz.* 1840.

The urchins grouped in this small genus constitute a type which is easily recognised, although in an organic point of view their most striking external character is not of the first value; in fact, the irregular structure of the tubercles of the inter-ambulacral areas essentially determines the peculiar physiognomy of the genus.

The small urchins composing this genus have a thick test, and a round, depressed, or sub-conoidal form; the ambulacral areas are narrow and straight, with two rows of marginal tubercles, very regularly arranged throughout.

The poriferous zones are narrow, and the pores strictly unigeminal.

The inter-ambulacral areas have two rows of well-developed tubercles at their base; but on the upper part of the area the regular tubercles disappear, and the surface of the plates is deeply sculptured with remarkable figures, which, in some species, resemble hieroglyphical characters.

The apical disc is large; the genital plates are prominent, and have their surface sculptured; the eye-plates are large, and the lines of division between the plates are very strongly marked.

The mouth opening is wide; the peristome decagonal, the notches are shallow, and the lobes very unequal.

The tubercles of both areas are imperforate, and have smooth, uncrenulated bosses.

Glypticus resembles *Temnopleurus* in having the plates sculptured on the upper surface; but in the former the figures are in relief, whilst in the latter the sculpture consists of deep impressions corresponding to a portion of the sutures of the ambulacral and inter-ambulacral areas. In *Temnopleurus* the tubercles form regular rows, and are crenulated, but imperforate; the poriferous zones are undulated, and the pores are in threes. *Glypticus* was thought to resemble *Cyphosoma*, but I cannot detect any two characters in common in these genera. In *Cyphosoma* the tubercles are regular, and well developed throughout; they have distinct areolas, crenulated bosses, and imperforate summits; the poriferous zones are undulated, and the pores bigeminal in the upper part of the zones.

Glypticus is distinguished from *Pseudodiadema* in the irregular sculpture on the inter-ambulacra, and in having the tubercles imperforate and uncrenulated. It resembles one section of that genus, however, in having the pores unigeminal in the zones.

Glypticus resembles *Stomechinus* in having the tubercles imperforate, and with smooth bosses; but is easily distinguished by the sculpture on the plates, and the size and foliated character of the apical disc.

The species of this genus have been hitherto found only in the Oxford Clay, Coral Rag, and Portland beds, so that it belongs to the Middle and Upper divisions of the Oolitic group.

GLYPTICUS HIEROGLYPHICUS, *Goldfuss*.   Plate XIII, fig. 3 *a, b, c, d, e, f*.

GLYPTICUS HIEROGLYPHICUS.	Bourguet, Traité des Pétrifications, pl. 51. fig. 377 (1742).
— —	Knorr, Recueil des Monuments des Catastrophes, que le globe terrestre a essayées, contenant des Pétrifications, tabl. E ii, No. 35, fig. 3, 1775.
ECHINITES TOREUMATICUS.	Leske, Additamenta ad Kleinii dispositionem Echinodermatum, p. 156, pl. 44, fig. 2, 1778.
ECHINUS HIEROGLYPHICUS.	Goldfuss, Petrefacta Germaniæ, p. 126, tabl. 40, fig. 17, 1829.
ARBACIA HIEROGLYPHICA.	Agassiz, Prodrome d'une Monographie des Échinodermes, p. 23, 1836.
ECHINUS HIEROGLYPHICUS.	Desmoulins, Tableaux synonymiques des Échinides, No. 60, p. 292, 1837.
— —	Lamarck, Animaux sans vertèbres, 2de édit., tome iii, p. 372, No. 43, 1840.
GLYPTICUS HIEROGLYPHICUS.	Agassiz, Catalogus systematicus ectyporum Echinodermatum fossilium, p. 13, 1840.
— —	Agassiz, Échinodermes Fossiles de la Suisse, 2e partie, p. 96, pl. 23, figs. 37—39, 1840.
— —	Bronn, Index Palæontologicus, p. 186, 1849.
— —	Agassiz and Desor, Catalogue raisonné des Échinides. Annales des Sciences Nat., 3me série, tome vi, p. 360, 1846.
	Bronn, Lethæa Geognostica, 3tte Auflage, Band ii, p. 146, pl. 17, fig. 4, 1851.
— —	D'Orbigny, Prodrome de Paléontologie stratigraphique universelle, tome ii, p. 26, 14th étage, No. 420, 1850.
— —	Quenstedt, Handbuch der Petrefactenkunde, p. 581, 1852.
— —	Cotteau, Études sur les Échinides Fossiles, p. 166, pl. 20, figs. 12—15, 1852.
— —	Desor, Synopsis des Échinides Fossiles, p. 95, tabl. 16, figs. 1—3.

Test thick, circular, flattened at the base, hemispherical on the upper surface, sub-depressed; ambulacral areas narrow, straight, with two rows of marginal tubercles; poriferous zones deeply sunk, pores unigeminal; inter-ambulacral areas wide, with two rows of tubercles at the base; on the upper surface the plates are deeply sculptured with irregular, hieroglyphic-like figures; tubercles imperforate and bosses smooth; apical disc large; genital plates elongated, with a sculptured surface, holes near the outer fourth; ocular plates triangular, holes marginal; mouth opening large, peristome decagonal, lobes unequal, notches small.

Dimensions.—Height, eleven twentieths of an inch; transverse diameter, one inch and one tenth.

Description.—This urchin has been long known to naturalists, having been figured by Bourguet and Knorr in their respective works, cited in the synonymy. Leske, however, in his 'Addimenta ad Kleinii Echinodermatum,' first described this curious form under the name *Echinus toreumaticus*, and gave a figure of it in his pl. xliv, fig. 2. Some authors have confused this species with Klein's *Cidaris toreumatica*, which is quite a different urchin, and forms the type of Agassiz's genus *Temnopleurus*. Moreover, *Temnopleurus toreumaticus*, Klein, is living in the Persian Gulph, whilst *Echinus toreumaticus*, Leske, is found fossil only in the Coral Rag.

The ambulacral areas are narrow, and very well defined by the straight poriferous zones, which form deep furrows at their sides; they are about one fourth the width of the inter-ambulacral, and have two rows of round tubercles, about fifteen in each row, very regularly arranged on the margins of the areas (fig. 3 *f*); the tubercles at the base are much larger, and more prominent; and on the upper surface, a zigzag line of granules separates the marginal tubercles (fig. 3 *f*).

The poriferous zones are narrow and straight, and, in consequence of the thickness of the sculptured plates, lie deeply sunk; the pores are unigeminal and oblique (fig. 3 *f*), and there are seven pairs of pores opposite each inter-ambulacral plate; near the peristome they are in ranks of threes.

The inter-ambulacral areas are four times as wide as the ambulacral; from the peristome to the equator there are two rows of large, round tubercles, on smooth, prominent bosses (fig. 3 *a, b, c*); and around the areolas circles of granules are disposed (fig. 3 *f*), which, when viewed in profile, are rather conspicuous (fig. 3 *e*). On the upper part of the areas the plates exhibit a most singular structure (fig. 3 *b, f*); around the primary tubercles there are a number of irregular elevations, in relief, resembling hieroglyphic characters (fig. 3 *f*); and although there is a considerable uniformity in the general *facies* of the sculpture, still scarcely are any two figures alike (fig. 3 *b, f*). It is this structure which suggested Leske's name, *toreumaticus*,* and Goldfuss's name, *hieroglyphicus*. The plates are thick and broad, and there are only about seven in each column; the tubercles at the base (fig. 3 *a*) are round, regular, and prominent, and form a remarkable contrast to the toreumatic sculpture exhibited on the upper part of the areas.

The apical disc is large, and forms a prominence on the vertex; it is one third the diameter of the test at the equator (fig. 3 *b*); the genital plates have an elongated, hepta-gonal form, and are sculptured on the surface like the plates of the inter-ambulacral areas; the oviductal hole is pierced about the outer third (fig. 3 *d*); the ocular plates are promi-nent and heart-shaped, the eye-hole is marginal, and the sutural lines between all the elements of the disc are defined by deep depressions (fig. 3 *d*); the vent is sub-pentagonal, and the madreporiform body is represented by a few granules on the surface of the right antero-lateral genital plate.

---

* *Toreuma,* (τόρευμα,) any work raised in relief.

The mouth opening is one half the diameter of the test, the peristome is unequally decagonal, the notches are small, and the ambulacral are nearly twice as wide as the inter-ambulacral lobes.

Affinities and differences.—As this is the only species of the genus *Glypticus* found in the English Oolites, it is impossible to mistake it: the thick test, with the hieroglyphic-like markings on the plates of the inter-ambulacral areas, and of the genital disc, distinguish it sufficiently from all others.

Locality and Stratigraphical position.—The rich cabinet of the late Mr. Channing Pearce contains three specimens of this species, collected from the Coral Rag of Calne, Wilts; and in the Museums of York and Whitby I saw one specimen in each, which were collected from Coralline Oolite of Malton.

In France, M. Cotteau collected it at Châtel-Censoir and Druyes from the Inferior Coral Rag, and the calcareo-siliceous layers subordinate to it. According to M. Desor it is found in the Corallien (Terrain à chailles) of the Swiss and French Jura, of Burgundy, Württemberg, and Franconia. On the Continent it is a very abundant and characteristic fossil of the Coral Rag, and is now recorded for the first time as a British urchin.

History.—Leske appears to have considered this species identical with *Cidaris toreaumatica*, Klein, he says, " Convenit hic Echinites omnino cum ipsa naturali Cidari toreumatica, figura testae lineolis insculptis, ambulacris biporosis et ani structura, ut nulla mihi relinquatur dutitatio, cum hujus esse speciei. Testa est calcareo spatosa, coloris cinereo flavescentis; nucleus calcareus griseus. In *Lotharingia* inventus est."*

Goldfuss figured and described it in his ' Petrefacta Germaniæ' by the name *Echinus hieroglyphicus*. M. Agassiz, in his ' Prodrome,' placed it in the genus *Arbacia*, but afterwards, in his 'Échinodermes Foss. de la Suisse,' proposed for this and other congeneric forms, the genus *Glypticus*. In this work it is beautifully figured, and described in detail as *Glypticus hieroglyphicus*. It has lately been figured by Bronn in his ' Lethæa,' by M. Cotteau in his ' Études sur les Échinides Fossiles,' and by M. Desor in his ' Synopsis des Échinides Fossiles,' and is now recorded for the first time as a British fossil urchin.

* ' Additamenta ad Kleinii dispositionem Echinodermatum,' p. 156, pl. 44, fig. 2.

NOTES

On Foreign Jurassic species of the genus GLYPTICUS, nearly allied to the British form, but which have not yet been found in the English Oolites.

Glypticus Burgundiacus, *Michelin.* Revue et Magasin de Zool., No. 1, 1853. "A large and fine species: it has, below the irregular portion of the inter-ambulacral areas, three or four very large tubercles.

"*Formation.*—Oxfordien ferrugineux d'Estrochey près Chatillon-sur-Seine.

"*Collection.*—Museum of Dijon: very rare." *Desor.*

Glypticus sulcatus, *Goldfuss.* Syn. *Echinus sulcatus,* Goldfuss. Petrefact. Germaniæ, tabl. 40, fig. 18.
A small hemispherical urchin, with a flat base; tubercles very irregular; on each inter-ambulacral area there are two deep grooves; the apical disc is large.

Formation.—Coral Rag, Nattheim; Engelhardsberg (Franconia); the environs of Vendôme.

Collections.—Museums of Munich, Tübingen, all foreign collections: very common.

Genus— MAGNOTIA, *Michelin.* 1853.

This genus was established by M. Michelin* for small urchins, closely resembling *Polycyphus*, but which are distinguished from that genus by the arrangement of the pores in the zones. The genus is thus defined by its learned author:

Test elevated, inflated, with a concave base, and numerous small, equal-sized, imperforate and uncrenulated tubercles in both areas.

The pores are disposed in single pairs from the disc to the circumference; from thence to the peristome they become crowded, and form many series.

The mouth is very large, and occupies much of the base; the peristome is unequally decagonal, and, from the width of the ambulacral lobes, is almost pentagonal.

The apical disc is small, and the vent round.

The ambulacral areas are narrow; the inter-ambulacral areas have a deep median depression, so that the test is divided into fifteen unequal lobes, of which the five ambulacral are the smallest. In both areas the tubercles are arranged in oblique ranks.

The distinctive structural character between *Magnotia* and *Polycyphus* consists in the arrangement of the pores, which are in simple pairs in *Magnotia*, and form triple oblique pairs in *Polycyphus*: in all other respects these urchins closely resemble each other.

The only English species of this small group was referred by me to the genus *Arbacia* of Agassiz, when I first described it in my 'Memoirs on the Cidaridæ of the Oolites;'† but as there is much doubt about the true characters of that genus, I have placed it in M. Michelin's *Magnotia*, of which it forms a good type.

Arbacia was a small group of the ECHINIDÆ, proposed by Dr. J. E. Gray, in October, 1835,‡ for a section of the genus *Echinus*, now living, in which, according to this author, the body is depressed, the ambulacral areas are very narrow, the poriferous zones narrow and straight, the pores in simple pairs, the ovarian and inter-ovarian plates middle sized, and the anus covered by four valves. The types of the genus cited by the author are *Echinus pustulosus*, Leske, and *Echinus punctalatus*, Lamk.

M. Desmoulins,§ in 1834, had proposed the genus *Echinocidaris* for the same group, and which he thus defined: " General form perfectly regular, circular; upper surface depressed; under surface slightly concave; areas very unequal; ambulacra always less than one half the width of the inter-ambulacra; ambulacra complete, lanceolate, straight,

* 'Revue et Magasin de Zoologie,' No. 1, 1853.

† 'Annals and Magazine of Natural History,' 2d series, vol. viii, p. 278.

‡ 'Proceedings of the Zoological Society,' part iii, p. 58; and 'Philosophical Magazine,' 3d series, vol vii, p. 329, Oct., 1835.

§ 'Tableau analytique des genres d'Échinides,' July, 1834. 'Études sur les Échinides,' p. 14.

bordered on each side with a single pair of pores; spiniferous tubercles as in Echinus." The six species included in this genus are all living.

M. Agassiz's[*] definition of *Arbacia* differs very materially from that of the original author. He confines it to "small, sub-spherical urchins, having the test covered by numerous, small, smooth-based, imperforate tubercles, ranged in numerous rows on the inter-ambulacral, and sometimes on the ambulacral areas. Pores disposed in simple pairs. Mouth circular, without deep notches. Apical disc narrow and ring-like." All the species enumerated in the 'Catalogue raisonné' are Cretaceous and Tertiary fossils, and the *Arbacia* of Gray are transferred to the genus *Echinocidaris*, Desml., so that the *Arbacia* of Agassiz is not the *Arbacia* of Gray, but a new group which required revision and a distinct name. To avoid future confusion, one section now forms the genus *Cottaldia*, Desor, another the genus *Magnotia*, Michelin, whilst the original *Arbacia* of Gray are placed in the *Echinocidaris* of Desmoulins, as that genus was first established.

MAGNOTIA FORBESII, *Wright*, Pl. XIII, fig. 6 *a, b, c, d, e, f.*

ARBACIA FORBESII.	Wright, Annals and Magazine of Natural History, 2d series, vol. viii, p. 278, pl. 13, fig. 4.
ECHINUS FORBESII.	Morris, Catalogue of British Fossils, 2d edit., p. 79.
— —	Salter, Memoirs of the Geological Survey, Decade V, description of pl. 4.
MAGNOTIA FORBESII.	Desor, Synopsis des Échinides Fossiles, p. 115.

Test small, hemispherical, sub-pentagonal; ambulacral areas straight, narrow, and of uniform width throughout, with four rows of small, equal-sized, close-set tubercles; poriferous zones narrow, and sunk in a groove; pores unigeminal throughout; inter-ambulacral areas wide, each divided into two lobes by a median depression, the surface of the plates covered with numerous close-set, equal-sized tubercles; base concave; mouth-opening large, situated in a depression; apical disc small, prominent, and ring-shaped.

Dimensions.—Height, nine twentieths of an inch; transverse diameter, three quarters of an inch.

Description.—The test of this beautiful sub-pentagonal little urchin is divided into fifteen unequal lobes; five of these, forming the ambulacral areas, are narrow, and ten much wider, the divided inter-ambulacral areas, which have deep furrows corresponding to the centro-sutural line, and dividing each area into two equal convex conical lobes; the

---

[*] 'Catalogue raisonné des Échinides, Annales des Sciences Naturelles,' 3ᵐᵉ serie, tome vi, p. 355.

entire surface of the plates of all the areas is covered with small, smooth, polished, equal-sized tubercles, crowded closely together, and disposed in oblique lines (fig. 6 f).

The ambulacral areas are narrow and prominent, and nearly of the same width throughout; they are furnished with four rows of tubercles, except at the apex of the area, where there are only two rows (fig. 6 f).

The poriferous zones form narrow, depressed, well-defined lines on the surface of the test (fig. 6 d); the pores are small, and strictly unigeminal throughout, and there are four pairs of pores opposite each inter-ambulacral plate (fig. 6 f).

The inter-ambulacral areas are upwards of four times the width of the ambulacral; each of these spaces is divided into two convex lobes by a well-defined longitudinal depression, which extends from the base to the disc in the direction of the sutural line; the narrow ambulacra, bounded by the deep poriferous zones, and the wide inter-ambulacra, divided by median depressions, produce a remarkable lobed appearance on the test of this little urchin (fig. 6 a, b); the surface of the areas is crowded with small, smooth, equal-sized tubercles; at the circumference there are from twenty-five to thirty rows (fig. 6 f), but the number diminishes at the upper surface, where the areas contract, and likewise at the base, where the tubercles are larger; it may be stated, as a general character of this species, that the tubercles are crowded so close together in all the areas that the surface of the plates is rendered invisible, and the test, when examined with a lens, has a uniform granulated appearance (fig. 6 c, d).

The base is concave (fig. 6 c), and the tubercles are larger in this region; the mouth opening is wide, and lies in a depression; the peristome has a pentagonal form, from the unequal size of the lobes. The notches are not well exposed in the only good specimen I possess.

The apical disc is small and prominent (fig. 6 a, b); it has a ring-like shape, from the smallness of the ovarial plates and the size and position of the oculars (fig. 6 e); the anterior pair of ovarials are the largest, and the right plate carries a small spongy madreporiform body; the vent is transversely oblong (fig. 6 e), and the oviductal holes are large.

Affinities and differences.—This urchin, at first sight, might be mistaken for *Polycyphus Normannus*, Desor, from the Great Oolite, but it is more depressed and pentagonal, has a more concave base, deeper areal depressions, smaller, more numerous, and closer-crowded tubercles; the ambulacra are narrower, and the pores in the zones are unigeminal. These characters are sufficiently marked to prevent *Magnotia Forbesii* being mistaken for *Polycyphus Normannus*: *Magnotia* has the zones narrow and the pores unigeminal, *Polycyphus* has the zones wide and the pores trigeminal.

Locality and Stratigraphical position.—This very rare urchin was collected in the upper ragstones of the Inferior Oolite, at Dundry, near Bristol. The two specimens in my collection, and the fine series in the Bristol Museum, with one found by Mr. W. H. Baily, are the only specimens I know.

History.—First figured as *Arbacia Forbesii*, in my 'Memoir on the Cidaridæ of the Oolites.' It was afterwards referred by M. Desor, in his Synopsis, to the genus *Magnotia*, and appears to form a good type of that new group ; it is the only English example of the genus we at present possess.

NOTES

On Foreign Jurassic species of the genus MAGNOTIA, nearly allied to the British form, but which have not yet been found in the English Oolites.

Magnotia Nodoti, *Michelin.* Rev. et Mag. Zool., No. 1, 1853.

Test hemispherical; ambulacra narrow, with four rows of tubercles; inter-ambulacra with twelve rows; tubercles of both areas of a uniform size, and not sensibly increased at the circumference; mouth opening large, occupying the greater portion of the base.

Formation.—Inferior Oolite d'Avesne (Côte-d'Or).

Collection.—Museum of Dijon; a single specimen, which forms the type of this genus.

Magnotia nodulosa, *Münster.* Syn. *Echinus nodulosus*, Münst , in Goldfuss, Petrefact. Germ., tabl. 40, fig. 16.

Test nearly hemispherical, more or less sub-pentagonal; ambulacra with four rows of tubercles, diminishing to two; inter-ambulacra divided by a median depression, and furnished with from eight to ten rows, disposed in slightly arched horizontal lines; the tubercles are a little larger at the circumference, and much larger at the base; they are nearly uniform in size in both areas; mouth opening large; apical disc very narrow, in the form of a ring.

Formation.—Coral Rag of Baireuth, and Nattheim.

Collections.—Museums of Bonn, Tübingen.

Magnotia Jurassica, *Cotteau.* Syn. *Arbacia Jurassica*, Cotteau, Études sur les Échinides, pl. 20, figs. 6—11.

Test small, flat below, inflated above; ambulacra with four rows of tubercles; inter-ambulacra with ten rows; upper surface of the areas depressed, and naked in the middle; the horizontal series of tubercles in inclined rows; mouth opening large; peristome slightly notched into ten nearly equal-sized lobes. This species is less granular than the preceding, from which it likewise differs in having smaller tubercles at the base.

Formation.—Inferior beds of " l'étage Corallien" at Châtel-Censoir (Yonne), d'Eccomoy (Sarthe).

Collection.—M. Cotteau : very rare.

MAGNOTIA DECORATA, *Agassiz.* Syn. *Eucosmus decoratus*, Agassiz. Cat. raisonné, Ann. Sc. Nat., 3ᵐᵉ série, tome vi, p. 356, pl. 15, figs. 12, 13.

Test small, hemispherical, depressed ; ambulacra so extremely narrow that there is only space for a single row of tubercles, arranged in a zigzag line, thus $\begin{smallmatrix}&\circ\\\circ&\\&\circ\\\circ&\\&\circ\end{smallmatrix}$; inter-ambulacra wide, with eighteen rows of tubercles, which do not form vertical and horizontal series, but are arranged in oblique lines ; the zigzag row of tubercles in the ambulacra results from the alternate plates in the area carrying only one tubercle on each.

Formation.—Argovien (Formation γ of Quenstedt), Lægern, Baden near Zurich, Lochen (Württemberg).

Collections.—Museums Zurich, Neuchâtel, Bale.

Genus—POLYCYPHUS, *Agassiz*. 1846.

This genus consists of small urchins, with a sub-globular or hemispherical form; the sides and upper surface of the test are covered with numerous, small, equal-sized, imperforate tubercles, which form numerous vertical and horizontal rows; at the base of both areas the tubercles are disproportionately large, when compared with those on the upper surface.

The poriferous zones are wide and depressed, and the pores are arranged in triple oblique pairs; near the peristome the pores are more numerous, and crowded.

The mouth opening is large; the peristome is pentagonal, the notches are shallow, with wide ambulacral, and short inter-ambulacral lobes.

The apical disc forms a narrow ring, and the genital and ocular plates are small.

The wide inter-ambulacral areas are each divided into two lobes by a slight median depression, so that the surface of the test in this genus is divided into fifteen nearly equal-sized lobes.

Polycyphus very much resembles *Magnotia* in the general *facies* of the species, in the size and ornamentation of the test, in its division by depressions into lobes, in the number of small imperforate tubercles which crowd its surface, and the ring-like character of the apical disc; but *Magnotia* is distinguished from *Polycyphus* by the narrowness of the ambulacral areas, and especially by having the pores strictly unigeminal throughout the narrow sunken, poriferous zones.

Polycyphus resembles *Stomechinus* in having the pores in triple oblique pairs; but is distinguished by the number and general uniform character of the tubercles, arranged in vertical and horizontal rows, and by having those at the base of the areas disproportionately large when compared with the small tubercles on the upper surface. In *Stomechinus* there is always a primary row of tubercles, and secondary rows; but in *Polycyphus* they are in general uniformly of the same size. The peristome in *Stomechinus* is always very deeply incised by large notches, whilst in *Polycyphus* the division of the peristome is but feebly marked.

POLYCYPHUS NORMANNUS, *Desor.*   Pl. XIII, fig. 4 *a, b, c, d, e, f.*

POLYCYPHUS NODULOSUS.	Agassiz and Desor, Catalogue raisonné des Échinides, Annales des Sciences Naturelles, 3ᵐᵉ série, tome vi, p. 361, pl. 15, fig. 18.
ARBACIA NODULOSA.	Wright, Annals and Magazine of Natural History, 2d series, vol. viii, p. 279, pl. 13, fig. 3 *a, b.*
POLYCYPHUS NODULOSUS.	Wright, Annals and Magazine of Natural History, 2d series, vol. xiii, p. 178.
ECHINUS NODULOSUS.	Morris, Catalogue of British Fossils, 2d edit., p. 79.
— —	Salter, Memoirs of the Geological Survey, Decade V, pl. 4, p. 8.
POLYCYPHUS NORMANNUS.	Desor, Synopsis des Échinides Fossiles, p. 117, tabl. 19, figs. 4—6.

Test circular, hemispherical above, flattened at the base; ambulacral areas a little more prominent than the inter-ambulacral, with six rows of tubercles at the equator; inter-ambulacra with fourteen rows of tubercles disposed in vertical and transverse lines; tubercles of both areas smooth, round, and nearly of the same size; basal tubercles much larger; poriferous zones wide, pores in trigeminal ranks; inter-ambulacra divided by slight median depressions, and naked at the upper part of the centro-suture: apical disc small and ring-like; base flat; mouth opening large; peristome very unequally lobed.

Dimensions.—Height, seven twentieths of an inch; transverse diameter, eleven twentieths of an inch.

Description.—This pretty little urchin has been long confounded with *Magnotia nodulosa*, Münster, from the Coral Rag of Nattheim, which it resembles much in its general physiognomy, but is distinguished by the structure of its ambulacra and poriferous zones, the pores in *M. nodulosa* being unigeminal, whilst in *P. Normannus* they are trigeminal: for this reason M. Desor has properly described it as a distinct species. Its hemispherical test exhibits a disposition to assume a sub-pentagonal form, in consequence of the prominence of the ambulacral areas (fig. 4 *a, b*); the surface of the test is divided into fifteen nearly equal lobes by the ten wide poriferous zones, and a median depression in the centre of the inter-ambulacra (fig. 4 *e*); these lobular divisions are more marked in young and small specimens than in old and large ones; the ambulacral areas (fig. 4 *b, e*) are one half the width of the inter-ambulacral, they have nine large tubercles at their base (fig. 4, *e, f*), and six rows of small tubercles at their widest part (fig. 4 *e*), which gradually diminish to four and two rows above (fig. 4 *b*).

The inter-ambulacral areas are twice the width of the ambulacral (fig. 4 *b*), and are each divided by a median depression into two equal-sized lobes (fig. 4 *c*); they have from twenty to twenty-four large tubercles at their base (fig. 4 *c*), and from twelve to fourteen tubercles on the same line at their widest part (fig. 4 *e*), which gradually diminish by the disappearance of the lateral rows to ten, eight, six, four, and two (fig. 4 *b*); the row on the centre of the plates alone extending from the peristome to the disc; the tubercles on the sides and upper surface of the ambulacra and inter-ambulacra are nearly of a uniform size, and regularly arranged in a series of horizontal and vertical rows (fig. 4 *f*), so that the tubercles are opposite each other, and do not alternate as in *Magnotia*.

The poriferous zones are wide and straight (fig. 4 *b*, *c*); the pores are arranged in oblique trigeminal ranks (fig. 4 *f*); at the base they are more crowded (fig. 4 *e*), to fill up the space between the ambulacral and inter-ambulacral areas.

The base is flat (fig. 4 *e*); the large mouth opening is nearly pentagonal, and lies in a concavity, surrounded by the larger tubercles which occupy this region (fig. 4 *c*); the peristome is feebly notched (fig. 4 *c*), and very unequally lobed, the ambulacral being three times the length of the inter-ambulacral lobes.

The apical disc is small and prominent (fig. 4 *b*); the genital plates are all nearly of the same size (fig. 4 *d*), and perforated near their apex; the ocular plates are small, and the eye-holes marginal; the vent is transversely oblong (fig. 4 *d*), and the genital plates form a prominent ring around it; the spongy madreporiform body is rather elevated, and a few microscopic granules adorn the surface of the plates.

Affinities and differences.—In its general *facies*, this little lobed and nodulated urchin resembles *Magnotia Forbesii*, but in the details of its structure it is very distinct from that form; thus its ambulacral areas and poriferous zones are wider, the median inter-ambulacral sulcus is shallower and not so defined, the tubercles are larger and arranged in a series of vertical and horizontal rows, and the pores are disposed in oblique trigeminal ranks; the test is likewise more hemispherical and inflated, and wants the marked sub-pentagonal outline, the narrow furrowed zones, and deep median sulci which so well characterise *Magnotia Forbesii*.

Locality and Stratigraphical position.—This species is very rare in England. I have collected one specimen from the Trigonia grit, Inferior Oolite, near Hampen; Mr. Jones found another near Birdlip, in the same rock; the Rev. P. B. Brodie collected one specimen from a bed of clay resting on the Stonesfield slate at Sevenhampton, associated with *Acrosalenia spinosa* and *Pecten varians;* Mr. Lycett obtained several from the Great Oolite of Minchinhampton Common; these are the only specimens known to me which have been collected in Gloucestershire. Mr. William Buy found a beautiful specimen in the Cornbrash near Sutton-Benger, Wilts, which is in my collection. He informs me that it is the only example of this urchin he has found. It was associated with *Acrosalenia hemicidaroides, Acrosalenia spinosa,* and other Cornbrash fossils; so that the range of this species is from the upper ragstones of the Inferior Oolite through all the intermediate beds into the Cornbrash, where it becomes extinct.

On the Continent it is more abundant. It has been collected from the Great Oolite of Langrune and Luc by Professor Deslongchamps and M. Tesson; and from the Calcaire à polypiers at Ranville by M. Michelin. I have to thank each of these gentlemen for the beautiful series of type specimens they have sent me for comparison, and which are perfectly identical with our English forms.

History.—First entered in the 'Catalogue raisonné des Échinides' of Agassiz and Desor as *Polycyphus nodulosus*, under the supposition that it was identical with Münster's species from Nattheim; a more pentagonal form was afterwards figured by me in the 'Annals of Natural History,' under the name of *Arbacia nodulosa*. M. Desor has the merit of having shown that it is even generically distinct from the German urchin, which has narrow zones and unigeminal pores; whereas our species has wide zones, and the pores in oblique trigeminal ranks. M. Desor has therefore described and figured it as a distinct species, under the name *Polycyphus Normannus*.

POLYCYPHUS DESLONGCHAMPSII, *Wright.* Pl. XIII, fig. 5 *a, b, c, d, e, f.*

POLYCYPHUS DESLONGCHAMPSII. Wright, Annals and Magazine of Natural History, 2d series, vol. xiii, p. 179, pl. 12, fig. 4.
— — Morris, Catalogue of British Fossils; additional species of Echinodermata.

Test small, hemispherical, and circular; ambulacral areas with two rows of large, and two rows of smaller tubercles; inter-ambulacral areas divided into lobes by a median depression, and provided with two rows of large, and several rows of smaller tubercles, the small tubercles often degenerating into granules; basal tubercles large and prominent; apical disc prominent.

Dimensions.—Height, seven twentieths of an inch; transverse diameter, thirteen twentieths of an inch.

Description.—Amongst the many beautiful forms of ECHINIDÆ found in the Oolitic rocks, this pretty little species will bear comparison for neatness and symmetry with any of the family to which it belongs. I first found a solitary specimen of this species about four years ago, and since then have obtained an interesting series of different ages; but it is a very rare species.

The ambulacral areas are one half the width of the inter-ambulacral (fig. 5 *b, d*); they have two rows of tubercles set closely and regularly together on the extreme margins of the areas (fig. 5 *b, f*); between these are two shorter, inner rows, which do not extend more than two thirds the length; at the base of the area ten large tubercles are disposed in pairs (fig. 5 *c, d*), those on the right alternating with those on the left side of the area.

The inter-ambulacral areas are twice the width of the ambulacral (fig. 5 *b, d*) ; a single row of larger tubercles, about sixteen in each row, is placed in the centre of each of the two columns of plates (fig. 5 *f*) ; and numerous smaller tubercles, degenerating into granules, occupy the rest of their surface (fig. 5 *f*) ; a second row of tubercles extends from the base to the circumference, where it terminates (fig. 5 *c, d*) ; the tubercles at the base of the inter-ambulacra are about the size of those occupying the same region in the inter-ambulacra ; there are twelve of these larger tubercles in each area, so that the base (fig. 5 *c*) has a much more granulated appearance than the upper surface of the test (fig. 5 *b*) ; as the median sulcus in the centre of the inter-ambulacra is sharply defined, and the poriferous zones are much sunk, the surface of the test is thereby nearly equally divided into fifteen lobes.

The poriferous zones lie in considerable depressions (fig. *b, d*) ; the pores are very indistinct, and arranged in oblique trigeminal ranks ; between each rank a small tubercle is developed, which adds to the granulated aspect of this region, and renders the situation of the pores still more difficult to trace (fig. 5 *f*).

The apical disc is small and prominent (fig. 5 *a, b*), the genital plates are nearly of the same size (fig. 5 *e*), the spongy madreporiform body is small, and the oviductal holes are pierced near the margin ; the ocular plates are small and heart-shaped, and the eye-holes are distinctly marginal ; the vent is small, and transversely oblong.

The base is flat (fig. 5 *d*), and remarkable for the much greater size of the tubercles in this region than on the upper surface ; the mouth opening is one half the diameter of the test, and the peristome is divided by feeble notches into ten unequal lobes (fig. 5 *c*.)

Affinities and differences.—In its general *facies*, this species resembles *Polycyphus Normannus*, but is readily distinguished from it by the marginal tubercles on the ambulacra, the two central rows of larger tubercles on the inter-ambulacra, and the smaller tubercles crowded around their base (fig. 5 *f*) ; by the row of tubercles between the trigeminal ranks of pores, the comparatively small number of tubercles in the inter-ambulacra, and the greater size of the central rows. The tubercles at the base are likewise much larger in *Polycyphus Deslongchampsii*.

Locality and Stratigraphical position.—I collected this urchin in the Pea-grit at Crickley Hill, with *Diadema depressum* and *Acrosalenia Lycettii*. Mr. Gibbs, of the Geological Survey, found a few specimens in the same rock and locality.

I have dedicated this species to Professor Deslongchamps, of Caen, to whom palæontologists are indebted for many valuable memoirs on the Oolitic fauna, published in the ' Mémoires de la Société Linnéenne de Normandie.'

NOTES

ON

FOREIGN JURASSIC SPECIES OF THE GENUS POLYCYPHUS, NEARLY ALLIED TO BRITISH
FORMS, BUT WHICH HAVE NOT YET BEEN FOUND IN THE ENGLISH OOLITES.

POLYCYPHUS STELLATUS, *Agassiz.* Catalogue raisonné, Annal. Sc. Nat., tome vi, 3me série, p. 361.

It closely resembles *P. Normannus*, but has fewer tubercles in the inter-ambulacral areas. The type specimen, sent me by M. Michelin, appears to be only a variety of that species.

Formation.—Great Oolite of Ranville, Calvados.

Collections.—MM. Deslongchamps and Michelin : rare. My cabinet.

POLYCYPHUS CORALLINUS, *Cotteau.* Études sur les Échinides, pl. 21, figs. 1—7.

A large, fine species, hemispherical above, and flat below; the narrow ambulacra, with two very regular rows of marginal tubercles, and a third row in the middle; the inter-ambulacra large, with a naked median depression; ten rows of tubercles at the equator, diminishing to two above, the horizontal rows slightly inclined. The prominence and regularity of the ambulacral tubercles forms a distinctive character of this beautiful species.

Formation.—Jura moyen " (Calcaire à chailles) des environs de Druyes : cette espèce est fort rare." *Cotteau.*

Collection.—M. Cotteau.

Polycyphus distinctus, *Agassiz*.   Syn. *Echinus distinctus*, Agassiz.   Cat. raisonné, An. Sc. Nat., tome vi, 3me serie, p. 366.

" A very large depressed species ; inter-ambulacra with from ten to twelve rows of tubercles ; ambulacra with four rows of tubercles, the two internal rows much more irregular than the external.

" *Formation*.—Corallien d'Angoulin près la Rochelle.

" *Collection*.—M. d'Orbigny : rare."   *Desor*.

Genus—STOMECHINUS,* *Desor.* 1854.

M. Desor has separated from the genus *Echinus* all those urchins which have a large mouth opening, with the peristome nearly pentagonal, through the great development of the ambulacral, and the rudimentary size of the inter-ambulacral lobes; and which have two wide, deep notches at each of the five angles of the pentagon.

This new genus or section of the ECHINIDÆ is composed of urchins of moderate size, having a hemispherical, globular, or conoidal test, which is sometimes more or less depressed in different species.

The ambulacral areas are about one third the width of the inter-ambulacral, having two marginal rows of small tubercles, from twenty to thirty in each row, and sometimes two additional internal rows, in general smaller than those of the other segments.

The inter-ambulacral areas have two principal rows of tubercles in the centre of the plates, and several secondary rows, at their sides, often as large as the principal ones; the miliary zone is sometimes broad and granular, or narrow, depressed, and more or less naked.

The apical disc is small, the genital plates are nearly equal-sized, and the spongy madreporiform body is large and prominent on the surface of the right anterior genital plate.

The mouth opening is large, the peristome is deeply notched at the base of the inter-ambulacral areas; the pairs of notches approach each other so close, that they leave only a small triangular lobe between them; the length of the ambulacral lobes is such, that it produces a pentagonal-shaped mouth; the notches have the margin reflected at each of the angles.

The poriferous zones are moderately wide, and the pores are arranged in triple oblique pairs, as in the genus *Echinus.*

The spines are small, short, stout, and blunt pointed; their surface is covered with well-marked longitudinal lines.

This genus is extinct, and appears to be limited to the Oolitic formations. The species are found in the Inferior Oolite, Great Oolite, Cornbrash, Coral Rag, Kimmeridge Clay, and Portland Oolite; they, however, attained their greatest development in the lower division of the Oolites.

---

* From στόμα, mouth.

A. *Species from the Inferior Oolite.*

STOMECHINUS GERMINANS, (*Phillips.*) Pl. XIV, fig. 1, *a, b, c, d.*

ECHINUS GERMINANS.	Phillips, Geology of Yorkshire, pl. 3, fig. 15, p. 127.
— —	Murchison, Geology of Cheltenham, 2d edit., p. 73.
— —	Morris, Catalogue of British Fossils, p. 52.
— PERLATUS, var. GERMINANS.	Wright, Annals and Magazine of Natural History, 2d series, vol. viii, pl. 13, fig. 1 *a—d*, p. 274.
— DIADEMATA.	M'Coy, Annals and Magazine of Natural History, 2d series, vol. ii, p. 410.
— PERLATUS.	Forbes, in Morris's Catalogue of Brit. Foss., 2d edit., p. 79.
— —	Salter, Memoirs of the Geological Survey, Decade V, pl. 4.
STOMECHINUS GERMINANS.	Desor, Synopsis des Échinides Fossiles, p. 126.
— SUB-CONOIDEUS.	Desor, Synopsis des Échinides Fossiles, p. 125.

Test high, conoidal; marginal fold acute, circumference sub-pentagonal; ambulacral areas with two rows of primary tubercles, thirty-four in each row; poriferous zones wide, the trigeminal ranks lie very obliquely across the zone; inter-ambulacral areas with one central row of primary tubercles in the middle, two rows of secondaries on the zonal, and one on the centro-sutural side of each column of plates; miliary zone wide and finely granulated, with a naked median depression in the upper half; apical disc small, excentral; genital plates very narrow, vent transversely oblong, base concave; mouth opening very large; peristome pentagonal, with bifid-notched angles, the ten notches wide and deep.

Dimensions.—Height, one inch and a half; transverse diameter, two inches.

Description.—Since the publication of my Memoir on the Cidaridæ of the Oolites,' in which this urchin was first described, I have obtained specimens of *Echinus perlatus*, Desmarest, which has enabled me to make a critical comparison between our specimens and the species to which it was referred, and I have thereby been enabled to correct errors in my previous determination.

The test of *Stomechinus germinans*, especially in adult specimens, is always high and conoidal (fig. 1 *a*); the sides rise abruptly from an acute marginal angle; the prominence of the ambulacra, and the median depression in the inter-ambulacra, impart a lobed appearance to its upper surface.

The ambulacral areas are one third the width of the inter-ambulacral (fig. 1 *a, c*); they are very uniform in breadth throughout, and from being convex and prominent, give a sub-pentagonal form to the circumference; they have two rows of tubercles on their margins, from thirty to thirty-four in each row, a smaller tubercle often alternating with a larger one (fig. 1 c); the tubercles of the ambulacra are smaller than those of the inter-ambulacra; they are raised on small bosses, and surrounded by depressed, ring-shaped areolas (fig. 1 c), which are bounded on one side by the poriferous zones, and on the other by a row of small lateral granules; between the two rows of granules there is a naked space (fig. 1 c), in which the suture is distinctly seen; towards the base of the area this space is filled with other rows of granules, and from four to six large tubercles, which extend round the angle to the base (fig. 1 b), where the marginal tubercles are disposed with great regularity, surrounded on each side by trigeminal ranks of pores, which lie at angles of 20° across the wide zonal space (fig. 1 d).

The poriferous zones are of a uniform width on the sides; they are slightly contracted at the angle, and again expand at the base; the trigeminal ranks of pores are disposed very obliquely across the zones, at angles of from 30° to 40°; there are three small granules between each rank (fig. 1 c), and each pair of pores is surrounded by an oval elevation of the test; the pores of each pair are placed obliquely upwards, and there are about five trigeminal ranks opposite every two large plates, which, estimating twenty-eight plates in a column, gives about seventy triple oblique pairs of pores in each zone.

The inter-ambulacral areas are three times the width of the ambulacral; they are composed of long, narrow, pentagonal plates, of which there are twenty-eight in each column of an adult shell; each area has two rows of primary tubercles in the centre of the plates, which extend from the peristome to the apical disc (fig. 1 a); between this row and the poriferous zones there are two rows of secondary tubercles, which become rudimentary half way up the sides; between the primary row and the centro-suture there is another short row of six or seven secondaries (fig. 1 a, c); the tubercles are raised on small bosses, which are surrounded by depressed areolas, and the granules of the miliary zone form complete circles around them.

On the upper half of the area there is a well-marked median depression, which divides it into halves; the depression is naked, but the rest of the wide inter-tubercular space is covered with numerous granules (fig. 1 a).

The marginal fold in this species forms an acute angle; the base is concave, and crowded with large tubercles, the secondaries being as large as the primary rows in this region of the test (fig. 1 b).

The mouth opening is very large, nearly one half the diameter of the base; the peristome is pentagonal, with two wide and deeply incised notches at each angle, opposite the base of the inter-ambulacral areas (fig. 1 b); between these the short lobe is rounded, and the margins of the large ambulacral lobes form double curved lines (fig. 1 b, d).

The apical disc is small and excentral; it is placed backwards, and projects into the single inter-ambulacrum; the genital plates are very narrow, the anterior pair being larger than the posterior pair; the right antero-lateral plate, with the prominent spongy madreporiform body, is the largest, and the posterior single plate the smallest; the oviductal holes are pierced near the apex; the ocular plates are small and heart-shaped, and the minute eye-holes appear like slits near the border; the vent is transversely oblong, or has sometimes an irregular form (Pl. XIV, fig. 1 *a*).

The spines are small, delicate, and subulate, but are very seldom found in connection with the tubercles. I have seen them twice when cleaning specimens, but unfortunately they have fallen off in drying.

Affinities and differences.—This is the true *Echinus germinans*, Phillips. Through the kindness of Mr. Reed, of York, I possess two specimens of the type urchin, collected by him at Whitwell, which I have critically compared with our specimens, and found them to be identical. This species is certainly distinct from *Echinus perlatus*, Desmarest, which has smaller and more numerous tubercles in the inter-ambulacra, narrower poriferous zones, and wider ambulacra, than *Stomechinus germinans*.

It differs from *Stomechinus intermedius* (Pl. XIV, fig. 2), which is considered by some to be a mere variety of *St. germinans*, in having a more elevated and conoidal test, larger tubercles, narrower and more prominent ambulacra, the apical disc more excentral, the marginal angle more acute, the mouth opening larger, and the notches wider, with deeper incisions; but the two forms are unquestionably most nearly allied, and it is only when placed side by side, and closely compared, that the differences become evident.

It differs from *Stomechinus bigranularis* (Pl. XIV, fig. 3) in all the structural details already referred to, in the comparison with *St. intermedius*; but in addition to these, the globose test, with its inflated sides and convex base, so characteristic of *St. bigranularis*, added to the small mouth opening which this species possesses, enables us to distinguish it readily from *St. germinans*; as I shall have to return to these affinities and differences when treating of both forms, I reserve further details until the respective species are the subject of our special study.

Locality and Stratigraphical position.—I have collected this species in the Pea-grit of Leckhampton, Birdlip, and Crickley Hills, and it is found in the freestone beds of the Inferior Oolite at Nailsworth and Wallsquarry, Gloucestershire. In Yorkshire it is collected from the freestone beds of the Inferior Oolite at Whitwell, near Castle Howard. These beds have been considered to be Great Oolite, but I am of opinion they are true Inferior Oolite, as Mr. Reed obtained, at Whitwell, with *Stomechinus germinans*, *Gervillia Hartmanni*, Münst., the large quadrate Inferior Oolite variety of *Trigonia costata*, Sow., and other well-known Inferior Oolite shells. It is by mistake that this urchin has been said to be found in the Coralline Oolite at Malton, as I shall endeavour to show in the history of the species. The *Echinus diademata*, M'Coy is a small Whitwell specimen of *St. germinans*. From the facts before me, I conclude that this is a true Inferior Oolite species, and the first and most typical of the genus to which it is referred.

History.—This urchin was first figured by Professor John Phillips, in his 'Geology of the Yorkshire Coast,' in his plate of Coralline Oolite fossils, and was said to be collected from the Great Oolite at Whitwell, and the Coralline Oolite of Malton and Scarborough. The statement that it occurred in the Coral Rag led both Professor Forbes and myself to suppose that it might be a variety of *Echinus perlatus*, and under this impression it was described as such by us both. Since the publication of my Memoir, I have visited most of the typical collections of Yorkshire fossils, with the view of determining some doubtful points relative to the stratigraphical position of certain species, and from this examination I am satisfied that *Stomechinus germinans* has not been found out of the Inferior (Great?) Oolite of Whitwell and other Inferior Oolite localities. The statement that it came from Malton was first made by a collector who obtained the specimens at Whitwell, and *sold them as Malton fossils*. Mr. Reed, of York, who was acquainted with the facts, writes me as follows: "There cannot, I think, be a doubt as to the correctness of your previously strongly expressed opinion regarding the distribution of *Echinus germinans*, Phil., and *Clypeus semisulcatus*, Phil., in the Oolitic beds, very *exclusively in the lower series*, and not both in the latter and Coralline Oolite, as stated by Professor Phillips in his work on the 'Geology of Yorkshire.' From long experience I am decidedly of opinion that neither species have ever been found in the Coralline Oolite at Malton or the neighbourhood. The error originated from the fact of a local collector, named Larcum, having sold the specimens which he obtained at the different quarries, and from different strata, six or eight miles round Malton, as Malton fossils. He was in the habit of procuring his specimens from the Great Oolite of Whitwell and Weston, also from some quarries at or near Coneysthorpe, most probably Inferior Oolite, and from the Calcareous grit at Appleton, as well as from the Coralline Oolite of Malton. I may also remark that he was wholly ignorant of geology, and I know, from personal experience, that he was unacquainted with the distribution and names of the Oolitic beds in the neighbourhood, having for many years obtained fossils from him. He was, I believe, the exclusive dealer in Malton for more than half a century."

Professor M'Coy described* a specimen of this species as *Echinus diademata*, from the Coralline Oolite of Malton. Through the kindness of Professor Sedgwick I have been enabled to examine this specimen, belonging to the Cambridge Museum, and have compared it with my Whitwell specimens, and there cannot be a doubt of their identity.

This species was first figured by Professor Phillips, in the 'Geology of Yorkshire,' as *Echinus germinans*. It was afterwards figured and described, for the first time, in my 'Memoir on the Cidaridæ of the Oolites,' as *Echinus perlatus*, var. *germinans*. Professor Forbes, in Morris's 'Catalogue,' and in lettering Pl. IV, Decade V, of the 'Memoirs of the Geological Survey,' retained the same name. Mr. Salter, in his elaborate article on this urchin, in the same work, describes it as *Echinus perlatus*.

* 'Annals and Magazine of Natural History,' 2d series, vol. ii, p. 410.

STOMECHINUS INTERMEDIUS, *Agassiz.*   Pl. XIV, fig. 2 *a, b, c, d.*

ECHINUS INTERMEDIUS.   Agassiz, Catalogus systematicus Ectyporum Echinodermatum,
                                     p. 12.
    —   GRANULARIS.   Wright, Annals and Magazine of Natural History, 2d series,
                                     vol. viii, p. 277.

Test hemispherical, depressed, inflated at the sides, nearly circular at the circumference, sometimes inclining to a sub-pentagonal form; ambulacral areas with two rows of small tubercles on the margins, and a miliary zone between; poriferous zones wide; trigeminal ranks very oblique; inter-ambulacral areas with two entire primary, and four short secondary rows of small tubercles; miliary zone very wide, and covered with small granules; apical disc slightly excentral, of moderate size, genital plates fully developed; mouth opening large; peristome pentagonal, with two small notches at each angle; inter-ambulacral areas slightly depressed at the centro-suture.

Dimensions.—Height, one inch and one fifth; transverse diameter, one inch and nine tenths.

Description.—Through the kindness of Professor Deslongchamps, I possess a type specimen of *Echinus intermedius* from the Great Oolite of Ranville, and which fossil was determined by M. Agassiz. This specimen has enabled me to distinguish the urchin mentioned in the ' Catalogue raisonné,' and entered as var. *major* of *Echinus bigranularis,* Lamarck, in that memoir.

It is certainly very difficult to describe the nice distinctions between the first three species, beautifully and truthfully drawn in Pl. XIV, as they graduate into each other so insensibly, that it is almost impossible to seize their distinctive characters; so much so, that at one time I considered them varieties of one species; having, however, examined a great many specimens of these Echini, I am satisfied that if varieties, they are permanent ones, and as such require a separate description. Their structural characters, however, are sufficiently marked to justify their separation into distinct species.

The test of *Stomechinus intermedius* is hemispherical, but rather depressed above; the sides are a little inflated, the marginal angle is obtuse, and the base is flat (fig. 2 *c*); the ambulacral areas are one third the width of the inter-ambulacral; they have two marginal rows of small tubercles, about thirty in each row, arranged with regularity on the margins of the areas (fig. 2 *a, d*); these tubercles have small bosses, and narrow areolas, which are encircled by minute granules; there are two rows of granules between the tubercles and the suture (fig. 2 *d*), which extend through the entire length of the area.

The poriferous zones are wide, the trigeminal ranks incline at about 40°; three small tubercles are placed between each rank (fig. 2 *d*), and there are two ranks opposite each inter-ambulacral plate ; the zones are slightly contracted at the margin, and expand again at the base.

The inter-ambulacral areas are three times as wide as the ambulacral (fig. 2 c) ; each column consists of about twenty-five elongated, pentagonal plates (fig. 2 d), and a median depression divides the area from the margin to the disc into two lobes ; there are two rows of small primary tubercles in the centre of the plates (fig. 2 c, d), which extend from the peristome to the disc, and two short rows of secondaries on their zonal sides, which disappear a little way above the margin ; within the primary rows at the base are six or seven more ; from the circumference to the disc there is a wide miliary zone (fig. 2 a, c), which is filled with small tubercles and minute granules, arranged, however, with considerable regularity on the plates (fig. 2 d), and extending even over the median depression, which, when taken in connection with the smallness of the tubercles, about the same size in both areas, imparts a peculiar granular appearance to the test of this species.

The base is flat, and the tubercles, as usual, are larger in this region ; the mouth opening is eight tenths of an inch, the diameter of the test being one inch and nine tenths ; the peristome is pentagonal, with two notches at each angle, and a small lobe between, the large ambulacral lobes forming the sides of the pentagon (fig. 2 b).

The apical disc is well developed, and slightly excentral (fig. 2 c) ; the genital plates (fig. 2 a, e), are elongated, and perforated near their outer third, the right antero-lateral being much the largest, and supporting a prominent madreporiform body (fig. 2 e) ; the ocular plates are small pentagons, with slit-like eye-holes ; the disc is much more developed in this species than it is either in *Stomechinus germinans* or *bigranularis*.

Affinities and differences.—This urchin is distinguished from *Stomechinus germinans* by the following characters : The test is more regular in its form, less elevated and rounder, and never assumes a conoidal figure ; the tubercles are much smaller, and equal-sized ; the small tubercles and granules on the miliary zone are larger, and have a more regular arrangement, which, with the smallness of the primary tubercles, gives the test a more granular *facies*. The marginal angle is more obtuse, and the sides are inflated ; the apical disc is larger, and its elements are more developed ; the mouth opening is proportionally smaller, and the notches are neither so wide nor so deeply incised. *Stomechinus intermedius* is considered by MM. Agassiz and Desor to be a variety of *St. bigranularis;* but, between the structure of the apical disc, the arrangement of the trigeminal ranks in the poriferous zones, and in the size of the mouth opening, we find characters sufficient to show that the affinity between *St. germinans* and *St. intermedius* is greater than between *St. intermedius* and *St. bigranularis.*

Locality and Stratigraphical position.—I have collected this species in the upper rag-stones of Shurdington, Rodborough, and Dundry Hills, where it is extremely rare. The Dundry specimen I formerly described under the name of *Echinus granularis.* I found a beautiful specimen in the Trigonia grit of Hampen, Gloucestershire, with *Pedina rotata*, *Holectypus depressus*, and *Collyrites hemisphæricus*. The Rev. A. W. Griesbach collected one specimen in the Cornbrash, at Rushden, Northamptonshire; and Mr. Buy obtained another from the Cornbrash, near Sutton, Wilts.

STOMECHINUS BIGRANULARIS, *Lamarck.*   Pl. XIV, fig. 3 *a, b, c, d, e.*

ECHINUS BIGRANULARIS.	Lamarck, Animaux sans Vertèbres, tome iii, p. 50.
— —	Agassiz and Desor, Catalogue raisonné des Échinides, Annal. des Sciences Naturelles, 3ᵐᵉ série, tome vi, p. 365.
— SERIALIS.	Wright, Annals and Magazine of Natural History, 2d series, vol. viii, p. 276, pl. 13, fig. 2.
— —	Forbes, in Morris's Catalogue of British Fossils, 2d edit., p. 79.
— PERLATUS, var. FORBESII.	Salter, Memoirs of the Geological Survey, Decade V, pl. 4, fig. 6.
STOMECHINUS BIGRANULARIS.	Desor, Synopsis des Échinides Fossiles, p. 125, tab. xviii, fig. 5—7.

Test hemispherical, depressed; sides inflated, circumference more or less sub-pentagonal; ambulacral areas with two rows of small tubercles on the margins, placed wide apart, twenty-six in each row, and a very fine granulation between; poriferous zones narrow, trigeminal ranks not very oblique, becoming nearly parallel on the upper and under surfaces; inter-ambulacral areas with two rows of primary tubercles, twenty in each row, and two short rows of secondaries, which disappear at the circumference; miliary zone wide, and uniformly covered with small, equal-sized granules; mouth opening small; peristome pentagonal, with ten shallow, obtuse notches; apical disc large, excentral; all the genital plates, with the exception of the right antero-lateral, small; vent large, and encircled by a moniliform line of small granules.

Dimensions.—Specimen A, fig. 3 *a.*   Height, one inch and one fifth; transverse diameter, one inch and seven tenths.

„   B, fig. 3 *b.*   Height, one inch and seven twentieths; transverse diameter, two inches.

Description.—This species is supposed to be *Echinus antiquus*, Defrance MSS., and *Echinus bigranularis*, Lamarck, although this is not quite clear.   M. Desor says— "I have been for a long time in doubt as to the identity of this species, which was so much more difficult to define, as, among the originals in the Paris Museum, ticketed by the hand of Lamarck, there are found many species.   After much hesitation, I propose to restrict the name *bigranularis* to the Oolitic species, so well figured by M. Wright under the name *serialis*, and afterwards by M. Forbes."* I am the more disposed to accept this determination, as I have a beautiful specimen of this species from the Ferrugenous or Inferior Oolite of Croisilles, Calvados, kindly sent me by M. Michelin as *Echinus bigranularis*, Agassiz, and determined by him, which exactly corresponds with our English specimens.

I have been fortunate in obtaining a fine series of this urchin, which has enabled me to study the variations it presents at different periods of growth; with the exception of some forms being more globular, others more pentagonal, all its essential characters are extremely persistent in my specimens.

The test is sometimes globular, but in general hemisperical and depressed; the circumference is occasionally round, but oftener sub-pentagonal, and varies in size from one to two inches in diameter. The specimen figured is an extremely pentagonal variety, and was selected because the tubercles and sculpture of the test were finely preserved (fig. 3 *a, c*).

The ambulacral areas are narrow, and nearly of a uniform width throughout; they have two rows of small tubercles, placed wide apart on the margins, of which there are twenty-six in each row (fig. *a, c*); every alternate plate in the column carries a tubercle, which alternates with a plate covered with two rows of small granules (fig. 3 *d*); at the base the tubercles are larger (fig. 3 *b*), and in the middle two rows of small granules separate the marginal tubercles (fig. 3 *d*).

The poriferous zones are narrower than in *St. germinans* and *St. intermedius;* the trigeminal ranks incline at angles of from 55° to 60°; they become more sub-parallel on the upper part of the zones, and towards the base; there is one small tubercle at the angle between each file (fig. 3 *d*) of pores, and there are five trigeminal ranks opposite two tubercular plates.

The inter-ambulacral areas are three and a half times as wide as the ambulacral (fig. 3, *a, c*); the plates are broader than in the preceding species; and consequently there are only from twenty to twenty-two plates in a column. There are two rows of primary tubercles, which are placed nearer the poriferous rows than the centro-suture (fig. 3 *c*); they are a little larger than the ambulacral tubercles, and nearly of the same size from the circumference upwards. A row of smaller secondary tubercles, about twelve in number, occupies their zonal side, and extends half way up the test (fig. 3 *c*). The tubercles are raised on bosses (fig. 3 *d*), surrounded by narrow, ring-like areolas, a series of small granules encircle the base, and the remaining surface of the plates is dotted over with numerous small equal-sized granules (fig. 3 *d*). There is a very slight median depression, from the upper part of which the small granules are absent (fig. 3 *a*). At the base of the area six larger tubercles are disposed within the primary rows (fig. 3 *b*).

* 'Synopsis des Échinides Fossiles,' p. 125.

The base is flat, and, as all the tubercles are of an equal size, this region of the test presents a remarkable tuberculous character (fig. 3 *b*), when compared with its smooth upper surface (fig. 3 *a*); the mouth opening is much smaller than in *St. germinans* and *St. intermedius*; thus, in fig. 3 *b*, the diameter of the test is nearly two inches, whilst the width of the mouth is seven tenths of an inch; in fact, the smallness of this opening is one the most important diagnostic characters of the species, and becomes very evident when we compare fig. 2 *b* and fig. 1 *a*, with fig. 3 *b*; this comparative smallness of the mouth is a persistent character in all the specimens I have examined, the peristome is pentagonal, with two obtuse notches at each angle.

The apical disc is of moderate size and slightly excentral, projecting backwards into the single inter-ambulacrum (fig. 3 *c*); the anterior pair of genital plates are larger than the posterior pair. The right plate, with the madreporiform body, is the largest, and the posterior single plate the smallest (fig. 3 *a, e*). On the surface of each of these plates, the anterior right plate excepted, there are six or seven small tubercles arranged, which collectively form a moniliform ring about the margin of the vent (fig. 3 *e*), which is extremely large in this species, and projects slightly towards the left side of the body; the ocular plates are small, and the eye-holes like oblong slits (fig. 3 *e*).

Affinities and differences.—In the general form and physiognomy of the test, this species resembles *Stomechinus intermedius;* it is distinguished from that species, however, by having fewer ambulacral and inter-ambulacral tubercles, the poriferous zones narrower, the trigeminal ranks of pores more upright, and the mouth opening much smaller. The apical disc is larger, the anal opening much wider, with a circle of granules around its margin; on the miliary zones the granules are likewise smaller and more numerous. The same group of characters serve to distinguish it still more distinctly from *St. germinans.*

M. Desor and M. Agassiz consider *St. intermedius* as a variety of *St. bigranularis;* but from this opinion I must beg to differ, because, although *St. germinans* and *St. intermedius* may possibly be varieties of one species, still assuredly *St. bigranularis* is distinct from both; the smallness of the mouth opening, the structure of the apical disc, and the greater breadth of the inter-ambulacral plates, in my judgment, justify the distinction.

Locality and Stratigraphical position.—I have collected this species only from the Upper Ragstones of the Inferior Oolite near Bridport, in Dorsetshire, associated with *Holectypus hemisphæricus*, Desor; *Clypeus altus*, M'Coy; *Ammonites Parkinsoni*, Sow.; and *Terebratula sphæroidalis*, Sow.

The specimen I figured in the 'Annals of Natural History' as *Echinus serialis* was said to have been collected from Dundry Hill, but I find this was a mistake, as it turns out to be a Dorsetshire fossil. I have never seen *St. bigranularis* in the Inferior Oolite of Gloucestershire. In this county it is represented by *St. intermedius*.

The only authentic foreign locality that I am acquainted with is Croisilles, Calvados, where it occurs in a brown ferruginous Inferior Oolite, containing numerous large grains of the hydrate of iron, the " calc. à polypiers de Croisilles," Michelin.

History.—First figured in my 'Memoir on the Cidaridæ of the Oolites' as *Echinus serialis*, afterwards by Professor Forbes in the fifth decade, pl. 4, fig. 6, of the ' Memoirs of the Geological Survey ;' where it was well described by Mr. Salter as *Echinus perlatus*, var. *Forbesii*.

B. *Species from the Great Oolite*—11ᵉ *Étage, Bathonien*, d'Orbigny.

STOMECHINUS MICROCYPHUS, *Wright*, nov. sp. Pl. XV, fig. 1 *a, b*.

Test circular, hemispherical, with a flat base ; ambulacral areas with four rows of tubercles, diminishing to two on the upper surface ; inter-ambulacral areas with ten rows of tubercles at the equator, irregularly disposed on the plates, and a median depression in the line of the centro-suture ; poriferous zones wide, the pores in oblique ranks of threes, and between each file two small granules regularly disposed.

Dimensions.—Height, six tenths of an inch ; transverse diameter, nearly one inch.

Description.—The modern generic divisions of the family ECHINIDÆ often repose upon characters which undergo many phases of development in the different species ; and thus it sometimes happens, as in the urchin now under consideration, some of the species approach, in their ensemble, nearer to aberrant forms of an allied genus than to the one to which they are referred.

This form certainly very much resembles a large *Polycyphus Normannus*, Desor ; although the state of conservation of the test, and the concealment of the base by adherent rock, prevents that amount of examination so necessary for a critical diagnosis ; still, however, the specimen exhibits such a group of characters, that I have placed it in the genus *Stomechinus* for the following reasons : 1st, The size of the body ; 2d, the thickness of the test ; 3d, the irregular arrangement of the tubercles on the inter-ambulacral plates ; and 4th, the absence of a median depression in the inter-ambulacra.

The ambulacral areas (fig. 1 *b*) have four rows of tubercles at the equator, two marginal rows placed on the extreme borders of the area, and two inner rows near the sutural line ; the tubercles all alternate with each other (fig. 1 *b*), and the two inner rows disappear at the upper surface.

The poriferous zones are wide; the pairs of pores are disposed in oblique ranks of threes, rather widely asunder (fig. 1 *b*), and between each rank two small tubercles are developed, which separate them from each other throughout the zones (fig. 1 *b*), so that there is only one trigeminal rank of pores opposite each large inter-ambulacral plate.

The inter-ambulacral segments are twice the width of the ambulacral; the centro-suture is very distinctly marked, but there is no median depression, and the division of the test into fifteen lobes, so characteristic of *Polycyphus*, is not seen in this form ; each of the large plates in general supports four or five tubercles, which are rather irregularly disposed on its surface (fig. 1 *b*); they have narrow, ring-like areolas around their base, encircled by rows of small granules (fig. 1 *b*), invisible to the naked eye, and only seen with a lens; the inter-ambulacra have therefore ten rows of tubercles at the equator ; the third tubercle from the poriferous zones represents the principal row, which is continuous from the base to the disc; but the other lateral rows disappear at different points between the circumference and the vertex.

The apical disc is unfortunately absent, and the base is adherent to a hard shelly fragment of the Great Oolite limestone, which cannot be removed without, at the same time, detaching the shell.   In this remarkable formation, the Oolitic grains are indented into the plates of the test of the Echinodermata, as well as into the shells of the Mollusca ; and unless the specimen separates readily from the matrix, it is almost hopeless to expect to remove it without so much injury to the sculpture as will render its determination doubtful, or even impossible.   For this reason I have been unable to expose the base of this beautiful rare form.

Affinities and differences.—The nearest affinities of this species is with *Polycyphus Normannus*, Desor; but it has a much thicker test, the inter-ambulacral tubercles are proportionately smaller and more irregularly disposed, and they have circles of granules around their areolas; the segments are not divided into lobes by median depressions ; the poriferous zones are wide and only slightly depressed, and the trigeminal ranks of pores are separated by small tubercles : all these diagnostic characters show how distinct *Stomechinus microcyphus* is from *Polycyphus Normannus*, when a critical comparison is made between these two urchins, which are often found in the same bed.   I know of no other form for which our species could be mistaken.

Locality and Stratigraphical position.—This beautiful urchin is one of the many exquisite fossils collected by my friend Mr. Lycett, from the Great Oolite of Minchinhampton Common.   It occurs in the shelly beds of limestone at the large quarry, and is the only specimen found by him, after many years' diligent search, in his favorite localities.

I have one small specimen of the same urchin, collected from the Great Oolite of

Ranville, Calvados, by M. Tesson, in which the apical disc is preserved. It is small and prominent; the genital plates are nearly equal-sized, the right anterior plate is the largest, and the spongy madreporiform body occupies all the surface of that plate; the oculars are small, and deeply indented where they receive the apex of the ambulacra; the genital holes are perforated about the outer third of the plates; the vent is oblong, and its long diameter lies obliquely across the test. It is doubtless a rare urchin in France, as none of the systematic authors make mention of it; or it may be that it has hitherto been confounded with *Polycyphus Normannus*.

C. *Species from the Coral Rag*—14ᵉ *Étage Corallien*, d'Orbigny.

STOMECHINUS GYRATUS, *Agassiz.*   Pl. XIV, fig. 4 *a, b, c, d, e.*

ECHINUS GYRATUS.	Agassiz, Échinoderm. Fossiles de la Suisse, part ii, p. 87, pl. 23, figs. 34—46.
— PETALLATUS.	M'Coy, Annals and Magazine of Natural History, 2d series, vol. ii, p. 409.
— GYRATUS.	Agassiz and Desor, Catalogue raisonné des Échinides, Annales des Sciences Naturelles, 3ᵐᵉ série, tom. vi, p. 366.
— GYRATUS.	Wright, Annals and Magazine of Natural History, 2d series, vol. ix, p. 85.
— —	Forbes in Morris's Catalogue of British Fossils, 2d edit., p. 79.
— —	Salter, Memoirs of the Geological Survey, Decade V. Notes on Echinus, p. 8.
STOMECHINUS GYRATUS.	Desor, Synopsis des Échinides Fossiles, p. 126.

Test hemispherical, more or less elevated, inflated at the sides, round at the circumference; sides divided by depressions into fifteen unequal, slightly convex lobes; ambulacral areas half the width of the inter-ambulacral, with two complete rows of marginal tubercles extending from the peristome to the disc, and two incomplete rows of central tubercles occupying two thirds of the sides; inter-ambulacral areas divided into lobes by naked median depressions, which extend from the disc to the circumference; each lobe has one complete central row, and four incomplete lateral rows of secondary tubercles; apical disc central; vent small; mouth opening wide; peristome pentagonal.

Dimensions.—Height, one inch; transverse diameter one inch and four tenths.

Description.—This beautiful urchin was figured, with details, by M. Agassiz, in his ' Échinodermes Fossiles de la Suisse;' its specific characters are so well marked that it can scarcely be mistaken for any other species. The test is hemispherical, more or less elevated, and very regularly formed; its surface is divided into fifteen nearly equal-sized lobes grouped in five divisions, with three lobes in each, of which the ambulacra form the centre, and the half of the adjoining inter-ambulacra the two lateral lobes (fig. 4 a).

The distinctive character of this species consists in the wide, naked median depression in the centre of the inter-ambulacral areas, which extends from the circumference to the apical disc, and is throughout entirely destitute of granules or any other sculpture (fig. 4 a); near the circumference, however, small tubercles occupy the space, and at this point there are twelve tubercles, in a horizontal row, in a single inter-ambulacral area; each area is thus divided into two convex lobes by a median sulcus; in each lobe there is one complete principal row of twenty-five tubercles, which extends from the peristome to the disc, and two incomplete rows on each side of the principal row, with from twelve to eighteen tubercles in each, which disappear on the sides; besides these at the widest part of the area, a few small additional tubercles are introduced (fig. 4 c, d); the bosses are small, and closely surrounded by circles of granules (fig. 4 d); the tubercles are large and prominent, and nearly all of the same size, which gives the surface of this urchin very much the appearance of a large *Polycyphus*.

The ambulacral areas are nearly half as wide as the inter-ambulacral; they are furnished with two complete rows of tubercles disposed on the margins of the area, and two incomplete rows which occupy the central parts of the sides (fig. 4 d); the tubercles have small bosses, around which granules are disposed in circles, and similar moniliform granular rings surround the larger tubercles of the incomplete rows (fig. 4 d).

The tubercles of both areas are large, prominent, and highly polished; those at the base are a little larger than the tubercles on the sides; and it may be said of this species in general that the test is uniformly very granular, and forms a remarkable contrast to that of *Stomechinus bigranularis* (fig. 3 a).

The poriferous zones are narrow, the trigeminal ranks form angles of about 65°, which become nearly sub-parallel in the upper part of the zone and at the base (fig. 4 b, c); two small granules are dotted between each file of pores.

The apical disc is small and central (fig. 4 a); the genital plates are nearly equal-sized, with the exception of the right antero-lateral, which is the largest, and supports a fine spongy madreporiform body (fig. 4 e); a series of granules on the plates form a circle around the vent, which is of a moderate size. The ocular plates are small, pentagonal pieces, projecting from between the angles of the genitals; there are two holes in each plate, with a transverse slit between (fig. 4 e). The oviductal holes are conspicuous, and perforated near the outer third of the plate.

The mouth opening is very large, nearly one half the diameter of the test (fig. 4 b); the peristome is pentagonal, with two wide, obtuse notches at each angle, and a small, lip-like lobe between; the ambulacral lobes are four times as broad as the inter-ambulacral, and form the sides of the pentagon.

Affinities and differences.—M. Desor, by mistake, has placed *Stomechinus gyratus* as a synonym of *Stomechinus germinans*, and described the true *Stomechinus germinans* as a new species under the name *Stomechinus sub-conoideus*, a confusion which has doubtless arisen from the stratigraphical error already pointed out in the history of *St. germinans*. The following characters will show how entirely distinct *St. gyratus* is from the three closely allied forms of the Inferior Oolite. It differs from *St. germinans* in having the naked median sulcus wider and deeper, and extending further down the sides ; in having four rows of equal-sized tubercles in the ambulacra, whilst *St. germinans* has only two. The secondary tubercles of the inter-ambulacral areas are as large as the principal row, and there is no miliary zone in this species ; whereas, in *Stomechinus germinans*, *intermedius*, and *bigranularis*, the miliary zone is a very wide granular space. The tubercles are likewise larger and nearly of the same size in both areas. I have been able to examine *Echinus petallatus*, M'Coy, through the extreme kindness of Professor Sedgwick, and compared it with the urchin I have figured ; I can, therefore, state that it is an un-questionable specimen of *Echinus gyratus*, Agass., and not in any way a distinct form, as Professor M'Coy supposed.

Locality and Stratigraphical position.—*Stomechinus gyratus* is rarely found in the Clay beds of the Coral Rag, near Calne, Wilts ; I am not aware that it has been collected in any other locality in England. Its foreign locality is likewise limited to the "terrain à Chailles" or Corallian stage of Besançon. I have a specimen from the Coral Rag of the department of Haute Saone, France, kindly sent me by M. Michelin.

History.—First described by M. Agassiz in his 'Échinodermes Fossiles de la Suisse,' where it is beautifully and accurately figured. The English form was afterwards described by M'Coy as *Echinus petallatus*. It was first described in detail in my 'Memoir on the Cidaridæ of the Oolites,' and its distinction from our other Oolitic Echinidæ was therein indicated. It is probable, that Parkinson's figure, in vol. 3 of his 'Organic Remains,' described as an 'Echinite from France,' refers to this species.

STOMECHINUS NUDUS, *Wright*, nov. sp. Pl. XV, fig. 2 *a, b, c, d, e.*

Test circular, conoidal ; ambulacra flat, with two regular marginal rows of tubercles, twenty-four in each row, and two inner rows which occupy the middle half of the area ; poriferous zones wide ; pores oblique ; two small granules between each trigeminal rank ; inter-ambulacra with eighteen plates in each column ; equatorial plates with four or five tubercles disposed irregularly on the surface on each plate, the number diminishing on the upper surface; tubercles surrounded by small, depressed, ring-like areolas, scarcely any granules on the surface of the plates; apical disc small, genital plates narrow, ocular plates prominent, with transverse eye-holes; base flat, inter-ambulacra forming convex lobes around the peristome.

Dimensions.—Height, seven tenths of an inch; transverse diameter nine tenths of an inch.

Description.—This small conoidal Echinite in many respects resembles *Stomechinus gyratus*, Agass., and might at first sight be supposed to be a mere variety of that form, but a closer examination proves it to be distinct; the test is circular and elevated, the sides rising suddenly from the basal angle; the ambulacral areas are flat, having two rows of tubercles regularly arranged on their margins, about twenty-four in each row, with two irregular internal rows, which occupy about the middle half of the area; the tubercles are all nearly of the same size; they have narrow, sunken areolas around their base, but scarcely any granules on the surface of the plates (fig. 2 *d*).

The poriferous zones are wide, the trigeminal ranks form angles of 40° to 50°; between each rank there are two granules, and there are three ranks opposite two tubercular plates; as there are eighteen plates in each column, there are therefore about twenty-seven trigeminal ranks in each poriferous zone.

The inter-ambulacral areas are rather more than twice the width of the ambulacral, each column contains about eighteen plates, and on each plate, at the widest part of the area, there are from four to five tubercles, so irregularly disposed that the arrangement is different on every plate (fig. 2 *d*); the tubercles are all about the same size, they are surrounded by narrow, sunken areolas; a few smaller tubercles are scattered among them, but there are no circles of granules around their base, which gives the surface of the plates a naked appearance; there is no depression in the middle of the areas, and the region of the centro-suture between the five uppermost pairs of plates is naked but not depressed; the areas become convex and prominent below.

The apical disc is small (fig. 2 *a*); the genital plates are narrow, and the anterior larger than the posterior pair; the spongy madreporiform body occupies the surface of the right anterior plate (fig. 2 *e*); the ocular plates are small, but prominent, and the eye-holes form transverse slits on their sides; the vent is oblong (fig. 2 *e*), and around its margin there are three parts of a circle of granules; the oviductal holes are large, and perforated near the apex of the plates.

The base is much covered with adhering matrix (fig. 2 *b*), so that the form of the mouth opening is nearly concealed; the tubercles are all much larger in this region, and the wide depressed poriferous zones, and convex inter-ambulacral lobes, impart a lobed appearance to the circumference of the mouth.

Affinities and differences.—This species most nearly resembles *Stomechinus gyratus,* but is distinguished from it by the absence of the naked median depressions in the inter-ambulacral areas, which gives such a remarkable lobed character to the surface of its test ; by the absence of the circles of granulations around the base of the tubercles ; and the naked appearance which the surface of the plates presents, in consequence of the almost total absence of this granular ornamentation, and which is expressed by its specific name *nudus*.

Locality and Stratigraphical position.—This Echinite was collected in Wiltshire, but I have not been able to ascertain the correct locality ; it appears to have been obtained from the Coral Rag, although this is not certain.

NOTES

On Foreign Jurassic species of the genus STOMECHINUS, nearly allied to British forms, but which have not been found in the English Oolites.

Stomechinus multigranularis, *Cotteau.* Syn. *Echinus multigranularis*, Cotteau. Études sur les Échinides Fossiles, pl. 7, figs. 6—8.

Test conoidal, with inflated sides, and a sub-pentagonal circumference; ambulacra narrow and prominent, with four rows of tubercles; inter-ambulacra more than three times as wide, with a naked median depression in the upper part of the area, and from eight to ten rows of tubercles, irregularly disposed, in the widest part, diminishing to two rows above; tubercles nearly all of the same size, except at the base, where they are larger; intermediate space filled with fine granules. Apical disc moderately large; genital plates pentagonal, with a granular surface; vent sub-circular; base flat; mouth opening very large, half the width of the test; peristome decagonal, with ten wide, obtuse notches, and small inter-ambulacral lobes between.

Dimensions.—Height, one inch and seven twentieths; transverse diameter, one inch and three quarters.

Formation.—Bathonien, Great Oolite, of Grimaux, Yonne: very rare.

Collections.—M. Rathier; a plaster mould in my cabinet.

Stomechinus Vacheyi, *Cotteau.* Syn. *Echinus Vacheyi*, Cotteau. Études sur les Échinides Fossiles, pl. 3, figs. 12—16.

Test small, depressed, and pentagonal; ambulacra with two rows of tubercles; inter-ambulacra with two rows of primary tubercles, and a few scattered secondary tubercles; mouth opening large; peristome with small notches.

Dimensions.—Height, three tenths of an inch; transverse diameter, six tenths of an inch.

Formation.—Bathonien, " dans les couches calcaires du Forest-marble, de Montillot, Yonne." *Cotteau.*

Collection.—M. Cotteau, a single specimen.

STOMECHINUS POLYPORUS, *Agassiz.* Syn. *Echinus polyporus,* Agassiz. Catalogue syst., p. 12.

Test hemispherical, circular depressed; ambulacra with four rows of tubercles; inter-ambulacra with close-set tubercles at the base and circumference, which rapidly diminish on the upper surface. It is distinguished from *Stomechinus bigranularis* by having four rows of tubercles in the ambulacra.

Formation.—Bathonien de Ranville? Rare.

Collection.—M. Michelin.

STOMECHINUS CAUMONTI, *Desor.* Synopsis des Échinides Fossiles, p. 128.

Test sub-conoidal, inflated at the sides; ambulacra with four rows of tubercles, the two internal rows short and less regular; inter-ambulacra with from eight to ten rows of tubercles at the circumference, which diminish little in size at the upper surface; mouth opening very large. " This species is found to be one of the most characteristic of the Bathonien stage." *Desor.*

Formation.—" Marnes à *Ostrea acuminata* (Vesulien) de Herznach (Argovie). Kellovien de Châtillon-sur-Seine."

Collections.—" Hébert, Mæsch, Mus. Zurich. Assez abondante." *Desor.*

STOMECHINUS APERTUS, *Desor.* Synopsis des Échinides Fossiles, p. 127.

Test sub-conical, pentagonal; ambulacra with two rows of tubercles, a little less in size, but more numerous than in the inter-ambulacra, which have six rows of tubercles at the circumference; mouth opening very large. "This species was formerly confounded with *Echinus excavatus*, Goldf., which has the ambulacral tubercles much smaller, and set more close together than those of the inter-ambulacra, whilst in *Stomechinus apertus* their difference in size is scarcely perceptible." *Desor.*

Formation.—Kellovien de marolles près Mamers, Courgains, Nantua. Rare.

Collections.—M. Michelin, d'Orbigny, Paris Museum.

STOMECHINUS ROBINALDINUS, *Cotteau.* Syn. *Echinus Robinaldinus*, Cotteau. Études sur les Échinides Fossiles, pl. 22, figs. 1—6.

Test sub-conoidal, inflated at the sides; circumference sub-circular; ambulacra with four regular rows of tubercles, identical with those of the inter-ambulacra; poriferous zones wide, trigeminal ranks very oblique; inter-ambulacra double the width of the ambulacra, and covered throughout with equal-sized tubercles, regularly disposed in numerous series, six tubercles on each plate; apical disc composed of equal-sized genital plates; vent large; mouth opening moderate in width, in the proportion of one to two and a half; peristome decagonal, with wide, obtuse notches.

Dimensions.—Height, one inch and a half; transverse diameter, two inches and one fifth.

Formation.—" Cette espèce caractérise les couches supérieures de l'étage Corallien et n'a jamais été rencontrée dans le Coral-rag inférieur de Châtel-Censoir et de Druyes." *Cotteau.*

Collection.—M. Robineau-Desvoidy. Very rare; only three specimens known.

STOMECHINUS ORBIGNYANUS, *Cotteau.* Syn. *Echinus Orbignyanus*, Cotteau. Études sur les Échinides Fossiles, pl. 21, figs. 8—13.

Test hemispherical, depressed; circumference circular; ambulacra with two marginal rows of tubercles, and a naked zone between : inter-ambulacra with six rows of tubercles at the equator, diminishing to two rows above; base concave; mouth opening wide, half the diameter of the test; peristome pentagonal, deeply notched.

Dimensions.—Height, six tenths of an inch; transverse diameter, one inch and one twentieth.

Formation.—"Calcaires marneux et lithographiques de Commissey Corallien étage." *Cotteau.*

Collection.—M. Rathier. Very rare.

STOMECHINUS EXCAVATUS, *Goldfuss.* Syn. *Echinus excavatus*, Goldfuss. Petrefacta. Germaniæ, tabl. 40, fig. 12.

Test hemispherical, depressed, sub-pentagonal; ambulacra with two marginal rows of close-set tubercles, smaller than those of the other areas; inter-ambulacra with two rows of primary tubercles on the sides, and four other secondary rows at the base.

Dimensions.—Height, eleven twentieths of an inch; transverse diameter, one inch.

Formation.—"Findet sich in den obersten Schichten des Jurakalkes bei Regensburg und in Schwaben." *Goldfuss.*

Collection.—Munich Museum. Very rare.

STOMECHINUS SERIALIS, *Agassiz.* Syn. *Echinus serialis*, Agassiz. Échinod. Foss. Suisse, II, tabl. 22, figs. 10—12.

Test circular, much depressed; ambulacra with two regular rows of tubercles; poriferous zones very narrow; inter-ambulacra with six rows of tubercles at the circumference, diminishing to two rows above; mouth opening very large; peristome decagonal, with obtuse notches.

Dimensions.—Height, four tenths of an inch; transverse diameter, three quarters of an inch.

Formation.—"Corallien infér. (Terrain à Chailles) du Fringeli (Jura Soleurois)." *Desor.*

Collection.—M. Gressly. Very rare.

STOMECHINUS LINEATUS, *Goldfuss.*   Syn. *Echinus lineatus*, Goldfuss.   Petrefacta Germaniæ, tabl. 40, fig. 11.

Test hemispherical, depressed, circular, or slightly sub-pentagonal; ambulacra with two rows of tubercles on the margins, and two small rudimentary rows internal to them; inter-ambulacra with six rows of tubercles at the circumference, diminishing to two rows above; tubercles of both areas prominent, and nearly of the same size; bosses encircled with granules; the general surface is very tuberculous.

Dimensions.—Height, one inch; transverse diameter, one inch and three quarters.

Formation.—From the Coral Rag of Regensburg and Basel. The specimen kindly sent me by M. Michelin is from the Coral Rag of Niederdoff, Canton de Bâle.

Collections.—Museums of Tübingen, Besançon, Munich, collection of M. Michelin. Royal College of Surgeons (Hunterian Collection), my cabinet.

STOMECHINUS PERLATUS, *Desmarest.*   Syn. *Echinus perlatus*, Agassiz.   Échinoderm. Foss. Suisse, part ii, pl. 22, figs. 13—15.
Knorr.   Petrefact., ii, tab. E, figs. 1, 2.

Test conoidal, sub-circular, or sub-pentagonal; the transverse and bucco-anal diameters are sometimes nearly equal; ambulacra with two complete rows of tubercles on the margins, two incomplete rows within, and numerous small granules encircling their bosses; inter-ambulacra with ten rows of small tubercles at the circumference, two rows of which are larger, and extend from the peristome to the disc, the others disappear on the upper surface; the bosses of all the tubercles are surrounded by fine granules, which imparts a highly sculptured appearance to the test of this species; poriferous zones narrow, trigeminal ranks, form angles of 50° to 55°; apical disc moderate, ovarial plates large; mouth opening very large, half the diameter of the test; peristome pentagonal, the ten notches wide and obtuse, with small lobes between.

Dimensions.—Specimen figured by M. Agassiz: height, one inch and three quarters; transverse diameter, two inches and four tenths.
Specimen in my cabinet: height, one inch; transverse diameter, one inch and nine tenths.

Formation.—Corallien infér. (Terrain à Chailles) de la Combe d'Échert (du Val de Moutiers), de Salins. Corallien étage Commercey (Meurthe), Châtel-Censoir, et Druyes (Yonne).

Collections.—Museums Neuchâtel, Bâle, Porrentruy. Collection of M. Michelin. British Museum, my cabinet.

STOMECHINUS SEMIPLACENTA, *Desor*.　Cotteau, Études sur les Échinides Foss., pl. 45, fig. 5.

Test sub-circular, sub-inflated above, concave below; ambulacra with four rows of tubercles; inter-ambulacra with ten rows of tubercles at the circumference, irregularly disposed, diminishing on the upper surface; tubercles of both areas nearly equal-sized, and each surrounded by circles of granules; mouth opening very large; peristome decagonal, notches very wide and deep.

Dimensions.—Height, nine tenths of an inch; transverse diameter, one inch and seven tenths.

Formation.—" Kimméridge étage des environs de Chablis et Havre." *Cotteau.*

Collections.—MM. Rathier, Royer.

Family 5—SALENIADÆ.

This natural family nearly corresponds to the Salénies of MM. Agassiz and Desor, and is distinguished from other families of the Echinoidea endocyclica by the peculiar structure and great development of the apical disc, which, besides the five genital and five ocular plates, has an additional or sur-anal plate, developed in the centre of the disc, immediately before the anal opening; this plate is sometimes single, or more frequently is composed of from three to eight separate elements.

The test is thin, and in general small, spheroidal, hemispherical, or depressed: the ambulacral areas are always narrow, straight, or flexuous, with two rows of small tubercles which alternate with each other on the margins of the area.

The poriferous zones are narrow, the pores unigeminal, except near the peristome, where they fall into oblique ranks of threes.

The inter-ambulacral areas are wide, with two rows of primary tubercles, which have large bosses with crenulated summits; in one section the tubercles are perforated, in the other they are imperforated.

The mouth opening differs in size in the different genera; the peristome is more or less decagonal, and is sometimes deeply notched, or only feebly indented. The jaws are known in one genus, in which they resemble those of *Hemicidaris.*

The spines of one section (the *Acrosalenia*) are only known; in this genus, the stems are long, slender, angular, or flattened, and the surface, although apparently smooth, is covered with very fine longitudinal lines.

From a misconception of the true relative position of the elements of the apical disc, in this family, much confusion exists in the works of different authors in the description of this part of the test. "The great difficulty in the study of this group," says M. Desor,* "is to find the place of the madreporiform body; we are consequently embarrassed when we attempt to assign the lateral parts to the longitudinal axis of these animals; unless we admit that the sur-anal replaces the madreporiform body; but this would be contrary to all analogy, because in all the other Cidarides, the madreporiform body is an integral part of one of the genital plates. M. Agassiz had got rid of the difficulty by means of an hypothesis, by admitting that the sur-anal plate is invariably placed in the plane of the animal, that it therefore could only be anterior or posterior; hence his two divisions in the genus *Salenia,*—the first with a sur-anal plate posterior, and, consequently, with the *périprocte* excentral and before; the second with the sur-anal plate anterior, and, consequently, with the *périprocte* excentral and behind."†

* 'Synopsis des Échinides Fossiles,' p. 138.

† For further details on this subject, M. Agassiz's 'Monographies d'Échinodermes, première Monographie des Salénies,' may be consulted.

Professor Johannes Müller assigns the left posterior genital plate as the bearer of the madreporiform body in *Salenia personata*. "Dies wird auch durch die Salenien bestätigt, wo die Längsachse durch die plaque suranale vor dem After bestimmt wird. An einem im mineralogischen Museum aufbewahrten ausgezeichnet schönen Exemplar der *Salenia personata*, Ag., mit vorderem After, Taf. 1, Fig. 9, ist die linke hintere Genitalplatte porös und Madreporenplatte."[*]

I have selected fine specimens of *Hyposalenia Wrightii*, Desor, from the Lower Green Sand, *Salenia petalifera*, Desmarest, from the Upper Green Sand, and *Salenia Austeni*, Forbes, from the Lower Chalk, in all of which the madreporiform body occupies the surface of the right anterior genital plate, as in the CIDARIDÆ, HEMICIDARIDÆ, DIADEMADÆ, and ECHINIDÆ. The sur-anal plate is central, and the anal opening posterior, and inclined to the right side. In fact, the madreporiform body and sand canal, whatever their true functions may be, have the same position in all the Echinoidea, recent and fossil, which I have examined; and probably the same in all the Echinodermata. Professor Müller's mistake, therefore, may have arisen from his placing the *Salenia* in a false position before him.

The study of the apical disc in the *Acrosalenia* reveals the true relation of its elements to each other, and proves that the sur-anal plate has nothing in common with the spongy madreporiform body which occupies the surface of the right anterior genital plate (Pl. XV, fig. 4 *a, i*). I had the good fortune to make this discovery some years[†] ago, when figuring and describing *Acrosalenia hemicidaroides*, which urchin has furnished the key to the true relation of the bilateral parts to the longitudinal axis of the SALENIADÆ. On this point M. Desor observes:

"Nous devons en outre à M. Wright une autre découverte plus importante, celle du corps madréporiforme, qui fait partie intégrante de l'une des plaques génitales comme dans les autres Cidarides. Or comme nous savons maintenant que cette plaque a une position fixe dans tous les oursins, nous sommes par là même en mesure de déterminer l'avant et l'arrière de ces animaux; et puisque les plaques sur-anales sont situées en arrière de cette plaque, il s'ensuit que le périprocte se trouve réellement refoulé en arrière. Il ne peut dès-lors plus être question d'Acrosalénies à périprocte eccentrique en avant, comme on supposait que c'était le cas de l'*Acrosalenia tuberculosa*," &c.

"Il n'arrive que trop souvent que le disque apicial manque, et dans ce cas, il est très-difficile de distinguer les Acrosalénies du genre Hemipedina décrit ci-dessus. Cependant, comme par suite du refoulement du périprocte en arrière la plaque génitale impaire ou postérieure gagne plus que les autres sur le test, on peut encore, d'après M. Wright, reconnaître la place de cette plaque même dans les individus dépourvus d'appareil apicial."[‡]

[*] Joh. Müller, 'Über den Bau der Echinodermen,' p. 7.
[†] Wright, 'Annals and Magazine of Natural History,' 2d series, vol. viii, p. 261.
[‡] 'Synopsis des Échinides Fossiles,' p. 140.

A Table showing the Classification of the SALENIADÆ.

Family.	Sections.	Diagnosis.	Genera.
SALENIADÆ.	Section A. Tubercles perforated. Apical disc small.	Inter-ambulacral tubercles large; apical disc small and not prominent; sur-anal plate composed of one or many pieces; vent posterior and excentral.	Acrosalenia, *Agassiz*.
	Section B. Tubercles imperforate. Apical disc large, shield-like, and prominent.	Inter-ambulacral tubercles moderate; apical disc large and prominent; genital plates in the form of elongated lobes; sur-anal plate single; vent posterior and slightly excentral.	Peltastes, *Agassiz*.
		Inter-ambulacral tubercles large, few in number; apical disc large, and forming a regular pentagon, with elevated angular carinæ independent of the sutures; sur-anal plate angular; vent large, excentral, oblong, and posterior.	Goniophorus, *Agassiz*.
		Inter-ambulacral tubercles large; apical disc large, prominent, and shield-like; border undulated, with punctuations or sculpture in the lines of the sutures: sur-anal plate single, central; vent excentral, posterior, and inclined to the right side.	Hyposalenia, *Desor*.
		Inter-ambulacral tubercles very large; apical disc large and very solid, with an undulated circumference; surface of the large plates ornamented with punctuations, or sculptured figures along the line of the sutures; sur-anal plate single; vent excentral, and directed towards the right side.	Salenia, *Gray*.

229

Genus—ACROSALENIA. *Agassiz*, 1840.

This genus is composed of small or moderate-sized urchins with a thin spheroidal, hemispherical, or depressed test.

The ambulacral areas are narrow, straight, or slightly undulated, with two rows of small crenulated and perforated tubercles on their margins, which diminish gradually in size from the base to the apex.

The inter-ambulacral areas are wide, having two rows of primary perforated tubercles, raised on large prominent bosses with crenulated summits.

The apical disc is proportionally smaller than in other genera of the SALENIADÆ, and forms no prominence on the surface of the test; the sur-anal plate, sometimes single, is in general composed of many separate pieces placed before the anal opening, which renders it excentral and displaces the vent backwards; the anterior pair of genital plates are larger than the posterior pair, and the single plate is small and crescentic; the spongy madreporiform body occupies the right anterior genital plate.

The mouth opening is large, often one half the diameter of the test; the peristome is decagonal, with ten wide notches which have the margin reflected over their border.

The primary spines are long, circular, flattened, or angular; and they are sometimes twice the length of the diameter of the body; although apparently smooth, still with a lens, their surface is seen to be covered with fine longitudinal lines. The secondary spines are short, regular, round, and striated.

The *Acrosalenias* are found in the different stages of the Oolitic rocks from the Lias to the Portland, but they are most numerous in the lower division of the Oolites; recent researches have shown that this genus contains a much larger number of species than was originally supposed, when many of them were erroneously referred to other genera, in consequence of the apical disc being absent in a very great number of specimens.

When the apical disc is absent, an *Acrosalenia* may be easily mistaken for a *Hemicidaris*, but the aperture in *Acrosalenia* is always larger, and one of its angles projects further into the single inter-ambulacrum than into the others; the posterior pair of ambulacra are more curved backwards than the anterior pair, and the single anterior area is always straight; it requires a considerable practice of the eye and the handling of many specimens, before we can determine accurately by these characters; but practice has convinced me of their value, and I have now no difficulty in distinguishing an *Acrosalenia*, whether the disc be present or not. The absence of true semi-tubercles from the base of the ambulacral areas, likewise assists in the diagnosis.

A. *Species from the Lias.*

ACROSALENIA MINUTA, *Buckman.*   Pl. XV, fig. 3 *a, b, c ;* Pl. XVII, fig. 2 *a, b, c, d, e.*

ECHINUS MINUTUS.	Buckman, in Murchison's Geology of Cheltenham, 2d ed., p. 95.
ACROSALENIA CRINIFERA.	Wright, Annals and Magazine of Natural History, 2d series, vol. xiii, p. 168, pl. 12, fig. 1.
— —	Forbes, in Morris's Catalogue of British Fossils, 2d edition. Additional species of Echinodermata.
— MINUTA.	Oppel, die Jura Formation Englands, Frankreichs, und des südwestlichen Deutschlands, p. 110.

Test circular, depressed ; ambulacral areas very narrow, with two rows of microscopic tubercles placed at some distance apart on the sides of the area, those on the right side alternating with those of the left ; inter-ambulacral areas with two rows of primary tubercles, from nine to ten in each row, so disposed that the test appears, from the narrowness of the ambulacra, to possess only ten rows of tubercles, nearly equidistant from each other ; spines long, numerous, and hair-like.

Dimensions.—Height, three twentieths of an inch ; transverse diameter, six twentieths of an inch.

Description.—This beautiful little urchin has been long known to our local geologists, as it was obtained in great numbers when cutting through the Oxynotus bed of the Lower Lias in the formation of the Birmingham and Bristol Railway ; it has often been a palæontological puzzle, for although a few specimens have been found in a tolerable state of preservation, still, for the most part, the test is so much injured by pyrites, that it requires a good lens, and much patient study, to make out the details of its structure.   I lately found a very good specimen in the Oxynotus shales near Lansdown, which forms the subject of fig. 2, Pl. XVII, and is the most perfect example I have seen.   In my 'Memoir on the Lias Echinodermata,' I figured and described this species as *Acrosalenia crinifera,* Quenst., but my friend Dr. Oppel, of Stuttgart, having kindly sent me the type of Quenstedt's species, I am satisfied, from the comparative shortness of the spines of our urchin, that it is distinct from that form, and have, therefore, restored its original specific name, *minuta.*

The test is nearly circular, and is more or less depressed ; the ambulacral areas are extremely narrow (Pl. XVII, fig. 2 *d*), with two rows of minute marginal tubercles, not much larger than the common granulation of the test ; these tubercles are placed in each row at some distance apart (fig. 2 *e*), and the tubercles of the one side alternate with those of the opposite (fig. 2 *b*); between these two rows of tubercles there is a narrow, zigzag line of granulations ; the tubercles are very uniform in size throughout the area, but to see them satisfactorily it is necessary to examine the test with a microscope under an inch object-glass.

The poriferous zones are extremely narrow; the pores are small, there being four pairs opposite each tubercular plate; the septa are slightly elevated, and form a microscopic moniliform line between the pores (fig. 2 e).

The inter-ambulacral areas are wide (fig. 2 b, c), with two rows of primary tubercles, from nine to ten in each row; they are situated near the zonal sides of the plates, and have a wide miliary zone between them; the bosses have deeply crenulated summits, and the tubercles are small and widely perforated; well-defined areolas encircle the bosses (fig. 2 e), which are confluent above and below; the miliary zone consists of an elevated band, composed of from four to six rows of small, unequal-sized granules (fig. 2 e), which extend from the peristome to the disc. When viewed with the naked eye, at the equator, this tiny urchin appears to possess only ten rows of tubercles, nearly equidistant from each other (fig. 2 a), but when examined with a microscope its true structure is seen; the extreme narrowness of the ambulacral areas, with their close, alternate rows of microscopic tubercles, and the width of the miliary zone, with its unequal-sized granules, alike contribute to make the deception almost complete (fig. 2 b).

The opening for the disc is nearly one half the diameter of the test, but I have never seen the trace of a plate in any specimen I have examined (fig. 2 c).

The mouth opening is small, about one third the diameter of the test, and lies in a concave depression (fig. 2 b); the peristome is feebly notched, and the lobes are of unequal size (fig. 2 b).

The most remarkable portions of the structure of this tiny fossil are the spines, which, in some crushed specimens, are preserved *in situ* between the laminæ of the Lias shales; they are long, slender, and hair-like, with a well-developed head; on some slabs these spines resemble so many fine bristles, laid down in all directions, upon the surface of the laminated Oxynotus shales; in one crushed test, four tenths of an inch in diameter, the spines measured upwards of an inch in length.

Affinities and differences.—The smallness of the test, and the length and hair-like character of the spines, are sufficient to distinguish *Acrosalenia minuta* from all other Lower Lias urchins. In the shales of the Upper Lias, there is another small urchin with long, hair-like spines; but I have never yet succeeded in obtaining a specimen in sufficient preservation to enable me to institute a comparison between it and *Acrosalenia minuta;* the Upper Lias form reminds me very much of *Cidarites crinifera*, Quenstedt, from the " Posidonienschiefer von Pliensbach bei Boll in Würtemberg," a bed of the Upper Lias.

Locality and Stratigraphical position.—*Acrosalenia minuta* was found in the laminated Oxynotus shales of the Lower Lias at Lansdown, Cheltenham, and likewise at Gloucester, whilst excavating the same bed to form a new dock. It was associated in both places with

Ammonites oxynotus, Quenstedt, and *Ammonites raricostatus,* Zieten; these Ammonites characterise the horizon of this species in a very definite manner. I have lately received the spines of this urchin, which were collected from the Lower Lias near Stratford-on-Avon.

Dr. Oppel states that *Acrosalenia minuta* is found in the Lower Lias of Württemberg. "Die flachgedrückten Körper mit den feinen Stacheln füllen eine ganze Schichte in der Region des *Pentacrinus tuberculatus.* Sie liegen häufig verkiest in den bituminösen Schiefern des untern Lias an der Steinlach bei Dusslingen, und wurden von meinem Freund Dr. Rolle zuerst darin aufgefunden."*

History.—First described by Professor Buckman, in the ' Geology of Cheltenham,' as *Echinus minutus.* Afterwards, in my ' Memoir on Lias Echinodermata,' it was figured for the first time, and described as *Acrosalenia crinifera,* Quenst.; a comparison of our species with the German urchin has induced me to restore its original specific name. It has been subsequently found by Dr. Rolle in the *Tuberculatus* bed at Steinlach.

B. *Species from the Inferior Oolite.*—10ᵉ *Étage Bajocien,* D'Orbigny.

ACROSALENIA LYCETTII, *Wright.* Plate XVI, fig. 1 *a, b, c, d, e, f.*

ACROSALENIA LYCETTII. Wright, Annals and Magazine of Natural History, 2d ser., vol. viii, p. 263, pl. 11, fig. 2
　—　　　—　　Forbes, in Morris's Catalogue of British Fossils, 2d ed., p. 70.
　—　　　—　　Desor, Synopsis des Échinides Fossiles, p. 142.

Test hemispherical, much depressed on the upper surface, and flat at the base; circumference circular or sub-pentagonal; ambulacral areas narrow, with two rows of prominent tubercles on the margins; inter-ambulacral areas wide, the two rows of tubercles with very large, prominent bosses; miliary zone with two rows of granules at the equator, and a naked median depression between the three upper pairs of tubercles; apical disc small and prominent; sur-anal plate single; vent small, and transversely oblong.

Dimensions.—Height, half an inch; transverse diameter one inch.

Description.—This urchin at first sight so much resembles a *Hemicidaris* that it might readily be mistaken for one, in consequence of the size of the ambulacral, and the prominence of the bosses of the inter-ambulacral tubercles; but a more careful study of the test soon discloses its true generic character.

A transmutationist will doubtless find it a difficult matter to discover the progenitor of this urchin; it makes its appearance in the lowest beds of the Inferior Oolite, with all its Acrosalenian characters so strongly developed, that it may be taken as a good type of the

* Dr. Oppel, die Jura Formation, p. 110.

genus. This is not the only example among fossil Echinodermata illustrative of the fact, that the first created forms of new types of life are often the most characteristic of the group they represent; the figure and description of the first species of every genus contained in this Monograph is, in fact, a practical commentary on this great natural law.

The ambulacral areas are narrow, straight, and prominent; they have two rows of small, well-developed tubercles, from twelve to fourteen in each row, disposed on the margins of the areas, which gradually diminish in size from the base to the apex (fig. 1 *a*, *b*); a zigzag line of small granules descends down the centre, sending lateral branches between every two tubercles; this granular network encircles three parts of the areolas, and leaves them open only to the poriferous zones (fig. 1 *d*); the tubercles at the base of the area (fig. 1 *b*, *e*) are large, and remind us of the semi-tubercles in this region in *Hemicidaris*.

The poriferous zones are narrow; the pores are unigeminal, except at the base, where they fall into oblique ranks of threes (fig. 1 *e*); there are eleven pairs of pores opposite two large plates (fig. 1 *d*).

The inter-ambulacral areas are three times and a half as wide as the ambulacral; there are eight tubercles in each row, which, at the circumference, are raised on very large prominent bosses (fig. 1 *a*, *c*); one of these mammæ, drawn in profile (fig. 1 *f*), shows its conical form; it is surrounded by a wide, oval areola (fig. 1 *d*), which is confluent with its fellow above and below; the tubercles are large at the base, but suddenly diminish in size in the upper part of the area; there are ten crenulations on the summits of the bosses, and the tubercles are small in proportion to the magnitude of these eminences; the two rows of tubercles are separated by a zigzag granular band, or miliary zone, in the centre of the area (fig. 1 *c*), consisting of two rows of granules, and a few additional smaller ones dotted here and there in the interspaces; similar crescentic bands of granules separate the areolas from the poriferous zones; each plate, therefore (fig. 1 *d*), has a semi-circular row of granules, with a few granulets, on its central side, and a similar row on its zonal side, whilst the upper and lower borders are destitute of sculpture. On the upper part of the area the granules disappear, and leave a small, naked, median depression between the three uppermost pairs of small tubercles (fig. 1 *a*).

The base is flat, or slightly concave (fig. 1 *c*); the mouth opening is large, one half the diameter of the test; the peristome is decagonal, and divided by wide notches into unequal-sized lobes, the ambulacral being one half larger than the inter-ambulacral lobes (fig. 1 *b*), and its margin is folded over at the angles of the notches (fig. 1 *c*).

The small apical disc is very seldom preserved; fortunately I lately found one specimen with most of the plates *in situ* (fig. 1 *a*), and another with the sur-anal plate, which is single, or more probably composed of three plates soldered together; it is the only specimen I have seen with this portion of the disc, the coarse character of the Pea-grit in which they are found being unfavorable for the preservation of this fragile and complex part of the shell.

Affinities and differences.—This species resembles *Acrosalenia hemicidaroides*, Wright, but is distinguished from it by having the test more depressed, the ambulacral tubercles larger and more prominent, the three upper pairs of the primary inter-ambulacral tubercles smaller, with a naked median depression in the centre of the upper part of the miliary zone; the apical disc is likewise smaller. The size and prominence of the ambulacral tubercles form a distinctive character between *Acrosalenia Lycettii* and *Acrosalenia pustulata* on the one side, and *Acrosalenia Wiltonii* on the other, both of which have small and widely spaced-out tubercles in the ambulacral areas.

Locality and Stratigraphical position.—I have collected this urchin from the lower ferruginous beds of the Inferior Oolite, the Pea grit, at Crickley, Leckhampton, and Cooper's Hills; from a sandy seam of the lower beds of the Inferior Oolite at Stinchcombe; and likewise from the lower beds of the Inferior Oolite at Brockhampton, Cleeve, and Sudely Hills. It occurs likewise in the shelly beds of the Roe-stone at Leckhampton, and has been catalogued in the Rev. P. B. Brodie's paper* as *Acrosalenia Hoffmani*, Römer. Although an abundant species, it is seldom found in good preservation, the shell being always more or less crushed; it is commonly associated with *Pseudodiadema depressum*, Agass., which it somewhat resembles, but is easily distinguished from that form by the disproportion between the size of the ambulacral and inter-ambulacral tubercles.

History.—It was first figured and described in my 'Memoir on the Cidaridæ of the Oolites,' and is dedicated to my friend John Lycett, Esq., one of the learned authors of the 'Monograph on the Mollusca of the Great Oolite,' which has already enriched the volumes of the Palæontographical Society.

C. *Species from the Great Oolite, Bradford Clay, Forest Marble, and Cornbrash.*—
11e *Étage Bathonien*, D'Orbigny.

ACROSALENIA HEMICIDAROIDES, *Wright*. Pl. XV, fig. 4 *a, b, c, d, e, f, g, h, l, m*.

ACROSALENIA HEMICIDAROIDES.	Wright, Annals and Magazine of Natural History, 2d ser., vol. viii, p. 161, pl. 11, fig. 1.
— —	Forbes, Memoirs of the Geological Survey, Decade IV, pl. 2.
— —	Forbes, in Morris's Catalogue of British Fossils, 2d edit., p. 70.
— —	Desor, Synopsis des Échinides Fossiles, p. 144.

* 'Quarterly Journal of the Geological Society,' vol. vi, p. 247.

Test hemispherical, more or less depressed, sometimes elevated ; ambulacral areas narrow, undulated, with two rows of small, perforated tubercles, fourteen to sixteen in each row, which alternate on the borders of the area, and gradually diminish in size from the base to the apex ; inter-ambulacral areas with large prominent tubercles on the sides, and small tubercles near the disc ; areolas sub-confluent ; miliary zone narrow, with two rows of granules ; apical disc very large ; sur-anal plate composed of six elements ; mouth large, decagonal ; peristome deeply notched ; primary spines twice or more in length the diameter of the test; stem sub-angular, tapering, or slightly compressed ; extremity sometimes bifid or trifid.

Dimensions.—This species varies so much in size and figure, that I have selected four specimens on account of their differences, the comparative dimensions of which are shown in the following table :

Acrosalenia hemicidaroides. Wright.	Largest form.	Elevated form.	Depressed form.	Common small form.
	Inch.	Inch.	Inch.	Inch.
Transverse diameter of test	$1\frac{4}{10}$	$1\frac{3}{20}$	$1\frac{1}{10}$	$0\frac{6}{10}$
Height of test	$0\frac{9}{10}$	$0\frac{7}{10}$	$0\frac{6}{10}$	$0\frac{1}{3}$
Diameter of mouth opening	$0\frac{7}{10}$	$0\frac{11}{20}$	$0\frac{1}{2}$	$0\frac{1}{3}$
Length of apical disc	$0\frac{5}{10}$	$0\frac{9}{20}$	$0\frac{4}{10}$	$0\frac{3}{10}$

Description.—This is the most common and best preserved of all our fossil sea-urchins ; it has long been known as a Cornbrash species, but was neither named, figured, nor described, until I gave its history, with figures and details, in my ' Memoir on the Cidaridæ of the Oolites ;' since then it has been figured and described by Professor Forbes in the fourth Decade of the ' Memoirs of the Geological Survey ;' and by M. Desor, in tabl. XX, figs. 19—23, of his valuable ' Synopsis des Échinides Fossiles.' This species exhibits much variation in size and figure, but its diagnostic characters are preserved with remarkable uniformity throughout these different phases of form and magnitude.

The test is sometimes elevated and globular, like a *Hemicidaris;* indeed, the large forms, the dimensions of which are given in the first column of the table of measurements, are commonly so named ; the absence of semi-tubercles at the base of the ambulacra, and the large size of the elongated discal opening, are the only characters by which they can be distinguished from that genus. In the more common form (fig. 4 *c*), the body is spheroidal and depressed on the upper surface ; and, when the flattening is excessive, it produces the depressed form of the third column.

The ambulacral areas are narrow and moderately prominent (fig. 4 *a*); they are nearly of a uniform width, gradually expanding in the lower half, and tapering in the upper; they exhibit a very slight undulating contour, and have two rows of small, nearly equal-sized secondary tubercles, from sixteen to eighteen in each row (fig. 4 *c*); the tubercles at the basal angle are the largest (fig. 4 *b, e*), and they imperceptibly diminish in size from the circumference to the disc (fig. 4 *a*), they are all perforated and raised on bosses, which have ten crenulations on their summits (fig. 4 *d*); the tubercles of each row alternate, and in the centre of the area there are one or two rows of granules, which send off small lateral branches to encircle the areolas (fig. 4 *d, e*).

The inter-ambulacral areas are nearly four times as wide as the ambulacral; they consist of two columns, each composed of seven or eight plates, each plate bearing a primary tubercle (fig. 4 *d*); the two plates nearest the peristome are very small (fig. 4 *b*), the four on the sides are very large (fig. 4 *c*), and the two near the disc are small (fig. 4 *a*); the tubercles are elevated on very large bosses (fig. 4 *f*), which have ten or more deep crenulations on their summits, the tubercles themselves are perforated; around the base of the boss there is a wide, smooth areola; the plates are bordered by a single row of granules (fig. 4 *d*), which, however, is sometimes absent from the upper and lower borders, the areolas then being confluent; the miliary zone is narrow, and is formed of two rows of granules, with the intermediate angles having a few smaller additional granules introduced; the areolas are separated from the poriferous zones by a row of granules on the zonal side of the plates (fig. 4 *d*).

The poriferous zones are narrow, the pores unigeminal throughout, except at the base, where they fall into triple oblique rows; the septa form small elevations on the surface, and a beaded line thereby passes down the zone between the pores forming a pair (fig. 4 *d*), and there are nine or ten pairs of pores opposite each tubercular plate.

The apical disc is often admirably preserved in this beautiful urchin; the study of its curious structure first enabled me to correct M. Agassiz's erroneous supposition, that the genital plate, which carries the madreporiform body, is the single plate, and represents the posterior side of the animal, instead of the right antero-lateral plate which occupies the same relative position in all the ECHINOIDEA.

The disc is one third the width of the test, and in consequence of the projection of the genital plates has a pentagonal form; it is slightly convex and prominent, the anterior and posterior pair of genital plates are nearly of the same size (fig. 4 *a, i*), the right anterior is the largest, and carries on its front part the madreporiform body; the small crescentic-shaped single plate forms the posterior boundary of the vent (fig. 4 *i*); the oviductal holes are all near the apices; the sur-anal plate is composed of six or seven pieces arranged like mosaic before the anal opening; the ocular plates are small and firmly wedged between the genital and sur-anal elements. All the plates of the apical disc have numerous granules scattered on their surface.

The mouth opening is wide, nearly one half the diameter of the test; the peristome is decagonal; the ambulacral being larger than the inter-ambulacral lobes, the ten deep notches, with reflected edges, indent the bases of the inter-ambulacra, and extend as far as the areolas of the second or third tubercles.

The primary spines are finely preserved *in situ* in fig. 4 *k*; they are variable in size in the same and in different specimens, and are proportionally shorter in young than in old urchins; sometimes they are three times the length of the diameter of the body (fig. 4 *b*), and are sometimes nearly three inches and a half in length; the head is conical, with a truncated extremity marked by deep crenulations (fig. 4 *m*), the ring is prominent, and the milling is angular and sharp; the stem (fig. 4 *b*) swells out beyond the ring, it is more or less irregularly sub-angular, with the angles rounded; a transverse section of one of the spines exhibits an irregularly elliptical figure; sometimes the spine tapers to a conical point, or the distal end terminates in a bifid or trifid extremity. The secondary spines articulating with the ambulacral tubercles (fig. 4 *n*) are short, about three tenths of an inch in length, they are round, and taper gently from the ring to the point, their surface being covered with fine longitudinal lines (fig. 4 *n*).

The jaws are preserved in one or two specimens (fig. 4 *g*); the lantern is strong (fig. 4 *h*), the teeth conical, and in its general appearance the dental apparatus closely resembles that of an *Echinus*.

Affinities and differences.—This urchin very much resembles a *Hemicidaris*; in fact, *Acrosalenia* and *Hemicidaris* have so many characters in common, which are always well preserved, and so few that are diagnostic, and which for the most part are either broken or absent, that it requires considerable practice, when the apical disc is wanting, to determine the genus; the absence of semi-tubercles at the base of the ambulacra and the magnitude and pentagonal form of the discal opening serve as good guides to the genus. *Acrosalenia hemicidaroides* is distinguished from *A. Lycettii* by having a larger and more spheroidal test, larger bosses on the sides, and smaller ambulacral tubercles; the apical disc is larger, and the sur-anal plate is composed of a greater number of pieces. *Acrosalenia hemicidaroides* is distinguished from *A. pustulata* in having regular rows of secondary tubercles in the ambulacral areas, which are larger and more fully developed; the miliary zone is narrower, with only two rows of granules; the primary tubercles are more developed at the upper surface, and the apical disc is likewise larger.

The same characters distinguish it from *A. Wiltonii*; but besides those already enumerated, that species has four rows of fine granules in the miliary zone, smaller tubercles, suddenly diminishing at the upper surface, and a much smaller mouth opening. The complex character of the sur-anal plate, the size of the test and of its primary tubercles, readily separate it from *A. spinosa*.

Locality and Stratigraphical position.—The finest specimens of this urchin with their spines attached (fig. 4 *k*) have been collected by Mr. William Buy, from the Forest Marble, near Malmesbury, in a thin vein of clay, which, according to that acute and careful collector, separates the Cornbrash from the Forest Marble. The beauty of these specimens, and the admirable state of preservation in which they are found, forms one of the marvels of the Oolitic fauna; a great quantity of fine tests are collected from the gray brashy beds of the Cornbrash near Chippenham. Mr. Bristow obtained it from the Cornbrash near Wincanton; Mr. Pratt from the Forest Marble at Hinton Abbey; Mr. Hull found very large specimens in a cream-coloured, calcareous, semi-indurated, marly bed of the Great Oolite near Burford, in a quarry not far from the Bird-in-hand Inn. I discovered two specimens in the Great Oolite limestone of Minchinhampton Common. The specimens said to have been found in the Inferior Oolite have been ascertained to be *A. Lycettii*. The Rev. A. W. Griesbach collected quinquefid spines of this urchin in the Great Oolite at Wollaston, Northamptonshire, along with *Acrosalenia pustulata*. This species, therefore, belongs to the Bathonian stage, and is a very characteristic urchin of this great zone of life.

ACROSALENIA SPINOSA, *Agassiz.*   Pl. XVII, fig. 3 *a, b, c, d, e, f.*

ACROSALENIA SPINOSA.		Agassiz, Catalogus systematicus Ectyporum Echinodermatum fossilium, p. 9, 1840.
—	LÆVIS.	Agassiz, Catalogus systematicus Ectyporum Echinodermatum fossilium, p. 9, 1840.
—	SPINOSA.	Agassiz, Échinodermes Fossiles de la Suisse, ii, tab. 18, figs. 1—5, p. 39.
—	—	Cotteau, Études sur les Échinides Fossiles, pl. 3, figs. 6—11, p. 58.
—	—	Agassiz et Desor, Catalogue raisonné, Annales des Sciences Naturelles, 3ᵐᵉ série, tome vi, p. 343.
—	—	Forbes, in Morris's Catalogue of British Fossils, p. 70.
—	—	Wright, Annals and Magazine of Natural History, 2d series, vol. viii, pl. 12, fig. 3, p. 265.
—	—	Bronn, Lethæa Geognostica, dritte Auflage, Band ii, tabl. xvii, fig. 7, p. 144.
—	—	Desor, Synopsis des Échinides Fossiles, tabl. 20, figs. 14—16.
—	—	D'Orbigny, Prodrome de Paléontologie, tome i, 11ᵐᵉ étage, Bathonien, p. 320, No. 417.
—	RADIATA.	Forbes, Memoirs of the Geological Survey, IV Decade. Note on new species of British Acrosalenias.

Test sub-pentagonal, depressed; ambulacra straight, prominent, with two marginal rows of small spaced-out tubercles ten to twelve in each row; inter-ambulacra with eight to nine primary tubercles in each row, large at the circumference, gradually diminishing towards the peristome, suddenly so on the upper surface; miliary zone narrow at the circumference, wide and naked in the middle at the upper surface; apical disc large, sur-anal plate single;

vent large, excentral; mouth large; peristome decagonal, lobes nearly equal; notches slightly incised.

Dimensions.—Height, nearly four tenths of an inch; transverse diameter, seven tenths of an inch.

Description.—This beautiful little urchin is so admirably preserved in the Cornbrash of Wiltshire, that it forms one of the most characteristic fossils of that formation; the test is small, sometimes circular, oftener sub-pentagonal, and always much depressed (fig. 3 *c*).

The ambulacral areas are narrow and prominent (fig. 3 *a*, *b*); as the poriferous zones are wide, for so small an urchin, the ambulacra appear isolated, and the upper surface of the test has a radiated appearance, which is increased by the size and smoothness of the apical disc (fig. 3 *a*); the ambulacra converge in straight lines from the base to the disc; they are furnished with two rows of small tubercles, which, although microscopic in size, have crenulated bosses and perforated summits; the tubercles are disposed with great regularity on the borders of the areas, at a moderate distance apart (fig. 3 *c*, *d*), so that there are only from ten to twelve in each row; in the middle of the area a double row of granules separates the tubercles, and lateral rows encircle the areolas, which are only open on the sides (fig. 3 *d*). The poriferous zones are wide and straight; the pores are large, and disposed obliquely in single pairs (fig. 3 *d*), forming a rectilinear file on each side of the ambulacral segments; there are from five to six pairs of pores opposite each large tubercular plate.

The inter-ambulacral areas are three times the width of the ambulacral; the two rows of tubercles, about eight in each row, are nearly equidistant throughout (fig. 3 *a*); the two basal pairs are small (fig. 3 *b*), the three equatorial pairs large (fig. 3 *c*), and the three upper pairs suddenly diminish in size and become dwarfed as they approach the disc (fig. 3 *a*, *c*); two or three pairs of tubercles at the equator have very large bosses (fig. 3 *e*), whilst all the others are of moderate size; they have narrow elliptical areolas, of which the larger are confluent (fig. 3 *d*); the miliary zone consists of two rows of granules, which form scrobicular circles around the areolas (fig. 3 *d*); as the rows diverge above, a small triangular space on the upper part of the segment is left naked (fig. 3 *a*); the areolas are likewise separated from the poriferous zones by a single row of granules (fig. 3 *d*).

The apical disc is large and pentagonal, about two fifths the diameter of the test (fig. 3 *a*); the anterior and posterior pair of genital plates are nearly alike in size and form (fig. 3 *f*); the oviductal holes are perforated near the external third of the plates; the posterior single plate projects a little further into its corresponding segment than the others, it is largely excavated for the vent, of which it forms the posterior wall (fig. 3 *f*); the sur-anal plate is central, single, and pentagonal, and forms the anterior wall of the vent (fig. 3 *f*), its sides being formed by the posterior pair of ocular plates, which are much larger than the three others (fig. 3 *f*) ; the madreporiform body is very small, and occupies as usual the right anterior genital plate; a few small granules form central clusters on the pairs of genitals, and similar granules dot the surface of the small oculars (fig. 3 *f*).

The mouth opening is nearly half the diameter of the test (fig. 3 *b*) ; the peristome is decagonal, and divided by small notches into nearly equal-sized lobes; the incisions are not deep, and the margin is reflected over all the angles (fig. 3 *f*).

Affinities and differences.—This species is distinguished by its small size and depressed test, the extreme regularity of its tubercles, the radiated appearance of the upper surface, the apical disc having the sur-anal plate single, and the vent of a triangular form ; these characters serve to distinguish it from all its other English congeners. It resembles *A. aspera* in many of its general characters, but that species has undulated ambulacral areas with close-set tubercles and the sur-anal plate composed of two pieces; it is so entirely distinct from all other forms of the genus at present known, that it is unnecessary to pursue a comparison with them.

Locality and Stratigraphical position.—I have collected two specimens of this urchin in the Pea grit, Inferior Oolite, at Crickley Hill, and in the yellow clay resting on the Stonesfield Slate at Sevenhampton, with *Anabacia orbulites, Pecten vagans, Ostrea acuminata,* and other Great Oolite fossils. I have likewise found it in the Great Oolite at Sham Castle, near Bath. Mr. Lycett has collected it from the Great Oolite of Minchinhampton Common, where it is small in size and rare. It is abundant in the Forest Marble and Cornbrash near Chippenham, with *Avicula echinata,* Sow., whence my finest specimens were collected by Mr. William Buy. It is found, likewise, in the Forest Marble and Cornbrash near Cirencester ; indeed, it may be considered an abundant fossil in the English Cornbrash, but I have seen no specimens at all equal in their preservation and beauty to those found in Wiltshire.

In Switzerland it was collected by M. Strohmeyer from the marls containing *Ostrea acuminata,* in the canton of Soleure. In France it has been found in the Great Oolite of Ranville, Calvados, by Professor Deslongchamps, and in the upper beds of the Bathonien étage in the environs of Châtel-Censoir, by M. Cotteau.

Acrosalenia Loweana, *Wright,* nov. sp. Pl. XVII, fig. 4 *a, b, c, d.*

Test circular, much depressed ; ambulacral areas narrow and sinuous, with marginal rows of very small wide-set tubercles ; inter-ambulacral areas with six primary tubercles in each of the two rows, of which the two equatorial pairs are very large ; apical disc of moderate size, sur-anal plate composed of three pieces ; primary tubercles near the disc very small.

Dimensions.—Height, seven twentieths of an inch; transverse diameter, three quarters of an inch.

Description.—This small depressed urchin was kindly communicated to me by my friend Mr. S. P. Woodward, and is the only specimen of the form which has come under my notice. The test is circular, much depressed, and well characterised by the disproportionate magnitude of the equatorial tubercles, when compared with those on the upper and under surfaces of the test.

The ambulacral areas are narrow, and slightly undulated; they have two rows of very small tubercles on the margins, from twelve to fourteen in each row, placed wide apart from each other, with a few granules in the interspace (fig. 4 *c*); the poriferous zones are narrow and undulated, winding round the bulging areolas of the large equatorial tubercles; there are from eight to nine pairs of pores opposite the large tubercular plates (fig. 4 *c*), and the septa form mammillated elevations on the surface.

The inter-ambulacral areas are more than four times as wide as the ambulacral (fig. 4 *a*); they have only six plates in each column (fig. 4 *b*), each plate bearing a primary tubercle; the two equatorial plates support tubercles with very large prominent bosses (fig. 4 *b*), which are likewise surrounded by wide, smooth areolas; on three parts of the margin of these equatorial plates there is a continuous series of small granules, which are absent, however, from the lower border (fig. 4 *c*); on the smaller of the larger plates the bosses become suddenly less, and between the areola and the centro-suture there are three rows of granules (fig. 4 *c*); the uppermost plate has a mere rudimentary tubercle on its surface, surrounded with numerous granules (fig. 4 *a*); the basal tubercles likewise suddenly diminish in size towards the peristome; the zones expand in consequence, and the pairs of pores are disposed in oblique ranks of threes closely laid together.

The apical disc is of moderate size for the genus *Acrosalenia* (fig. 4 *a*); the four ovarial plates are shield-shaped, the anterior pair being a little larger than the posterior pair (fig. 4 *d*); the single ovarial plate is small and crescentic, the oviductal holes are perforated near their apices, and the small spongy madreporiform body occupies the centre of the right anterior (fig. 4 *d*); the sur-anal plate is composed of three pieces, one central pentagonal piece, and two small latero-posterior rhomboidal pieces, which form the anterior wall of the vent (fig. 4 *d*); the sides of this opening are formed by the posterior genital and ocular plates, and the posterior wall by the single crescentic genital (fig. 4 *d*); the three anterior ocular plates are heart-shaped, the posterior pair are rhomboidal, and the eye-holes are all marginal (fig. 4 *d*); all the plates are moderately large.

The base is flat, the mouth opening large, and the peristome decagonal; the notches are deep and have reflected margins; the opening is one half the diameter of the test, and the ambulacral are larger than the inter-ambulacral lobes.

Affinities and differences.—This urchin at first sight resembles some of the depressed forms of *Acrosalenia hemicidaroides* (Pl. XV, fig. 4), but from these it is distinguished by having fewer primary tubercles in each row, the two equatorial pairs are disproportionately larger than the others, and those on the upper surface are rudimentary; the apical disc likewise is smaller, and the sur-anal plate is composed of three pieces, whilst in *A. hemicidaroides* it has six elements.

It is distinguished from *Acrosalenia spinosa* in having fewer primary tubercles in the inter-ambulacral areas, in having the ambulacral areas sinuous, the sur-anal plate with three elements, *A. spinosa* having only one (Pl. XVII, fig. 3 *f*); it is so entirely distinct from other congeneric forms that it is unnecessary to make a comparison with them.

Locality and Stratigraphical position.—This urchin was collected from the Forest Marble at Malmesbury, and belongs to the British Museum; the honour of detecting the species is due to Mr. Woodward, who has kindly communicated it for this work.

I have much pleasure in associating the name of my friend Josiah Graham Lowe, Esq., of Kensington Park, with this Forest Marble species, as a tribute of gratitude for his kindness in presenting me with several rare and valuable specimens for this Monograph, and as an acknowledgment of the valuable service he has rendered to the palæontology of the Forest Marble and Cornbrash formations, so well developed in Wiltshire, in making the finest collection extant from these rich fossiliferous deposits

ACROSALENIA PUSTULATA, *Forbes*.   Pl. XVI, fig. 2 *a, b, c, d, e, f, g.*

ACROSALENIA PUSTULATA.	Forbes, Memoirs of the Geological Survey, Decade IV. Notes to pl. 3.
— —	Forbes, in Morris's Catalogue of British Fossils, 2d ed., p. 70.
— —	Desor, Synopsis des Échinides Fossiles, p. 143.

Test hemispherical, depressed; ambulacral areas narrow, with two rows of small tubercles disposed alternately, rather wide apart, on the sides of the area; inter-ambulacral tubercles large at the equator, gradually diminishing towards the peristome, suddenly diminishing on the upper surface; miliary zone with four rows of granules; apical disc small, sur-anal plate composed of many pieces; mouth opening large; peristome nearly equally decagonal, notches wide; spines long, slender, tapering; stem flattened, ring prominent, striæ wide apart.

Dimensions.—Height, half an inch; transverse diameter, one inch.

As the size of this urchin varies very much, I have selected four specimens, from different localities in Northamptonshire, and have given their measurements in the following table—

Acrosalenia pustulata, Forbes.	Height.	Diameter.	Mouth opening.	Apical disc.
	Inch.	Inch,	Inch.	Inch.
A. Large variety, Great Oolite	$\frac{8}{10}$	$1\frac{3}{10}$	$\frac{6}{10}$	$\frac{1}{2}$
B. Yardley ditto, Great Oolite	$\frac{6}{10}$	$1\frac{1}{10}$	$\frac{9}{20}$	$\frac{4}{10}$
C. Yardley ditto, Great Oolite	$\frac{1}{2}$	1	$\frac{4}{10}$	$\frac{7}{20}$
D. Oundle ditto, Forest Marble	$\frac{3}{10}$	$0\frac{11}{20}$	$\frac{1}{4}$	$\frac{1}{5}$

Description.—I have had much difficulty in making out the history of this species; and had it not been for the zeal, industry, and perseverance of my friend, the Rev. A. W. Griesbach, in obtaining, for the sake of comparison, a fine series of specimens from Oundle, Yardley, and Wollaston, Northamptonshire, I should have been unable to clear up the doubts. For several years I regarded the large, inflated forms of this urchin, with broad miliary zones, as granulated varieties of *Acrosalenia Wiltonii*; the only perceptible difference consisting in the size of the mouth opening, which is smaller in that species (Pl. XVI, fig. 3 *b*) than in *Acrosalenia pustulata* (Pl. XVI, fig. 2 *b*). The specimens which had been collected in Gloucestershire, Oxfordshire and Northamptonshire, were, on an average, about the size of the one figured in Pl. XVI, fig. 2, whilst Professor Forbes's type specimen was only half the size of our urchins; moreover, his diagnosis, always so correct in reference to the object he described, "areolis disjunctis, areâ centrali angustissimâ, bigranulatâ, granulis sparsis (diam. $\frac{1}{2}$ unc., alt. $\frac{1}{4}$ unc.)," did not agree with those I had hitherto collected, for their areolas were confluent, and the area centralis was broad, with four rows of granules. It occurred to me, therefore, that Professor Forbes's specimen, from the Forest Marble of Malmesbury, was possibly an immature form of the species. I, therefore, requested my friend, the Rev. A. W. Griesbach, to obtain specimens of the same age as those described by Professor Forbes, which he fortunately discovered, and these small urchins I found agreed very well with his diagnosis.

The ambulacral areas are narrow and straight, they have two rows of small tubercles, from eighteen to twenty in each row, disposed at some distance apart on the sides of the area, those at the base (fig. 2 *b*) are a little larger than those on the sides (fig. 2 *c*); a double series of granules takes the direction of the zigzag suture, and sends lateral branches between their small areolas, so that on the upper part of the area these granules are nearly as large as the small tubercles placed between them (fig. 2 *c, d*).

The poriferous zones are narrow, the pairs of pores are oblique, and the septa form moniliform elevations on the surface (fig. 2 *d*), which produce a bead-like line down the centre of the zone between the pores; and there are from eight to nine pairs of pores opposite each large tubercular plate (fig. 2 *d*).

The inter-ambulacral areas are nearly four times as wide as the ambulacral; they have eight plates in each column, of which the three equatorial pairs support large tubercles

32

(fig. 2 c); the three uppermost pairs of plates have mere rudimentary tubercles; the sudden diminution in size between the fifth and sixth forms a remarkable feature in the character of this species (fig. 2 a); at the base, the tubercles gradually diminish towards the peristome (fig. 2 b). In young shells, and in some adult ones, the miliary zone consists of two rows of granules; but, in other varieties, it is formed of four rows, closely set together, on the upper surface; the plates are covered with numerous equal-sized granules, among which the rudimentary tubercles are placed (fig. 2 a); the areolas of the three equatorial tubercles are circular and confluent, the uppermost has a distinct scrobicular circle of granules, but the others are confluent (fig. 2 d), and one row of granules separates the areolas from the poriferous zones; this description applies to the variety figured, but does not agree with others. The Rev. Mr. Griesbach, on this point, has made the following observations: "Among the very numerous specimens I found at Oundle, there is about an *equal* number of two dissimilarly constructed individuals. I pointed this out to you before (though you had seen it), and it is the same with the Wollaston specimens. When I found that age would not account for the difference, I was set to thinking about sex, making inquiries on this subject. I see from Professor Owen's lectures on the Invertebrate Animals, that the 'sexes are in distinct individuals in the Echinoids as in star-fishes;' and Professor Forbes, in Decade III, 'Mem. of the Geol. Survey,' accounts for the difference in form between some individuals of *Micraster coranguinum*, on the supposition that it may possibly be due to sex. This was all I wanted, to confirm an impression I had already entertained that the tall, ample form of this *Acrosalenia pustulata*, Forbes, with its broad granulated space between the primary tubercles, is the *female*, and the depressed one, with a narrow central area, the *male*. They are, without doubt, the same species; and I quite believe my sexual hypothesis to be true." As my excellent friend has carefully examined a greater number of specimens of this urchin than any other naturalist, it affords me much pleasure to record his explanation of a difficult problem, and to add that it affords a physiological reason for an admitted fact.

The apical disc is seldom preserved in this species (fig. 2 a); in one or two small specimens the genital and ocular plates are *in situ*, but only a portion of the sur-anal remains; the disc is pentagonal; the anterior and posterior pair of genital plates are shield-shaped; the right plate is the largest, and supports in its centre a spongy madrepori-form body; the single plate is crescentic, and its body is absent, to give place to the vent; the three anterior oculars are heart-shaped, and the two posterior pairs are elongated, and form the sides of the anal aperture; the sur-anal is composed of many pieces; the three anterior, however, are alone *in situ*—one central, and two latero-posterior plates, behind these there was evidently a series to form the anterior boundary of the vent. The surface of all the plates is covered with numerous granules, similar to those on other parts of the test.

The base is flat; the mouth opening is large, nearly, but not quite, one half the width of the equatorial diameter; the peristome is decagonal, with wide notches (fig. 2 b); the ambulacral, however, being larger than the inter-ambulacral lobes.

The beautiful specimen, on a slab of Great Oolite from Yardley (fig. 2 *f*), was collected by Mr. Griesbach, and presented to me for this work; it shows the spines *in situ*. The primary spines are long, slender, and smooth; in length, once and two thirds the diameter of the test; they have a small, conical head, with a prominent milled ring (fig. 2 *e*); the stem tapers gently to the point, a transverse section shows it is somewhat triangular, with flattened or rounded angles; the surface, although apparently smooth, is covered with microscopic longitudinal lines. The secondary spines are small, stout, dagger-like bodies (fig. 2 *g*), with a flattened stem, and covered with longitudinal lines (fig. 2*).

Affinities and differences.—This species closely resembles *Acrosalenia Wiltonii*, but is distinguished by the following characters: it has larger equatorial tubercles, with more prominent bosses and wider areolas; the three superior tubercles more suddenly diminish in size; the miliary zone, even in the large granulated varieties, is narrower, the mouth is larger, the notches are wider and deeper, the test is more depressed, and the sides less inflated; but the most marked character resides in the mouth opening which is small, being about two fifths the diameter in *Acrosalenia Wiltonii*, and nearly one half the diameter in *Acrosalenia pustulata*.

Locality and Stratigraphical position.—This species has been collected by the Rev. A. W. Griesbach, from the Forest Marble at Oundle; the Great Oolite at Yardley; and Wollaston; and also from Strixton, Wimmington, Blisworth, and Kingsthorp, Northamptonshire. It was collected from the Great Oolite, near Woodstock, by Mr. Gavey; and from near Kiddington, Oxon, by Mr. Dominicke Brown, who kindly sent me a fine slab, with fourteen urchins on its surface, which he obtained from a Great Oolite quarry. In describing this specimen, Mr. Brown observes: "On examining the quarry where the *Cidaris* (*Acrosalenia pustulata*) is found, I can see at once the reason why the shells are not in a better state of preservation. The *Cidarites* appear only in a thin layer of rock not more than a foot or two below the surface, then comes a thick bed of marl, and below this the solid rock in which I occasionally find spines of *Cidaris*, but not often." This species has been found by Mr. Bravender, in the Great Oolite and Bradford Clay near Cirencester, and I have collected two specimens from the Forest Marble near Naunton Downs, which are both depressed and highly granulated varieties; I have seen separate portions of the test on slabs of the Forest Marble near Upper Cubberley, Gloucestershire, and in the Bradford Clay near the Tetbury-road Station, Great Western Railway, but I have never seen even a fragment of a specimen in the Great Oolite, of Minchinhampton.

ACROSALENIA WILTONII, *Wright.*   Pl. XVI, fig. 3 *a, b, c, d, e.*

ACROSALENIA WILTONII.   Wright, Annals and Magazine of Natural History, 2d series, vol. ix, p. 83, pl. 3, fig. 4 *a—e.*
—   —   Forbes, in Morris's Catalogue of British Fossils, 2d ed., p. 70.
—   LAMARCKII. Desor, Synopsis des Échinides Fossiles, p. 141.

Test hemispherical, depressed, sometimes elevated; sides always tumid; ambulacral areas narrow, straight, with two rows of small perforated tubercles on the margins, set wide apart, and a miliary zone of three rows of fine granules between the marginal tubercles; inter-ambulacral areas four times the width of the ambulacral; ten tubercles in each row—of these the three middle pairs only are developed, the basal are small, and those at the upper surface rudimentary; miliary zone wide, composed of four rows of small granules in the middle, and six rows above; apical disc convex and prominent; sur-anal plate formed of two large and five small pieces; basal angle obtuse, from the tumidity of the sides; base concave; mouth opening small; peristome unequally decagonal.

Dimensions.—Height, half an inch; transverse diameter, one inch.

Description.—This beautiful little urchin has almost always a more or less elevated hemispherical test, which is rarely as much depressed as in fig. 3 *c*; the sides are tumid, and the base is concave; the ambulacral areas preserve a very uniform width throughout; two rows of small perforated tubercles placed wide apart occupy the sides of the area (fig. 3 *d*), with from eighteen to twenty in each row; the eight basal pairs are larger, and the lateral and dorsal pairs smaller, they are arranged closer together, and gradually diminish in size until they become quite microscopic (fig. 3 *d*); two or three rows of fine granules (fig. 3 *d*) form a miliary zone, which sends lateral branches to divide the small marginal tubercles.

The poriferous zones are narrow and undulated, the pores are unigeminal, except near the base, where they are trigeminal (fig. 3 *b*); opposite each of the large lateral plates there are eight pairs of pores (fig. 3 *d*).

The inter-ambulacral areas are four times as wide as the ambulacral; they have ten pairs of tubercles; the three basal pairs are small (fig. 3 *b*), gradually increasing in size from the peristome upwards, to blend in with the three lateral pairs, which are the largest (fig. 3 *c, d*); the eighth pair are suddenly smaller than the seventh pair, and the ninth and tenth pairs are quite rudimentary (fig. 3 *a*) : the miliary zone is wide (fig. 3 *c*), and occupied by four or five rows of small close-set granules on the sides, increasing to six or seven rows on the upper surface; between the areolas and the poriferous zones there are likewise two rows of granules : the bosses of the four middle pairs of tubercles are moderately large (fig. 3 *e*), they have small areolas, which are nearly divided from each other by horizontal lines of granules ; in some they are absent, and the areolas are then confluent.

The apical disc (fig. 3 *a*) is seldom preserved ; in one specimen however (Pl. XVII, fig. 5), it is seen to consist of an anterior and posterior pair of ovarial plates, and a single rudimentary ovarial ; the sur-anal plate is composed of two unequal-sized pentagonal pieces, united with the anterior and posterior ovarials, and six or seven small pieces which form an arch, extending from the right to the left posterior ovarials, and completing the anterior wall of the anus (Pl. XVII, fig. 5) ; the posterior pair of ocular plates form the lateral, and the single ovarial the posterior boundary of the vent, which is transversely oblong, slightly excentral (fig. 3 *a*), and projects into the single inter-ambulacrum. The ocular plates are heart-shaped and of moderate size, and the eye-holes are very minute ; the surface of all the discal elements is covered with small close-set granules (Pl. XVII, fig. 5).

The tumid sides are gently rounded towards the base, which is concave ; the mouth opening is small, being less than two fifths the diameter of the test ; the peristome is unequally decagonal, the ambulacral being one third larger than the interambulacral lobes.

The fragment of a primary spine shows that it was smooth and cylindrical, and, judging from its thickness, must have been long ; the secondary spines are short and prickle-shaped, and are sculptured with fine longitudinal lines.

Affinities and differences.—This species so very closely resembles *Diadema* (*Acrosalenia*) *Lamarckii*, Desmoulins, that at one time I thought them identical, and stated as much in a letter to M. Desor ; but from a more attentive study of the form, I now consider them distinct. *A. Wiltonii* has a more concave base, and a smaller mouth opening than any other form of the genus at present known. It differs from *A. pustulata* in having a wider miliary zone, a greater number of much smaller granules, a more concave base, and a smaller mouth opening ; the same group of characters serve to distinguish it from *A. hemicidaroides* and *A. Lycettii*.

Locality and Stratigraphical position.—This urchin was collected many years ago by Dr. William Smith, from the Cornbrash of Wiltshire. My type specimens were found by Mr. William Buy, in the Cornbrash, near Sutton-Benger, Wilts, where it appears to be very rare, and is known as " the small-mouthed Cidaris." Mr. Bravender obtained it from the Bradford Clay, near Cirencester.

I dedicate this species to my esteemed friend, John Wilton, Esq., of Gloucester, with whom I have spent many hpapy days exploring the Natural History of different parts of Gloucestershire.

ACROSALENIA HEMICIDAROIDES, JUNIOR? *Wright.*   Pl. XVI, fig. 4 *a, b.*

The small urchin, with spines attached, which forms the subject of fig. 4 *a, b,* was collected by Mr. William Buy, from the same bed of clay, between the Forest Marble and Cornbrash, from whence *Acrosalenia hemicidaroides* (Pl. XV, fig. 4) was obtained. At first, this specimen was thought to be distinct from that species, in consequence of the spines being more cylindrical; but as the test agrees with undeniable tests of *A. hemicidaroides,* about the same size, I am disposed to consider the difference in the form of the spines as depending on age.   It may, however, hereafter be found to be a distinct species, although in the mean time, until better evidence is obtained, I regard it as the young of *A. hemicidaroides,* the materials at my disposal not enabling me to draw up a diagnosis. Fig. 4 *a* shows the upper surface of the test, with the primary spines *in situ,* the longest of which is once and two thirds the length of the diameter of the test; the head and milled ring agree in form with the same parts of the spine in *A. hemicidaroides*; the stem swells out a little in the middle, and tapers gently to the point; the sides are slightly flattened (fig. 4 *b*), and the surface is covered with fine, longitudinal lines: the secondary spines are short, stout, dagger-like bodies, on the surface of which the longitudinal lines are more distinctly marked, and the milled ring is proportionally more prominent than in the primaries.

Locality and Stratigraphical position.—From the band of clay between the Forest Marble and Cornbrash, near Malmesbury, where it was associated with *A. hemicidaroides* and numerous stems and fragments of *Pentacrinus.*

ACROSALENIA RADIATA, *Forbes,* nov. sp.

" A. ambulacris angustis, tuberculis parvis seriebus, duobus approximatis alternatis subdivergentibus dispositis; inter-ambulacrorum tuberculis numerosis regulariter graduatis, superne decrescentibus centro-lateralibus mediocribus, areolis disjunctis; areâ centrali angustâ pauci granulatâ, granulis sparsis.

" Diameter, seven twelfths of an inch; altitude, three tenths of an inch.

" It has affinities with *A. spinosa,* but differs in having the primary tubercles regularly diminishing instead of suddenly decreasing above." (Mus. Pract. Geol.)

Locality.—Collected by Mr. Lycett in the Great Oolite of Minchinhampton.

I have made the above extract entire from Professor Forbes's note on undescribed species of *Acrosalenia*, Decade IV, pl. 3, of the 'Memoirs of the Geological Survey.' After a careful comparison of the type specimen in the Museum of Practical Geology, in Jermyn Street, with a large series of *Acrosalenia spinosa*, I have in my collection, from the Great Oolite of Minchinhampton, and of Ranville, Calvados, and from the Cornbrash of Wiltshire, I can detect no persistent specific character in *A. radiata*; the upper inter-ambulacral tubercles are larger than in many specimens, but in others, the same gradual diminution observed in *A. radiata* prevails in a large series; in fact, the links between a sudden diminution and a gradual diminution are abundantly supplied; for these reasons I consider *A. radiata*, Forbes, as a synonym of *A. spinosa*; the structure of the apical disc, and of its single sur-anal plate, are the same in both.

ACROSALENIA ASPERA, *Agassiz.*

Professor M'Coy has entered this species in his list of Mesozoic Radiata,* contained in the Geological Collection of the University of Cambridge, giving as the localities of the species "Great Oolite, Minchinhampton, Inferior Oolite, Dundry." Having seen this specimen, I can state that I believe it to be a small *A. hemicidaroides* from Minchinhampton. The original *Acrosalenia aspera*, Agass., was found by M. Gressly in the Kimmeridge Clay of Banné, near Porrentruy, only a single specimen of which is known. I have never seen an urchin from Minchinhampton, nor Dundry, which resembled this Swiss species.

D. *Species from the Coral Rag.*

ACROSALENIA DECORATA, *Haime.*   Pl. XVII, fig. 1 *a, b, c, d, e, f, g, h, m.*

MILNIA DECORATA.	Haime, Annales des Sciences Naturelles, 3me série, tom. xii, Zoologie, 1849, pl. 2, figs. 1—3, p. 217.
ACROSALENIA DECORATA.	Wright, Annals and Magazine of Natural History, 2d series, vol. ix, p. 81, 1851.
— —	Forbes, Memoirs of the Geological Survey, Decade IV, pl. 3, 1852.
— —	Cotteau, Notes sur les Échinides de l'étage Kimméridgien du dép. l'Aube, Bulletin Soc. Géol. de France, 2d série, tom. xi, p. 355, 1854.
— —	Forbes, in Morris's Catalogue of British Fossils, 2d edit., p. 69, 1854.
— —	Desor, Synopsis des Échinides Fossiles, p. 143, 1856.
— —	Cotteau, Études sur les Échinides Fossiles, p. 322, 1856.

* 'Annals and Magazine of Natural History,' 2d series, vol. ii, p. 419.

Test sub-pentagonal, depressed on the upper surface, concave at the base; ambulacral areas convex and prominent, the anterior and posterior pairs slightly sinuous, with two rows of small marginal tubercles, and the intermediate space filled with very small close-set granules; inter-ambulacral areas with from eight to ten tubercles in each row, the four equatorial tubercles very large, those on the upper surface very small; apical disc large, sur-anal plate composed of eight pieces; vent elongated and extremely excentral; base concave; mouth large; peristome equally decagonal; primary spines long, smooth, slender, and tapering; secondary spines small, hair-like.

Dimensions.—Height, four tenths of an inch; transverse diameter, nine tenths of an inch.

Description.—This elegant little urchin is remarkable among its congeners for its pentagonal form, arising from the flatness of the inter-ambulacral areas and the convexity and prominence of the ambulacral, and likewise for exhibiting in a most remarkable manner the bilateral symmetry of the Saleniadæ (fig. 1 *a*).

The ambulacral areas are one fourth the width of the inter-ambulacral; the anterior single area is quite straight, and the anterior and posterior pairs are sometimes slightly curved (fig. 1 *a*); the apices of the anterior pair curve gently backwards, and those of the posterior pair upwards and inwards (fig. 1 *a*); two rows of small secondary tubercles, from twenty to twenty-four in each row, occupy alternately the margins of the area (fig. 1 *d*), those at the base and circumference being much larger (fig. 1 *e, h*) than those on the upper surface (fig. 1 *d*); the intermediate space is occupied with four rows of very small close-set granules (fig. 1 *d, e*); both the tubercles and granules are extremely regular in their size and arrangement throughout the areas.

The poriferous zones are narrow and depressed, which increases the prominence of the ambulacra; the pores are unigeminal throughout, except just at the base, where they fall into indistinct ranks of threes (fig. 1 *b, h*); there are from eight to nine pairs of pores opposite each large tubercular plate (fig. 1 *e*); and the septa are slightly elevated on the surface.

The inter-ambulacral areas are four times as wide as the ambulacral (fig. 1 *c, e*), they are so much flattened that they form nearly straight lines at the circumference; each segment is composed of two rows of primary tubercles, about eleven in each row (fig. 1 *c*), which are unequally developed in different regions of the area; the four basal pairs (fig. 1 *b*) are small, the three equatorial pairs large (fig. 1 *c*), and the four dorsal pairs dwarfed and rudimentary (fig. 1 *d, e*); from the first to the eighth pair, the areolas are oval and confluent (fig. 1 *e*), whilst those on the upper surface are surrounded by clusters of granules (fig. 1 *d*); the miliary zone is broad, and composed at the equator of six rows of small close-set granules (fig. 1 *c*); on the upper surface the granules cover all the surface of the plates, the eight rudimentary tubercles appearing as only larger granules in their midst (fig. 1 *c, d*);

between the areolas and the poriferous zones there is another band of granulations, composed of three rows of larger granules, among which a few secondary tubercles rise at intervals (fig. 1 *e*), so that this species has wider granulated inter-tubercular bands than any other ACROSALENIA.

The large apical disc is oblong (fig. 1 *a*), with the vent excentrical, and placed so far back that it encroaches considerably on the single inter-ambulacrum (fig. 1 *g*) ; the two anterior pairs of genital plates are the largest (fig. 1 *g*), the posterior pair are smaller, and the single plate is extremely elongated ; the body is absent, and the plate is represented by a thickened semi-lunar border with a reflected margin (fig. 1 *g*) ; this plate extends far down the single segment, and appears as if excavated for the passage of the vent (fig. 1 *a*) ; the ovarials are shield-shaped, and have the perforation for the ducts near their apex. The ocular plates (fig. 1 *g*) are small heart-shaped bodies, wedged between the genitals and the apices of the ambulacral areas, the three anterior ones are symmetrical, and the two posterior unsymmetrical, and slightly produced backwards ; the eye-holes are lodged in the marginal sinus ; the sur-anal plate is large (fig. 1 *f*), and composed of several pieces ; namely, one central pentagonal plate, two smaller lateral rhomboidal plates behind the first, another pentagonal posterior to these two, and four smaller lateral pieces, making in all eight plates (fig. 1 *f*) ; these five small plates form the anterior wall of the opening for the vent, which occupies nearly the entire area of the single ovarial plate (fig. 1 *g*) ; the surface of the discal elements is covered with the same-sized delicate granules (fig. 1 *f*) as those which adorn the miliary zones (fig. 1 *d*).

The base is concave, the mouth opening large, one half the diameter of the body ; the sides are tumid and rounded, so that the mouth appears to lie in a deep depression (fig. 1 *b*) ; the peristome is decagonal and nearly equal-lobed, the notches are wide and have the margin reflected all round the rim (fig. 1 *h*) ; the equatorial tubercles are extremely prominent, and their bosses are conoidal, with deeply crenulated summits (fig. 1 *i*).

The primary spines (fig. 1 *h*) are seen *in situ* in the specimens on the slab (fig. 1 *k*), some of them are one and a half the diameter of the test ; they are sub-cylindrical, subulate, and, to the naked eye, appear smooth, but with the lens are seen to be finely striated ; they are often slightly curved near their bases, and have their milled rings set obliquely ; the head is small and conical, and the milling of the ring not well marked (fig. 1 *l*) ; a transverse section shows that the stem is slightly flattened (fig. 1 *l*).

The secondary spines (fig. 1 *m*) are very small, rather stouter in proportion to their length, and more distinctly striated than the primary spines (fig. 1 *m*) : the head is larger, and the milled ring more prominent.

Affinities and differences.—This species differs so much from all its congeners, that it cannot be mistaken for either of them. In its general character it approaches *Acrosalenia spinosa*, but is distinguished from that species by its pentagonal form, the smallness of the superior inter-ambulacral tubercles, the concavity of the base, the width of the notches, the equal size of the peristomal lobes, the magnitude of the apical disc, the compound character of the sur-anal plate, and the great excentricity of the vent; the fineness and abundance of the granulation which adorns the test fully entitles it to the name *decorata*.

Locality and Stratigraphical position.—I have collected this beautiful Acrosalenia from the seams of yellow clay which traverse the Coralline Oolite near Calne, Wilts, and from the limestone of the same locality associated with *Hemicidaris intermedia, Pseudodiadema mamillanum,* and *Echinobrissus scutatus.* The specimens in the Museum of Practical Geology were collected by the officers of the Geological Survey, from the Coral Rag at Steeple Ashton, Wilts, and near Abbotsbury Castle, Dorset; some of the specimens in the British Museum were obtained from the Coralline Oolite at Malton, Yorkshire.

It has been recently collected by M. Cotteau, "dans les couches Kimméridgiennes inférieures des environs de Bar-sur-Aube."*

History.—This species was first figured and described in the twelfth volume of the ' Annales des Sciences Naturelles,' by M. Jules Haime, as a remarkable urchin, " which he had seen in the collection of the British Museum, on the supposition that it exhibited characters not met with in any known genus of sea-urchins, and that it combined the anal arrangements of the CASSIDULIDÆ with the usual characters of the CIDARIDÆ, an union of structures not hitherto observed." For this apparent anomaly, he proposed the genus *Milnia,* which he considered as the type of a new family, designated by him PSEUDOCIDARIDES, and mistaking Malton, the locality from which this type specimen came, for Malta, he considered it as probably a tertiary species. Mr. Woodward, who was previously aware that the urchin thus erroneously described was an *Acrosalenia,* directed Professor Forbes's attention to the type specimen which was selected as the subject for plate 3, Decade IV, of the 'Memoirs of the Geological Survey,' where it is beautifully figured, with full details of structure. My lamented friend, knowing that I had collected this urchin in an excursion I made into Wiltshire, and being informed that I was engaged in writing a description of it in my memoirs, kindly communicated a proof impression of that plate to me : finding that it was deficient both in the anatomy of the apical disc and in the structure of the spines, I supplied Professor Forbes with the materials for both, which were then incorporated in his plate. It was fully described in my 'Memoirs on the Cidaridæ of the Oolites,' and afterwards by Professor Forbes in his description of plate 3, Decade IV, of the 'Memoirs of the Geological Survey of Great Britain.' As a foreign species, it was first found by M. Cotteau, and recorded in his 'Note sur les Échinides de l'étage Kimméridgien. du dép. l'Aube,' and afterwards was described in his valuable work on Fossil Echinoderms.†

* 'Bulletin Soc.Géol. de France,' 2ᵉ série, tom. xi, p. 355, 1854.
† Cotteau, ' Études sur les Échinides Fossiles du département de l'Yonne.'

NOTES

On Foreign Jurassic species of the genus ACROSALENIA nearly allied to British forms, but which have not yet been found in the English Oolites.

Acrosalenia Lamarckii, *Desmoulins*. Syn. *Diadema Lamarckii*, Desmoulins, Tabl. Synon., p. 316.
Hypodiadema Lamarckii, Desor, Synop. Échin. Foss., tabl. 10, figs. 1—5.

Test small, inflated at the sides; ambulacra straight, with two marginal rows of small, distant, marginal tubercles, and a miliary zone of fine granulations between the rows; inter-ambulacral tubercles large at the equator, the three uppermost pairs rudimentary; miliary zone wide, composed of six rows of granules at the equator, which cover the whole surface of the plates on the upper surface; mouth opening small, base flat, peristome pentagonal, apical disc convex, sur-anal composed of several (six to eight) pieces.

Dimensions.—Height, half an inch; transverse diameter, nine tenths of an inch. Spines, according to M. Desor, " en forme de petits bâtons cylindriques, d'apparence lisse, à bouton haut."

Formation.—Cornbrash (Bathonien) de Marquise; abundant.

Collections.—In all English and Foreign public collections. My cabinet.

Acrosalenia Bouchardii, *Desor*. Synopsis des Échinides Fossiles, p. 142.

Test globular, depressed on the upper and under surfaces; ambulacra narrow, with two marginal rows of small tubercles, and a double row of granules between them; inter-ambulacral tubercles, nine in each row, diminishing gradually in size from the equator towards the disc and peristome; miliary zone moderate in width, and composed of four rows of granules; mouth opening large, peristome decagonal; discal opening large.

Dimensions.—Height, eight tenths of an inch; transverse diameter, one inch and four tenths.

Formation.—Cornbrash (Bathonien) Marquise.

Collections.—British Museum, Neuchâtel Museum, M. Bouchard, M. Michelin, my cabinet. M. Desor states that this species is often ticketed in Continental collections by mistake, *Hemicidaris Luciensis,* d'Orbigny; the latter is altogether a different urchin.

ACROSALENIA GRANULATA, *Merian.* Syn. *Hemicidaris granulata*, Merian, in Agassiz's Catalogue raisonné, Annal. Sc. Naturelles, tome vi, p. 339, 3me série.

A small, depressed species. The ambulacral tubercles are so small that they resemble miliary granules; the tubercles of the inter-ambulacral areas are disposed at some distance apart.

Formation.—" Grande Oolite (Vesulien) de Gensingen (Argovie), avec les *Dysaster analis, Holectypus depressus,*" &c. *Desor.*

Collections.—Museum Bâle, Collection M. Mæsch. Very rare.

ACROSALENIA ELEGANS, *Desor.* Synopsis des Échinides Fossiles, p. 143.

This species, according to M. Desor, resembles the preceding. It is larger, but has a small mouth opening, and the ambulacral tubercles are excessively small.

Formation.—" Grande Oolite (Vesulien) du Kornberg (Argovie), avec la précédente.

" *Collection.*—Mæsch." *Desor.*

ACROSALENIA LENS, *Desor.* Synopsis des Échinides Fossiles, p. 143.

According to M. Desor, this is the smallest species of the genus. It is about the size of a lentil, and is well characterised by its close-set inter-ambulacral tubercles, of which there are from nine to ten in a row.

Formation.—" Grande Oolite (Vesulien) de la cluse de Pfeffingen près Bâle.

" *Collection.*—Museum Bâle. Exemplaire unique." *Desor.*

ACROSALENIA RADIANS, *Agassiz.* Syn. *Hemicidaris radians*, Agassiz. Catalogue raisonné des Échinides, Annal. Sc. Naturelles, 3me série, tom. vi, p. 339.

Test hemispherical, depressed; ambulacra narrow, with two rows of small, prominent tubercles; inter-ambulacral tubercles, nine in a row, large at the equator, small above; miliary zone with two rows of granules in the middle; the median suture naked, and depressed at the upper surface; mouth opening large; peristome nearly equally decagonal, with wide notches; discal opening small.

Formation.—Kellovien de Vivoin, Courgains (Sarthe), Saint-Aubin (Calvados).

Collections.—MM. Michelin, Rouault; my cabinet. The type specimen kindly sent me by M. Michelin.

ACROSALENIA INTERPUNCTATA, *Quenstedt.* Handbuch der Petrefactenkunde, pl. 49, figs. 3 and 4.

Test small and much depressed; ambulacra very narrow; inter-ambulacra wide, with large, spaced-out tubercles; apical disc large; sur-anal plate single, smaller than the other plates; eight punctuations at the angles between the genitals, oculars, and sur-anal plate; five of these are at the inner angles of the eye-plates, and three around the circumference of the sur-anal plate; the five oviductal holes are very small.

Formation.—The White Jura ε of Nattheim = Coral Rag.

Collection.—Museum of Tübingen.

ACROSALENIA KŒNIGII, *Desmoulins.* Syn. *Diadema Kœnigii*, Desmoulins. Études sur les Échinides, Tabl. Synon., p. 312, No. 10.

Hemicidaris Königii, Agassiz. Catalog. raisonné des Échinides, Annal. Sc. Naturelles, 3ᵐᵉ série, tome vi, p. 337.

Hemicidaris Boloniensis, Cotteau, in Desor's Synopsis des Échinides Fossiles, p. 53.

Test hemispherical, base flat, sides inflated; ambulacra narrow, straight, with two rows of small, marginal tubercles, and a double row of granules within; inter-ambulacra more than three times as wide as the ambulacra, with two rows of large primary tubercles, eight to nine in each row; the bosses of the tubercles large, with wide, oval areolas; miliary zone composed of two or three rows of granules; apical disc small, the right anterior genital plate supports a large, prominent, spongy madreporiform body; single genital plate small and crescentic; sur-anal compound, number of pieces unknown; mouth opening large; peristome unequally decagonal; primary spines long, cylindrical, and tapering.

The discovery of a portion of the sur-anal plate, *in situ*, added to the excentral position of the vent, and the absence of semi-tubercles from the base of the ambulacra, are my reasons for removing this urchin from the genus *Hemicidaris*, and placing it in *Acrosalenia*. I have, at the same time, retained M. Desmoulin's specific name.

Dimensions.—Height, one inch (?); transverse diameter, one inch and three quarters. Mr. Davidson's specimen exceeds these dimensions.

Formation.—Étage Kimméridgien, Ningles, near Boulogne-sur-Mer.

Collections.—British Museum, MM. Michelin, Desmoulins, Cotteau, Bouchard-Chantereaux, Davidson, my cabinet. The two largest specimens I possess were collected, and kindly given me, by M. Bouchard. The species is rare, these two specimens being all that eminent palæontologist collected in thirty years. Mr. Davidson has figured, in his original manuscript plates, a very beautiful specimen contained in his collection, which has the spines *in situ*.

ACROSALENIA WOODWARDI, *Wright*, nov. sp.

Test spheroidal, depressed; ambulacral areas narrow, with two rows of small tubercles placed very obliquely in the upper part, a single row of granules between them; inter-ambulacral areas four times as wide as the ambulacral, with few tubercles in each row; bosses large and rather prominent; tubercles widely perforated; apical disc moderate, four and a half lines in diameter; genital plates unequal, anterior pair largest; ocular plates small; sur-anal plate composed of several pieces.

Primary spines three inches and a quarter long, and one line and a half thick, some three inches long, and less than one line in their greatest thickness; tapering, and sometimes forked, at the extremity; very finely striated, or granulo-striated on the surface.

Secondary spines three lines long, slender, cylindrical, pointed, and striated.

Dimensions.—Height (?); transverse diameter, thirteen lines.

Formation.—Cornbrash? British?

Collection.—British Museum.

ACROSALENIA HUNTERI, *Wright*, nov. sp.

Test conoidal; ambulacral areas narrow, with two rows of small tubercles; inter-ambulacra with two rows of moderate-sized tubercles; test very much elevated; apical disc absent; its form, and excentral position, with the absence of semi-tubercles, at the base of the ambulacra, indicate its acrosalenian character.

Formation.—Unknown. I found this urchin in the celebrated John Hunter's collection. Not having access at present to the specimen, I am unable to give a more correct diagnosis of its specific characters.

Collection.—Hunterian Museum, Royal College of Surgeons, London.

Family 6—ECHINOCONIDÆ.

GALÉRITES (pars), *Desor.* 1842.
ECHINONÉIDES (pars), *Agassiz* and *Desor.* 1847.

This natural family includes urchins which have the mouth opening central, or sub-central, and the vent excentral. Their test is thin, with a circular or sub-pentagonal circumference; the upper surface in general is elevated, and sometimes even conical; the ambulacral areas are simple and lanceolate; the poriferous zones extend without interruption from the mouth to the apical disc, and the pores are unigeminal, except near the mouth, where they lie in triple oblique pairs. The inter-ambulacral areas are wide, the tubercles are small and perforated; they are arranged with more or less regularity on the plates, and supported on bosses which have either smooth or crenulated summits. In some genera, they form regular vertical and concentric rows, and the inter-tubercular surface is covered with small granules, which form complete circles round the base of the tubercles.

The mouth opening is circular, central, or sub-central, and the peristome is more or less divided by notches into ten lobes. The organs of mastication consist of five jaws, which appear to resemble those of the ECHINIDÆ.

The anal opening is always large and excentral; it has an oblong or pyriform shape, and is either dorsal, marginal, infra-marginal, or basal, sometimes occupying the entire space between the mouth and the border.

The apical disc is mostly central and vertical, composed of five ovarial and five ocular plates; the right antero-lateral plate is very large, and extends backwards into the centre of the disc; it supports on its surface a prominent, spongy madreporiform body; the anterior and posterior pairs of ovarials are perforated, whilst the single posterior plate is imperforate; and the five small ocular plates are perforated near the margin.

The spines are small, short, conical appendages, with a smooth head, and having the stem covered with longitudinal microscopic lines.

I include in this extinct family the genera *Holectypus, Discoidea, Echinoconus, Pygaster,* and *Hyboclypus.* The *Holectypi* are found mostly in the Oolitic rocks, the *Discoideæ* and *Echinoconi* are true Cretaceous forms, the *Hyboclypi* are Oolitic, and the *Pygasters* are common to the Oolites and Chalk, although they had their greatest development during the Oolitic age. The oldest types are the *Pygasters* and *Hyboclypi,* which were created about the same time, at the commencement of the deposition of the basement beds of the Inferior Oolite.

Genus—HOLECTYPUS, *Desor*, 1847.

Discoides (pars), Klein. 1734.
Echinites (pars), Leske. 1778.
Galerites (pars), Lamarck. 1816.
Discoidea (pars), Gray. 1835.

The genus *Holectypus* was established by M. Desor for the reception of those *Discoideæ* which are deprived of ribs, or projecting processes, on the inner wall of the test; the species referred to this group constitute one of the oldest types of the Echinoconidæ, and are met with chiefly in the Oolitic rocks. They form, according to the views of the late Professor Forbes, " a section or sub-genus of the *Galerites*, more valuable on account of their palæontological merits, and limited distribution in time, being in the main characteristic of the Oolitic period, than for the zoological importance of the characters of their organization, which are rather transitional than distinctive."*

The test is thin, circular, or sub-circular, more or less hemispherical, conical or sub-conical, always tumid at the sides, and flat or concave at the base.

The ambulacral areas are narrow, straight, and lanceolate, with six or eight rows of small tubercles, of which the marginal rows only extend from the base to the apex.

The poriferous zones are narrow, and the pores are unigeminal throughout.

The inter-ambulacral areas are three times the width of the ambulacral, the large pentagonal plates support numerous small, perforated tubercles, which are very regularly arranged in vertical and concentric rows (Pl. XVIII, fig. 1 *d*). They are raised on bosses with crenulated summits, and surrounded by ring-like areolas; numerous minute granules are scattered over the surface of the plates, and form circles around the tubercles (Pl. XVIII, fig. 1 *e*).

The mouth opening is circular, and situated in the centre of the base; the peristome is divided by obtuse notches into ten equal lobes (fig. 1 *b*). The organs of mastication consisted of five jaws, which are preserved *in situ* in the specimen figured at 1 *g*.

The anal opening is large, inferior, infra-marginal, rarely marginal, sometimes occupying the entire space between the mouth and the border.

The apical disc is nearly central and vertical, composed of five ovarial and five ocular plates; the right antero-lateral ovarial is much the largest, and extends into the centre of the disc; it supports a prominent, convex, madreporiform body; the anterior and posterior pairs of ovarials are perforated, whilst the single plate is imperforate; the five ocular plates are small, triangular bodies, with marginal perforations (fig. 1 *i*).

* 'Memoirs of the Geological Survey,' Decade III, pl. 6, *Holectypus hemisphæricus.*

34

The internal moulds of *Holectypus* want those depressions occasioned by ribs projecting from the inner walls of the test, which so well characterise *Discoidea*.

The spines are short, with a smooth head, and milled ring; and they have the surface sculptured with fine, longitudinal lines.

The *Holectypi* are distinguished from the *Echinoconi* by having a larger mouth and vent, a concave base, and a less-elevated dorsal surface; and from the *Discoideæ* in having tumid sides, a larger mouth and vent, and the absence of ribs from the internal wall of the test.

The small crenulated tubercles, and basal vent, with the absence of any aperture in the upper surface of the inter-ambulacrum, distinguishes the *Holectypi* from the *Pygasters;* and the want of a longitudinal valley in the inter-ambulacrum separates them from the *Hyboclypi.*

The *Holectypi* are abundant chiefly in the Oolitic rocks; two of the species are found in the Neocomian and the Chalk. They make their first appearance in the Inferior Oolite, in the zone of the *Ammonites Parkinsoni*, Sow.

A. *Species from the Inferior Oolite.* = 10ᵉ *Étage Bajocien*, D'Orbigny.

HOLECTYPUS DEPRESSUS, *Leske.*   Pl. XVIII, fig. 1 *a, b, c, d, e, f, g, h, i.*

	Jacob. a Melle de Echinitis Wagricis, tab. 1, fig. 2, 1718.
	Kundman, Rariora Naturæ, tab. 5, fig. 12, 1737.
	Brückner, Merkwürdigkeiten der Landschaft. Basel, tab. 22, figs. G, H, 1748-63.
	Van Phelsum, Brief. de Gewelvslekken of Zee-egelen, p. 31, No. 16 (*Egelsteen tienband plattop*), 1774.
	Knorr, Petrefactions, vol. ii, tab. E, ii, figs. 6, 7, 1775-78.
	Favanne, pl. 67, figs. 1, 2.
ECHINITES DEPRESSUS.	Leske, Additamenta ad Kleinii Echinodermata, p. 164, pl. 41, figs. 5, 6, 1768.
ECHINUS DEPRESSUS.	Linné, Systema Naturæ, Gmelin, p. 3182.
GALERITES DEPRESSUS.	Lamarck, Animaux sans Vertèbres, tom. iii, p. 21, 1816.
— —	Deslongchamps, Enc., tom. ii, p. 432; Encyl. Méthod., tab. 152, figs. 7, 8.
— —	Defrance, Dict. des Sciences Naturelles, tome xviii, p. 86, 1820.
ECHINITES ORIFICIATUS.	Schlotheim, Jahrb., 1813; Petrefactenkunde, p. 317, 1822.
GALERITES DEPRESSUS.	Goldfuss, Petrefacta Germaniæ, Band i, p. 129, tabl. 41, fig. 3, 1826.
— —	Phillips, Geology of Yorkshire, tab. 7, fig. 4, 1829.
— —	De Blainville, Zoophytologie, p. 204, 1830.
— DEPRESSA.	Leonhard and Bronn, Jahrb., 1834, p. 135.
— —	Desmoulins, Études sur les Échinides, p. 254, No. 6, 1835.
— DEPRESSUS.	Grateloup, Mémoire sur les oursins Fossiles (Dax), p. 56, No. 9, 1836.

GALERITES RADIATUS.	Valenciennes, Encyclopéd. Methodique, tab. 153, figs. 1, 2 (Explication des Planches).
— DEPRESSA.	Koch and Dunker ? Norddeutschen Oolithgebildes, p. 40, tab. 4, fig. 2 a, b, 1837.
HOLECTYPUS DEPRESSUS.	Bronn and Römer, Lethæa Geognostica, 3^{tte} Auflage, Band ii, p. 148, tab. 17, fig. 5 a, b, 1851.
DISCOIDEA DEPRESSA.	Agassiz, Prodromus, p. 86, 1837.
— —	Agassiz, Échinodermes Foss. de la Suisse, part i, p. 88, tab. 13, figs. 7—13, 1839.
— —	Desor, Monographie des Galérites, p. 65, tab. 10, figs. 4—12, 1842.
HOLECTYPUS DEPRESSUS.	Desor, Catalogue raisonné, Ann. des Sc. Nat., 3me série, tom. vii, p. 145, 1847.
— —	Albin Gras, Oursins Fossiles du département de l'Isère, p. 41, 1848.
— —	Wright, Annals and Magazine of Natural History, 2d series, vol. ix, p. 94, 1851.
— —	Forbes, in Morris's Catalogue of British Fossils, 2d edit., p. 82, 1854.
	Desor, Synopsis des Échinides Fossiles, p. 169, 1857.

Test thin, hemispherical, sometimes conoidal, more or less depressed; circumference circular, or sub-pentagonal, and slightly contracted posteriorly; base flat, or a little concave; mouth central; peristome nearly equally decagonal, with reflected margin; anal opening large, pyriform, occupying nearly all the space between the mouth and the border; ambulacral areas lancet-shaped, with from six to eight rows of tubercles at the circumference, arranged in two oblique rows; inter-ambulacral areas with from sixteen to twenty rows of tubercles, which form a single row on the centro-sutural half of each plate, and two or three rows on the zonal half of the same; apical disc small, four genital plates nearly equal sized, the right antero-lateral large, with the madreporiform body projecting into the centre of the disc.

Dimensions.—Cornbrash specimen. Height, nine tenths of an inch; transverse diameter, two inches and two tenths of an inch; antero-posterior diameter, two inches and two tenths.

Inferior Oolite specimen. Height, seven tenths of an inch; transverse diameter, one inch and a quarter; antero-posterior diameter, one inch and seven tenths.

Description.—This is one of the most ubiquitous Echinites of the Oolitic rocks, and as its synonymy shows, has been long known to naturalists. In different formations it attains various degrees of development, being small in the Inferior Oolite, but large in the Cornbrash, as shown by the measurement of the specimens from these two stages.

The form is sub-conoidal, or more or less depressed; the circumference is circular, or sub-pentagonal, with the postero-lateral border a little compressed, and the single inter-ambulacrum slightly produced and truncated (fig. 1 *a, b*). The ambulacral areas are one third the width of the inter-ambulacral; they have a lanceolate form, and taper gradually from the border to the disc; at the circumference there are eight rows of tubercles, which are disposed so as to form double oblique

rows, thus— (fig. 1 *d*). In the upper part of the areas the inner

rows disappear, and the test is depressed in the line of the median suture (fig. 1 *a*).

The inter-ambulacral areas are three times as wide as the ambulacral, each plate supports a number of tubercles, one row represents the principal range, and extends from the peristome to the disc, and the tubercles of this row are a little larger than the others on the upper surface, but nearly of the same size on the sides and base. Between this primary row and the centro-suture, the tubercles are arranged in a single line on the same plane; but between it and the poriferous zones they form three super-imposed rows (fig. 1 *d*), so that the zonal side of the plates contains many more tubercles than the central half. The tubercles are surrounded by sunken areolas; they have crenulated bosses and perforated summits, and the intermediate surface of the plates is covered with a fine, abundant, miliary granulation. The tubercles at the base are much larger, they have deeper areolas, and are all disposed in single, concentric rows (fig. 1 *b*).

The poriferous zones are extremely narrow, and the pores are unigeminal throughout. The septa are elevated above the surface, and form a moniliform line between the pores (fig. 1 *d*), of which there are eight pairs opposite each inter-ambulacral plate.

The apical disc is well preserved in many of our specimens. Fig. 1 *i*, is an accurate drawing of this part (fig. 1 *a*). The anterior and posterior pairs of genital plates are perforated, but the single posterior plate is imperforate; the antero-lateral plate is very large, and extends into the centre of the disc; it supports a prominent, spongy, madreporiform body, which forms the summit of the test, and occupies the centre of the disc; not, however, as a new element introduced into the centre of the genital circle, but formed simply by the development of the right antero-lateral plate. The ocular plates are small, heart-shaped bodies, inserted between the genitals (fig. 1 *i*), with the eye-holes perforated near their margin (fig. 1 *i*); all the discal plates are covered with small tubercles, similar to those which form the miliary granulation. In M. Desor's 'Monograph on the Galerites,' tab. 10, fig. 7 *a*, the eye-holes are represented as marginal, and the madreporiform body is figured and described as distinct from the genital plates. It is probable that the imperfection of his specimens led the learned author into these errors, which we have now the pleasure of correcting by our better examples. The under surface is slightly concave, the mouth opening is central, and upwards of one third the diameter of the test (fig. 1 *b*). The peristome is nearly equally decagonal, and the notches are wide, with reflected margins. In one specimen, collected by Mr. William Buy, from the Forest Marble of Wiltshire, the jaws are preserved *in situ* (fig. 1 *g*).

Although the structure of the peristome in the ECHINOCONIDÆ led us to infer the existence of jaws in the entire family, still we had only feeble traces of them in *Echinoconus albo-galerus*, Klein, and *Pygaster umbrella*, Agass. Our discovery of these organs in *Holectypus depressus*, Leske, is a fact new to science; each jaw consists of two slender branches united near the apex, and supports a long, slender, prismatic tooth (fig. 1 *g*).

The anal opening is large and pyriform, with the apex directed towards the mouth; it occupies nearly all the space between the mouth and the border, and is therefore situated entirely at the base of the shell (fig. 1 *b*); the direction of the apex of the vent inwards, and its basal position, form important diagnostic characters, by which this species is clearly distinguished from its congeners. The tubercles situated at the base, are much larger and more fully developed than those on the upper surface (fig. 1 *f*), but instead of having a circle of granules around the areolas, they are separated laterally from each other by sharp elevations of the test, the upper and under sides of the areolas being bounded by crescents of granules (fig. 1 *f*).

Many spines are preserved *in situ* on the shell (fig. 1 *g*); they have a long, conical head, with a prominent ring, and a slender, tapering stem, which is sculptured with numerous longitudinal lines (fig. 1 *h*); the apex is blunted. Besides these primary spines, which were articulated with the largest tubercles, I find the miliary granulation at the base likewise possessed fine, hair-like spines, as they are observed in connection with their granules among the larger spines at the under surface (fig. 1 *g*).

Affinities and differences.—*Holectypus depressus*, Leske, may be regarded as the best type of this genus. It is distinguished from *H. hemisphæricus*, Desor, by its more depressed form, basal vent, and more acute marginal fold. The apex of the pyriform anal opening is directed towards the mouth, whilst in *H. hemisphæricus*, it is just the reverse.

It is distinguished from *H. oblongus*, Wright, by the form of the test, which is circular or sub-pentagonal, whereas in that Coral Rag species it has an oblong shape; the tubercles on a portion of the inter-ambulacral plates form a single line of tubercles, whilst in *H. oblongus* they are disposed in double lines on the same plate (fig. 3 *d*).

Locality and Stratigraphical position.—This urchin is found for the first time in the Upper Ragstones of the Inferior Oolite, in the zone of *Ammonites Parkinsoni*, Sowerby. I have collected it from the Trigonia Grit, wherever that bed is present, along the whole range of the Cotteswold Hills, as at Dundry, Wotton-under-Edge, Stinchcombe, Rodborough, Coopers, Birdlip, Shurdington, Leckhampton, and Sudeley Hills; and at Hampen, Naunton, and Stow-in-the-Wold, likewise in the Cornbrash near the Kemble Tunnel, and in the neighbourhood of Cirencester, Gloucestershire. It is found rarely in the Cornbrash near Chippenham, but abundantly in that formation near Trowbridge, Wilts, whence I have a very fine series of large specimens in the finest preservation, collected by Mr. Macniel.

The Rev. A. W. Griesbach has collected the finest specimens I have seen in the Cornbrash near Rushden, Northamptonshire; the beautiful specimens fig. 1, *a*, and *b*, are two of these, kindly presented for this work. The Cornbrash specimens are always larger and finer than those collected from the Inferior Oolite. Mr. Gavey obtained it from the Cornbrash near Woodstock, Oxon.; and it is found abundantly at Scarborough, in the Cornbrash near the Castle Hill.

HOLECTYPUS HEMISPHÆRICUS, *Agassiz*.   Pl. XVIII, fig. 2 *a*, *b*, *c*, *d*, *e*, *f*, *g*, *h*, *i*.

DISCOIDEA HEMISPHÆRICA.	Agassiz, Catalogus Systematicus, p. 7.
— —	Desor, Monographie des Galérites, pl. 8, figs. 4—7, p. 71.
HOLECTYPUS HEMISPHÆRICUS.	Agassiz et Desor, Catalogue raisonné des Échinides, Annal. des Sc. Nat., 3ᵐᵉ série, vol. vii, p. 146.
GALERITES HEMISPHÆRICUS.	Forbes, Memoirs of the Geological Survey, Decade III, pl. 6.
DISCOIDEA MARGINALIS.	M'Coy, Annals and Magazine of Natural History, 2d ser., vol. ii, p. 413.
HOLECTYPUS DEVAUXIANUS.	Cotteau, Études sur les Échinides Fossiles, pl. 2, figs. 7—9, p. 46.
— HEMISPHÆRICUS.	Wright, Annals and Magazine of Natural History, 2d ser., vol. ix, p. 96.
— —	Forbes, in Morris's Catalogue of British Fossils, 2d edit., p. 82.
— —	Desor, Synopsis des Échinides Fossiles, p. 172.

Test hemispherical, more or less depressed; sides tumid; margin rounded; posterior half of the test longer than the anterior; single inter-ambulacrum slightly produced; anal opening marginal, large, and pyriform, with the apex directed upwards; base nearly flat; mouth opening small.

Dimensions.—Height, eight tenths of an inch; transverse diameter, one inch; antero-posterior diameter, one inch and a quarter.

Description.—The marginal vent of this urchin sufficiently distinguishes it from its congeners. It is very abundant in some localities, but rare in others; it appears to have had a very limited life in time, having been hitherto only found in a marly vein, about an inch in thickness, which traverses the Trigonia bed of the Inferior Oolite, in the zone of *Ammonites Parkinsoni*, Sow. I have examined many hundreds of specimens of this Echinite, which all came from the same bed.

M. Desor has given a very good figure of this species in his 'Monograph on the Galerites,' which appears to have been the small conical variety.

Professor Forbes has given most beautiful figures, with full details, of *Holectypus hemisphæricus*, in Decade III, pl. 6, which leave nothing more to be desired.

The general outline of this urchin is sub-hemispherical, but it is more or less convex in different individuals; some varieties are depressed, and others are conoidal, but few specimens are regularly convex; most commonly they have a slight obliquity, from the test in a majority of specimens being slightly elongated in the antero-posterior diameter, and declining on the side towards the vent; the vertex is therefore not quite central, and the apical disc is nearer the anterior than the posterior border; the sides are a little tumid, and the margin is gently rounded thereby.

The ambulacral areas are about one third the width of the inter-ambulacral; from the border to the disc they are quite conical, and taper gradually between these two points; there are six rows of tubercles at the margin, which gradually diminish to four and two on the upper surface; the two outer rows alone extend from the peristome to the disc; each pair of the small, narrow ambulacral plates (fig. 2 *d*) supports one tubercle, which occupies the same relative position thereon on every fourth plate, so that the areas are adorned at their widest part with six rows of tubercles, arranged obliquely in V-shaped lines, thus— ⋮ ⋮ ⋮ ⋮ In a specimen one inch and a quarter in diameter I counted one hundred and twenty plates in each column. The poriferous zones are narrow, the pores are strictly unigeminal, and there are from four to five pairs of pores opposite each inter-ambulacral plate.

The inter-ambulacral areas are rather more than three times the width of the ambulacral (fig. 2 *a*); the number of plates in each column varies with the age of the urchin; in the one before me there are thirty-two plates; those on the sides are slightly bent upwards in the middle, whilst the basal plates are nearly straight; each plate supports one or two tiers of tubercles, the number and arrangement of which varies exceedingly in different individuals; fig. 2 *d* shows a common distribution of the tubercles on the plates near the margin; besides these spinigerous tubercles, the entire surface of the plates is covered with a fine, close-set, miliary granulation, from the midst of which the tubercles appear to arise; near the margin, in one specimen there are sixteen tubercles abreast in one inter-ambulacral area; the tubercles become crowded towards the margin; they increase in size at the base (fig. 2 *h*), but the number on the plates in this region is inconsiderable. The perforated tubercles are raised on crenulated bosses, which are surrounded by a sunken areolas; at the base many of these are encircled by perforated granules (fig. 2 *i*).

The apical disc is small (fig. 2 *a*), and formed of five genital, and five ocular plates; the pairs of genital plates are shield-shaped and perforated; the single plate is imperforate; the right anterior genital plate is the largest, and extends into the centre of the disc, supporting on its surface the spongy madreporiform body. The ocular plates are small pentagonal bodies, wedged between the angles formed by the ovarials and the summits of the ambulacra; the eye-holes are large, and pierced near the lower border (fig. 2 *e*), their axis having a slanting direction upwards. The surface of all the discal elements is covered with close-set, miliary granules, similar to those on the surface of the plates.

The vent occupies the margin of the single inter-ambulacrum (fig. 2 *c*); its form is pyriform, with the rounded base extending half way into the ventral surface (fig. 2 *b*), and its apex directed upwards towards the apical disc (fig. 2 *c, g*). The widest or basal portion of the vent occupies the outer half of the space between the peristome and the margin (fig. 2 *b*), and the apex extends one fourth of the distance between the base and the disc (fig. 2 *g*), but these proportions vary considerably in every individual I have examined; the marginal opening for the vent, however, is a constant diagnostic character of the species.

The mouth opening is small, being rather more than one fourth the diameter of the test; the peristome is equally decagonal, the notches are wide and rounded, with the margin reflected (fig. 2 *b*); the base is concave; and the oral opening lies in a central depression of the ventral surface.

Among the many hundreds of specimens which I have examined, the specific characters above described are retained in all with remarkable persistence; still there are certain points, such as differences in the amount of convexity, the proportion between the height and the breadth, the distance between the vent and the mouth, and the number and regularity of the tubercles, which vary in certain specimens; they may all, however, be reduced to three principal types.

Var. *a*.   *Hemisphæricus*, represented by our fig. 2 *a, b, c*.

Var. *β*.   *Conicus*, represented by our fig. 2 *g*.

Var. *γ*.   *Depressus*, with the upper surface much depressed, but the vent marginal.

In the following table I have embodied the comparative dimensions of six specimens from six different localities in Dorsetshire:

No.	Locality.	Height. Inch.	Breadth. Inch.	Anus to Mouth. Inch.
1	Stoke Knaps	$\frac{5}{12}$	1	$\frac{3}{12}$
2	Loders	$\frac{1}{12}$	$1\frac{1}{13}$	$\frac{2}{12}$
3	Crewkerne	$\frac{8}{12}$	$1\frac{1}{12}$	$\frac{0}{12}$
4	Castle Cary	$\frac{7}{12}$	$\frac{8}{12}$	$\frac{3}{12}$
5	Bridport Harbour	$\frac{5}{12}$	$\frac{8}{12}$	$\frac{1}{12}$
6	Walditch	$\frac{9}{12}$	$1\frac{4}{12}$	$\frac{4}{12}$

Affinities and differences.—The tumid sides, convex upper surface, and marginal vent of *Holectypus hemisphæricus* serve to distinguish this species from all its congeners, and by these characters it is readily separated from all other *Holectypi*.

Locality and Stratigraphical position.—In the stratigraphical position of this species, I am at issue with the statement made by Professor Forbes, who remarks : " During the examination of the Inferior Oolite strata in Somersetshire and Dorsetshire, by the members of the Geological Survey, this species was collected abundantly, chiefly in the sands of the Inferior Oolite, in numerous places, associated in most instances with *Dysaster ringens,* and often *Dysaster bicordatus.*" The following is a list of the principal localities in which it was found : Hazelbury, Crewkerne, Lyttelton Hill, near Cadbury, Whatley, near Frome, Little Windsor, Loders (top beds), Stoke Knaps, Greenland, Compton Pauncefoot (bottom beds), Pilcombe, Bruton, Shipton Gorge, Burton Bradstock, and near Burton Castle (top beds), Bridport Harbour, Chideock Hill, Mapperton, West Swillets, Beaminster."*

I have never seen this urchin lower than the marly vein which traverses the upper ragstones in the zone of *Ammonites Parkinsoni,* Sow. ; where I have found it with *Cidaris Bouchardii,* Wright ; *Stomechinus bigranularis,* Lamarck ; *Clypeus Agassizii,* Wright ; *Clypeus altus,* M'Coy ; *Hypoclypus gibberulus,* Agassiz ; *Ammonites subradiatus,* Sow. ; *Ammonites Parkinsoni,* Sow. ; *Terebratula sphæroidalis,* Sow. ; *Rhynchonella plicatella,* Sow. It is true, that in many of the localities in the above list, the ragstones rest on the sands of the Inferior Oolite ; a collector, therefore, might readily obtain specimens from the sands which had dropped out of the marly seam, and thereby conclude that they belonged to that formation ; but I have shown† that the so-called sands of the Inferior Oolite belong to the upper region of the Upper Lias, to the zone of *Ammonites Jurensis,* Zieten, and *Ammonites variabilis,* d'Orbig., and that, as far as we at present know, urchins have not been found in this bed.

I have collected *Holectypus hemisphæricus* from the Trigonia grit at Shurdington, Leckhampton, and Hampen, in Gloucestershire, associated with *Pedina rotata,* Wright ; *Holectypus depressus,* Leske ; *Echinobrissus clunicularis,* Llhwyd ; *Echinobrissus Hugii,* Agass. ; *Clypeus Plottii,* Klein ; *Ammonites Parkinsoni,* Sow. ; and *Terebratula globata,* Sow.

B. *Species from the Coral Rag* = 14ᵉ *Étage Corallien,* d'Orbigny.

HOLECTYPUS OBLONGUS, *Wright,* nov. sp. Pl. XVIII, fig. 3 *a, b, c, d.*

Test oblong, inflated at the sides, depressed at the upper surface ; anterior half of the test shorter, broader, and rounder than the posterior half, which becomes narrow,

---

* ' Memoirs of the Geological Survey,' Decade III, pl. 6, pp. 4, 5.

† ' Quarterly Journal of the Geological Society,' vol. xii, p. 292.

and has the single inter-ambulacrum produced; anal opening at the base; tubercles on the upper surface very small and indistinct; those at the base and margin moderately large; mouth opening concealed.

Dimensions.—Height, seven tenths of an inch; transverse diameter, one inch and a quarter; antero-posterior diameter, one inch and four tenths.

Description.—The few specimens of this urchin I have examined were not in good preservation, the surface of the test having been more or less rubbed in them all, so much so, that a detailed description, with the materials at my disposal, is at present impossible.

The test is oblong, with tumid sides; it is rounded before, contracted and elongated behind, so that the anterior half is an eighth of an inch shorter than the posterior half, when measured from the disc to the border (fig. 3 *a*).

The ambulacral areas are one third the width of the inter-ambulacral; they have six rows of small tubercles, which form V-shaped lines throughout the area; the poriferous zones are narrow; and there are five pairs of pores opposite each inter-ambulacral plate (fig. 3 *d*).

The inter-ambulacral spaces are of equal width, and three times as broad as the ambulacral (fig. 3 *a*); from the disc to the border there are twenty plates in each column, which are all slightly bent in the middle; on some of the plates there is only one row of tubercles, but in the others there are two rows, arranged as in fig. 3 *d*, which are small and indistinct from friction; the basal tubercles of both areas are larger, and they are arranged in single concentric lines.

The base is much concealed in all the specimens I have seen; the mouth is entirely, the vent partially so, in those before me; the anal opening appears to be a large aperture, situated between the mouth and the border, but encroaching a little upon the latter; and has the apex directed towards the mouth.

The apical disc is small, vertical, and excentral, being nearer the anterior than the posterior border; the ovarial plates are small, with the exception of the right antero-lateral, which is nearly twice as large as the others, and supports an oblong, madreporiform body; the ocular plates are very small, and perforated near their border.

Affinities and differences.—This species very much resembles *H. depressus*, Leske; but it has a more oblong form, more tumid sides, and the anal opening is nearer the border than in that species; by the smallness of the tubercles it is distinguished from *H. corallinus*, d'Orbigny, and by the arrangement of the tubercles on the plates from *H. depressus*; the basal position of the anus separates it from *H. hemisphæricus*, Agassiz. The imperfect condition of all the specimens I have seen renders a more critical comparison at present impossible.

Locality and Stratigraphical position.—The four or five specimens I have examined were all collected from the Coralline Oolite of Malton, Yorkshire.

NOTES

On Foreign Jurassic species of the genus HOLECTYPUS, nearly allied to British forms, but which have not yet been found in the English Oolites.

Holectypus planus, *Desor*.   Monogr. des Galérites, tab. 9, figs. 1—3.

Test small, flat, pentagonal; dorsal surface convex, base concave; tubercles on the upper surface small, about ten rows in the inter-ambulacra at the circumference, basal tubercles much larger; mouth small, central; vent small, occupies only half the space between the mouth and the border.

Dimensions.—Height, seven twentieths of an inch; transverse diameter, three quarters of an inch.

Formation.—Oxford Clay, Vaches-Noires. Normandy.

Collections.—M. Michelin, my Cabinet.   Very rare.

Holectypus Mandelslohi, *Desor*.   Monogr. des Galérites, tab. 9, figs. 14—16.

Test sub-conical; tubercles small on the upper surface, large at the base, arranged in vertical and horizontal rows; mouth small; vent at the base nearly as wide as the mouth; miliary granulation abundant, and close-set.

Dimensions.—Height, nine twentieths of an inch; transverse diameter, eight tenths of an inch.

Formation.—Corallien inférieur (terrain à chailles), Albe Wurtembergeoise, Lusberg (Canton du Soleure).   "Argovien (avec *Dysaster granulosus*) du Randen de Birmansdorf, de Baden."   *Desor*.

Collections.—Count Mandelslohe, M. Gressly, M. Moesch.   Mus. Bâle, Mus. Neuchâtel.

HOLECTYPUS PUNCTULATUS, *Desor*. Monogr. des Galérites, tab. 9, figs. 17—19.

Test hemispherical, sides tumid; tubercles form regular, horizontal, and vertical series, ten rows in the inter-ambulacra, and six rows in the ambulacra, which are much more developed at the base; mouth small, vent large, between that opening and the border. This species has many important affinities with the preceding, and is only distinguished from it, according to M. Desor, by having fewer tubercles, which are arranged with greater regularity.

Dimensions.—Height, four tenths of an inch; transverse diameter, eight tenths of an inch.

Formation.—Corallien inférieur de Largue (Canton de Berne), Dettingen, Württemberg, Mont-de-Bregille (près Besançon). Oxfordien de Chambery. *Renevier*.

Collections.—MM. Gressly, Parandier; Count Mandelslohe. Abundant.

HOLECTYPUS CORALLINUS, *d'Orbig*. Cotteau, Études sur les Échinides Foss., pl. 32, figs. 1—8.

Test sub-pentagonal, sub-conical above, flat or sub-concave below; tubercles disposed in regular, vertical, and concentric series, a single longitudinal row on each plate, small and few in number on the upper surface, large and numerous at the circumference and base, about twelve in the widest part of the inter-ambulacra. Granules unequally scattered on the surface of the plates. Mouth central, peristome deeply notched; vent large, elliptical, between the mouth and the border.

Dimensions.—Height, eleven twentieths of an inch; transverse diameter, one inch and a quarter.

Formation.—Corallien 14ᵉ étage, d'Orbig.; Druyes, Châtel-Censoir (Yonne). Kimméridgien 15ᵉ étage. Pointe-du-Ché (Yonne). *Cotteau*.

Collections.—MM. Cotteau, d'Orbigny. Abundant.

HOLECTYPUS ARENATUS, *Desor*.   Monogr. des Galérites, tab. 9, figs. 11—13.

The tubercles, although very numerous, are disposed with great regularity; the miliary granules form continuous horizontal series; mouth opening small, with rows of oblong granules near the peristome; in its general form it resembles *H. punctulatus*, Des., and *H. Mandelslohi*, Des., and is distinguished from these species only when the details of the test are well preserved.

Dimensions.—Height, four tenths of an inch; transverse diameter, eight tenths of an inch.

Formation.—Oxfordien du Canton de Soleure.   Very rare.

Collection.—M. Gressly.

HOLECTYPUS SPECIOSUS, *Münster*.   Syn. *Galerites speciosus*, Goldfuss, Petref. German., tab. 41, fig. 5 *a*, *b*.
Discoidea speciosa, Agass., Echin. Foss. Suisse, tab. 6, fig. 16. (*non.*)
— — Desor, Monogr. des Galérites, tab. 10, fig. 13. (*non.*)

Test large, depressed; the plates of the ambulacral and inter-ambulacral areas very narrow in proportion to their length; tubercles on the upper surface small and irregularly disposed; tubercles at the base larger, and arranged in double horizontal rows, closely placed together on each plate; mouth opening small, peristome decagonal; anal opening large, pyriform, occupying half the space between the border and the peristome.

Dimensions.—Height, one inch and a half; transverse diameter, three inches and three quarters.

Formation.—Kimmeridge Clay of the valley de La Birse, near Lanfon (Greifel); Upper Stage of the Calcaire Jurassique de Heidenheim (Würtemberg).

Collections.—Count Münster, M. Gressly.

HOLECTYPUS INFLATUS, *Agassiz.* Syn. *Discoidea inflata,* Agassiz, Échinoderm. Foss.
Suisse, tab. 6, figs. 4—6.

—   —   Desor, Monogr. des Galérites,
tab. 9, figs. 7—10.

Test thin, small, and depressed, with tumid, inflated sides; tubercles small, numerous, twelve rows in the inter-ambulacra, and four in the ambulacra; miliary granules do not form horizontal series; mouth opening small and decagonal; anal opening very large and pyriform, extending from the peristome to the border, with the apex directed towards the mouth.

Formation.—" Portlandien inférieur (Astartien) du Jura Neuchâtelois.

"*Collections.*—Museum of Neuchâtel.   Coll. M. Gressly." *Desor.*

HOLECTYPUS GIGANTEUS, *Desor.*   Syn. *Discoidea speciosa,* Agassiz, Échinoderm. Foss. de
Suisse, tabl. 6, fig. 16.

—   —   Desor, Monogr. des Galérites,
tabl. 10, fig. 13.

" A very large, depressed species.   The base is furnished with numerous close-set tubercles, but without a regular arrangement; mouth opening proportionally very small; anal opening pyriform, hardly occupying half the space between the peristome and the border."

Formation.—" Corallien de la Vallée de la Birse.

" *Collection.*—M. Gressly; very rare."   M. Desor adds, in a note to the above description, in his ' Synopsis des Échinides Foss.,' that it is by mistake this species has been confounded by M. Agassiz and himself with *H. speciosus,* Münst.   The arrangement of the tubercles is very different in *H. giganteus ;* the regular concentric disposition observed in *H. speciosus* is absent in *H. giganteus ;* their regular distribution in the one, and irregular distribution in the other, form a diagnostic character between these two gigantic *Holectypi.*

Genus—PYGASTER, *Agassiz.* 1834.

GALERITES? (pars), Lamarck. 1816. (?)
CLYPEUS (pars), Phillips. 1829.
ECHINOCLYPEUS (pars), De Blainville. 1830.
NUCLEOLITES (pars), Desmoulins. 1837.

It is by no means certain that Lamarck was acquainted with the urchins now included in the genus *Pygaster*, although he is invariably cited as the author of one of the most typical species of the group, for reasons which will be given in the article on *Pygaster umbrella;* it appears that Lamarck's reference to Klein's tab. 12 was an error. In his genus *Echinoclypeus*, De Blainville associated three distinct types of Echinides,—*Clypeus, Pygaster,* and *Conoclypeus;* whilst Desmoulins placed the *Pygasters* with his *Nucleolites*, urchins which have limited petaloidal ambulacra, wide poriferous zones, small and irregular-disposed tubercles, with a five-lobed edentulous mouth; characters which are quite opposite to those possessed by the *Pygasters*. The only explanation that can be given for these errors of arrangement by authors of such eminence, is the fact that neither De Blainville nor Desmoulins had seen a *Pygaster*, as appears from a note by the latter author, appended to his description of the species in his 'Synonymie générale.'[*]

M. Agassiz has the merit of having first detected and pointed out the leading characters of the genus *Pygaster*, which he established from a figure of the only species then known, and which has subsequently been proved to be one of the most natural and best-defined of all the genera of the ECHINOIDEA EXOCYCLICA.

The test is sub-pentagonal, more or less elevated and convex on the upper surface, and concave at the base; the ambulacral areas are narrow, with four or six rows of small tubercles, the marginal rows only extending from the base to the apex of the areas.

The poriferous zones are narrow, simple, and complete; and the pores are strictly unigeminal throughout.

The inter-ambulacral areas are in general four times the width of the ambulacral; each of the large pentagonal plates supports numerous tubercles, and those at the border have from six to eight. The tubercles are more or less regularly arranged in vertical and horizontal rows; it is only the representatives of the two primary rows of each area which extend from the mouth to the disc, all the others disappear in succession on the sides, and the length of each is in proportion to its proximity to the two primary rows. The tubercles are small, and nearly equal-sized; they are perforated, and raised on bosses with smooth, uncrenulated summits; depressed, ring-like areolas surround their base; and the inter-

[*] 'Études sur les Échinides,' p. 354, No. 2.

tubercular surface of the plates is covered with numerous small miliary granules, which form circles around the areolas. The basal tubercles are larger than those on the upper surface, their areolas are more excavated, and they have a quadrate or hexagonal figure, which makes a striking contrast to their circular form on the upper surface.

The apical disc is large, composed of five ovarial and five ocular plates; the right antero-lateral ovarial is the largest, it extends into the centre of the disc, and supports a large madreporiform body on its surface. The five ocular plates are small, and wedged between the angles left by the ovarials.

The anal opening is situated on the upper surface of the single inter-ambulacrum; it is a very large, oblong, or pyriform aperture in the test, and in the living urchin appears to have been closed by a tegumentary membrane; in all the species at present known, *P. pileus* excepted, the discal and anal openings are continuous; but in that species a portion of the test separates the apical disc from the vent.

The mouth opening is circular, and situated in a depression in the centre of the base; the peristome is divided by deep notches into ten equal-sized lobes. Remains of jaws have been found in some species; and imprints of ten carinæ, radiating from the centre to the periphery, are seen in the moulds of others. The *Pygasters*, therefore, possessed masticating organs probably allied in structure to those I have already described in *Holectypus depressus*, Leske.

The spines are small, short appendages, which resemble the same parts in *Echinus*, and, like them, their surface is covered with well-marked longitudinal lines.

The *Pygasters* form a type of structure which nearly approaches the *Hemipedinas*, of the family DIADEMADÆ; like them, the tubercles are perforated, their bosses have smooth, uncrenulated summits, and they are arranged in regular vertical and horizontal rows. The poriferous zones are narrow and unigeminal; the mouth opening is circular and decagonal; and they possess organs of mastication.

Unlike the DIADEMADÆ, however, the *Pygasters* are true *exocyclous urchins*, and have a large vent, placed without the circle of the apical disc. They differ from *Holectypus* in having the tubercles larger, and the vent at the upper surface; and from *Hyboclypus* in the absence of the dorsal sulcus in which the vent is situated in that genus. They are distinguished from *Clypeus*, *Echinobrissus*, *Catopygus*, *Pygaulus*, and other ECHINOBRISSIDÆ, in having perforated tubercles arranged in rows; narrow, complete poriferous zones; and the mouth armed with jaws. They form, in fact, a well-defined group of urchins, separated by prominent organic characters from all the others.

The *Pygasters* first appeared in the lowest beds of the Inferior Oolite; and their different species are discovered in the Cornbrash, Calcareous Grit, Coral Rag, and Kimmeridge Clay. They have been found likewise in the Gault.

A. *Species from the Inferior Oolite* = 10ᵉ *Étage Bajocien,* d'Orbigny.

PYGASTER SEMISULCATUS, *Phillips.*   Pl. XIX, fig. 1 *a, b, c, d, e, f, g.*

CLYPEUS SEMISULCATUS.	Phillips, Geology of Yorkshire, vol. i, p. 104, pl. 3, fig. 17 (two thirds nat. size), 1829.
— ORNATUS.	Buckman, in Murchison's Geology of Cheltenham, 2d ed., p. 95, 1845.
NUCLEOLITES SEMISULCATA.	Desmoulins, Études sur les Échinides, Tabl. Synoptique, p. 362, 1837.
PYGASTER SEMISULCATUS.	Agassiz, Prodrome d'une Monogr. des Échinodermes, p. 185, 1837.
— —	Dujardin, in Lamarck's Hist. Naturelle des Animaux, 2ᵐᵉ édit., tome iii, p. 353, 1840.
— BREVIFRONS.	M'Coy, Annals and Magazine of Natural History, 2d series, vol. ii, p. 414.
— SEMISULCATUS.	Wright, Annals and Magazine of Natural History, 2d series, vol. ix, p. 89, 1852.
— —	Forbes, in Morris's Catalogue of British Fossils, 2d ed., p. 88, 1855.
— —	Desor, Synopsis des Échinides Fossiles, p. 165.

Test sub-pentagonal, depressed, sometimes conoidal; ambulacral areas prominent and convex, with four rows of tubercles at the margin, diminishing to two marginal rows above; inter-ambulacra four times the width of the ambulacra, with from eighteen to twenty vertical rows of tubercles at the margin, diminishing to two rows in the upper part; marginal fold acute, base flat, towards the centre very concave; mouth sunk in a depression, peristome nearly equally decagonal; discal opening wide, central; anal opening large, semisulcate, not contracted towards the disc, extending one half the distance between the centre of the apical disc and the posterior border.

Dimensions.—Height, one inch and a half; transverse diameter, three inches and one eighth.

Description.—I have omitted from the list of synonyms the *Galerites umbrella,* Lamarck, because that species is clearly the *Clypeus sinuatus,* Leske; likewise the *Pygaster umbrella,* Agassiz, of the 'Échinodermes Fossiles de la Suisse,' and of Desor's 'Monographie des Galérites,' inasmuch as the figures given in these works represent, probably, another British species. It is extremely doubtful whether *Pygaster semisulcatus* has been yet found out of the English Inferior Oolite; it was said to have been discovered in the Bajocien stage of the department of the Sarthe, but M. Cotteau informs me that the specimen supposed to have been collected therefrom, turns out to be a Gloucestershire fossil, which had got by mistake into a collection of Chauffour urchins.

When Professor Phillips figured *Pygaster semisulcatus* in his 'Geology of Yorkshire,' he was not aware that two distinct species of the genus *Pygaster* existed in the Oolites of that county, one collected from the Inferior Oolite of Whitwell, and the other from the Coralline Oolite of Malton; as the type specimen is now lost, it is uncertain whether the Whitwell or Malton urchin was the one sketched in tab. 3, fig. 17 of that work. When Professor Forbes was studying this species, he applied to Professor Phillips for the type of *P. semisulcatus*, and the one sent was a Whitwell specimen, which I compared with my Inferior Oolite urchins, and ascertained the identity of the species.

It was assumed, therefore, by Professor Forbes, that the Whitwell urchin was *P. semisulcatus*, and his beautiful plate of this species, in the fifth Decade of the 'Memoirs of the Geological Survey,' was lettered accordingly.

The general form is sub-hemispherical, more or less depressed on the upper surface; the base is flat or slightly concave, and the mouth is lodged in a considerable central depression; the circumference is sub-pentagonal, and the bilateral symmetry of the test is very evident in specimens which are free from distortion (fig. 1 *a*); the upper surface is in general convex (fig. 1 *c*), but sometimes in large examples the sides are flattened, and the test then assumes a conoidal figure, when it becomes the *P. brevifrons*, M'Coy.

The ambulacral areas are equidistant, and of equal width, and about one fourth the breadth of the inter-ambulacral areas; the anterior single area is straight and lancet-shaped; the anterior pair are gently bent backwards, and the posterior pair in large individuals have a slight *f*-shaped flexure, occasioned by the great width of the vent (fig. 1 *a*). They are composed, according to the size of the test, of from one hundred and twenty to one hundred and sixty pairs of plates, as weathered specimens show that every third plate carries a primary tubercle (fig. 1 *f*). There are two complete rows of tubercles on the margins of the areas, which extend without interruption from the peristome to the disc; in each row there are about fifty-eight tubercles; of these, twenty-two belong to the base and margin, and thirty-six to the sides and upper surface (fig. 1 *a*, *b*); within these, two other rows commence about half an inch from the peristome, and extend from thence three quarters of an inch up the sides, about which point they disappear; at the marginal angle sometimes a fifth, or in some cases a sixth, row is introduced for a very limited extent, and the miliary zone in the upper part of the area is filled with small granulations. The poriferous zones are narrow; the pores are strictly unigeminal throughout, one pair of pores nearly corresponding to each of the ambulacral plates. The pores constituting a pair are placed slightly oblique; in some specimens the inner hole is round and the outer is oval, but this character is individual, and not general. The septa are narrow, and form very inconsiderable elevations on the surface (fig. 1 *f*, *h*).

The inter-ambulacral areas are four times the width of the ambulacral; they are composed of about thirty-five pairs of plates, of which about fifteen occupy the base, and twenty the sides; these plates, according to their breadth, support a variable number of tubercles, those at the margin having from nine to ten tubercles on the same horizontal line (fig. 1 *d*). Two of these rows, the fifth on each side from the centro-sutural line, represent the primary tubercles, and extend from the peristome to the disc; the other rows, as the plates become narrower, disappear at various points on the sides, so that, whilst at the margin of the specimen (fig. 1 *d*) there are twenty rows of tubercles, in the vicinity of the apical disc there are only two rows (fig. 1 *a*). The tubercles are nearly of the same size, and form very regular horizontal and vertical rows on both areas; those of the primary rows are larger in the upper surface, and all the basal tubercles are so likewise. The tubercles, which are perforated, are raised on small bosses, with smooth summits (fig. 1 *e, f*); their base is encircled by narrow, sunken, ring-like areolas, which are much more developed around the ventral than on the dorsal tubercles; the circumference of the areolas is surrounded by a circle of small granules, and the intervening portion of the plates is covered with a like-sized miliary granulation (fig. 1 *d, e, f*).

The opening for the apical disc is very large, half an inch in diameter, but in only one of the hundreds of specimens of this species which I have examined has a vestige of the plates remained. These consist of four ovarial, and three ocular; the ovarial plates are small, and dove-tail with the angular incisions in the discal opening; the spongy madreporiform body is large, extending inwards and backwards, and the ovarial holes are perforated near the apex; the ocular plates are very small, situated at the summit of the ambulacral areas, and the eye-holes are perforated near their margin; as the centre of the disc is absent; the form of its posterior boundary is therefore unknown.

The anal aperture is a very large, oblong-oval opening, which occupies nearly the upper half of the single inter-ambulacrum (fig. 1 *a*). Its shape forms an important diagnostic specific character, and ought to be carefully noted in making determinations of the species. The borders of the opening are incurved, and the vent appears to have lain in a depression; at the point where the plates incline towards the lateral parts of the vent, two tumid ridges extend downwards to the posterior border, having a slight concave depression between them (fig. 1 *a, c*). In young specimens, the anal aperture is proportionately smaller.

The base is concave, and the mouth central, and placed in a considerable depression. It is of moderate size, about one fifth the diameter of the test; the peristome is nearly equally decagonal, the ambulacral being larger than the inter-ambulacral lobes; the notches are wide, and the margin is everted (fig. 1 *b*). Although I have searched diligently for the teeth, I have never yet seen a vestige of one, although there cannot be a doubt that the *Pygasters* possessed jaws like those of *Holectypus depressus*, Leske.

The spines adhering to the fine specimen I have figured are short and needle-shaped, and delicately striated longitudinally.

Affinities and differences.—*Pygaster semisulcatus* resembles in many points *Pygaster umbrella*, Agassiz, but is distinguished from that urchin by the following characters: The tubercles, especially those on the upper surface, are disposed in much more regular horizontal and vertical rows, the ambulacral areas are not so lanceolate, the anal opening is wider in the upper part, and does not descend so far down the inter-ambulacrum as in *P. umbrella,* in which it has a well-marked pyriform shape, and is much contracted in the upper part like a key-hole. (Compare Pl. XIX, fig. 1, with Pl. XX, fig. 2.)

Locality and Stratigraphical position.—This species has been collected from the Pea-grit, Inferior Oolite of Crickley, Birdlip, Shurdington, Leckhampton, Cleeve, and Sudeley Hills, Gloucestershire, where it is abundant, although good specimens are rare. I possess a series of all sizes, from half an inch to three inches and a half in diameter. It is found likewise in the shelly freestone at Leckhampton, and I have extracted small specimens from the planking beds of the Great Oolite at Minchinhampton Common. In Yorkshire it is collected only from the Inferior (Great?) Oolite at Whitwell. In the Pea-grit at Crickley Hill it is associated with *Ammonites Murchisonæ,* Sow., *Nautilus truncatus,* Sow., *Terebratula simplex,* Buck, *Terebratula plicata,* Buck; and *Thecidium triangulare,* d'Orbig., is often adherent to its test. *Pseudodiadema depressum,* Agas., *Hyboclypus agariciformis,* Forb., *Cidaris Fowleri,* Wright, *Cidaris Bouchardii,* Wright, and *Acrosalenia Lycetti,* Wright, are its usual associates.

PYGASTER CONOIDEUS, *Wright.* Pl. XIX, fig. 2 *a, b, c, d, e, f.*

PYGASTER CONOIDEUS.	Wright, Annals and Magazine of Natural History, 2d series, vol. ix, p. 91, pl. 3, fig. 1.
— —	Forbes, in Morris's Catalogue of British Fossils, 2d ed., p. 88.
— —	Salter, Memoirs of the Geological Survey, Decade V, pl. 8.
— —	Desor, Synopsis des Échinides Fossiles, p. 166.

Test pyramidal, pentahedral; posterior border sub-acute; ambulacral areas narrow and prominent, with two rows of small marginal tubercles, and two imperfect, incomplete rows within; inter-ambulacral areas upwards of four times the width of the ambulacral; tubercles of both areas very small, and scattered without much order on the surface of the plates, which are covered with minute spaced-out granules; vent small, occupying rather more than the upper third of the single inter-ambulacrum; marginal fold acute, sides of the pentahedron rising abruptly therefrom; base flat, mostly concealed by the matrix.

Dimensions.—Height, one inch and three tenths; antero-posterior diameter, two inches and nine twentieths; transverse diameter, two inches and four tenths.

Description.—This very rare *Pygaster* is remarkable for its pyramidal form; the sides, which are pentahedral, rise abruptly from the marginal fold; the anterior and lateral pairs of inter-ambulacra form angles of from 50° to 55° with the base, and the single inter-ambulacrum makes an angle of about 42° (fig. 2 a). The ambulacral areas are narrow and prominent, and placed nearly equidistant from each other; the single anterior area is straight; the anterior pair curve forwards, upwards, and backwards; the posterior pair rise forwards and upwards for three parts of their course, then curve inwards towards the anal opening, and terminate near the posterior part of the disc (fig. 2 a). There are two complete rows of tubercles extending from the peristome to the disc, arranged on the margins of the areas, a primary tubercle being developed on every third plate (fig. 2 d); two inner rows commence near the mouth, pass round the angle, and rise half way up the sides, but the tubercles thereof are smaller, less regular in their arrangement, and more incomplete than the marginal rows; like them, however, one tubercle rises from every third plate; at the widest part of the areas, near the border, a few tubercles form a fifth row of very limited length (fig. 2 b).

The poriferous zones are narrow, and sunk in slight depressions (fig. 2 c); the pores are strictly unigeminal throughout; the pores are nearly equal, and the septa develop a small ridge between them (fig. 2 d); one pair of pores corresponds to a pair of ambulacral plates; as there are thirty marginal tubercles on the upper surface of one area, and every third plate supports one tubercle, it follows that there are ninety pairs of pores in the poriferous zones on the upper surface alone; as the areas are only partially exposed at the base, the pores cannot be counted in that region.

The inter-ambulacral areas are more than four times as wide as the ambulacral; there are eighteen pairs of plates in each area, between the margin and the disc; each plate is bent in the middle (fig. 2 d), and supports numerous small tubercles, their number on the plates varying according to their length; they are disposed in single transverse rows on the centro-sutural half, and in double rows on the zonal half of the plates; near the margin, there are four tubercles in the single row, and three pairs of tubercles in the double rows, but on the shorter plates they are much fewer and less regular (fig. 2 d). On the under surface the tubercles are larger, and arranged in concentric rows; there are from six to eight tubercles on each plate, the smooth areolas of which nearly touch (fig. 2 b).

The tubercles on the upper surface of this species are extremely small, and form a remarkable contrast to those in the same region of the test in *Pygaster semisulcatus*; their areolas are scarcely sunk, and the granulations on the surface of the plates are very small, and placed rather widely apart (figs. 2 a, c, e).

The single inter-ambulacrum is concave between the disc and the border, and the anal opening occupies its upper third (fig. 2 e); this aperture is of the same shape, but smaller than in *P. semisulcatus* (fig. 2 a); the portion of the area below the vent is flattened, and the tumid ridges are absent.

The base is partly concealed by firmly adhering matrix, which cannot be removed without the risk of splintering the test; enough, however, is exposed, to show that the base is slightly concave, and that the tubercles in that region are larger than those on the upper surface (figs. 2 *b* and *f*).

Affinities and differences.—This species resembles *P. semisulcatus* in its pentagonal form, but it is pyramidal and pentahedral, and is neither hemispherical nor depressed; it is distinguished from that species by a greater prominence of the ambulacra, the smallness of the tubercles, the superficiality of the areolas, the microscopic character of the granules, and the smallness of the vent. It is a very rare species, as I have only seen a specimen in Mr. Lycett's cabinet, besides the one now figured in detail.

Locality and Stratigraphical position.—I collected this urchin from the Pea-grit at Crickley Hill. Mr. Lycett's specimen was found in the lower beds of the Inferior Oolite, near Stroud.

B. *Species from the Cornbrash* = 11ᵉ *Étage Bathonien*, d'Orbigny.

PYGASTER MORRISII, *Wright.* Pl. XX, fig. 1 *a, b, c, d, e, f.*

> PYGASTER MORRISII. Wright, Annals and Magazine of Natural History, 2d ser., vol. ix, p. 92, pl. 4, fig. 1.
> — — Forbes, in Morris's Catalogue of British Fossils, 2d edit., p. 88.
> — — Wright, Memoirs of the Geological Survey, Decade V. Notes on British species of Pygasters.
> — — Desor, Synopsis des Échinides Fossiles, p. 166.

Test pentagonal, depressed; marginal fold very tumid; single inter-ambulacrum much truncated; ambulacral areas wide, convex, and prominent, with six rows of tubercles; inter-ambulacral areas wide, with rather large tubercles, in very regular vertical and horizontal rows, from twenty to twenty-two in each space on the same horizontal line at the equator; base flat, concave towards the mouth opening, which is small; anal opening long, pyramidal, occupying two thirds of the upper surface of the inter-ambulacrum.

Dimensions.—Height, eight tenths of an inch; antero-posterior diameter, two inches and three twentieths of an inch; transverse diameter, two inches and a quarter of an inch.

Description.—This is one of our rarest *Pygasters*, and the specimen figured is the only one I know. It has a thick test, with a very pentagonal outline, is much depressed on the dorsal surface, has a flat base, tumid sides, is remarkable for the size of its tubercles, and for the regularity of their arrangement in vertical and horizontal rows.

The ambulacral areas are wide, convex, and prominent, and form an exception to the generalisation of M. Agassiz, that in the genus *Pygaster* the ambulacra are furnished with only four rows of tubercles, for in this species at the widest part of the area there are six rows of well-developed tubercles (fig. 1 *a, b*); the two marginal rows, with about forty-five tubercles in each row, extend from the peristome to the disc; the second rows commence at a short distance from the mouth, and extend two thirds of the length of the upper surface; the third rows commence at the base, about half an inch from the border, and extend to about the same distance up the sides of the area (fig. 1 *b*).

The poriferous zones are straight, the pores have moderately thin septa, with only slight elevations on the surface, and there are four pairs of pores opposite each inter-ambulacral plate (fig. 1 *d*).

The inter-ambulacral areas are three times the width of the ambulacral; at the circumference they are furnished with twenty-two rows of tubercles, which attain a greater size and are arranged with more regularity than in any other English species; each plate above the circumference has from eight to nine tubercles developed on its surface (fig. 1 *d*); those on the half of the plate nearest the centro-suture are arranged horizontally on the same line, whilst the tubercles nearest the zones form double rows in oblique pairs. The tubercles are raised on prominent bosses, and surrounded by wide, sharply defined, sunken areolas (fig. 1 *d*); each areola is surrounded by a circle of granules (fig. 1 *e*); there are thirty-six pairs of plates in each area, twenty-one of which are dorsal and fifteen are basal. The tubercles attain their greatest development at the base, where they are so uniformly arranged, and so closely set together, that they appear to arise from hexagonal spaces (fig. 1 *b*); the areolas are deeply excavated (fig. 1 *f*), and have a square or hexagonal figure, two sides being bounded by granules, the other two by sharp elevations of the areolar border (fig. 1 *f*); of the twenty-two rows of tubercles which occupy the area at the circumference, only six, and these the three central rows of each column, extend from the peristome to the disc, the others disappear at shorter distances, the length of the rows being in proportion to their distance from the margin of the columns.

Although the upper surface of this urchin is much depressed, its sides are tumid, as is well seen in the profile (fig. 1 *c*). The single inter-ambulacrum is much truncated (fig. 1 *a*), and the large anal opening, which has an oblong shape, occupies the upper three fourths of this area; the discal opening is small, and the plates are all absent.

The marginal angle is obtuse, the outer half of the base is convex, and the inner half concave; the small mouth opening, which is about one sixth the diameter of the test, lies in a deep depression.

The spines are short, stout, needle-shaped bodies, marked with fine longitudinal lines.

Affinities and differences.—This species resembles *P. laganoides*, Agassiz, in its depressed form, obtuse basal angle, and truncated single inter-ambulacrum; but

P. Morrisii has a greater number of tubercles in the ambulacral and inter-ambulacral areas, *P. laganoides* having four rows in the ambulacral and fourteen in the inter-ambulacral areas, whilst *P. Morrisii* in the same region of the corresponding area possesses six rows and twenty-two rows.

P. Morrisii resembles *P. Gresslyi*, Desor, in its general form, in the size and disposition of its tubercles, and in their surrounding granulation, but *P. Morrisii* is more depressed, has a larger anal opening, and more rows of tubercles in both areas.

Locality and Stratigraphical position.—This rare species was collected by Mr. W. Buy from the Forest Marble or Cornbrash near Stanton, Wilts. The specimen figured (Pl. XX, fig. 1) is the only one I know. I have seen a *Pygaster* from the Great Oolite near Cirencester, which resembles my urchin; but the specimen was crushed, and not otherwise determinable; it had the same number of tubercles in the areas as *P. Morrisii*. I dedicate this species to my friend Professor John Morris, to whose valuable labours British palæontologists are under so many lasting obligations.

C. *Species from the Coral Rag = 14ᵉ Étage Corallien*, d'Orbigny.

PYGASTER UMBRELLA, *Agassiz*.　　Pl. XX, fig. 2 *a, b, c, d, e, f.*

PYGASTER UMBRELLA.		Agassiz and Desor, Catalogue raisonné, Ann. des. Scienc. Nat., 3ᵉ série, tom. vii, p. 144, 1847.
—	—	Bronn, Index palæontologicus, Band i, p. 1066, 1848.
—	—	D'Orbigny, Prodrome de Paléontologie, tom. i, 13ᵉ étage, No. 510.
—	EDWARDSEUS.	Buvignier, Statistique géologique, paléontologique depart. de la Meuse Atlas, p. 46, pl. 32, figs. 31—33, 1852.
—	UMBRELLA.	Wright, Memoirs of the Geological Survey, Decade V, pl. 8. Note on British Pygasters, 1856.
—	—	Cotteau, Études sur les Échinides Foss., p. 194, pls. 27, 28, fig. 1, 1856.
—	—	Desor, Synopsis des Échinides Fossiles, p. 165, 1857.

Test large, more or less elevated, sometimes circular, oftener sub-pentagonal; ambulacral areas narrow, with two complete marginal rows of tubercles, and two incomplete rows, which commence near the peristome, and extend only half way up the sides; inter-ambulacral areas wide, with from sixteen to eighteen rows of tubercles at the equator, which are small and rather irregularly disposed on the upper surface, but large and arranged in regular horizontal and vertical series at the circumference and base; anal opening distinctly pyriform, occupying rather more than one half the length of the upper surface of the single inter-ambulacrum; base concave; mouth opening small, peristome equally decagonal, apical disc large, composed of unequal-sized genital plates; the right antero-lateral very large, and extending backwards beyond the centre; ocular plates small, eye-holes marginal.

Dimensions.—In the following table I have given the dimensions of four English specimens from different localities enumerated in this article, and have added the measurements of M. Cotteau's urchin from Châtel-Censoir.

Pygaster umbrella, Agass.	Height.	Transverse diameter.	Antero-posterior diameter.	Length of discal and anal opening.
	Inches.	Inches.	Inches.	Inches.
Specimen from Malton . .	2	$4\frac{1}{4}$	$4\frac{1}{10}$	$1\frac{1}{4}$
,, ,, Headington .	$1\frac{1}{2}$	$4\frac{1}{4}$	$4\frac{1}{10}$	indeterminable
,, ,, Lyneham .	$1\frac{1}{10}$	3	$2\frac{9}{20}$	$1\frac{3}{20}$
,, ,, Farringdon .	$1\frac{7}{10}$	$2\frac{8}{10}$	$2\frac{7}{10}$	$1\frac{1}{10}$
,, ,, Malton . .	$1\frac{2}{20}$	$2\frac{7}{10}$	$2\frac{6}{10}$	$1\frac{1}{4}$
,, ,, Châtel-Censoir .	$1\frac{6}{10}$	$3\frac{6}{10}$	$3\frac{7}{10}$	$1\frac{3}{4}$

Description.—The greatest confusion exists regarding the synonyms of this species. MM. Agassiz, Desor, and Cotteau, who have each figured and described this urchin, cite *Galerites umbrella*, Lam., as its type; now, Lamarck * does not say that he had seen the fossil, but refers to "Leske apud Klein, p. 157, t. 12, and 'Encycl. Méthod.,' pl. 142, figs. 7, 8," as the species intended. After consulting the original figure in Klein, which represents *Clypeus sinuatus*, Leske, in the Dresden Museum, I do not see how such a reference can be sustained, as the *Clypeus* in question belongs not only to a different genus, but even to a distinct family of the Echinoidea. As Deslongchamps,† Defrance,‡ De Blainville,§ and Desmoulins‖ all founded their synonyms on this mistake, I have not cited these authors in my list, for the obvious reason, that their references were made under an erroneous impression. It is strange that MM. Agassiz and Desor, when correcting the synonyms of the *Pygasters*, in their 'Catalogue raisonné,' should not have de-

---

* *Galerites umbrella.* Lamarck, 'Animaux sans Vertèbres,' tom. iii, p. 23, 1816.

† *Galerites umbrella.* Deslongchamps, 'Encyclopédie méthodique Hist. Nat. des Zoophytes,' tom. ii, p. 434.

‡ *Nucleolites umbrella.* Defrance, ' Dict. des Sciences Naturelles,' art. Galérites, tom. xviii, p. 87, 1825.

§ *Echinoclypeus umbrella.* Blainville, 'Dict. des Sciences Naturelles,' art. Zoophytes, tom. ix, p. 189, 1830.

‖ *Nucleolites umbrella.* Desmoulins, 'Études sur les Echinides,' p. 354, No. 2, 1837.

tected this primary error; so true is it, that unless we compare typical specimens with each other, the highest authorities may be misled by imperfect figures.

In the Échinodermes Foss. de la Suisse,' M. Agassiz* figured as *Pygaster umbrella,* an urchin found by M. Gressly in "le Portlandien du Jura Soleurois." This figure was reproduced by M. Desor,† in his 'Monographie des Galérites.' In the 'Catalogue raisonné,' however, these authors have changed the name of the Swiss urchin to *P. dilatatus,* and have given that of *P. umbrella* to a species from the Coral Rag of Châtel-Censoir (Yonne).

In the "Études sur les Échinides Foss. du département de l'Yonne," M. Cotteau has given for the first time good figures with details, and a description of this species; and he has likewise kindly sent me type specimens thereof, so that I have ample materials for comparing our specimens with *P. umbrella,* Agassiz (1847). Unfortunately M. Cotteau's figured specimen is partly denuded of its shell, and my specimen is only an interior mould, but the general form of the test, and the figure of the vent opening, are well preserved.

The test of *Pygaster umbrella* is thick; in some specimens it has nearly a circular, in others a sub-pentagonal, form; in some the upper surface is moderately convex (fig. 2 c), in others it is more or less depressed; in fact, its general outline varies much with age, sex, and external conditions, so that it is difficult to give a general description of its form that will be true for three or four specimens from different localities.

The test (fig. 2 a, c) was kindly given me by my friend Mr. J. G. Lowe; it exhibits the var. a, with a circular outline, and is selected in consequence of the fine preservation of the shell on its upper surface. The ambulacral areas are narrow and lanceolate (fig. 2 a), the single area and the anterior pair are straight, and the posterior pair are only slightly curved inwards at their upper part. They have two marginal rows of tubercles, which extend uninterruptedly from the peristome to the disc, there are thirty tubercles on each row between the margin and the apex, and from eighteen to twenty at the base; midway between the peristome and the margin two inner rows of tubercles commence, which extend only one third the distance up the sides, and at the widest part of the areas a few additional tubercles forming a fifth row exist (fig. 2 d). As the marginal tubercles are developed on every third plate, it follows that there are $48 \times 3 = 114$ plates in each area. The poriferous zones are narrow (fig. 2 d), one pair of pores corresponding to one ambulacral plate; the septa are narrow, and slightly elevated on the surface; and there are from six to seven pairs of pores opposite each inter-ambulacral plate (fig. 2 d).

* *Pygaster umbrella.* Agassiz, 'Description des Echinodermes Fossiles de la Suisse,' 1" partie, p. 8, tab. 12, figs. 4—6, 1839.

Pygaster umbrella. Agassiz, 'Catalogus Systematicus Ectyporum Echinodermatum Fossilium,' p. 7, 1840.

† *Pygaster umbrella.* Desor, 'Monographie des Galérites,' p. 77, tab. 12, figs. 4—6, 1842.

The inter-ambulacral areas are four times as wide as the ambulacral; there are about thirty-two pairs of plates in each column, of which eighteen belong to the upper surface, and fourteen to the base; the plates are bent in the middle, and at the widest part of the area, each plate supports seven tubercles; the four nearest the zones are arranged in oblique pairs, and the three nearest the centro-suture in horizontal lines (fig. 2 *d*); at the margin there are from eighteen to twenty tubercles in one horizontal series.

The tubercles on the upper surface are small (fig. 2 *a*); they have slightly sunken areolas, and are raised upon small bosses with smooth summits (fig. 2 *f*); the areolas are surrounded by six small granules (fig. 2 *f*), and others are scattered sparingly on the surface of the plates; these granules, like the tubercles, are all perforated (fig. 2 *d*).

The tubercles at the base are much larger (fig. 2 *b*), and they are arranged in close-set, horizontal lines; the areolas are more excavated, and rather square-shaped; they are bounded by a single row of granules (fig. 2 *g*); and the tubercles of both areas are about the same size.

The base is concave, slightly so at the sides, but much depressed at the centre; the mouth opening is small, and the peristome is equally decagonal. " Its circumference is armed with ten prominent carinæ, which, in the interior moulds, have left on the borders of the inter-ambulacral areas very apparent imprints." * These carinæ, which M. Desor† had already observed in a specimen of *Pygaster costellatus*, are doubtless destined in the *Pygasters* to replace the auricles of the Cidaridæ and Clypeasteridæ. M. Michelin‡ has recently discovered in a *Pygaster umbrella* a masticating apparatus, which resembles that of a *Clypeaster*, the united pieces of which form a very acute pyramid.

The anal opening is very large, and occupies more than the upper half of the single inter-ambulacrum (fig. 2 *a*); it has a pyriform figure, contracted above and swelling out below, which, with the discal space, produces a keyhole-like opening in the test. In most of the specimens this aperture is not symmetrical, and bulges more to the left than to the right side; in fig. 2 *a* a portion of the test is broken, which gives the vent in this specimen a more symmetrical form than in all the others I have examined (fig. 2 *c*).

The greater portion of the apical disc is preserved in one of the specimens from Malton, kindly given me by Dr. Murray (fig. 2 *b*, *e*); the genital plates have an irregular rhomboidal form, with pointed apices, and perforations for the canals near the apex; the right antero-lateral plate is disproportionately large, and the shield-like madreporiform body extends backwards into the centre of the disc (fig. 2 *e*); the ocular plates are small triangular bodies wedged into the angles formed by the genitals; the eye-holes are marginal and lodged in a depression between the plate and the apex of the ambulacral area (fig. 2 *e*); the surface of

---

* Cotteau, ' Études sur les Échinides Foss.,' p. 197.

† ' Monogr. des Galérites,' p. 76.

‡ Hardouin Michelin, Déscription de quelques nouvelles espèces d'Echinodermes fossiles, ' Revue et Magasin de Zoologie,' 2ᵉ série, tom. v, p. 36, 1853.

all the disc plates is covered with numerous small granules; and still smaller granules are scattered over the surface of the madreporiform body (fig. 2 *e*).

Affinities and differences.—This species very much resembles *Pygaster semisulcatus*, Phil., and was for a long time mistaken for that species; Professor Phillips having stated that *P. semisulcatus* was common to the Great (?) Oolite of Whitwell and the Coralline Oolite of Malton; this, however, I have proved, by an examination of the specimens, to be an error, the Malton *Pygasters* being quite distinct from the Whitwell species.

In *P. umbrella* the tubercles on the upper surface are always smaller, and not very regularly arranged in vertical and horizontal rows; the bosses are less elevated, the areolas less defined, and the intervening granules smaller and fewer in number than in *P. semisulcatus*. The discal opening is smaller; the anal opening is longer, more pyramidal, and seldom symmetrical, occupying sometimes two thirds of the area; the basal tubercles of both species are nearly alike in size; in *P. umbrella*, however, there is a much greater disproportion between the size of the tubercles on the upper and under surfaces of the test than exists in *P. semisulcatus*.

P. umbrella is easily distinguished from *P. Morrisii* by the size of the tubercles in the latter, its more decided pentagonal form, larger ambulacra, and tumid sides.

P. dilatatus, as stated by MM. Agassiz and Desor, "Se distingue du *P. umbrella* par son bord plus tranchant et sa forme plus dilatée." The shape of the anal opening is different, and it wants the pyramidal figure so characteristic of *P. umbrella*; it belongs likewise to a different stratigraphical horizon, having been collected from the " Portlandien du Jura Soleurois, carrière de Greifel (vallée de la Birse) ;" whilst *P. umbrella* is found only in the Coral Rag.

Locality and Stratigraphical position.—I have collected this urchin from the lower calcareous grit at Headington, near Oxford; from the Coralline Oolite of Malton, Yorkshire; from the Coral Rag near Farringdon, Berks; and from the Coral Rag of Lyneham and Calne, Wilts; in the two latter localities it was associated with *Cidaris florigemma*, *Pseudodiadema versipora*, *P. mamillanum*, and *Echinobrissus scutatus*—all true Corallian forms.

On the Continent it has been collected by M. Cotteau from the " Calcareo-siliceuses des environs de Druyes, à Châtel-Censoir, et à Montillot (Yonne) ;" by M. Buvignier from the Coral Rag environs of Saint Mihiel (Meuse); and M. Sæmann, of Paris, kindly sent me a specimen which was obtained from the Coral Rag of Commercey (Meuse).

NOTES

On Foreign Jurassic species of the genus PYGASTER nearly allied to British forms, but which have not yet been found in the English Oolites.

Pygaster laganoides, *Agassiz.* Échinoderm. Foss. Suisse, tabl. 12, figs. 13—16.
Desor. Monogr. des Galérites, tabl. 11, figs. 5—7.

Test sub-circular, depressed; four rows of tubercles in the ambulacra, which are one third the width of the inter-ambulacra; twelve rows of tubercles in the inter-ambulacra; the tubercles are large, and nearly all of the same size; base concave, mouth opening large, peristome feebly notched; anal opening large, of an elongated form, and occupying nearly two thirds of the single inter-ambulacrum.

Dimensions.—Height, six tenths of an inch; transverse diameter, one inch and six tenths; antero-posterior diameter, one inch and a half.

Formation.—Bathonien, "Calcaire à polypiers de Ranville," Normandy.

Collections.—Deslongchamps, Michelin, d'Orbigny; my cabinet.

Pygaster Gresslyi, *Desor.* Monogr. des Galérites, p. 80. Synopsis des Échinides Foss., tabl. 22, figs. 1, 2.
Cotteau. Études sur les Échinides Foss., pl. 28, figs. 2—6.

Test thin, small, sub-pentagonal, depressed; inflated at the sides, flat at the base, ambulacra with four rows, inter-ambulacra with twelve rows of tubercles at the circumference; the tubercles are large, equal-sized, and surrounded by granules; anal opening pyriform, contracted near the disc; base concave, mouth opening moderate in size, peristome deeply notched; tubercles at the base surrounded by hexagonal areolas.

Dimensions.—Height, six tenths of an inch ; transverse diameter, one inch and a half; antero-posterior diameter, nearly one inch and a half.

Formation.—Corallien, " Couches supérieures de l'étage corallien, et recueillis dans les carrières de Vauligny près Tonnerre (Yonne)." *Cotteau.* " Dans un banc à coraux du Portlandien, à Rædersdorf (Haut-Rhin)." *Agassiz.*

Collections.—M. Rathier, M. l'abbé Bellard, M. Gressly, M. Cotteau.   Rare.

PYGASTER PATELLIFORMIS, *Agassiz.*   Échinoderm. Foss. Suisse, tabl. 13, figs. 1—3.
                           Desor.   Monog. des Galérites, tabl. 11, figs. 11—13.

Test thick, hemispherical, more or less depressed ; circumference sub-pentagonal ; tubercles large, equal-sized, and arranged in regular rows ; ambulacra with four rows of tubercles ; inter-ambulacra four times the width of the ambulacra, with fourteen rows of tubercles ; all the tubercles surrounded by small granules ; anal opening large, pyriform, contracted above, wide below, occupying two thirds the length of the area ; mouth opening large, peristome deeply notched.

Dimensions.—Height, one inch and a quarter ; transverse and antero-posterior diameters, three inches.

Formation.—" Kimmeridge de Lauffon dans la vallée de la Birse (Berne)." *Agassiz.*

Collections.—Museum of Neuchâtel, M. Gressly.   Very rare.

PYGASTER DILATATUS, *Agassiz.*   Syn. *Pygaster umbrella*, Agassiz.   Échinoderm. Foss. Suisse, part. 1e, tabl. 13, figs. 4—6.
                           Pygaster umbrella, Desor.   Monogr. des Galérites, tabl. 12, figs. 4—6.

This is the urchin which was described by MM. Agassiz and Desor in their respective works as the type of *Pygaster umbrella*.   In their ' Catalogue raisonné,' however, it is separated from *P. umbrella* of the Coral Rag, under the name *P. dilatatus*, with this remark—" Se distingue du *P. umbrella* par son bord plus tranchant et sa forme plus dilatée." The original specimen is an interior mould ; M. Gressly has, however, found one with a portion of the test preserved.   The fragment of the test is very thick, and has large tubercles, disposed nearly as in *P. patelliformis*.   The general form of the mould, and the shape of the anal opening, which is pyriform, likewise resemble that species. M. Desor, in his 'Synopsis,' remarks—"Grande espèce subpentagonale, étalée, à bord plus aminci que dans aucune autre espèce." It is probable that *P. patelliformis* and *P. dilatatus* may be different conditions of the same species, as they are both Kimmeridge forms.

Dimensions.—Height, two inches; transverse diameter, four inches and a half.

Formation.—" Kimmeridge du Jura Soleurois, carrière de Greifel, vallée de la Birse Canton de Berne." *Agassiz.* Very rare.

Collection.—M. Gressly.

PYGASTER PILEUS, *Agassiz.* Syn. *Pygaster pileus,* Cotteau. Études sur les Échinides Foss., pls. 29 and 30.
   Pileus hemisphæricus, Desor. Syn. Échinod. Foss., tabl. 22, fig. 6.

Test large, sub-pentagonal; upper surface sub-conical; base flat or sub-concave; ambulacra convex and prominent, with six rows of small, irregularly disposed tubercles; inter-ambulacra wide, with numerous small tubercles disposed partly in single, and partly in double, horizontal rows; middle of the areas depressed, and deprived of tubercles; anal opening pyriform, situated in the lower half of the single inter-ambulacrum; between the apex of the vent and apical disc the test is entire, so that a large triangular space exists in this species, filled with the plates of the test, not found in other *Pygasters,* in all of which, with this exception, the discal opening and vent conjoin; mouth small, situated in a deep depression in the centre of the under surface.

Dimensions.—Height, one inch and three quarters; transverse and antero-posterior diameters, four inches.

Formation.—" Coral-rag inférieur; dans les couches blanches et pisolithiques de Coulanges-sur-Yonne et de Châtel-Censoir; calcaire à chailles de Druyes." *Cotteau.*

Collections.—Muséum d'histoire naturelle de Paris, M. Salomon, M. Cotteau. Very rare. Plaster mould in my cabinet.

In consequence of this *Pygaster* having the anal opening in the lower part of the inter-ambulacrum, and a considerable portion of the test between its apex and the disc, M. Desor has erected it into a new genus, under the name *Pileus hemisphæricus.*

PYGASTER MACROCYPHUS, *Wright*, nov. sp.   Davidson's MSS., pl. 2, figs. 1, 2, 3.

Test thick, large, sub-pentagonal; upper surface convex, or more or less depressed; base concave; ambulacra narrow, with two irregular rows of small marginal tubercles, and two short rows, equally irregular, internal to them, which disappear from the upper two thirds of the areas; inter-ambulacra five times the width of the ambulacra; each of the plates near the border supports about seven unequal-sized tubercles; those representing the primary rows are larger than the others; the tubercles, however, are in general large, and very irregularly disposed; the inter-tubercular surface of the plates is covered with an abundance of miliary granules; the bosses of the tubercles are prominent, and surrounded by ring-like areolas, and the granules form circles around each; there is a median depression in the centre of the upper part of the areas; the anal opening is small in proportion to the size of the urchin, and occupies nearly the upper two thirds of the area; it appears to have had a pyriform shape, much of the upper surface of the test is broken in my specimen, so that its precise form is indeterminable; base concave; mouth opening small, one sixth the diameter of the test; peristome deeply notched, lobes equal.

> *Dimensions.*—Height, one inch and a half; transverse diameter, four inches and a half.

> *Formation.*—Kimmeridge clay, from a cliff between Boulogne-sur-Mer and Portel. Very rare; I only know two specimens.

> *Collections.*—Thomas Davidson, Esq., F.R.S.   The specimen in my collection was found by M. Bouchard-Chantereaux in the same rock and locality, and was generously given me by that gentleman for this work.

PYGASTER TENUIS, *Agassiz.*   Échinoderm. Foss. Suisse, p. 83.
Desor.   Monogr. des Galérites, tabl. 12, figs. 1—3.

" A very large, sub-pentagonal, depressed species, with small and numerous tubercles, twenty rows in the inter-ambulacral areas, and six in the ambulacral areas, which are not very regular.   This species is distinguished by its very thin test.

" *Formation.*—Corallien inférieur (Terrain à chailles) de Fringeli, Canton de Soleure.

" *Collection.*—M. Gressly.   Very rare." *Desor.*

Genus—HYBOCLYPUS. 1839.

The urchins grouped in this genus present an assemblage of characters which belong to so many distinct forms of ECHINOIDEA, that they may probably hereafter be found to constitute a separate family, rather than a section of the ECHINOCONIDÆ, with which they are now provisionally placed for want of sufficient materials to justify such a separation.

Their test is thin, and in general as wide as it is long; the anterior half is more elevated than the posterior half, and sometimes rises into a prominent ridge. The surface of the plates is covered with numerous concentric rows of small perforated tubercles, set close together, and raised on low crenulated bosses, which are encircled by sunken areolas, and the intermediate surface is crowded with microscopic miliary granules.

The ambulacral areas are narrow, flexuous, and disjointed at the summit by the length of the apical disc. The three anterior areas terminate at the front of the disc, and the posterior pair at some distance behind them. The single area is lodged in a depression of the anterior border.

The poriferous zones are very contracted; the pores are simple and unigeminal; they are placed close together on the upper surface, and wide apart at the base.

The inter-ambulacral areas are very wide; the single inter-ambulacrum is traversed superiorly by a deep longitudinal valley, which commences behind the apical disc; in the upper part of this channel the wide vent opens on the dorsal region.

The apical disc is central, but not vertical; it is narrow and elongated, in consequence of the length and singular disposition of its component elements; the anterior pair of ovarial plates are shield-shaped; the right plate is the largest, and supports the madreporiform body; the posterior pair are much larger and longer than the anterior pair, and both pairs are perforated; the single ovarial plate is composed of two or more long, narrow, imperforate pieces, placed end to end in the centre of the disc; the single ocular plate is small; the anterior pair are large, and disposed side by side in a line in the middle of the disc, *between* the anterior and posterior ovarials, instead of being lodged in angles formed *by* them; the posterior pair of oculars are situated at the end of the posterior ovarials (Pl. XXI, fig. 2 *e*), and thereby give lengthened extension to the disc.

I have discovered this singular arrangement of the discal elements in three different species, and now regard it as a valuable generic character (Pl. XX, fig. 2 *e*; Pl. XXII, fig. 1 *e*, and fig. 2 *h*).

The small mouth opening, in general, is situated near the anterior third of the base, which in some species is concave, and much undulated; the peristome is sub-pentagonal or elongated in the antero-posterior diameter.

The *Hyboclypi*, like other ECHINOCONIDÆ, have simple poriferous zones, perforated

38

and crenulated tubercles, and manifest a disposition to a conical elevation of the upper surface of the test; they are unlike the true type forms of this family, however, in the excentral position of the mouth, the absence of notches from the peristome, in the possession of a longitudinal dorsal valley, and in having the opening of the vent therein; the elongation of the apical disc, the singular disposition of its elements, and the disjunction of the posterior pair from the three anterior ambulacra, which converge around the front of the disc, form a group of negative characters that sufficiently justify our doubts as to the propriety of placing this genus in the family ECHINOCONIDÆ.

The *Hyboclypi* resemble in some respects the COLLYRITIDÆ, in possessing an elongated apical disc, and having two ambulacral summits, consequent on the disjunction of the anterior from the posterior pair of ambulacra. They have affinities with the ECHINOBRISSIDÆ in having the test contracted before and expanded behind, the mouth small and excentral, and the vent opening into a longitudinal valley on the dorsal surface; but in the structure of the tubercles, the apical disc, and poriferous zones, they are very different from all the forms of that family.

The *Hyboclypi* form an aberrant type of the ECHINOIDEA, having close affinities with the ECHINOCONIDÆ on the one side, and the COLLYRITIDÆ on the other; they appear to form a transition link between these two families, and are probably entitled to rank as a separate sub-family.* Our imperfect knowledge of the intimate structure of the mouth, and the presence or absence of jaws in these urchins, make it uncertain how far this separation would be justified by anatomical characters, concealed or unknown.

The different species of this genus have hitherto been only found in the Oolitic rocks; they had their greatest development in the lower division of that series, as all the English species are found chiefly in the Inferior Oolite.

A. *Species from the Inferior Oolite.*

HYBOCLYPUS AGARICIFORMIS, *Forbes.*   Pl. XXI, fig. 1 *a, b, c, d, e, f, g.*

PYGASTER SUBLÆVIS?	M'Coy, Annals and Magazine of Natural History, 2d series, vol. ii, p. 413.
HYBOCLYPUS AGARICIFORMIS.	Wright, Annals and Magazine of Natural History, 2d series, vol. ix, p. 99.
— —	Forbes, Memoirs of the Geological Survey, Decade V, pl. 4.

---

* The genus *Hyboclypus* is placed by M. Desor, in his 'Synopsis des Échinides Fossiles,' p. 192, in his 2d type of the family Galéridées—" *Genres à appareil apicial alongé, sans plaque génitale impaire.*" I have already shown that the *Hyboclypi* not only possess a single imperforate ovarial plate, but that it is composed of two or more pieces. I must therefore regard M. Desor's definition as a mistake, which has probably arisen from the imperfect condition of the disc in the specimens he examined.

Hyboclypus agariciformis.	Forbes, in Morris's Catalogue of British Fossils, 2d edition, p. 82.
Nucleolites decollatus.	Quenstedt, Handbuch der Petrefactenkunde, tab. 50, fig. 6, p. 585.
Galeopygus agariciformis.	Cotteau, Mém. lu à la Soc. Géol. de France, le Juin, 1856.
— —	Desor, Synopsis des Échinides Fossiles, p. 167.

Test in general disciform, sometimes convex or conical, with a pentagonal circumference; ambulacral areas narrow, the three anterior straight, the posterior pair sinuous; inter-ambulacral areas wide, unequal, the plates covered with a great number of microscopic tubercles; apical disc small, central, and vertical; anal valley deep, with parallel sides, which gradually expand about the middle of the single inter-ambulacrum; mouth opening small, sub-central, nearer the anterior border; peristome feebly decagonal; poriferous zones narrow, pores unigeminal in the upper surface, at the base wide apart and trigeminal.

Dimensions.—The measurements given in the following table are made from four specimens which represent the four varieties this species assumes.

No.	*Hyboclypus agariciformis*, Forbes.	Height.	Transverse diameter.	Antero-posterior diameter.
		Inches.	Inches.	Inches.
1	Disciform var. α . . .	$0\frac{7}{10}$	$3\frac{3}{4}$	$3\frac{6}{10}$
2	Type var. β . . .	1	$2\frac{8}{10}$	$2\frac{3}{4}$
3	Convex var. γ . . .	$1\frac{2}{10}$	$2\frac{4}{10}$	$2\frac{4}{10}$
4	Conical var. δ . . .	$1\frac{3}{10}$	$2\frac{3}{10}$	$2\frac{3}{10}$

Description.—In the above table I have given the measurements of four specimens of this urchin, with the view of showing the diversity which exists in the relative proportions of different forms of the same species. Were an observer to find only Nos. 1 and 4, he would be almost justified in considering them distinct species; but when he discovers a number of intermediate forms, by which the depressed, hemispherical, and conical varieties are seen to blend into each other, he is convinced that they all belong to one and the same species, and at the same time is taught the important lesson, that mere difference in form and bulk alone do not constitute a specific character.

This fine urchin, the oldest of the genus, is the largest *Hyboclypus* we are acquainted with; it has a sub-orbicular or sub-pentagonal shape (Pl. XXI, fig. 1), and is the most typical form of the discoidal variety No. *a*. It is expanded and depressed above,

with an acute and sinuous margin arising from the convexity of the inter-ambulacra
(fig. 1 c); the base is flat, or moderately concave, and is slightly undulated by the
depressions formed by the ambulacra, and the convexities by the inter-ambulacra.   In
general the antero-posterior equals the transverse diameter, but sometimes the transverse
exceeds the antero-posterior diameter (fig. 1 a, b).

The ambulacral areas are very narrow and of unequal width; the anterior single area
is the narrowest; the antero-laterals are a little broader than the odd one, and narrower
than the postero-laterals (fig. 1 a); the three anterior areas converge around the front of
the apical disc, whilst the postero-laterals curve sinuously up to its posterior border
(fig. 1 a); each area has four or six rows of small tubercles, so arranged on the plates that
they form oblique rows, which meet in the median line, and branch upwards and
outwards, thereby forming v-shaped figures (fig. 1 d).

The poriferous zones are narrow; on the dorsal surface the pores are unigeminal,
and placed close together (fig. 1 d); at the base, in consequence of the plates being broader,
they are wider apart, and form slightly oblique ranks of threes; whilst nearer the mouth
they lie in very oblique trigeminal ranks (fig. 1 e); the oral portions of the ambulacra form
a radiate rosette around the mouth.

The inter-ambulacral areas are very wide, but of unequal width; on an average they
are eight times as broad as the ambulacrals (fig. 1 a, b).   In one specimen the anterior
pair measure at the circumference one inch and three tenths, the posterior pair one inch
and eleventh twentieths, and the single area is one tenth of an inch wider than the
posterior pair; the margin of all the areas is convex, except the posterior single inter-
ambulacrum, which is slightly truncated.   The dorsal surfaces of the anterior and posterior
inter-ambulacral and of all the ambulacral areas are gently convex; but the single inter-
ambulacrum, which is somewhat wider and longer than the others, has a deep valley with
parallel vertical sides in its dorsal half; these gradually decline and expand into a concave
depression at the lower half (fig. 1 a, c).

The anal aperture opens into the upper part of this valley immediately below the
apical disc; the basal portion of the area is slightly produced and truncated, and forms a
lip-shaped process, which imparts a considerable prominence and convexity to it (fig. 1 c).

The base is concave and undulated (fig. 1 b, c), the ambulacra lie in nearly straight,
depressed valleys, whilst the inter-ambulacra form gentle convex eminences at the circumference; the mouth opening is small and excentral, being nearer the anterior border;
when well exposed, the peristome is seen to be unequally decagonal, the ambulacral being
larger than the inter-ambulacral lobes (fig. 1 b).

The apical disc is small and central; the margin of the opening is notched with four
inter-ambulacral and five ambulacral notches, the former corresponding to the external
angle of the ovarial plates, the latter to the margins of the oculars (fig. 1 a); the disc
appears to have been lodged in a depression; but in none of the hundreds of specimens
which I have examined have I detected any of its elements.

The whole of the plates on the dorsal surface are covered with numerous, minute, regular, and nearly equal-sized, perforated tubercles (fig. 1 *f, g*), which are raised on bosses with crenulated summits, and surrounded by depressed, ring-like areolas; on one plate near the border I have counted as many as one hundred tubercles; the inter-tubercular surface of the plates is, besides, strewed with microscopic granules, which form circles around the areolas, and fill up the intervening spaces (fig. 1 *f*); the tubercles at the base are much larger and better developed than those on the dorsal surface; at the border and outer third they are very numerous and set close together; the areolas here are diamond-shaped, or hexagonal, and are more excavated, a single row of granules only separates the areolas from each other; towards the mouth they are larger, and not so regularly arranged, having several rows of granules between them (fig. 1 *e*).

Affinities and differences.—The adult forms of *H. agariciformis*, Forbes, differ so widely from its other Oolitic congeners, that this urchin cannot be mistaken for either of them; its sub-orbicular shape and depressed dorsal surface distinguish it from *H. gibberulus*, Agassiz, and *H. ovalis*, Wright; the central position of the apical disc, and the absence of any elongation of the single inter-ambulacrum, are diagnostic distinctions between it and *H. caudatus*, Wright. From the *Pygasters*, with which some of the species have been erroneously grouped, it is distinguished by the microscopic character and greater number of its tubercles; the deep anal valley, with vertical walls, in which tho vent opens, tho small sub-central mouth opening, and narrow apical disc, form a group of generic characters by which the *Hyboclypi* are distinctly separated from the *Pygasters*; whereas the narrow poriferous zones, complete and continuous from the mouth to the apical disc, without any petaloidal expansion on the dorsal surface, distinguish *Hyboclypus* from *Clypeus*, *Echinobrissus*, and *Pygurus*.

Locality and Stratigraphical position.—This urchin is very abundant in the lower ferruginous beds of the Inferior Oolite, "the Pea Grit" of Leckhampton, Crickley, Cooper's, Cleeve, and Sudely Hills, and at Camlong Down, near Uley Bury, in Gloucestershire. It was collected from the Inferior Oolite by the Geological Surveyors between Wayford and Seaborough in Dorsetshire, where it was accompanied by *Collyrites ringens*, Agassiz, and *Holectypus hemisphæricus*, Desor. I have found two specimens in the Great Oolite of Minchinhampton, Gloucestershire, along with *Purpurina Morrisii*, *P. nodulata*, Young, and other species of Mollusca characteristic of that formation.

History.—This species was first described in my Memoir on the CASSIDULIDÆ of the Oolites, and afterwards figured by Professor Forbes in the IV Decade of the 'Memoirs of the Geological Survey.' In consequence of the opening for the apical disc having a denticulated border, M. Cotteau has proposed for it the new genus *Galeopygus*; but as I have never seen any of the elements of the disc, I am unable to judge of the value of the character on which the separation of this urchin into a new genus is proposed to be made.

HYBOCLYPUS CAUDATUS, *Wright.*　　Pl. XXII, fig. 2 *a, b, c, d, e, f, g, h, i, j, k.*

HYBOCLYPUS CAUDATUS.	Wright, Annals and Magazine of Natural History, 2d series,
— —	vol. ix, p. 100, pl. 3, fig. 2 *a—e.*
— —	Forbes, in Morris's Catalogue of British Fossils, 2d ed., p. 82.
— —	Desor, Synopsis des Échinides Fossiles, p. 193.

Test small, oblong, much depressed; the single inter-ambulacrum produced into a caudal prolongation; mouth very near the anterior border; apical disc and vertex excentral; anterior border rounded and elevated, the posterior produced, and truncated.

Dimensions.—Height, nine tenths of an inch; antero-posterior diameter, one inch and two tenths; transverse diameter, one inch and one tenth. The great majority of the specimens, however, have the following measurement: Height, seven twentieths of an inch; antero-posterior diameter, fifteen twentieths of an inch; transverse diameter, thirteen twentieths of an inch.

Description.—The test of this elegant little urchin has an oblong shape, rounded and elevated before, produced and truncated posteriorly, and having the mouth and apical disc excentral, placed much nearer the anterior than the posterior border; the surface of the plates is covered with very small tubercles, which require the aid of a good lens to discover; without this the observer might suppose the test was altogether destitute of sculpture (fig. 2 *a, b*); in consequence of the excentricity of the mouth opening and disc, the single and antero-lateral ambulacral areas are straight and short, and terminate at the anterior border of the disc (fig. 2 *a*); the posterior pair are one seventh longer than the anterior pair; they curve upwards, inwards, and forwards on the dorsal surface, and terminate by the margin of the longitudinal valley, at a short distance from the posterior border of the disc (fig. 2 *a*).

The poriferous zones are narrow; the pores are situated some distance apart on the dorsal portion of the zones, and much wider apart at their basal region (fig. 2 *b, f*).

The inter-ambulacral areas are of unequal width; the anterior pair are the shortest and narrowest, the posterior pair the widest, and the single area the longest; it is likewise considerably produced into a lip-like process, which curves gently downwards, and is abruptly truncated posteriorly (fig. 2 *a, b*); the anterior and posterior pairs of the inter-ambulacral, and all the ambulacral areas, are convex on the upper surface; but the single inter-ambulacrum is traversed superiorly by a deep, broad, longitudinal valley, with vertical and parallel walls in the upper part, gradually expanding into two ridges at the lower half, and which form the lateral boundaries of the anal valley (fig. 2 *a*).

The anterior border is bluntly rounded (fig. 2 *f*), with a slight depression in the middle, formed by the single ambulacral area; the base is concave, and slightly undulated (fig. 2 *c*), in consequence of the basal portions of the inter-ambulacral areas being convex, and those of the ambulacral forming straight valleys between them (fig. 2 *b*). The small mouth opening is sub-pentagonal, and (fig. 2 *b*) situated near the anterior third; the peristome is undulated, and the inter-ambulacral are smaller than the ambulacral lobes; the tubercles at the base are larger than those on the dorsal surface; they are likewise fewer in number, and arranged with much irregularity on the plates (fig. 2 *b*).

The apical disc is fortunately preserved in one of my specimens (fig. 2 *d*); it is formed of two small anterior ovarial plates, and two larger posterior ovarials; the right anterior plate supports the spongy madreporiform body; between the posterior ovarials the single imperforate ovarial is situated, composed of two pieces, an anterior and a posterior half (fig. 2 *h*). This arrangement of the genital plates is common to all the *Hyboclypi* with preserved discs I have examined; I have therefore noted it as a character common to the genus. The five ocular plates are small, and wedged between the depressions formed by the ovarials; their eye-holes are marginal, whilst the perforations in the ovarials are situated near the border.

The tubercles are small and numerous, each plate having three or four concentric rows arranged in diagonal lines on its surface (fig. 2 *g*). As the dorsal portion of this pretty little urchin varies considerably, I have given outlines in figs. *c, f, i, k*, of some of the most remarkable deviations from what I consider as the typical form, fig. 2 *d*.

Affinities and differences.—*Hyboclypus caudatus* differs from the other Oolitic species in its more oblong form, and especially in having the single inter-ambulacrum developed into a kind of caudate process; by this character it is readily distinguished from *H. gibberulus*, Agassiz, and *H. ovalis*, Wright, which it otherwise resembles; the mouth and vent are likewise placed nearer the anterior border than in these allied forms. The sub-orbicular shape which *H. agariciformis* invariably retains throughout its numerous varieties of elevation and depression of the upper surface readily distinguish it from all the forms of *H. caudatus* I have met with.

Locality and Stratigraphical position.—This is rather a rare urchin. I have found it occasionally in the upper beds of the Inferior Oolite, "the Gryphæa Grit" of Leckhampton, Birdlip, Shurdington, and Ravensgate Hills, associated with *Gryphæa sublobata*, Desh., *Lima pecteniformis*, Scloth., *Myopsis punctata*, Buck., *Cercomya rostralis*, Wright, *Terebratula impressa*, V. Buch. It occurs likewise in the Trigonia Grit at Hampen, associated with *Ammonites Parkinsoni*, Sow., and the numerous other Mollusca and Echinida which characterise that rich zone of life.

It is found occasionally in the planking beds of the Great Oolite at Minchinhampton Common. The Great Oolite specimens, however, are small, and not well preserved. M. Deslongchamps kindly communicated a specimen which he collected from the "Oolite ferrugineuse de Bayeux." On the ticket which accompanied it was written, "Seul exemplaire que j'ai trouvé," so that it is extremely rare in Normandy. This urchin very much resembled our small common examples from the Inferior Oolite. I know it from no other foreign locality.

HYBOCLYPUS GIBBERULUS, *Agassiz.*   Pl. XXI, fig. 2 *a, b, c, d, e, f, g.*

HYBOCLYPUS GIBBERULUS.		Agassiz, Échinodermes Fossiles de la Suisse, part i, p. 75, pl. 13, figs. 10—12.
—	—	Desor, Monographie des Galérites, p. 84, pl. 13, figs. 12—14.
—	—	D'Orbigny, Prodrome de Paléontologie, tom. i, p. 290, étage 10*, Bajocien.
—	—	Bronn, Lethaea Geognostica, 3* Aufl., tabl. 17[1], fig. 11.
—	—	Agassiz and Desor, Catalogue raisonné, Annales des Sciences Naturelles, tom. vii, p. 152, série 3*.
—	—	Desor, Synopsis des Échinides Fossiles, tabl. 26, figs. 11—13, p. 192.
NUCLEOLITES EXCISUS.		Quenstedt, Handbuch der Petrefactenkunde, tabl. L, fig. 3, p. 585.

Test oblong, elevated above, and contracted on the sides before; enlarged, depressed, produced, and truncated behind; single ambulacral area the highest, and, with the antero-lateral inter-ambulacral areas, form a gibbous crest; longitudinal valley wide and deep; single inter-ambulacrum slightly produced, deflected, and truncated; base much undulated, a depression in the anterior border; apical disc elongate, nearly central, but not vertical; vent very wide, opening at the inner extremity of the valley; mouth opening small, oblong, sub-central, near the anterior border.

Dimensions.—Height, nearly one inch; transverse diameter, two inches and one tenth of an inch; antero-posterior diameter, two inches and one tenth of an inch.

Description.—This urchin has so singular a form, that when once seen, it is not likely to be mistaken for any other, being remarkable for a prominent gibbous crest (fig. 2 *d, e*), which rises from the anterior half of the test, formed by the elevation of the single ambulacral area, and the two anterior inter-ambulacral areas; from the anterior border to the mouth a depression extends (fig. 2 *b, d*); the anterior lateral are more contracted than the posterior lateral borders (fig. 2 *a, b*); the posterior half of the test is less elevated than the anterior half (fig. 2 *e*), and gradually declines from the vertex to the posterior border, which is a little produced and truncated (fig. 2 *a*, *b*).

The ambulacral areas are of unequal width; the single area is the narrowest, the anterior pair are a little wider, and the posterior pair are the widest; the single area makes a straight line from the mouth to the vertex; its upper half forms the ridge of the anterior gibbous crest (fig. 2 *d*); where it turns round the border a depression is formed, both there and at the base (fig. 2 *b*, *d*), by the bulging out of the inter-ambulacra; the anterior pair, between the border and the disc, are gently curved backwards (fig. 2 *e*), and the posterior pair are sinuous (fig. 2 *a*); the single and anterior pair terminate around the front of the disc (fig. 2 *a*), which is elongated in the longitudinal direction; the posterior pair terminate in the large specimen (fig. 2 *a*) within the longitudinal valley, one quarter of an inch behind the anterior pair; in each area there are six rows of tubercles, arranged so that they form oblique ᴠ-shaped lines (fig. 2 *f*).

The poriferous zones are narrow, and, from the border to the disc, the pores are placed close together, whilst, from the border to the mouth, they are wide apart (fig. 2 *a*); this arises from the ambulacral plates on the dorsal surface being narrow, whilst those at the base are broad; there are six pairs of pores opposite each of the large fourteen plates on the upper surface, which makes eighty-four pairs of pores on the dorsal portion of the zones, whilst there are only about fifteen pairs of pores in the base.

The inter-ambulacral areas are of unequal width; the anterior pair are the narrowest; the posterior pair are one third wider than the anterior, and the single inter-ambulacrum is the widest (fig. 2 *a*); the middle of its dorsal portion is occupied by the longitudinal valley, which is wide and deep above, and shallow and expanded below; the small and numerous tubercles are arranged very closely and regularly together in four or five concentric rows (fig. 2 *f*); and I have counted upwards of one hundred tubercles on one large plate near the margin.

The tubercles are very numerous at the border and external third of the base, whilst they are fewer, larger, and less regular between that point and the mouth; they are all perforated, and raised on low, crenulated bosses, surrounded by sunken circular areolas (fig. 2 *g*); the inter-tubercular surface of the plates is likewise covered with close-set miliary granules (fig. 2 *g*).

The base is concave, and much undulated (fig. 2 *b*, *c*), the basal portions of the inter-ambulacra being very convex (fig. 2 *d*), whilst the ambulacra lie in narrow valleys between them; the small mouth opening is sub-central, and situated nearer the anterior border, almost opposite to the vertex (fig. 2 *b*); the peristome has a sub-pentagonal form, but its minute structure is more or less concealed in all my specimens.

The apical disc is nearly central; it lies behind the summit of the crest, which rises above it (fig. 2 *e*), so that the vertex is before the disc in this remarkable species; it has a narrow, elongated form, and is composed of six ovarial and five ocular plates (fig. 2 *e* gives a faithful representation of the disc); the anterior are smaller than the posterior ovarials; the right is larger than the left plate, and supports, as usual, a spongy madreporiform body; the posterior ovarials are large and elongated, and separated from the anterior pair by the excessive development of the anterior pair of ocular plates, which form an important part of the disc; the large oviductal holes always indicate the true relative position of the genital plates; the single imperforate ovarial is composed of two plates, placed end to end, and behind the posterior ovarials the posterior pair of ocular plates are placed. M. Desor's figure of the disc of this species does not agree with our specimens, as the single imperforate ovarial, which is central and double, is entirely wanting in his figure, probably through an oversight on the part of the artist. I have verified Mr. Bone's beautiful drawing in three different individuals by a careful examination with the lens.

Affinities and differences.—This remarkable urchin, which formed the type of the genus, is so well characterised by its gibbous crest that it cannot be mistaken for any other species. *H. ovalis* has a considerable resemblance to it; but the absence of the crest, its oval form, and the want of those prominent traits which distinguish *H. gibberulus*, clearly distinguish them from each other.

Locality and Stratigraphical position.—I have collected this species in the upper beds of the Inferior Oolite, in the zone of *Ammonites Parkinsoni*, Sow., at Burton Bradstock, and Walditch Hill, near Bridport, Dorset; a few specimens have been found in the Inferior Oolite in the parish of Charlcomb, near Bath. The Dorsetshire specimens were associated with *Holectypus hemisphæricus*, Desor, *Stomechinus bigranularis*, Lamk., *Clypeus altus*, M'Coy, *Collyrites ringens*, Agass., *Terebratula sphæroidalis*, Sow., *Rhynchonella plicatella*, Sow., *Terebratula Phillipsii*, Mor., *Ammonites Parkinsoni*, Sow., *Ammonites sub-radiatus*, Sow.

It appears to be a rare species on the Continent. M. Desor observes, "Suivant les indications que m'a fournies M. Gressly, on la trouve dans une couche particulière de l'Oolite inférieure, la marne à *Ostrea acuminata*, la même qui contient aussi *Discoidea* (*Holectypus*) *depressa*, et le *Dysaster analis*, c'est au moins dans cette couche que l'a rencontrée M. Strohmeyer."* In his 'Synopsis des Echinides Fossiles,' M. Desor gives, as the other foreign localities of this species, "Vesulien (Marnes à Discoidées du Jura Soleurois et Argovien, du Hummel, près Waldenburg. Grande Oolite de Macon (Hébert), du départ. de l'Ain."

M. De Loriére has sent me a specimen which he collected at "Nogent, étage Callovien, département de la Sarthe;" and from M. Sæmann I have received another from the étage Callovien of the same department. I have already expressed a doubt as to the age

---

* 'Monographie des Galérites,' p. 85.

of the beds which have yielded *Hyboclypus gibberulus*, *Pygurus depressus*, and *Collyrites ringens*, in the department of the Sarthe; my friend M. Cotteau is at this moment engaged on a work on the Echinoderms of that department, and my friend M. Triger has undertaken a review of the stratigraphical distribution of the species. From this monograph we shall therefore learn the true age and position of these strata.

History.—This species was first figured and described by M. Agassiz, in his 'Échinodermes Fossiles de la Suisse,' he only knew two specimens from the Inferior Oolite of Switzerland. It was afterwards figured by M. Desor in his valuable 'Monograph on the Galerites,' and was described by me as a British fossil for the first time.* It has lately been figured by M. Bronn in his 'Lethaea Geognostica,' and by M. Desor in his 'Synopsis des Échinides Fossiles.'

HYBOCLYPUS OVALIS, *Wright*, nov. sp.   Pl. XXII, fig. 1 *a, b, c, d, e, f.*

Test oval or suborbicular, upper surface convex, rather more elevated anteriorly; ambulacral areas narrow, nearly of equal width; the single and anterior pair straight; posterior pair curved gently upwards, inwards, and forwards; apical disc small, nearly central, rather nearer the anterior border; longitudinal valley of moderate width and depth; base concave and undulated, from the convexity of the basal inter-ambulacra; mouth opening small, excentral, nearer the anterior than the posterior border.

Dimensions.—Height, seven tenths of an inch; transverse diameter, one inch and eleven twentieths; antero-posterior diameter, one inch and thirteen twentieths.

Description.—This species very much resembles *H. gibberulus*, but it wants the striking features of that urchin; the contracted anterior border, expanded posterior border, wide anal valley, and prominent gibbous crest are absent; although in the minute structure of the test there is much resemblance between them. It is collected, moreover, from the same stratigraphical horizon, the zone of *Ammonites Parkinsoni*, Sow., and may be regarded as the representative of *H. gibberulus* in the Cotteswold Hills, as that species, so far as I know, has never yet been found in this district.

The test is in general oval, but it has sometimes a sub-orbicular shape (fig. 1 *a*). The upper surface is very uniformly convex, and the anterior part is rather more elevated than the sides (fig. 1 *c*); most *Hyboclypi* manifest a disposition to the formation of

* 'Annals and Magazine of Natural History,' 2d series, vol. ix, p. 120.

an anteal elevation of the test, which attains its greatest development in *H. gibberulus*. (See fig. 2 *f, k.*)

The ambulacral areas have all nearly a uniform width; the single and anterior pair are nearly straight, and converge around the front part of the disc (fig. 1 *a*); the posterior pair are a little bent, and curve upwards, inwards, and forwards, to terminate at the posterior lateral part of the disc, their apices just falling within the longitudinal valley.

The apical disc is large, and formed of seven ovarial and five ocular plates; the two pairs of ovarials are perforated; the single plate is imperforate, and composed of three pieces (fig. 1 *e*); the pairs of ovarials expand between the apices of the ambulacra, and the posterior elements of the single plate bend over the upper part of the longitudinal valley (fig. 1 *a*); the madreporiform body is round, button-shaped, and placed between the anterior and the right antero-lateral ambulacra; it therefore rests on the surface of the right ovarial plate, as in other ECHINIDÆ. The single ocular plate is small; the anterior pair are large, and placed side by side in the centre of the disc, between the anterior and posterior pairs of ovarials; but the posterior oculars are not seen *in situ* in any of my specimens.

The base is concave, and much undulated, from the convexity of the basal portions of the inter-ambulacral areas; the mouth opening is small and excentral, it has an oval form, its long diameter being in the direction of the antero-posterior diameter of the body; around the border of the opening the poriferous zones become closely crowded with pores, which lie in triple oblique pairs, and form a radiate rosette around the orifice. These areas are well seen in fig. 1 *b*, and a portion of one of them, magnified, is drawn in fig. 1 *f*, where each of the broad plates are seen to be perforated with a pair of pores, having one, two, or three large tubercles on their surface.

There are four or five rows of small tubercles, so arranged on the plates that they form, as in most of the ECHINOCONIDÆ, oblique V-shaped lines (fig. 1 *d*). The poriferous zones are narrow, and on a level with the general surface of the test; the pores are placed close together on the dorsal surface, there being six pairs of pores opposite one inter-ambulacral plate; at the base they are wide apart, from the increased breadth of the ambulacral plates in this region (fig. 1 *b* and *f*).

The inter-ambulacral areas are five or six times the width of the ambulacral (fig. 1 *a* and *b*); they are, however, of unequal width; the anterior pair are the narrowest, the posterior are wider, and the single area is the widest; in its upper part is the longitudinal valley, which is very deep above, with vertical walls, that expand in the lower half, and form a concave depression (fig. 1 *a*); the posterior border, which is very little produced, is rounded, or only slightly flattened (fig. 1 *a* and *b*), and not truncated, as in *H. caudatus* and *H. gibberulus*. The tubercles are small, and very numerous (fig. 1 *d*); they are arranged in concentric rows, and I have counted fifty tubercles on the third plate, above the border; they are surrounded by sunken areolas, and the inter-tubercular surface is crowded with small granules; the tubercles of both areas are the same size.

Affinities and differences.—This urchin in its general *facies* bears so much resemblance to *H. gibberulus*, Agass., that at one time I considered it a variety of that species; but the absence of the gibbous crest, the fulness of the anterior border, the straightness of the anterior ambulacra, the form and direction of the posterior pair, together with the difference observed in the shape of the longitudinal valley, have induced me to separate it from that species under the name *H. ovalis*. By its oval shape and elongated apical disc it is distinguished from *H. agariciformis*, and by the shortness of the single inter-ambulacrum from *H. caudatus*.

Locality and Stratigraphical position.—I have collected this species only from the marly fossiliferous vein which traverses the upper ragstones of the Inferior Oolite, in the zone of *Ammonites Parkinsoni*, Sow., near Hampen, Gloucestershire, where it is associated with *Holectypus depressus*, Leske, *H. hemisphæricus*, Desor, *Pedina rotata*, Wright, *Clypeus Hugi*, Agass., *Clypeus Plottii*, Klein, *Trigonia costata*, Sow., *Pectin symmetricus*, Morris, *Ammonites Parkinsoni*, Sow. I have found one specimen in the Trigonia bed at Cold Comfort, where it was associated with the large *Perna isognomonoides*, Stahl.

B. *Species from the Coral Rag.*

HYBOCLYPUS STELLATUS, *Desor.*

HYBOCLYPUS STELLATUS.	Desor, Catalogue raisonné Annales des Sciences Naturelles, 3ᵉ série, tome vii, p. 152.
— —	Desor, Synopsis des Echinides Fossiles, p. 193.

This urchin is described by M. Desor as " Espèce intermédiaire par sa forme, entre les *H. canaliculatus* et *H. marcon*, mais différant de l'un et de l'autre par ses ambulacres postérieurs qui sont rectilignés an lieu d'être arqués. T. 76 (type du l'espèce).

" *Formation.*—Corallien du Wiltshire. Rare.

" *Collection.*—M. Le Viscomte d'Archiac."

I only know this species from the above notice, as I have never been so fortunate as to see a specimen of it in any of the collections of Coral-rag urchins which I have examined.

40

Family 7—COLLYRITIDÆ, *Wright*, 1856.

Famille 7me—Dysastéridées, *Albin Gras*, 1848.
Famille 1re—Collyritidæ, *D'Orbigny*, 1853 (pars).
Famille des Dysastéridées, *Desor*, 1856.

The urchins forming this family have a thin test which, in general, has an ovoid, elongated, or cordiform shape. The ambulacral areas converge at two points on the upper surface, which are more or less apart, whilst in all other Echinoidea, they converge towards one point, the apical disc, around which are the genital apertures, and the holes for the eyes. In the Collyritidæ, on the contrary, the ambulacra are divisible into two groups, the anterior, composed of the single area, and the anterior pair, meet at the apical disc, which is composed of the three anterior ocular, and four genital plates; the posterior, composed of the posterior pair of ambulacra, meet at some distance from the former, and form an arch over the anal opening. This disjunction of the ambulacra occasions a corresponding separation of the elements of the apical disc, as three of the ocular plates are placed before, and two behind, at the apices of the ambulacral areas; the four genital and three anterior ocular plates are intimately soldered together at the union of the three anterior ambulacral areas, and which junction forms the true vertex of the test.

The poriferous zones are narrow, the pores are unigeminal, and the tubercles in general small, numerous, perforated, and crenulated.

The mouth-opening is small, and placed near the anterior border; the peristome is oval or circular, and its margin is entire; the anal opening is round or oval, and is situated in the region of the posterior border, over which the two posterior ambulacra form an arch, the summit of which is more or less distant from the vent.

This family forms a small, natural group of urchins, connected with the Echinocoridæ on the one side, and with the Echinoconidæ and Spatangidæ on the other. Like the Echinoconidæ, the ambulacral areas in the Collyritidæ are narrow, the poriferous zones complete, the pores unigeminal throughout, and the tubercles perforated and crenulated. The genus *Hyboclypus* connects these two families by the elongated arrangement of the elements of the apical disc, and the partial disjunction of the apices of the three anterior from the two posterior ambulacra in that small group.

In the ovoidal form of the test, the structure of the ambulacra, and poriferous zones, the excentrical position of the mouth, and the supra-marginal situation of the vent, the Collyritidæ resemble the genus *Holaster*, by which they are connected with the family Echinocoridæ, but their two distinct ambulacral summits, placed at some distance apart, and the consequent disjunction of the posterior ambulacra from the apical disc, form a diagnostic character between them, and afford a good zootomical reason why they should not be united in one family, as proposed by the late M. A. d'Orbigny.

The petaloidal character of the ambulacra, circumscribed for the most part by *fascioles*, the bilabiate structure of the mouth, the completeness of the apical disc, and the union of the ambulacra around the same, form a group of characters by which the SPATANGIDÆ are distinguished from the COLLYRITIDÆ, which they only feebly resemble in external form, being widely separated from that family by organic structure.

M. Desor, who has carefully studied the species of this family, and figured them in his excellent 'Monograph on the Disasters,'* has made the following remarks on the affinities of the COLLYRITIDÆ, in his latest work now passing through the press.†
"On nous objectera peut-être qu'il existe certains types, les Ananchydées par example, chez lesquels les ambulacres ne sont qu'imparfaitement bornés, et qui n'en sont pas moins très voisins des Spatangoïdes. Nous répondrons à ceci que cette ressemblance des Ananchydées avec les Dysastéridées est plus apparent que réelle. Sans doute que vus par en haut, les ambulacres des Ananchydées n'ont pas cette apparence pétaloïde qui charactérise à un si haut dégré les Clypéastroïdes, les Cassidulides et les vrais Spatangoïdes. Mais qu'on les regarde par la face inférieure et l'on retrouvera le véritable type des Spatangoïdes, savoir de très larges plaques ambulacraires en général lisses et percées d'une seule paire de pores, tandis que chez les Dysaster les plaques ambulacraires de la face supérieure ne subissent aucun changement sensible, et sont beaucoup plus petites que les plaques inter-ambulacraires. Il y a donc sous ce rapport une bien plus grande variété de structure ambulacraire chez les Ananchydées. De plus ces derniers ont en général le péristome bilabié et, ce qui est plus significatif, l'ambulacraire impair est différent des ambulacraires pairs, tandis que chez les Dysastéridées tous les ambulacres sont égaux. Ces considérations suffiront, je l'espère, pour expliquer pourquoi nous n'avons pas suivi l'example de M. d'Orbigny qui réunit les Dysaster et les Ananchydées en une seule famille, sous le nom de Collyridées.

"D'une autre côté, il importe également qu'on ne confonde pas les Dysastéridées avec les Galéridées, comme on a pu être tenté de le faire à une certaine époque, alors que la structure intime du test n'était pas suffisamment étudiée. Le fait que chez certains Galéridées particulièrement ceux du second type, les ambulacres ne convergent pas complètement ne constitue pas encore un démembrement de l'appareil apicial. L'écartement plus ou moins considérable des ambulacres n'est ici que la conséquence d'une disposition particulière des plaques ocellaires qui, au lieu d'être rejettées dans les angles externes des plaques génitales, se placent avec elles sur le même rang. Mais l'unité de l'appareil apicial n'est pas rompu pour cela. C'est une combinaison qui se produit également dans la famille des Ananchydées, dans celle des Galéridées et même chez certains Dysastéridées."

* Monographie des Dysaster, 1842.
† 'Synopsis des Échinides Fossiles,' p. 200.

The family COLLYRITIDÆ includes at present four genera *Disaster, Collyrites, Metaporhinus,* and *Grasia,* which are limited to the Oolitic and Cretaceous rocks. The most ancient form at present known was found in the Lias; the species attained their greatest development in the middle division of the Oolites, are sparingly found in the Neocomian, and Lower Cretaceous rocks, and finally disappear with the White Chalk.

The following table exhibits the stratigraphical distribution of the Oolitic species of this family.

STRATIGRAPHICAL DISTRIBUTION OF THE OOLITIC SPECIES OF THE COLLYRITIDÆ.

List of Species.	Lias.	Inferior Oolite.	Great Oolite.	Kelloway.	Oxfordian.	Corallian.	Kimmeridge.	Type localities.
DISASTER.								
Disaster granulosus, Desml.	—	—	—	—	*	—	—	Amberg, Streitberg, Würtemberg.
„ *anasteroïdes,* Leym.	—	—	—	—	—	—	*	Bar-sur-Aube (Aube).
„ *Moeschii,* Desor	—	—	—	*	—	—	—	Kornberg (Argovie).
COLLYRITES.								
Collyrites prior, Desor	*	—	—	—	—	—	—	Frick, (Canton d'Argovie).
„ *ringens,* Desml.	—	*	*	—	—	—	—	England, France, Switzerland.
„ *ovalis,* Leske	—	*	*	—	—	—	—	England, France, Switzerland.
„ *elliptica,* Desml.	—	—	—	*	—	—	—	Chaufour, Mamers (Sarthe).
„ *castanea,* Desor	—	—	—	*	—	—	—	St. Croix, de Pouillerel.
„ *transversa,* D'Orbig.	—	—	—	*	—	—	—	Escragnolles (Var).
„ *bicordata,* Leske	—	—	—	—	*	*	—	England, France, Switzerland.
„ *carinata,* Desml.	—	—	—	—	*	—	—	Gunsberg (Canton Soleure.)
„ *capistrata,* Desml.	—	—	—	—	—	*	—	Würtemberg, Switzerland.
„ *Buchii,* Desor	—	—	—	—	—	*	—	Sirchingen, Würtemberg.
„ *Voltizii,* Agass.	—	—	—	—	*	—	—	Voirons près Genève.
„ *Desoriana,* Cotteau	—	—	—	—	—	*	—	Chatel-Censoir, Druges (Yonne).
„ *pinguis,* Desor	—	—	—	—	*	—	—	Bötzberg, Geissberg (Argovie).
„ *Orbigniana,* Cotteau	—	—	—	—	*	—	—	Stigny (Yonne).
„ *Loryi,* Alb. Gras.	—	—	—	—	—	*	—	Echaillon (Isère).
„ *acuta,* Desor	—	—	—	—	*	—	—	Villiers-les-Hauts (Yonne).
„ *conica,* Cotteau	—	—	—	—	*	—	—	Pacy, Ancy-le-Franc (Yonne).
„ *faba,* Desor	—	—	—	*	—	—	—	Ueken près d'Effingen (Argovie).
METAPORHINUS.								
Metaporhinus Michelinii, Cott.	—	—	—	—	—	*	—	Châtel-Censoir (Yonne).
„ *Censoriensis,* Cott.	—	—	—	—	—	*	—	Châtel-Censoir (Yonne).
GRASIA.								
Grasia elongata, Alb. Gras.	—	—	—	—	—	*	—	Echaillon (Isère).

Genus—COLLYRITES,* *Deluc,* 1831 ; *Desmoulins,* 1835.

DISASTER, *Agassiz,* 1836.
DISASTER, *Desor,* 1842.
COLLYRITES, *d'Orbigny,* 1853.
COLLYRITES, *Desor* (pars), 1857.

The generic characters of this group have been already indicated in our description of the family of which it is the type. The Collyrites, in general, are urchins of moderate size, although some species attain a considerable magnitude. They have an ovoid, oblong, cordiform, or triangular shape, more or less depressed on the dorsal surface, and have the anterior border slightly grooved, with an anteal sulcus, in which the single area is lodged. The ambulacral areas are disjoined, and form two distinct summits on the dorsal surface ; the anterior, composed of the single area and anterior pair, occupies the anterior third of the dorsal surface ; and the posterior, composed of the posterior pair, meet over the vent near the posterior third. The poriferous zones are narrow, equal, and complete ; the pores unigeminal, the holes round, oval, or oblong ; and the zones visible throughout from the mouth to the summits.

The apical disc, situated at the anterior summit, at the junction of the three anterior ambulacra, is composed of four perforated genital plates, between which are interposed two large; ocular plates. The right antero-lateral genital plate is the largest, and supports a prominent, spongy, madreporiform body ; the anterior ocular is a very small plate, which is lodged between the anterior genitals. The two posterior ocular plates are seen only in well-preserved specimens at the apices of the posterior ambulacra.

The mouth-opening is situated in the anterior third of the base ; the peristome is obtusely pentagonal, approaching a circular form.

The oval or elliptical anal opening is situated in the middle of the posterior border, and is in general destitute of a distinct anal area, and always without a fasciole.

The tubercles are small, uniform in size, and are perforated and crenulated ; the miliary granules are small and numerous.

The first known species of this genus were placed by systematic authors in different genera, with which they had few characters in common. Leske grouped them with the *Spatangites,* Lamarck with the *Ananchytes,* Defrance, Goldfuss, and Münster, with the *Nucleolites.* In 1831, M. Deluc, in a letter to M. Desmoulins,† proposed for the urchin which Lamarck named *Ananchytes elliptica,* that of *Collyrites sub-elliptica,* but it was not until August, 1835, that M. Desmoulins established definitely the genus *Collyrites.*‡

* From the Greek *Collyra,* a little loaf.
† Desmoulins ' Études sur les Échinides,' 1er Mémoire, p. 47.
‡ Desmoulins, loc. cit., p. 46.

About the same time M. Agassiz, was actively engaged in collecting materials for his great work on the Echinodermata, and in the course of 1836 appeared his Prodrome,* in which an important reform was proposed in the arrangement of that class, and many new genera were therein described for the first time; among these, was the genus *Disaster*, which nearly corresponded to the genus *Collyrites* of Desmoulins. The publication of this work formed an era in the History of the Echinidæ, and the author's classification and nomenclature was soon adopted by the naturalists of England, France, and Germany. In a note appended to the first page of the Prodrome, it was stated, that the memoir had been read before the Society of Natural Sciences at Neufchâtel, on the 10th of January, 1834,† which gave the Prodrome an apparent priority to the Mémoires of M. Desmoulins, and thus the genus *Disaster*, Agass. was adopted to the exclusion of the genus *Collyrites*, Desml. A closer investigation into the question of priority, however, showed that the memoir of M. Desmoulins was published in August, 1835. The Prodrome of M. Agassiz had been read in 1834, but was not published until July, 1836. In the interval which had elapsed between the reading and publication of the memoir, M. Agassiz had introduced into his work many important modifications, so much so, that the work published in 1836, was no longer the work which had been read in 1834; according, therefore, to the laws of nomenclature, the date of publication, and not that of the reading of a paper, after subsequent alterations and modifications, must serve to decide disputed points of priority in all questions affecting the natural history sciences.

In contending for the priority of his genus, M. Desmoulins‡ says—

" Le Prodrome de M. Agassiz a été publié en 1836, dans le 1ᵉʳ tome des ' Mémoires de la Société d'Histoire Naturelle de Neuchâtel:' j'ai donc un an d'antériorité sur lui. Il est vrai que ce Prodrome, d'après une note placée à la première page, a été lu le 10 Janvier, 1834, à la ' Socìètìè d'Hist. Nat. de Neufchâtel,' ce qui semblerait, dans un certain sens, faire tourner l'antériorité au profit de M. Agassiz; mais comme, dans ce travail, M. Agassiz adopté aux genres *établis par M. Gray en* 1835 (publiés d'après les renseignemens que j'ai pu recueillir en *Octobre*, 1835), il s'ensuit : 1ᵉ Que le travail de M. Agassiz n'a pas pu être, en 1836, *imprimé tel qu'il avait été lu en* 1834, et que des details de genres et d'espèces ayant été modifiés entre la lecture et l'impression, celle-ci seule prend une date authentique pour les noms de genres. 2ᵉ que j'ai une antériorité d'un an sur le genre *Disaster* de M. Agassiz, et une antériorité de deux mois sur l'*Arbacia* de M. Gray. Donc j'ai pu et dû conserver ma propre nomenclature, comme la plus ancienne. J'avais soumis cette question, avec tous ses détails, à un juge éclairé, parfaitement expert en ces sortes de matières, et j'ai agi d'après la décision motivée.

These historical facts are sufficient to justify the restoration of genus *Collyrites*, Desml.

* ' Prodrome d'une Monographie des Radiaires ou Échinodermes,' dans le premier tome des ' Mémoires de la Société des Sciences Naturelles de Neufchâtel,' 1836.

† ' Mémoires de la Soc. d'Hist. Nat. de Neufchâtel,' tome i, p. 168.

‡ ' Études sur les Échinides,' première Mémoire, p. 207.

in preference to that of *Disaster* Agass., by which name the urchins included in this group have been long known to English Geologists through the classical monograph on this genus published by M. Desor, in 1842.

A. *Species from the Inferior Oolite.*

COLLYRITES RINGENS, *Agassiz.*   Pl. XXII, fig. 3 *a, b, c, d, e, f, g, h, i.*

DYSASTER RINGENS.	Agassiz, Prodromus, 1ᵉʳ vol., des Mém. de la Société des Sciences Naturelles de Neufchâtel, 1836.
COLLYRITES RINGENS.	Desmoulins, 3ᵉ Mémoires sur les Echinides, p. 368, 1837.
DYSASTER RINGENS.	Agassiz, Echinoderm. Foss. de la Suisse, 1ʳᵉ partie, p. 5, tab. 1, figs. 7—11, 1839.
— —	Agassiz, Catalogus Systematicus, Ectyp. foss. p. 3, 1840.
— —	Desor, Monographie des Dysaster, p. 24, tab. 1, figs. 13—17, 1842.
— —	Agassiz and Desor, Catalogue raisonné des Échinides, Annales des Sciences Naturelles, 3ᵉ série, tome viii, p. 33, 1848.
— EUDESII.	Agassiz, Catal. System, Ectyp. foss. p. 3, 1840.
— —	Desor, Monographie des Dysaster, p. 23, tab. 1, figs. 5—12, 1844.
— SUBRINGENS.	M'Coy, Annals Nat. Hist., 2d series, vol. ii, p. 415, 1848.
— RINGENS.	Forbes, Mem. of the Geol. Survey, decade 3, pl. 9, figs. 1—10, 1850.
— —	D'Orbigny, Prodrome de Paléontologie, tome i, p. 289, 1850.
— —	Wright, Annals of Nat. Hist., 2d series, vol. ix, p. 207, 1851.
— —	Cotteau, Études sur les Échinides Fossiles, p. 46, pl. ii, figs. 10—13, 1852.
— —	Forbes, in Morris's Catalogue of British Fossils, 2 ed., p. 78, 1854.
COLLYRITES RINGENS.	Desor, Synopsis des Échinides Fossiles, p. 207, 1857.
— —	Cotteau and Triger, Échinides du Département de la Sarthe, pl. viii, figs. 5, 6, p. 48.
— EUDESII.	D'Orbigny, Paléontologie Française Ter. Cretacés, t. vi, p. 49, 1853.
— —	D'Orbigny, Note rect. sur divers genres d'Echid., Rev. Mag. de Zool., 2ᵉ série, t. vii, p. 26, 1854.

Test sub-orbicular or sub-pentagonal, rounded anteriorly, rostrated posteriorly; upper surface convex, more or less depressed; sides tumid; vertex nearly central; apices of the ambulacra widely disjoined, posterior pair forming an arch over the anal opening; vent pyriform, situated in a sulcus on the posterior margin; base concave, very much undulated, inter-ambulacra extremely tumid, single posterior area very prominent and much deflected; mouth-opening small, sub-central, and sub-pentagonal.

Dimensions.—Height seven tenths of an inch; antero-posterior diameter one inch and one tenth; transverse diameter one inch and one twentieth.

As this urchin presents very variable proportions, I subjoin a table, by Professor Forbes, showing the dimensions of eight specimens from the inferior Oolite of Dorsetshire, measured in inches and twelfths.

	A	B	C	D	E	F	G	
Length .	$1\frac{1}{12}$	$1\frac{1}{12}$	$0\frac{11}{12}$	$0\frac{8}{12}$	$0\frac{10}{24}$	$0\frac{9}{12}$	$0\frac{10}{12}$	$0\frac{10}{12}$
Breadth .	1	$1\frac{1}{24}$	$0\frac{11}{12}$	$0\frac{8}{12}$	$0\frac{8}{12}$	$0\frac{9}{12}$	$0\frac{10}{12}$	$0\frac{10}{24}$
Thickness .	$0\frac{6}{12}$	$0\frac{8}{12}$	$0\frac{7}{12}$	$0\frac{5}{12}$	$0\frac{5}{12}$	$0\frac{5}{12}$	$0\frac{5}{12}$	$0\frac{6}{12}$

Description.—The preceding table shows how much the general outline of this curious urchin varies in different individuals, so much so, indeed, that out of the varieties of *Collyrites ringens*, no less than three other species, *C. Eudesii*, Agass., *C. Agassizii*, d'Orb., *C. subringens*, M'Coy, have been described and proposed as distinct species. Having carefully examined and compared upwards of one hundred specimens of this urchin, I can confidently state that the orbicular, sub-pentagonal, and oblong varieties met with in the Inferior Oolites of Dorsetshire, are all referable to one and the same species, the extreme forms in different individuals being blended together by numerous intermediate gradations of structure.

The dorsal surface is uniformly smooth and convex, and more or less depressed, it is elevated posteriorly, and declines gently anteriorly, the vertex in general is situated nearer the anterior than the posterior border (fig. 3 *c*); the sides are tumid, the anterior border is flattened, and the posterior is produced and truncated (fig. 3 *c*); the antero-lateral is in general narrower than the postero-lateral region, and the prominence of the inter-ambulacral spaces in some individuals produces the sub-pentagonal varieties. The base is very unequal from the convexity of the inter-ambulacra, which form five nodulose eminences around the mouth; the posterior single area in particular is extremely prominent, gibbous (fig. 3 *b*, *c*), and much deflected; its posterior border is truncated and channelled to form the anal valley (fig. 3 *c*, *d*), which is bounded by two ridges, commencing at the apices of the posterior pair of ambulacra, and passing downwards and outwards towards the base, where they may be traced on the summit of the single area as far as the mouth (fig. 3 *b*). The anal opening has a pyriform shape, with its apex directed upwards; it is situated in the upper part of the valley, nearly on a level with the dorsal surface, and immediately below the ambulacral arch (fig. 3 *d*, *c*).

The ambulacral areas are all complete, at the dorsal surface they are on a level with the inter-ambulacral plates, but at the base, they lie in depressions. They are of unequal width, the single area being the narrowest, the posterior pair the widest, and the anterior pair of intermediate breadth; the single area is straight, and the anterior pair describe three curves in their course between the mouth and the disc (fig. 3 *c*); the three anterior ambulacra converge nearly in the centre of the back, at the front, and sides of the apical disc (fig. 3 *a, i*). The posterior pair are somewhat wider than the others, they curve gracefully round the single inter-ambulacrum (fig. 3 *c*), form an arch round its produced and truncated border, and converge above the anal opening (fig. 3 *d*); the ambulacral areas are formed of small plates, of which, on the dorsal surface, there are four opposite one inter-ambulacral plate (fig. 3 *e*), but at the base there are only about three in the same space (fig. 3 *f*). Each ambulacral plate is perforated at its outer side with two very small pores, placed obliquely across the zones (fig. 3 *e*), which are narrow, and observed with difficulty. Near the mouth the ambulacra widen, and the pairs of pores are disposed in about three oblique series of three pairs in each (fig. 3 *b*); as the plates are closely soldered together, their relation to the pores in this region is very indistinct.

The inter-ambulacral areas are of unequal width; on the upper surface they are on a level with the ambulacra, and with them form a uniform, convex back, but at the base they are extremely prominent and nodulated; so much so, that the ventral is as remarkable for its undulations as the dorsal is for the smoothness of its surface. The single inter-ambulacrum differs from the others in being produced posteriorly, and is extremely prominent and gibbous inferiorly (fig. 3 *b, c*).

All the plates are covered with minute, perforated tubercles, raised on uncrenulated bosses, and surrounded by sunken areolas (fig. 3 *g*); the inter-tubercular surface is so finely granulated that the tubercles appear conspicuous only when examined with a lens. The tubercles are most numerous at the base; the spines are unknown.

The apical disc is a curious structure, it is composed of two pairs of perforated ovarial plates (fig. 3 *h, i*), disposed in pairs at some distance apart, and separated by three largely developed ocular plates, which extend into the centre of the disc; the anterior ovarials are of an irregular shape, and separate the single ambulacrum from the antero-lateral ambulacra. On the surface of the right plate is the spongy, prominent, madreporiform body; behind and between them in the median line, is a small diamond-shaped plate, its anterior angle unites with the apex of the single ambulacrum, and its posterior border with the anterior ovarials; behind these are two rhomboidal-shaped plates, which articulate before with the anterior ovarials, laterally with the apices of the antero-lateral ambulacra, and behind with the posterior ovarial plates. Near the points of junction of these plates with the ambulacra, the three small eyeholes are situated, and behind the rhomboidal ocular plates the small, oblong, posterior ovarials are placed. I have failed to discover ocular plates at the summits of the posterior ambulacra in this species. The specimen which furnished these details has been mislaid, and could not be found in time for the artist, but a very similar type of structure exists in *Collyrites ovalis* (Pl. XXIII, fig. 2 *f*), where it is accurately drawn. The disc, it is right to state, was studied with the microscope under an inch object-glass.

The mouth is more or less sub-central, and lodged in a concavity formed by the prominent nodulated inter-ambulacra. It is obscurely decagonal, and appears as if round; in some specimens it is much nearer the anterior border than in others. The mouth and anus are nearly of the same size, and about one eighth of the length of the shell; no traces of jaws have yet been found, nor has the structure of the peristome been sufficiently made out.

Affinities and differences.—Many specimens of this urchin agree with M. Desor's figures of *Disaster Eudesii*, Agass., whilst others have the depressed dorsal surface, and angular outline of *D ringens*, Agass. As I have many series of intermediate forms connecting these two extremes, I have referred them all to one species. On this subject, M. Cotteau observes that he collected, with M. Moreau, from the "Oolite ferrugineuse" of Tour du Pré, a suite of specimens of *D. ringens;* these presented various degrees of tumidity and more or less circularity of outline; among them were all the gradations conducting to *D. Eudesii*, Agass., from this he concluded that the urchin figured in his excellent work, and which may be taken as a fair representation of many of our specimens, is a small and more elongated variety of *D. ringens*, Agass.* This conclusion, Professor Forbes† admits, agrees with the experience of the collectors of the Geological Survey.

Professor M'Coy, in his memoir 'On some new Mesozoic Radiata,'‡ enumerates *D. Eudesii* as a British species from the Inferior Oolite of Dundry and Bridport, and has described another form under the name *D. sub-ringens*. As he has favoured me with a sketch of this urchin, I can state with certainty that it is only a large individual of *D. ringens*. The characters which Prof. M'Coy regarded as specific, namely, the "greater gibbosity, and less prominence of the ridges on the under side," and, the "disproportionate narrowness of the three anterior ambulacra, as in *D. ringens*," vary almost in every one of the many individuals I have collected, I therefore do not hesitate to include *D. sub-ringens* among the synonyms of *Collyrites ringens*, Agass., which is distinguished from its congeners by the convexity of the basal portions of the inter-ambulacra, and especially by that of its single inter-ambulacrum.

Locality and Stratigraphical position.—All my specimens have been collected from the marly vein which traverses the upper ragstones of the Inferior Oolite of Dorsetshire, in the zone of *Ammonites Parkinsoni*, Sow. From this stratum I have collected it between Sherborne and Yeovil, at Burton Bradstock, Walditch Hill, and Chideock Hill, near Bridport. It is generally associated with *Collyrites ovalis*, Leske, *Holectypus hemisphæricus*, Agass., *Clypeus altus*, M'Coy, and *Stomechinus bigranularis*, Lamck.

* 'Etudes des Échinides Fossiles,' p. 48.
† 'Memoirs of the Geological Survey,' decade iii, pl. 9.
‡ 'Annals of Natural History,' 2d series, vol. ii, p. 120.

I have only seen one specimen (an oblong variety) which was said to have been found in the Cornbrash, near Fairford (Gloucestershire). Professor M'Coy states that it is not uncommon in the Inferior Oolite of Leckhampton; but this is a mistake, as it is not within the memory of any of our local collectors that any Collyrite has been found in the Inferior Oolite of that locality.

On the continent, it is found in the Marnes à Discoidées (Vesulien) of Goldenthal, Mont-Terrible, Salins, Besançon, Neufchâtel, the Inferior Oolite (Bajocien) of Tour du Pré (Yonne), and of St.-Vigor, Port-en-Bessin, Moutiers, Bayeux (Calvados). It was collected by M. Triger, from the Forest Marble, Ass. No. 4, of his table, at Pécheseul, Noyen, Saint-Pierre-des-Bois, Chemiré-le-Gaudin, département de la Sarthe, where it is abundant.

History.—This species was recorded for the first time by Agassiz, in his 'Prodrome,'[*] and it has been successively figured and described in his 'Échinodermes Fossiles de la Suisse,' M. Desor's 'Monographie des Dysaster,' M. Cotteau's 'Études des Échinides Fossiles,' 'Échinides du département de la Sarthe,' and Professor Forbes's 'Memoirs of the Geological Survey.' M. Desmoulins entered it in his 'Tableaux Synonymiques des Échinides.' The history of this species is so intimately connected with its previous description, that it is unnecessary to enter upon it further in detail under this head.

COLLYRITES OVALIS, *Leske.*   Pl. XXIII, fig. *a, b, c, d, e, f, g.*

	Van Phelsum, p. 32, sp. 3 (*Egelschuitje twee-top*) i. e., *Echinoneus bivertex.*
	Knorr, Petrefactions, ii, p. 182, tab. E iii, No. 6.
	D'Annone, Acta Helvetica, vol. iv, p. 275, sq.; tab. 14, figs. 1, 2, 3.
	D'Annone, Miner. Belust., v, p. 161, tab. iv, figs. 1, 2, 3.
SPATANGITES OVALIS.	Leske, Additamenta Kleinii ad Disposition. Nat. Echinoderm., p. 253, tab. 41, fig. 5, 1778.
COLLYRITES ANALIS.	Desmoulins, Études sur les Échinides, p. 368, No. 14, 1837.
DISASTER ANALIS.	Agassiz, Échinodermes Fossiles de la Suisse, ii, p. 6; pl. 1, figs. 12—14, 1839.
DISASTER AVELLANA.	Agassiz, Catal. Syst. Ectyp. Foss., p. 3, 1840.
DISASTER BICORDATUS.	Agassiz, Catal. Syst. Ectyp., Suppl., non Leske, non Goldfuss.
COLLYRITES ELLIPTICA.	Desmoulins, Tableaux Synonymiques, p. 366, 1835.
DISASTER BICORDATUS.	Desor, Monographie des Dysaster, p. 9, tab. 2, figs. 1—4, 1842.
— —	Agassiz and Desor, Catalogue raisonné des Échinides, Annales des Sciences Naturelles, 3ᵐᵉ série, tome viii, p. 31.
DISASTER ANALIS.	Desor, Monographie des Disaster, p. 10,. pl. 2, figs. 8—10, 1842.

* 'Soc. d'Hist. Nat. de Neufchâtel,' tome i, p. 168, 1836.

Disaster avellana. — Desor, Monographie des Disaster, p. 23, pl. 1, fig. 1—4, 1842.

—   —   Agassiz and Desor, Catal. rais. des Échinides, Ann. des Sc. Nat., 3ᵐᵉ série, t. vii, p. 32, 1847.

— bicordatus. — Agassiz and Desor, Catal. rais. des Échinides, Ann. des Sc. Nat., 3ᵐᵉ série, t. vii, p. 31.

Disaster Robinaldinus. — Cotteau, Études sur les Échinides Fossiles, p. 73, tab. vii, figs. 1—5, 1849.

— symmetricus. — M'Coy, Annals of Natural History, 2d series, vol. ii, p. 415, 1848.

— avellana. — Agassiz, Catal. Syst., p. 3.

— — — D'Orbigny, Prodrome de Paléontologie, t. 1, p. 289, No. 43, 1856.

— — — Desor, Monographie des Dysaster, p. 23, tab. 1, figs. 1—4.

— — — M'Coy, Annals of Natural History, 2d series, vol. ii, p. 420, 1848.

Disaster bicordatus. — Wright, Annals of Natural History, 2d series, vol. ix, p. 210, 1851.

— — — D'Orbigny, Prod. de Pal., t. i, p. 318, No. 399, 1850.

— — — Forbes, in Morris's Catalogue of British Fossils, 2d ed., p. 77, 1854.

— Agassizii. — D'Orbigny, Prod. de Pal., t. i, p. 290, No. 494, 1850.

Collyrites bicordata. — D'Orbigny, Pal. Franç. Ter. Cretacés, t. vi, p. 49, 1853.

— analis. — D'Orbigny, ibid., p. 48, 1853.

— avellana. — D'Orbigny, ibid., p. 48.

— Agassizii. — D'Orbigny, ibid., p. 48.

— — — D'Orbigny, Note rect. sur divers genres d'Échinides, Rev. et Mag. de Zool., 2ᵈᵉ série, t. vi, p. 27, 1854.

— bicordata. — D'Orbigny, ibid., p. 27.

— analis. — D'Orbigny, ibid., p. 27.

— analis. — Desor, Synopsis des Échinides Fossiles, p. 206, 1857.

— ovalis. — Cotteau, Note sur quelques Oursins de la Sarthe, Bull. de la Soc. Géologique de France, 2ᵈᵉ série, p. 649, 1856.

— — — Cotteau et Triger, Échinides du Départment de la Sarthe, pl. vii, figs. 7—9, p. 45, 1857.

Plaster moulds R 15, R 16, type of *bicordatus*, Agassiz; Q 82, type of *analis*, X 76, type of *avellana*.

Test thin, oval, anterior and posterior borders nearly uniform in convexity; sides tumid; dorsal surface convex, sometimes flattened; vertex excentral, situated near the anterior third; anal opening pyriform, supra-marginal; postero-lateral ambulacra terminate by the sides or immediately above the anus; base smooth, convex, without undulations; mouth-opening small, situated at the junction of the anterior with the middle third.

Dimensions.—Eight tenths of an inch; antero-posterior diameter, one inch and one fifth; transverse diameter, one inch and one tenth.

Description.—After a careful examination of all the evidence on the subject, I have come to the conclusion that this is the urchin which was figured by Leske (tab. XLI, fig. 5), and described by that author as *Spatangites ovalis.* The *Ananchytes bicordata,* Lamk. to which it was formerly referred, is clearly the *Spatangites bicordatus* of Leske, and not the *S. ovalis* of that author; the confusion occasioned by this mistake, prevails throughout all the works which treat of the species, down to that of our friend M. Desor, in which its true synonyms, with one exception, according to our ideas on the subject, are given in his excellent 'Synopsis des Échinides Fossiles.' The *Spatangus ovalis* of Parkinson and Phillips is a Coral Rag urchin, and in my opinion is the *Spatangites bicordatus* of Leske. I reserve further details on this question until I give the description of the latter species.

M. Desor regards this urchin as an inflated variety of *Dysaster analis,* Agassiz, and preserves that author's specific name; but surely, if we admit the identity of the species, we should in justice to Leske retain his specific name *ovalis.* On comparing our urchins, however, with the very excellent figure given of *D. Analis* by M. Agassiz, I confess I have never met with so depressed a form as that figured in the 'Echinodermes Fossiles de la Suisse;' moreover, there is always an elevation before the apical disc, amounting in some individuals to a monstrosity, which does not exist in *D. analis;* the posterior ambulacra, likewise, terminate at a greater distance before the anal opening than in our specimens.

The regular oval outline of this urchin forms a striking contrast to the orbicular and sub-pentagonal figures of *Collyrites ringens;* the sides are tumid; the dorsal and lateral surfaces are smooth, and convex; and the test has a uniformly gibbous appearance, with a slight elevation on the upper surface; the ambulacra are all complete, passing, without interruption, from the mouth to their terminations on the dorsal surface, and on the same level as the inter-ambulacra; the three anterior ambulacra converge near the junction of the anterior with the middle third of the back, and the apex of the single area is separated from those of the antero-laterals by the anterior pair of ovarial plates; the single ambulacrum is the narrowest, the postero-lateral are the widest; and the antero-lateral of intermediate width; the anterior border is slightly flattened, and lies in a depression in the centre of this region; the single area passes in a straight line from the mouth to the vertex; the apex of this area rises into a small prominent triangular eminence, which forms the highest point of the test, so that the vertex is situated at the anterior third of the back (fig. 1 *c*), whilst in *Collyrites ringens,* it is at the posterior third.

The antero-lateral ambulacra (fig. 1 *c*), curve gently upwards, backwards, and forwards, from the mouth to the apical disc, thereby forming an undulated course. The postero-lateral pair take a long sinuous course from the mouth, passing backwards, outwards, and upwards, over the posterior border, and terminate by making rather a sharp curve at the upper end of the anal opening, or sometimes they form an arch immediately over that aperture (fig. 1 *d*); the point at which these areas terminate, in this species, is an important guide to its determination, and this character is very accurately delineated in Leske's original figure of his *Spatangites ovalis*; the apices of the postero-lateral ambulacra converge on the back at the distance of three fifths of an inch behind the antero-lateral pair.

The poriferous zones are narrow, and in consequence of the depth of the ambulacral plates the pores are wide apart; each plate is perforated with a pair of holes near its lower border (fig. 1 *e*), and there are from two to three ambulacral plates opposite one inter-ambulacral.

The pores are disposed in oblique pairs, they are very distinct on the sides and back fig. 1 *a, c, d*), but are small and indistinct at the base (fig. 1 *b*).

The inter-ambulacral plates are large and bent, each plate forming a double inclined (plane (fig. 1 *e*). The anterior are much narrower than the posterior areas; they are uniformly smooth and gibbous on the sides; the single inter-ambulacrum is bevelled obliquely, and slightly flattened at the upper part of the posterior border; at the extreme upper part of this region, and immediately beneath the centre of the arch formed by the posterior ambulacra, the anal opening is situated (fig. 1 *a, c,d*). It has a pyriform shape, with the apex directed upwards, and terminates between the apices of the posterior ambulacra. From its sides two obtuse ridges pass downwards and outwards, which gradually disappear about the middle of the area, near the anal opening; the tubercles are larger, and set closer together, than at any other part of the upper surface (fig. 1 *e*). The basal portion of the inter-ambulacrum is more gibbous and produced than the corresponding parts of the other inter-ambulacra.

The base is convex, with little or no undulation; anteriorly there is a slight concavity, and posteriorly an increased convexity, occasioned by the gibbosity of the single inter-ambulacrum (fig. 1 *b, d*). The small mouth-opening is lodged in a slight depression, situated near the anterior fourth of the antero-posterior diameter of the test, but the precise relative situation of this aperture appears to vary a little in different individuals. The peristome is sub-pentagonal, and in one specimen, appears to have rudimentary notches, which would imply that the *Collyritidæ* possessed jaws, but the organic evidence is too feebly developed to enable me to state that such is unquestionably the case, unless confirmed by a similar inequality of the margin in other specimens.

The apical disc is situated behind the sub-triangular vertex, and therefore occupies the anterior third of the back; it has a lengthened rhomboidal figure, and is formed very much like the disc of *C. ringens*, already described in detail. (See page 311.)

The single ocular plate is small, and occupies nearly the vertex, having behind it the two anterior ovarial plates, which are situated between the single and the antero-lateral ambulacra, the left plate being placed further forward than the right plate, which supports the madreporiform body. Behind the anterior ovarial are the two large anterior pair of ocular plates, and immediately behind them are the posterior ovarials (fig. 2 *f*). The position of the posterior pair of ocular plates I have not ascertained; if they exist at the summit of the posterior ambulacra, their true homological position, they are so intimately soldered to the surrounding structures that I have as yet failed to detect them with my compound microscope, provided with an inch object-glass; but the elements of the disc are, in general, united so entirely to the adjoining plates of the test in the COLLYRITIDÆ, that it is only in weathered specimens, or in those in which the shell has passed into the condition of calcareous spar, that we can distinguish the separate pieces of which it is composed. The tubercles on the sides and upper surface are small, and arranged in tolerably regular lines on the plates. There are, in general, three rows on each plate (fig. 1 e); those at the base are larger (fig. 1 f), and more prominent; they are raised on prominent bosses, which are surrounded by areolas, and the intermediate spaces are covered with close-set granules.

Affinities and differences.—The general outline of *C. ovalis* resembles *C. ringens*, but is distinguished from it by the following characters. In *C. ovalis*, the highest point of the back is near the anterior third, whilst in *C. ringens*, it is at the posterior third; in *C. ovalis*, the base is nearly uniformly convex, in *C. ringens* it is very much undulated; in *C. ovalis*, the apical disc is situated near the anterior third of the back, whilst in *C. ringens*, it is nearly central. The single inter-ambulacrum, likewise, is not so much developed; the anal opening is larger, and higher up, and the anal valley is more rudimentary in *C. ovalis* than in *C. ringens*.

This species is distinguished from *C. bicordatus*, Leske, by the posterior ambulacra always terminating in *C. ovalis* at or near the apex of the anal opening, whilst in *C. bicordatus*, the apices of these areas terminate at a point about one third of the distance between the vent and the disc. The same character, the proximity of the apices of the posterior ambulacra to the anal opening, serves to distinguish *C. ovalis* from *C. elliptica*. *C. ovalis* resembles *C. analis*, but in the former they are nearer the vent than in the latter. It is highly probable, however, that this is a mere variety, and not a specific difference.

Locality and Stratigraphical position.—I have collected this species in the marly vein which traverses the upper ragstones of the Inferior Oolite, in the zone of *Ammonites Parkinsoni*, Sow., at Walditch Hill, near Bridport, where it was associated with *Collyrites ringens*, and *Holectypus hemisphæricus*. It has been found by Mr. Walton in the same zone at Charlcomb, near Bath, and by Dr. Bowerbank in the Cornbrash of Wilts, where it is rare, as I only know of his solitary specimen from that formation.

History.—First figured by Knorr, and afterwards figured and described by Leske as *Spatangites ovalis*, in his edition of Klein's 'Echinodermata,' which figure was confused with the *Spatangites bicordatus* of the same author. It was beautifully figured as *Disaster bicordatus* by M. Desor, in his 'Monographie des Dysaster;' afterwards by M. Cotteau under the name *Disaster Robinaldinus*, in his 'Études des Echinides Fossiles,' and described under the name *D. symmetricus*, by Professor M'Coy. It was recorded by Professor Forbes as an abundant species, collected by the officers of the Geological Survey, from several Inferior Oolite localities in Dorsetshire.

M. Desor enumerates the following localities where this species is found. Knorr and Leske's type specimens came from the Marnes à Discoidées (Vesulien) de Muttenz près Bâle. It is found likewise in the Great Oolite (Bathonien) de Macon, de Bysé près Caen, la Latte près Nantua (Ain), environs de Besançon, Véseloy (Yonne), Calcaire à Polypiers (Bathonien) de Croisille, Charroux, and M. Triger collected it in abundance from the Bradford Clay, Ass. No. 1, 2, and 3, of his table, and from the Forest Marble bed No. 1 at Monné, La Jaunelière, Tassé Champfleur, Petit-Oiseau, and several other localities in the department of the Sarthe.

B. *Species from the Coral Rag.*

COLLYRITES BICORDATA, *Leske*.   Pl. XXIII, fig. 2 *a, b, c, d, e, f, g, h.*

	Andreae Briefe aus der Schweiz, p. 16, tab. 2, fig. c, 1763— 1776.
SPATANGITES BICORDATUS.	Leske, apud Klein, Dispositio Naturalis Echinodermatum, tab. 42, fig. 6, p. 244, 1778.
SPATANGUS OVALIS.	Parkinson, Organic Remains, vol. iii, tab. 3, fig. 3, 1811.
SPATANGITES OVALIS.	Young and Bird, Geology of the Yorkshire Coast, tab. 6, fig. 9, p. 215, 1828.
SPATANGUS OVALIS.	Phillips, Geology of Yorkshire, tab. 4, fig. 23, p. 127, 1829.
DISASTER OVALIS.	Agassiz, Prodromus, Mémoires de la Soc. des Sc. Nat. de Neufchâtel, tome i, 1836.
COLLYRITES OVALIS.	Desmoulins, Tableaux Synoptiques, p. 368, 1837.
DISASTER PROPINQUUS.	Agassiz, Échinoderm. Fossiles de la Suisse, part i, p. 2, tab. 1, figs. 1—13, 1840.
— OVALIS.	Agassiz, Catalogus Systematicus ectyporum Echinodermatum fossilium, p. 3, 1840.
— PROPINQUUS.	Desor, Monographie des Dysaster, p. 14, tab. 3, figs. 24—26, 1844.
— TRUNCATUS.	Dubois, Voy. au Caucase, (Ser. Geol.), tab. 1, fig. 1.
— —	Desor, Monographie des Dysaster, p. 17; des Galérites, tab. 13, figs. 8—11, 1844.
— —	Agassiz and Desor, Catalogue raisonné des Échinides, Annales des Sciences Naturelles, 3ᵐᵉ série, tome viii, p. 32, 1847.
— OVALIS.	Agassiz and Desor, ibid.
— —	Cotteau, Études sur les Échinides Fossiles, p. 86, tabs. 8 and 9, 1849.
— —	Wright, Annals and Magazine of Natural History, 2d series, vol. ix, p. 213, 1851.
COLLYRITES BICORDATA.	Desor, Synopsis des Échinides Fossiles, p. 204, 1857.

Test nearly oval, broader before than behind, upper surface convex, more or less depressed; sides rather tumid, front border slightly grooved, posterior border feebly truncated, ambulacral areas of unequal width, posterior pair the widest, anterior ambulacral summit nearly central, posterior situated nearly mid-way between the apical disc and the vent; base flat anteriorly and laterally, convex behind; mouth-opening small, circular, situated at the anterior fourth of the base, tubercles small, placed at some distance apart, three or four concentric rows on each large plate. Anal opening oval, situated in the upper part of the posterior border, about midway between the ambulacral arch and the basal angle.

Dimensions.—A. Height, one inch and one twentieth; transverse diameter, one inch and nine tenths; antero-posterior diameter, two inches and one tenth. B. Height, one inch and one twentieth; transverse diameter, one inch and three quarters; antero-posterior diameter, one inch and nine tenths. C. Height, eight tenths of an inch; transverse diameter, one inch and four tenths; antero-posterior diameter, one inch and a half.

Description.—It is rather remarkable that so much confusion should have arisen regarding the only two *Collyrites* figured by Leske, and that a complete transposition of his specific names should have been the result of this mistake. The urchin now under consideration is, doubtless, the *Spatangites bicordatus*, Leske, which was erroneously identified by Parkinson with the *Spatangites ovalis*, Leske, this formed the starting point of the error, which has been faithfully copied by succeeding authors until corrected by M. Desor, who has given, in his 'Synopsis,' the correct synonymy of this species. A single character, which has been well represented by Leske, determines this point, namely the position of the posterior ambulacral summit, in *Spatangites bicordatus* it is between the apical disc and the vent, whilst in *Spatangitis ovalis* it is immediately above the anal

The general form of *Collyrites bicordata* is nearly oval, the anterior half is more enlarged than the posterior half (fig. 2 *a, b*), the upper surface is uniformly convex (fig. 2 *c*), and the base is flat (fig. 2 *b, c*), the anterior border forms the segment of a much larger circle than the posterior border, and it has a median depression which extends to the mouth in which the single ambulacrum is lodged (fig. 2 *a*), the sides are slightly tumid and the posterior border feebly truncated, the greatest width of the test is behind the antero-lateral ambulacra, from which point it gradually tapers backwards (fig. 2 *a*).

The ambulacral regions are of unequal width, the three anterior areas are about the same diameter on the sides, but the posterior pair are one fourth larger, the single area is the shortest; in specimen c it measures $1\frac{1}{2}$ inch, the antero-lateral $1\frac{11}{20}$ inch, and the postero-lateral $1\frac{7}{10}$ inch in length from the peristome to the apical disc, the ambulacral plates are narrow, and on the sides of the anterior pair there are from six to seven plates opposite each large interambulacral plate (fig. 2 *e*); at the base they become much wider

42

and are nearly half as deep as the inter-ambulacral plates (fig. 2 *b*). In the posterior pair the plates on the sides are much deeper, for in these areas there are not more than two or three plates opposite one large inter-ambulacral plate. The left postero-lateral ambulacral area extends farther forward than the right.

The inter-ambulacral regions are large and of unequal width ; the antero-lateral areas are the narrowest, the single area is of the same width, and the postero-lateral are one third wider than the anterior pair ; the large pentagonal plates composing these areas are bent in the middle, and there are from fifteen to sixteen pairs in each ; the plates carry three or four concentric rows of small tubercles (fig. 2 *e*), numbering from fourteen to sixteen on each plate ; the tubercles which are perforated are surrounded by well-defined areolas, raised on crenulated bosses (fig. 2 *g*), and the inter-tubercular surface is crowded with microscopic miliary granules (fig. 2 *g*) ; the basal are larger than the dorsal tubercles, especially those situated on the prominent portion of the single inter-ambulacrum ; some of these have wide hexagonal areolas, closely set together (fig. 2 *h*).

The poriferous zones are narrow ; in the anterior zones on the sides there are about six pairs of pores opposite each large plate (fig. 2 *e*), whereas in the posterior zones in the same portion of the test there are only three or four (fig. 2 *d*). At the base the pores are small, very indistinct, and placed widely apart.

The base is flat at the sides and before, but is convex in the region of the single inter-ambulacrum. As the basal portions of the ambulacral areas are nearly destitute of tubercles, and the basal portions of the inter-ambulacral areas are furnished with larger tubercles, the course of the former is readily made out by the comparatively naked track formed by them, from the border to the mouth (fig. 2 *b*).

The small circular mouth-opening is situated in a depression at the anterior fourth of the base, the peristome is entire, and there is no trace of notches nor of any armature for the mouth (fig. 2 *b*).

The apical disc is placed rather nearer the anterior than the posterior border (fig. 1 *a*). It has an elongated figure, but its elements are so intimately soldered together that their relative anatomy is with difficulty made out. In most specimens the anterior ovarials project between the single and the antero-lateral ambulacra (fig. 2 *f*) and the right plate, which is the largest, supports the madreporiform body on the same line. Behind the anterior ovarials the anterior pair of ocular plates are placed, and behind them come the posterior pair of ovarials, with the single imperforate ovarial behind and between them ; the small single ocular plate is seen at the apex of the single ambulacrum, but the position of the posterior oculars I am unable to determine, notwithstanding the perfect conservation of one of my specimens.

The posterior border is slightly truncated ; at its upper part, and nearly midway between the basal angle and the apex of the posterior ambulacra, the anal aperture is situated (fig. 2 *d*). It is a small oval opening, the longest diameter of which is in the direction of the longitudinal axis of the test. The basal portion of the single inter-ambulacrum is convex and prominent, more especially so nearest the border (fig. 2 *d*).

Affinities and differences.—This species very much resembles *Collyrites elliptica*, Lamk. From the Great Oolite and Kellovian strata of the Department of the Sarthe I possess for comparison a very good series of this species, collected and sent me by my friend, M. Triger, besides other good types from MM. Michelin, Bouchard, and Cotteau. I find that in *Collyrites bicordata* the shell is more cordiform, and the anteal sulcus is well developed, whilst in *C. elliptica* that depression is absent. The left postero-lateral ambulacrum generally rises higher up on the dorsum than the right postero-lateral ambulacrum. The anterior summit is rather more forward, and the basal portion of the single inter-ambulacrum is more prominent than in *C. elliptica;* whilst in the relative position of the mouth and vent, and the distance of the apices of the posterior ambulacra from the latter opening, there is much similarity between these two species.

Locality and Stratigraphical position.—The Yorkshire specimens have been collected from the lower calcareous grit, near Scarborough; they are in general denuded of their test, and are mostly deformed. C. W. Strickland, Esq., has collected several specimens in the Coralline Oolite of Hildenley, where they are associated with *Cidaris florigemma*, Phil., *Hemicidaris intermedia*, Flem., and *Pseudo-diadema hemisphæricum*, Agas. James Carter, Esq., of Cambridge, has collected this species from blocks of Oolitic drift at Holywell, St. Ives, the rock containing these specimens is a ferruginous, coarse-grained oolite, which I suppose may have been derived from the base of the Coral Rag, or lower calcareous grit. This gentleman has likewise obtained fine specimens of the same species at Ely, from a rock which he conjectures to be Kimmeridge clay. One of these urchins is figured in Pl. XXIII, fig. 2. It is unfortunate that a doubt should still exist relative to the true age of these beds, and I feel under many obligations to Mr. Carter, which it affords me great pleasure to acknowledge, for the series of specimens of *C. bicordatus* he has kindly communicated and generously given me. Professor Sedgwick has likewise been most obliging in making an examination of one of these specimens, and in expressing his opinion relative to the age of the Oolitic drift at Holywell, containing *C. bicordatus*. I have much satisfaction in adding that distinguished geologist's opinion on the subject, which accords with my own conjectures made from a palæontological and not from a stratigraphical point of view:

"I have no doubt," says Professor Sedgwick, in a letter to me on the subject, "the specimen is what is commonly called our glacial drift, not, however, a drift brought on icebergs, but a drift caused by a great change of level about the end of the so-called glacial period, and it contains fragments innumerable of rocks belonging to the whole series, from the lias to the chalk inclusive. Our general order of super-position round about Cambridge is, 1st, *gravel* and *drift* of different ages, irregular in thickness, and capping all the other deposits in a disorderly manner, without absolute continuity.   2d, *chalk.*   3d, *upper-green-sand*—very thin, but full of fossils, and now worked much for the phosphatic nodules called (by mistake) coprolites; 3 *a, galt* 3 *b, lower-green-sand* resting, with some discordance on the beds below.   It is seen in one place resting on good Kimmeridge clay, in another immediately on Oxford clay.   4th, *Kimmeridge clay.*   5th, *Coral rag, or Middle Oolite*—no Portland Oolite in this country. The Coral rag is not continuous, and is only well seen in one spot, but it may exist in other places under the drift which conceals so many of our strata.   6th, *Oxford clay.*

"The lower Oolitic terrace, including everything from Cornbrash to Inferior Oolite, is far removed from us.

"There are good brick-pits in the upper part of the Oxford clay, and immediately over the clay are some stone bands, which may possibly represent the base of the Coral rag.   My belief is that your fossil (*Collyrites bicordata,* Leske,) has been drifted out of the Coral Rag or Middle Oolite, which will agree well with your idea of the true place of this species."

The foreign distribution of this fine species is, according to M. Desor, from the Terrain à Chailles, or Inferior Corallian, equivalent to the lower calcareous grit of English geologists ; in the Swiss Jura, it is found in that stage at Fringeli, Liesberg, Wahlen, Delémont, Porrentruy ; of the Salinois Jura, at Mont Bregille, near Besançon.   It is collected in the same stage in France, according to M. Cotteau, from the " Calcaires Oxfordiens " of Lucy-le-Bois, Villiers-les-Hauts (Yonne).

NOTES

On Foreign Jurassic species of COLLYRITIDÆ nearly allied to British forms, but which have not yet been found in the English Oolites.

DISASTER GRANULOSUS, *Münster*. Syn. *Nucleolites granulosus*, Goldf., tab. xliii, fig. 4.

Test small, elongated, obovate, convex on the upper surface, flat on the under side, rounded before, and obliquely truncated behind; anterior ambulacral summit excentral, nearer the anterior border; posterior ambulacra much arched over the vent; tubercles numerous, irregularly scattered over the surface; granules, small, numerous, close-set.

Dimensions.—Height, three quarters of an inch; breadth, one inch and one eighth; length, one inch and three tenths.

Formation.—Oxford-gruppe, (Oppel.) Amberg, Streitberg, Bavaria.

Collections.—British Museum, Continental collections. My cabinet. Common.

DISASTER ANASTEROÏDES, *Leymerie*. Syn. *Disaster anasteroïdes*, Cotteau, Échinides Fossiles, tab. xlvi, figs. 4—10.

This Disaster closely resembles the preceding species, but is larger and less cylindrical than it.

Dimensions.—Height, seven tenths of an inch; breadth, one inch; length, one inch and four tenths.

Formation.—Kimmeridge of Bar-sur-Aube.

Collections.—M. Leymerie, M. Cotteau.

DISASTER MOESCHII, *Desor*. Syn. *Disaster Moeschii*, Cotteau et Triger, Échinides du Département de la Sarthe, Pl. XIV, figs. 9—11, p. 51.

Test elongated, sub-cylindrical, enlarged and rounded before, truncated obliquely, and square behind; upper surface convex, base flat; ambulacra narrow; anterior summit removed to a considerable distance from the posterior summit; apical disc small, square, vent pyriform, supra-marginal; peristome extremely excentral.

Dimensions.—Height, six tenths of an inch; breadth, seven tenths; length, one inch.

Formations.—Great Oolite, Saint-Marceau, (Sarthe), Étage, Callovien, Hornussen, Kornberg (Argovia), Erlinsbach, and Pouillerel, near Chaux-de-Fonds, Switzerland.

Collections.—MM. Guéranger, Moesch, Cartier, Nicolet, Mus. Neuchâtel. Rare.

COLLYRITES ELLIPTICA, *Lamk.* Syn. *Disaster ellipticus*, Desor. Monographie des Disaster, tab. ii, figs. 5—7.

Test regularly elliptical, convex above, more or less depressed, plano-convex below; the apices of the posterior ambulacra converge at a point about one third of the distance between the vent and the anterior summit, which is more or less excentrally forwards.

Dimensions.—This species varies very much in size and relative dimensions, the following is M. Cotteau's estimate from abundant materials :—" A. *Circular variety.*—Height twenty millimètres; transverse diameter, forty millimètres; antero-posterior diameter equal to the transverse. B. *Elongated variety.*—Height twenty-three millimètres; transverse diameter, thirty-eight millimètres; antero-posterior diameter, forty-two millimètres. C. *Large variety.*—Height forty millimètres; transverse diameter, sixty millimètres; antero-posterior diameter, sixty-three millimètres. (This is the *Disaster malum*, Ag.)"

Formation.—Kellovian, Chaufour, Mamers, " Kelloway Ferrugineux." (Sarthe.)

Collections.—Very common in most collections. I have a fine series in my cabinet, sent by M. Triger, M. Cotteau, and M. Michelin, from this locality.

COLLYRITES PINGUIS, *Desor.* Synopsis des Échinides Fossiles, p. 205.

Test depressed, of an elliptical form, with the border inflated, resembling *C. bicordata* a little, but larger than that form; posterior ambulacra nearly straight and rising much higher than the vent.

Formation.—Oxfordian superior of Bötzberg, and Geisberg, near Brugg (Argovia).

Collections.—Mus. Bâle, Coll. MM. Gressly, Schmiedlin, and Cotteau.

COLLYRITES ORBIGNIANA, Cotteau. Échinides Foss. Pl. IX, figs. 3—5.

Test oval, cordiform, upper surface inflated, under surface flat ; the posterior ambulacra ascend in nearly a straight direction, and meet above the vent.

Dimensions.—Height, three quarters of an inch ; breadth, one inch ; length, one inch and one twentieth.

Formation.—Oxford-clay of Stingy, (Yonne). Rare.

Collection.—D'Ormois.

COLLYRITES LORYI, *Albin Gras.* Catal. des Foss. de l'Isère, tab. ii, figs. 1—3.

Test elongated, contracted behind ; the posterior ambulacra rise high up on the dorsal surface, as in *C. bicordata,* but the ambulacral areas are narrower.

Formation.—Corallian d'Echaillon (Isère). Rare.

Collection.—M. Albin Gras.

COLLYRITES DESORIANA, *Cotteau.* Échinides Foss. Pl. XXXIX, fig. 1.

Test large, oval, or sub-elongated ; anterior border with an anteal sulcus ; posterior border obtuse ; convex and depressed on the upper surface, flat on the under surface ; the three anterior ambulacra narrow, anterior pair flexed ; summit nearly central ; posterior ambulacra converge at a considerable distance above the vent.

Dimensions.—Height, thirty-three millimètres ; breadth, seventy-four millimètres ; length, eighty-seven millimètres=three inches four tenths. *Cotteau.*

Formation.—Inferior Corallian of Châtel-Censoir, and Druyes, (Yonne).

Collection.—M. Cotteau ; known only as a siliceous mould, and in general deformed.

Collyrites castanea, *Desor.* Synopsis des Échinides Fossiles, p. 207.

" Test very much inflated, short, almost spherical, contracted and sub-rostrated posteriorly; posterior ambulacra much arched, and converging immediately above the vent, which is only visible below; inferior surface convex and undulated; the ambulacra repose in depressions, and the inter-ambulacra form convexities between them; the single inter-ambulacrum is very prominent." *Desor.*

> *Formation.*—" Kellovien de St.-Croix, de Poillerel près Chaux-de-Fonds. Pas trop rare."

> *Collections.*—" Mus. Neuchâtel, coll. of MM. Campiche, Nicolet." *Desor.*

Collyrites Voltzii, *Agass.* Syn. *Disaster Voltzii*, Agassiz. Échinodermes Foss. de la Suisse, tab. iv, figs. 11—13. Desor, Monogr. des Disaster, tab. i, figs. 18—21.

Test large, circular, depressed, convex above, flat below; posterior ambulacra very much arched and converging above the vent, which is large, pyriform, and infra-marginal; mouth-opening central; peristome circular; base nearly flat, and only slightly cushioned; the ambulacra lie in depressions, and the inter-ambulacra form inconsiderable convex prominences; the ambulacra are all equal in width, and there is no anteal sulcus; the pores are closely approximated, and form double rows near the peristome.

> *Formation.*—" Oxfordien des Voirons près Genève. (Voltz)."

> *Collection.*—Museum of Strasbourg. Very rare; only two specimens known, both of which are in this Museum. One of them measures $3\frac{1}{2}$ inches in diameter.

Collyrites transversa, *d'Orbigny.* Paléontologie Française Ter. Crétacès. p. 50.

A large species, sixty-one millimètres in breadth. Under this name M. d'Orbigny notices a cordiform species, the length of which is only nine tenths of its breadth, and is as high as it is long. This species is remarkable for its transverse form, its nearly round figure, and deep anteal sulcus.

Formation.—" Du 12° étage Callovien des environs d'Escragnolles (Var.)"

Collection.—M. d'Orbigny.

COLLYRITES CARINATUS, *Leske.*   Syn. *Spatangites carinatus*, Leske, apud Klein, tab. 51,
figs. 2, 3.   Goldfuss, Petref. Germ., tab. xlvi, fig. 4.
Desor, Disaster, tab. iii, fig. 1.

Test small, inflated, cordiform; with an anteal sulcus, and a carina or ridge more or less marked, which extends from the middle of the upper surface to the vent.   Base convex and cushioned, single inter-ambulacrum very prominent at the base, anal opening marginal.

Formation.—Oxfordien de Gunsberg (Cant. de Soleure).   Oxfordthon; Thonkalke mit
Terebratula impressa, Weisser, Jura (*a*).   (Oxfordian.)   *Quenstedt.*

COLLYRITES CAPISTRATA, *Desmoulins.*   Syn. *Spatangus capistratus*, Goldfuss, Petref. Germ.
Agassiz, Echinid. Foss., tab. iv.   Desor, Disaster,
tab. iii.

Test cordiform, rounded anteriorly, with an anteal sulcus, tapering behind, and inflated at the sides, without a dorsal carina; vent in the middle of the posterior border; posterolateral ambulacra rise high up on the dorsal surface and converge at a point nearly equidistant from the anterior summit and the vent.

Dimensions.—Height, seven tenths of an inch; breadth, one inch; length, one inch
and one fifth.

Formation.—Corallien inférieur d'Urach, (Würtemberg), de Porrentruy, Ste. Croix.
Weisser, Jura *a* (Oxfordian).   *Quenstedt.*   Würtemberg.

Collections.—Museums Neuchâtel, Bâle, British.   Abundant.

COLLYRITES EXCENTRICA, *Münst.*   Syn. *Neucleolites excentricus*, Goldfuss, Petr. Germ.,
tab. xlix.   Desor, Disaster, tab. iv, figs. 1—3.

Test small, nearly circular, or elliptical, abruptly truncated posteriorly; posterior ambulacra converge above the vent, which opening is high up towards the dorsal surface.

43

Formation.—Jurassic limestone near Kehlheim; age not determined.

Collection.—Museum at Munich, coll. Count Münster.

COLLYRITES FABA, *Desor.*   Synopsis des Échinides Fossiles, p. 209.

"Test small, intermediate in form between *C. capistrata* and *C. bicordata*, Leske, less triangular than the former, but more contracted behind; posterior ambulacra converge at some distance above the vent, which is nearly entirely visible above."

Formation.—" Kellovien d'Ueken, près d'Effingen (Argovie)."

Collection.—" Moesch; very abundant." *Desor.*

METAPORHINUS MICHELINII, *Agass.*   Michelin, Revue et Mag. de Zool., No. 8, 1854.

COLLYRITES MICHELINII.   Cotteau, Échinides Foss., pl. xl, fig. 5.

" A large species, rounded before, truncated behind; the anterior summit very excentral, and occupying the most elevated part of the test, which is prolonged forwards in the form of a rostrum."

Formation.—" Corallien (Calcaire à chailles) de Druyes et Châtel-Censoir (Yonne)."

Collection.—" MM. Michelin, Cotteau." *Desor.*

METAPORHINUS CENSORIENSIS, *Desor.* Synopsis des Échinides Fossiles, p. 211.

COLLYRITES CENSORIENSIS.   Cotteau, Échinides Foss., pl. xl, figs. 6, 7.

This singular urchin resembles the preceding species; according to M. Cotteau, it is more inflated, and rounder before, but not so much sloped behind; the posterior ambulacra are more flexuous and unite nearer the vent.

Formation.—Corallien of Châtel-Censoir, (Yonne).

Collection.—M. Cotteau, a single specimen ; a plaster mould in my cabinet.

GRASIA ELONGATA, *Michelin.* Revue et Mag. de Zool., No. 8, 1854.

HYBOCLYPUS ELONGATUS. Albin Gras, Cat. des Fossiles de l'Isère, tab. ii, fig. 1—3.

A large urchin, very much elongated, measuring seven centimetres in length, by four in breadth ; the anterior ambulacral summit very near the anterior border ; the posterior ambulacra meet high up on the dorsal surface near the middle of the test, and at a considerable distance from the vent.

Formation.—Corallien d'Echaillon (Isère) ; very rare.

Collection.—Dr. Albin Gras.

Family 9.—ECHINOBRISSIDÆ, *Wright*, 1856.

Famille des CASSIDULIDES, *groupe des* NUCLEOLIDES, *Agassiz* and *Desor*, 1846.
Famille des NUCLÉOLIDÉES (in pars) *Albin Gras*, 1848.
Famille des ECHINOBRISSIDÉES (in pars) *D'Orbigny*, 1855.

The family ECHINOBRISSIDÆ, as I have defined it, comprehends only those urchins which have petaloidal or sub-petaloidal ambulacra; the vent lodged in a dorsal sulcus, or opening at the posterior border; the mouth edentalous; and the peristome pentagonal and sub-central.

The test, in all the smaller species, is thin, has a sub-quadrate, oblong, sub-pentagonal, circular, or clypeiform shape; is uniformly covered with small imperforate tubercles, raised on uncrenulated bosses, and surrounded by deep sunken areolas (Pl. XXIV, fig. 1 *h*); the spines are small, short, and slender (Pl. XLI, fig. 1); the tubercles at the base are always larger, and more fully developed, than those on the upper surface.

The ambulacral areas are narrow and lanceolate; they are enclosed by poriferous zones of unequal width, the pores being placed more or less wide apart in different regions of the zones; the holes of the inner row are circular, those of the outer are oblong or slit-like, and they often communicate with the inner series by transverse sulci. In consequence of this structure, the ambulacra present elegant petaloidal forms, more or less developed in different species, but always lanceolate above, expanded in the middle, and contracted and open below. The poriferous zones at the border and base are narrow, and the pores are small, equal, and placed wide apart (Pl. XXIV, fig. 1 *i*).

The apical disc is in general small, and composed of at least ten elements; of the five genital plates, four are perforated, and one is imperforate; the right antero-lateral plate is always the largest and carries the madreporiform tubercle, which is often very large, and nearly conceals the other elements of the disc; the five ocular plates are small, with marginal orbits (Pl. XXIV, fig. 1 *g*), placed opposite the apices of the ambulacra.

The vent is large, and in general opens into a valley situated in the upper surface of the inter-ambulacrum, or at the middle or margin of the posterior border; in the only living species at present known, *Nucleolites recens*, Edwards, from New Holland, figured in the illustrated edition of the 'Règne Animal,' this aperture is found to be closed by a series of small plates, in the same manner as the anal aperture is filled up in the ECHINIDÆ and SPATANGIDÆ. I have copied this figure in Pl. XLI, fig. 1 *a*, and a specimen of the urchin, without spines or anal plates, is now in the collection of living Echinoderms in the British Museum.

The mouth-opening is small, sub-central, and edentulous; the peristome in general is pentagonal, and in the genus *Clypeus* is surrounded by five oral lobes (Pl. XXVIII, fig. 1 *b*).

The inter-ambulacral areas are wide; the plates composing these portions of the test are large, and bent in the middle; each plate supports about three rows of small tubercles, which are arranged in lines, with much regularity (Pl. XXXII, fig. 1 c).

The ECHINOBRISSIDÆ have many organic characters in common with the ECHINO-LAMPIDÆ, but they are distinguished from that family by the position of the vent, and by the absence of the poriferous petals which in the latter surround the mouth.

The ECHINOBRISSIDÆ, thus limited, nearly corresponds with the group of Nucléolidées in the family CASSIDULÉES of Agassiz and Desor; it differs from the family NUCLEOLITIDÆ of Dr. Albin Gras, and the family ECHINOBRISSIDÆ of M. d'Orbigny, by the elimination of all the urchins therefrom which have a basal vent, peristomal lobes and petals, and an organization in accordance with the genus *Echinolampas*.

Genus—ECHINOBRISSUS, *Breynius*, 1732.

NUCLEOLITES, *Lamarck*, 1801.
— *Goldfuss*, 1826.
— *Agassiz*, 1837.
ECHINOBRISSUS, *D'Orbigny*, 1855.
— *Desor*, 1857.
— *Cotteau*, 1858.

This natural group is composed of small urchins, which have an oval, oblong, sub-quadrate, or sub-circular form, more or less convex on the upper surface, and slightly concave beneath; the test is rounded anteriorly, more or less produced, truncated, or lobed posteriorly, and is in general narrow before, and wider behind; the vent opens in the upper surface into a dorsal valley, which in one section extends from the apical disc to the posterior border, and in another is limited to the lower half thereof; the aperture is closed by a series of small anal plates, which are usually absent in fossil species, but are preserved in the only living example of the genus at present known (Pl. XLI, fig. 1).

The base is more or less concave; the small mouth-opening is excentral, and lodged in a depression, nearer the anterior than the posterior border; the peristome in many of the species is regularly pentagonal, in others it is oblique. M. d'Orbigny* has separated the latter into a distinct genus, under the name *Trematopygus*, but these oblique-mouthed species can at most be considered only as a section of the genus *Echinobrissus*; which are supposed to be special to the Cretaceous Rocks.

The apical disc is small, square, and compact; it is composed of four perforated genital plates, arranged in pairs, and one single imperforate plate; the right antero-lateral is the largest, and supports the madreporiform tubercle; the five small

* 'Paléontologie Française Ter. Cretacés,' tome vi, p. 374.

ocular plates are triangular, and occupy depressions between the larger genital plates, opposite the summits of the petaloid ambulacra.

The tubercles are small on the upper surface, and larger at the base.

The genus Echinobrissus was established by Breynius in 1732, in his important memoir ' De Echinis et Echinites,' in which it was thus characterised : " *Echinobrissus est Echinus, cujus oris apertura centrum basis fere occupat, ani vero in vertice conspicitur, a centro aliquantulum remota, et in sinu quodam ori oblique opposita.*"*

Klein, who published two years afterwards, unfortunately did not preserve any of the well-defined genera of his learned contemporary, and his commentator, Leske, in 1778, placed the *Echinobrissus* of Breynius among the *Spatangus* of Klein. When Lamarck, in 1801, instituted his genus *Nucleolites*, he appears not to have been aware that the same group had been well figured and accurately defined sixty-nine years before by Breynius; in his great work, " Animaux sans vertèbres," he reproduced his genus Nucleolites, and for the first time refers to Breynius's work for figures of the species.

M. Agassiz, in dismembering Lamarck's Nucleolites, did not restore Breynius's genus, although that author was most careful to reproduce the genera established by the older authors; thus, *Nucleolites columbaria*, Lamk., was repeated as the type of his new genus *Catopygus*, and *N. ovalus*, Lamk., became the type of *Pyrina*, whilst *N. scutatus*, Lamk., remained the type of *Nucleolites*. M. d'Orbigny restored the genera so well established by Breynius, and proved beyond all question that author's claim to *Echinobrissus*; M. Cotteau and M. Desor have both admitted the justice of his views, and the genus *Echinobrissus* now occupies the position from which it ought never to have been displaced.

A. *Species from the Inferior Oolite.*

Echinobrissus clunicularis, *Llhwyd.*   Pl. XXIV, fig. 1 *a, b, c, d, e, f, g, h, i, k.*

Echinites clunicularis.	Llhwyd, Lithophylacii Brittanici Ichnograph., p. 48, 1698.
Echinobrissus planior.	Breynius, Schediasma de Echinis, tab. vi, figs. 1, 2, 1732.
Nucleolites Sowerbyi.	Defrance, Dic. Sciences Naturelles, tome xxxv, p. 213, 1825.
Clypeus lobatus.	Fleming, British Animals, p. 479, 1828.
— clunicularis.	Phillips, Geology of Yorkshire, tab. vii, fig. 2, 1829.
Nucleolites clunicularis.	Blainville, Dic. Sciences Naturelles, tome lx, p. 188, 1830.
— latiporus.	Agassiz, Échinoderm. Foss. de la Suisse, p. 43, tab. vii, figs. 13—15.
— clunicularis.	Desmoulins, Tableaux Synonymiques des Échinides, p. 358, 1837.

---

* ' De Echinis et Echinites, sive Methodica Echinorum Distributione, Schediasma,' 1732.

NUCLEOLITES CLUNICUNARIS.		Dujardin, in Lamarck's Animaux sans vertèbres, 2ᵉ ed., tome iii, p. 345, 1840.
—	TERQUEMI.	Agassiz, Catalogue raisonnée des Echinides, Revue et Mag., 1847.
—	THURMANNI.	Desor, Catalogue raisonnée des Échinides, Annales des Sciences Naturelles, 3ᵉ série, tome vii, p. 154, 1847.
—	PYRAMIDALIS.	M'Coy, Annals and Magazine of Natural History, 2d series, vol. ii, p. 116, 1848.
—	CLUNICULARIS.	Forbes, Memoirs of the Geological Survey, decade 1, pl. ix, 1849.
—	EDMUNDI.	Cotteau, Échinides Fossiles, p. 67, tab. v, figs. 1—3, 1850.
—	CONICUS.	Cotteau, Échinides Fossiles, p. 64, tab. iv, figs. 4—6, 1850.
—	CLUNICULARIS.	Wright, Annals and Magazine of Natural History, 2d series, vol. ix, p. 297, 1851.
—	—	Bronn, Lithaea Geognostica, 3ᵉ auflage, Band. ii, p. 152, 1852.
—	—	Guéranger, Essai d'un rept. Paléont. de la Sarthe, p. 25, 1853.
—	—	Forbes, in Morris's Catalogue of British Fossils, 2d ed., p. 84. 1854.
ECHINOBRISSUS CLUNICULARIS.		D'Orbigny, Note rectif. sur divers genres d'Échinides, Rev. et Mag. Zool., 2ᵉ série, t. vi, p. 24, 1854.
—	TERQUEMI.	D'Orbigny, ibid., p. 24, 1854.
—	LATIPORUS.	Ibid.
—	THURMANNI.	Ibid., p. 25, 1854.
—	CLUNICULARIS.	Desor, Synopses des Échinides Fossiles, p. 263, pl. xxx, figs. 18—20, 1857.
—	TERQUEMI.	D'Orbigny, Paléontologie Française Ter. Cretacés, t. vi, p. 390.
—	LATIPORUS.	Ibid., p. 391, 1857.
—	SARTHASENSIS.	Ibid. ,, ,,
—	CLUNICULARIS.	Ibid. ,, ,,
—	CONICUS.	Ibid. ,, ,,
—	EDMUNDI.	Ibid. ,, ,,
—	THURMANNI.	Ibid. ,, ,,
—	CLUNICULARIS.	Cotteau and Triger, Échinides de depart. de la Sarthe, pl. x, fig. 7, p. 12.
NUCLEOLITES CLUNICULARIS.		Pictet, Traité de Paléont., t. iv, p. 217, Atlas, pl. xciv, fig. 10, 1857.
—	SARTHASENSIS.	D'Orbigny, Prodrome de Paléontologie, tome i, p. 220, 1858.
AGASSIZ MOULDS.		P. VII, type. T. LXXXIII, *Nucl. Terquemi.* S. XLVI, *Nucl. latiporus.*

Test sub-quadrate, anterior border rounded, posterior bilobed; dorsal surface convex, declining abruptly anteriorly, more gently posteriorly; apical disc and vertex sub-central; ambulacral areas narrowly lanceolate above, converging below; dorsal valley deep, lanceolate, extending to the posterior border of the apical disc; posterior lobes gently tapering, not tumid; base concave and grooved by the ambulacra; mouth-opening sub-pentagonal, situated nearer the anterior border.

Dimensions.—*Large Cornbrash specimen* (fig. 1 *k*). Height, eighteen twentieths of an inch; anterior posterior diameter, one inch and thirteen twentieths; length from the border to the posterior sulcus, one inch and eleven twentieths; breadth across the apical disc, one inch and eight twentieths.

Cornbrash specimen (fig. 1 *a*).—Height, eight tenths of an inch; antero-posterior diameter, one inch and one quarter of an inch; transverse diameter, one inch and one quarter of an inch.

Inferior Oolite specimen.—Height, six tenths of an inch; antero-posterior diameter, one inch and three tenths of an inch; transverse diameter, one inch and three tenths of an inch.

Description.—This is doubtless the urchin to which Llhwyd* gave the name *Echinites clunicularis*, and described "Echinites è lapide selenite, quinis radiis è duplice serie transversarum lineolarum conflatis," he refers to the figures in Plott † and Lister, ‡ but as these exhibit no anal valley, it is impossible to say whether they represent this species or the Coral Rag form. The figure given by Breynius, § and described as *Echinobrissus planior*, represents this species as shown by the length of the anal valley, which reaches as high as the disc, whilst *Echinobrissus elateor*, fig. 3 of the same plate, represents *E. scutatus* from the Coral Rag. This most acute observer had therefore clearly distinguished and figured a diagnostic specific character which has been overlooked by subsequent authors, and led to much confusion in the synonyms of *E. clunicularis*. *Spatangus depressus*, of Leske, ‖ most probably represents a large quadrate variety of *E. scutatus*, and is consequently omitted from our list of synonyms. Dr. William Smith¶ figured in his plate of characteristic Coral Rag fossils an urchin to which he gave no name. This Fleming** refers to *E. clunicularis*, whilst he calls *E. lobatus* the species figured by Lister, forgetting that Llhwyd referred to Lister's figure as the type of *E. clunicularis*. Professor Phillips†† distinguishes the two species from each other, and restricts the name *E. clunicularis* to the Cornbrash form, whilst he figures the Coral Rag nucleolite as *E. dimidiatus*. Continental authors have made several species out of the simple varieties of *E. clunicularis;* but when we recollect the many varieties of figure and outline which this urchin exhibits, any erroneous multiplication of species is readily explained.

Professor Edward Forbes‡‡ considered *E. scutatus*, Lamk., a variety of *E. clunicularis*. This error, however, he afterwards corrected, when he became acquainted with the true

* 'Lithophylacii Britannici Ichnographia,' p. 48, No. 988.
† 'History of Oxfordshire,' table ii, fig. 12, 1698.
‡ 'De Lapidibus Turbinatis,' cap. 11, titulus xxvi, 1678.
§ 'Schediasma de Echinis,' tab. vi, figs. 1, 2.
‖ 'Dispositio Naturales de Echinodermatum,' tab. li, figs. 1, 2.
¶ 'Strata Identified by Organized Fossils.'
** 'British Animals.'
†† 'Geology of the Yorkshire Coast.'
‡‡ 'Memoirs of the Geological Survey,' decade i, pl. ix.

types of Lamarck's species. In fact, *Echinobrissus clunicularis*, Llhwyd, may be regarded as the type of that section of nucleolites in which the anal valley extends from the border to the apical disc, and *Echinobrissus scutatus*, Lamk., of that smaller section in which the anal valley never extends from the border to the disc.

Professor M'Coy* described three nucleolites under the names *planulatus, pyramidalis,* and *æqualis.* The two latter forms appear to be varieties of *E. clunicularis,* and *planulatus* a variety of *E. scutatus.* I have not seen the original specimens which belong to the Cambridge Museum; but the author having kindly furnished me with outlines of these urchins to assist me in identifying his species, I have formed that opinion from his figures.

This urchin exhibits much diversity as regards size, outline, height, and tumidity. Its most typical forms are found in the Inferior Oolite and Cornbrash, and one of the best specimens I know is that from the cabinet of my friend, the Rev. A. W. Griesbach, which forms the subject of our figure. The suite of specimens before me from these *terrains* vary from a sub-orbicular to a sub-quadrate outline, and present nearly all the intermediate forms. They are rounded anteriorly, a little contracted before, enlarged at the sides, and more or less bilobed posteriorly. The upper surface is convex, and exhibits various degrees of elevation; in some it is much depressed, in others it rises into a sub-conical form. In a series of specimens before me scarcely two have the same proportional height. The vertex is almost always excentral and inclined towards the anterior border; but the amount of inclination, like the height, varies considerably in different individuals.

The ambulacral areas have a petaloid or lanceolate figure, with sub-parallel sides; the single area and the anterior pair are nearly of the same length and width, and the posterior pair are the longest and widest (fig. 1 *a, c*). The poriferous zones vary in structure on the dorsal and ventral surfaces; in the petaloid portion of the dorsal surface the pores of the inner row are round, those in the outer row are oblong, and the furrow uniting the pores varies in depth in different examples (fig. 1 *f*); in a large specimen there are about forty pairs of pores in the petaloid portion of the dorsal zone; between each pair there is a short row of small microscopic granules which separates the outer oblong pores from each other (fig. 1 *f*). In all the non-petaloid portion of the zones, the pores are small, round, and very indistinct (fig. 1 *e*); at the sides and outer third of the base, near the mouth, however, they become more distinct.

The inter-ambulacral areas are of unequal width; the posterior pair are nearly one fourth wider than the anterior pair, and the single area is the widest (fig. 1 *a*). The anal valley extends from the posterior part of the disc to the border; it is narrow above, wider in the middle, and expanded below, in the excavated portion; it has vertical parallel walls, which gradually diverge, then approximate, and afterwards expand outwards, forming a well-defined groove (fig. 1 *a*). The pyriform anal opening is situated at the extreme end of the

* 'Annals of Natural History,' 2d series, vol. ii, p. 416.

valley (fig. 1 *d*). The surface of all the plates is covered with small, close-set, spinigerous tubercles, surrounded by circular areolas, and having the interspaces minutely granulated (fig. 1 *e*).   At the base the tubercles are larger, their summits are perforated, and they are surrounded by wider and deeper areolas (fig. 1 *h*).   The diagram of the dorsal structure (fig. 1 *i*) shows the relative disposition of the plates in both the areas and poriferous zones, and fig. 1 *k* is a correct figure of the Rev. A. W. Griesbach's unique and largest specimen from the Cornbrash of Rushden.

The apical disc (fig. 1 *g*) is composed of four perforated ovarial plates, and a small imperforate ovarial.   The right antero-lateral plate is the largest, and extends into the centre of the disc; it supports on its surface a prominent spongy madreporiform body. The left antero-lateral and postero-lateral ovarials are about the same size; the oviductal holes in each are near the apex of the plates; the single imperforate plate is much smaller than the others.   The five ocular plates, with marginal eye-holes, are wedged in the angles of the ovarials, and the surface of all the discal plates is granulated.   Fig. 1 *g* represents these discal elements in the most perfect development I have ever met with them in this species, and it is, I believe, the first correct figure which has been given of the apical disc of this common and well-known urchin.

The base is more or less concave, most so in the pyramidal varieties (fig. 1 *b*); the ambulacral areas radiate in depressed furrows from the margin of the mouth-opening, becoming shallower and wider as they approach the border.   The poriferous zones are more apparent near the peristome, and the pores lie close together in oblique pairs.   The mouth-opening is excentrical, being situated in a deep depression nearer the anterior than the posterior border.   The tubercles at the base are larger and set wider apart than those on the dorsal surface; their summits are likewise perforated, and their bosses surrounded by wider and deeper areolas (fig. 1 *h*), the intermediate space being filled with larger granules.

Affinities and differences.—This species resembles very much *E. scutatus*, Lamk., and *E. dimidiatus*, Phil.; it is distinguished from these forms by the length of the anal valley, which extends from the disc to the margin, whilst in *E. scutatus*, Lamk., and *E. dimidiatus*, Phil., there is an undepressed portion of test between the disc and the commencement of the valley.   The posterior lobes are likewise more tapering and less tumid than in the Coral Rag species.   *E. clunicularis* is distinguished from *E. orbicularis*, Phil., by its sub-quadrate shape, by its sides increasing in diameter towards the posterior part, and by its posterior bilobed border.   The same group of characters serve to distinguish it from *E. Woodwardii*, Wright, a species remarkable for its tumid sides and long narrow anal valley.

Locality and Stratigraphical position.—This urchin makes its first appearance in the *Clypeus Plottii* bed of the Inferior Oolite, in the zone of *Ammonites Parkinsoni*.   The specimens from this rock often attain a considerable size; the flattened varieties are the largest, and the pyramidal the smallest. I have collected this species from the upper zone of the Inferior Oolite at Rodborough, Birdlip, and Shurdington Hills, and from Hampen, Naunton, and Stow-in-the-Wold, Gloucestershire; from the Stonesfield Slate at Seven-hampton, Eyeford, and near Pewsdown; from the Great Oolite of Minchinhampton, Salperton, and near Cirencester; from the Forest Marble near Trowbridge; from the Cornbrash near Cirencester, Gloucestershire; and from Chippenham and Trowbridge, Wilts. Mr. Hull has collected it from that formation near Shilton, and Mr. Gavey near Woodstock. The Rev. A. W. Griesbach has collected the finest specimens I have seen from the Cornbrash at Rushden, Northamptonshire, where they attain a very large size (fig. 1 *a*, *k*), and are found in excellent preservation. Messrs. Bristow and Gapper, of the Geological Survey, collected specimens in abundance from the Cornbrash of Dorset, and I have likewise found it in the same rock at Gristhorpe Bay, and Scarborough, Yorkshire. It forms, therefore, a very characteristic urchin of the upper portion of the lower division of the Oolitic series, extending without interruption from the zone of *Ammonites Parkinsoni* to the uppermost beds of the Cornbrash; a practised eye, however, can readily detect the varieties which are gathered from the Inferior Oolite, Great Oolite, and Cornbrash; the Inferior Oolite forms are usually large and depressed, the Great Oolite small and dwarfed, and the Cornbrash the largest and best developed of the series.

The foreign distribution of this species, according to M. Desor, is Calcaire à Polypiers (Bathonien) de Ranville, Forest-Marble de Châtel-Censoir (Yonne), Vesulien du Hornussen, Kreisacker, Wolfliswyl et Frick (Argovie), Meltingen. (Canton de Soleure), Maiche (Doubs).

It was collected in abundance by M. Triger from the Great Oolite (department of the Sarthe) at "La Jaunelière (tuilerie), Domfront (four à chaux), Conlie, Monnè, Saint-Christophe, route de Mamers à Marolette, Aubigné (ferme de Gesnes-le-Gaudelin), Pécheseul, Noyen, Saint-Pierre-des-Bois, route de Contilly à la Perrière."

Table of M. Triger, Bradford Clay, Ass. Nos. 1, 2, and 3, and Forest Marble, Ass. No. 4.

B. *Species from the Great Oolite.*

ECHINOBRISSUS WOODWARDII, *Wright.* Pl. XXIV, fig. 2 *a, b, c, d, e.*

NUCLEOLITES WOODWARDII.	Wright, Annals and Magazine of Natural History, 2d series, vol. 13, p. 161, pl. xii, fig. 5 *a—e*, 1852.
— —	Forbes, in Morris's Catalogue of British Fossils, 2d ed., p. viii, ad. sp.
ECHINOBRISSUS WOODWARDII.	Desor, Synopsis des Échinides Fossiles, p. 268.

Test thin, sub-quadrate; sides tumid; dorsal surface flatly convex; anal valley deep, narrow, and spear-shaped, extending from the apical disc to the posterior border; ambulacral areas narrowly lanceolate; posterior lobes short, and truncated; base flat; inter-ambulacral areas slightly inflated at the base; single inter-ambulacrum scarcely produced; mouth-opening pentagonal, excentral; apical disc small, central.

Dimensions.—Height, six tenths of an inch; antero-posterior diameter, one inch and one tenth; transverse diameter, one inch and one fifth; the larger specimens are so much deformed that their proportional dimensions cannot be accurately given.

Description.—This species was formerly considered a variety of *E. orbicularis*, Phil., but a careful study of many specimens of the former, compared with good typical examples of the latter, convinces me that these two urchins are specifically distinct.

The test is thin, and not often well preserved; the specimen drawn (fig. 2 *a*, *b*) is a small but tolerably perfect individual; the outline is sub-quadrate, the shell is one tenth of an inch more in breadth than in length; it is a little narrower anteriorly, than posteriorly, and the posterior border is broadly truncated; the sides are very tumid (fig. 2 *c*), sometimes irregularly so; and the test is higher across the apices of the postero-lateral ambulacra than at any other point (fig. 2 *c*); the tumidity of the sides produces a greater flatness of the upper surface than is observed in any of its congeners. The ambulacral areas are nearly all of the same width; they have a narrow, graceful, lanceolate form (fig. 2 *a*), from the mouth to about midway between the margin and the apical disc they are nearly all of equal width; at this point the pores gradually change their form, they are slightly separated for a short distance, and again converge as they approach the disc; the pores of the inner row are round (fig. 2 *d*), those of the outer row form oblique slits; the widest part of which is external; the round pores are formed by notches in the upper part and sides of the small plates forming the avenues, and the oblique pores by uncalcified portions of the margins of the same plates; from the termination of the petaloidal portion of the zones to the mouth, the pairs of pores are small, and set wider apart, whilst the diameter of the areas remains the same; near the peristome they are crowded close together and form arches, the convexity of which is towards the mouth-opening. The inter-ambulacral areas are of unequal width; the anterior pair are the narrowest, the posterior are wider than the anterior pair, and the posterior single inter-ambulacrum is the widest (fig. 2 *a*). The anal valley forms a long narrow depression, extending from the disc to the border; it has perpendicular sides, and a small vent opens into it about the middle; the base is flat, becoming more or less concave near the mouth (fig. 2 *b*); the anterior and posterior pairs of inter-ambulacra are moderately convex in this region, and the basal portion of the inter-ambulacrum is very slightly produced; the mouth-opening is excentral, situated in a depression nearer the anterior than the posterior border; the peristome has a pentagonal shape, with five rudimentary lobes. The surface of the plates is covered with microscopic tubercles (fig. 2 *d*), arranged in tolerably regular oblique rows; on each plate there are from twenty to thirty, disposed in three or four rows (fig. 2 *d*) ; and those at the base are somewhat larger than the tubercles on the upper surface. The apical disc is small, and nearly central ; its elements are so intimately soldered together, that its general form alone can be made out ; none of the specimens I have examined display the separate plates ; the eyeholes are marginal, at the apices of the ambulacra, and the ovarial holes are further outwards, and between them, whilst the madreporiform tubercle occupies the centre of the disc. The test is very thin, and almost always deformed by tumidity, and the upper surface is often irregular from this cause. The beauty and regularity of the small specimen I have figured, forms an exception to almost all the other specimens I have seen.

Affinities and differences.—This species resembles most *E. orbicularis*, Phil., which is the only form amongst its Oolitic congeners for which it can be mistaken. The following diagnostic characters will enable the geologist to distinguish these allied forms from each other. In *E. Woodwardii* the sides are very tumid, and the dorsal surface is flat, whilst in *E. orbicularis* the sides taper, and the upper surface is convex. In *E. Woodwardii* the base is flat, and the inter-ambulacra are slightly produced, whilst in *E. orbicularis* the base is concave, and the inter-ambulacra convex and prominent. In *E. Woodwardii* the anal valley is narrow, in *E. orbicularis* it is wide. In *E. Woodwardii* the outline is sub-quadrate, in *E. orbicularis* it is circular ; the petaloidal disposition of the dorsal portion of the poriferous zones extends farther down the sides in *E. orbicularis* than in *E. Woodwardii*. It differs from *E. Brodiei* and *E. scutatus* in the anal valley extending from the disc to the border, whereas in these species there is always an undepressed portion of test between the disc and the valley. Between *E. Woodwardii* and *E. clunicularis* the difference in the general shape and development of the posterior lobes is so great, that they cannot be mistaken for each other.

Locality and Stratigraphical position.—I have collected this urchin from the Great Oolite at Minchinhampton, near Cirencester, at Salperton tunnel, Great Western Railway, at Highgate, and near Pewsdown, Gloucestershire, and at Burford, Oxon ; as far as I know, it has not been found out of the Great Oolite. Mr. John Bravender, of Cirencester, collected this species at Tetbury, and between Tetbury and Bourton-on-the-Water, embracing a distance of twenty-five miles, in ten or twelve different localities ; the specimens were found in a hard marly rock, at the upper part of the Great Oolite ; the test is unfortunately not often well preserved. Mr. Frederick Bravender has kindly sent me the following note on the distribution of this urchin, which I herewith give entire :

" *Echinobrissus Woodwardii* is generally found in a rubbly bed, nearly, if not quite, at the top of the Great Oolite ; and although this stratum covers a considerable tract of elevated country in the neighbourhood of Cirencester, this urchin has not been collected very plentifully, except in one or two localities. It has been found at Perrimore Quarry, on the Royal Agricultural College Farm, which is the best locality; at Coates, at the Wood-house in Earl Bathurst's Park, at Downs Farm, and Stowell Park, on the Stow Road, and at Stratton, and North Cerney, near Cirencester. The specimens obtained from the Woodhouse were from some light excavations to mend a private road, where they were abundant. In one instance (near Stowell Park) the rubbly bed containing the *E. Woodwardii* was covered over with Bradford Clay, about two feet in thickness, when the characteristic fossil *Terebratula digona*, Sow., occurred in abundance. In two or three instances *E. Woodwardii* was found associated with *Trigonia Phillipsii*, Lyc., which has not been very long recognised as a Gloucestershire shell; it is not improbable that the two may be found together in Northamptonshire. I am not aware that this species has been found out of the Great Oolite.

In three specimens collected from the same white marly bed of the Great Oolite, whilst this sheet was passing through the press, the small apical disc is better preserved than in those I had previously examined. The ovarial plates are long, narrow, and lanceolate, with large, oblique, oviductal holes, perforated near their apices, the madreporiform body covers so much of the disc, that I cannot discern whether there are supplementary plates in its centre as in *E. orbicularis*. The ocular plates are small, and have minute marginal orbits.

ECHINOBRISSUS GRIESBACHII, *Wright*, nov. sp. Pl. XXV, fig. 1 *a, b, c, d, e, f.*

Test quadrate, elevated; sides tumid; posterior lobes small; apical disc large, expanded; anal opening large, adjoining the disc without any intermediate depressed space between the disc and the vent; poriferous zones narrow; pores approximated throughout; anal valley wide.

Dimensions.—Height, four tenths of an inch; antero-posterior diameter, three quarters of an inch; transverse diameter, nearly equal to the length.

Description.—I am indebted to my friend, the Rev. A. W. Griesbach, for calling my attention to this new form, which has hitherto, doubtless, been considered a small variety of *E. clunicularis*. After a careful study of this urchin, I feel disposed to adopt my friend's view, and in justice to Mr. Griesbach, I subjoin his notes on this species, in which he has most accurately pointed out its diagnostic characters:

"There is a small nucleolite found in the Great Oolite round about here, which I have for a long time heedlessly confounded with *N. clunicularis*, but which I yesterday discovered to be quite a distinct form, and I do not know any other species to which it can be referred. I have seen it at Wimmington, Higham Ferrers, and Blisworth, and have received a specimen from Mr. Brodie from 'Fuller's Earth Rock, Gloucestershire.' The specimens I have seen and collected have been, with only one exception, either entirely denuded of the test, or else so eroded as to leave no structure visible. I send you the one exception, which I found at Blisworth about four years ago (fig. 1 *a*, *c*). It is narrower than the usual form, which is quadrate—as broad as it is long. Mr. Brodie's specimen is considerably larger, but, though having the test, it has been scraped and spoiled. Compared with *N. clunicularis* (of which I have seven beautiful specimens from the Cornbrash of Rushden before me, four of these are depressed, and three high and conical), this nucleolite, among minor differences, presents the following main distinctions:

Affinities and differences.—" 1st. Whereas in *E. clunicularis* there is always a narrow depressed space (as long as the anal opening itself) between the apical disc and the vent; in *E. Griesbachii* there is *no such space*, but the anal aperture *immediately adjoins the apical disc* (fig. 1 *a*, *c*).

" 2d. The apical disc is proportionally *much* larger than in *E. clunicularis*, in consequence of which the apices of the ambulacra do not approach each other so nearly, and the perforations in the disc are wider apart (fig. 1 *b*).

" In this small urchin (fig. 1 *a*) the disc is fully as broad as in a fine specimen of *E. clunicularis*—one inch and three tenths in antero-posterior diameter.

" If several false species have been made out of *E. clunicularis*, that is no reason why we should too readily unite different forms under that name. I think this i a distinct species, and hope, therefore, to have avoided what might otherwise have been said, ' Incidit in Scyllam cupiens vitare Charybdim.' I have no idea that this form is *rare* in the Great Oolite, but it appears to have been overlooked."

ECHINOBRISSUS ORBICULARIS, *Phillips*.   Pl. XXV, fig. 2 *a*, *b*, *c*, *d*, *e*, *f*, *g*.

CLYPEUS ORBICULARIS.	Phillips, Geology of Yorkshire, tab. vii, fig. 3, 1829.
— —	Agassiz, Prodrome d'une Monogr. des Radiaires Mem. Soc. des Sc. Nat. de Neuchâtel, t. i, p. 136, 1836.
— —	Dujardin, in Lamarck, Animaux sans vertèbres, 2d ed., t. iii, p. 348, 1840.
— —	Morris, Catalogue of British Fossils, p. 50, 1843.
NUCLEOLITES ORBICULARIS.	Forbes, Memoirs of the Geological Survey, decade first, description of pl. vii, 1849.
— —	Wright, Annals and Magazine of Natural History, 2d series, vol. ix, p. 301, 1851.
— —	Forbes, in Morris's Catalogue of British Fossils, 2d ed., p. 84, 1854.
	Davoust, Note sur les Foss. Spéciaux, à la Sarthe, p. 25, 1856.
ECHINOBRISSUS ORBICULARIS.	Desor, Synopsis des Échinides Fossiles, p. 265, 1857.
— —	Cotteau and Triger, Échinides du departement de la Sarthe, pl. ix, figs. 5—8, p. 57, 1858.

Test orbicular; upper surface irregularly convex and depressed; sides tumid; apical disc complex and central; ambulacral areas lanceolate; poriferous zones narrowly petaloid between the border and disc; posterior lobes obsolete; anal valley broad, extending from the disc to the border; vent large, situated in the upper part, near the disc; base flat or slightly concave; mouth sub-central, situated in a depression nearer the anterior border.

Dimensions.—Height, eight tenths of an inch; transverse diameter, one inch and six tenths; antero-posterior diameter, one inch and six tenths.

Description.—The tumid sides, obsolete lobes, orbicular circumference, its length and breadth being equal, with the broad, flat, and somewhat irregular dorsal surface, serve to distinguish this species from its congeners.

The ambulacral areas are narrow, and nearly all of the same width (fig. 2 *a*); they have a more petaloidal form than in the preceding species, and are furnished with two rows of small tubercles arranged in zig-zag lines on the areas, the external row being the largest and most regular (fig. 2 *e*), whilst the middle of the area is covered only with minute granules.

The poriferous zones are moderately wide, and with the ambulacral areas form well-developed petaloidal figures on the upper surface of the test; the pores of the inner row are round, those of the external row are oblong (fig. 2 *e*); there are about six pairs of holes opposite each inter-ambulacral plate; on the upper surface the pores are wider apart (fig. 2 *g*), on the sides they are close together (fig. 2 *c*), at the margin they are very indistinct (fig. 2 *f*), at the base they are scarcely visible and still more widely apart, and continue so to the peristome (fig. 2 *b*).

The inter-ambulacral areas are of unequal width; the antero-lateral pair are the narrowest, the postero-lateral are wider, and the single area is the widest (fig. 2 *a, b*); the sides are more or less tumid (fig. 2 *c*), and the upper surface is irregularly convex (fig. 2 *f*), in consequence of the antero- and postero-lateral areas being slightly depressed in the centre (fig. 2 *c, f*); the inter-ambulacral plates form long pentagons, bent upwards in the middle (fig. 2 *e*). Each plate is covered with about twenty very small tubercles, arranged in three rows; in a large majority of specimens they are so minute on the sides, margin, and upper surface that it requires the aid of a good lens to discover them, at the base they are larger and crowded closer together; each tubercle is surrounded by a sunken areola (fig. 2 *e*), and the inter-tubercular surface of the plates is closely crowded with microscopic granules which encircle the areolas or fill up all the intervening spaces. The single inter-ambulacrum is not at all produced, and the lobes are very small. The anal valley is large; it commences at the posterior border of the disc and extends to the margin (fig. 2 *a. g*). The large vent has an elliptical shape, and opens near the surface in the upper third of the valley, whilst the lower two thirds of that depression form a considerable furrow in the middle of the area (fig. 2 *a, g*).

The apical disc is large, and presents a very remarkable modification of the usual type of structure observed in other ECHINOBRISSIDÆ. As I have seen this structure in many specimens collected from localities widely apart, it must be regarded as a normal form in this species. Mr. Bone has figured this curious disc with great accuracy at fig. 2 *d*; the two pair of perforated ovarial plates have a pyriform shape, the right antero-lateral is the largest, and supports the spongy madreporiform body; between these there are introduced into the middle of the disc four small plates, and posterior to them three other smaller plates. The front rows are bounded before by the madreporiform plate, and behind by the left posterior ocular plates, and the three posterior plates lie between the posterior oculars. The surface of all the elements of the disc is covered with numerous close-set tubercles. The five ocular plates are cordate in form, and have marginal orbits. The abnormal deviation from the usual structure consists in the introduction of two rows of small plates behind and between the normal discal elements.

The base is concave and undulated (fig. 2 *b*), from the prominence of the inter-ambulacra (fig. 2 *f*) and the depressions formed by the ambulacra which radiate in straight furrows from the peristome to the margin. The mouth lies at the bottom of a subcentral depression. The peristome is central and slightly five-lobed; the tubercles are more conspicuous at the base than on the upper surface, and are closely placed upon the plates.

The test is very thin, and the spines, which were preserved on one specimen, are moderately long and needle-shaped.

Affinities and differences.—The orbicular form and long anal valley liken this species to *Echinobrissus Woodwardii*, Wr., but the convexity of the upper surface, the wideness of the anal valley, the concavity of the base, and the slight tumidity of the sides, form a group of characters of sufficient value to distinguish it from that congeneric species. It resembles some of the small round varieties of *Echinobrissus Hugii*, Ag., but is readily distinguished from them by the anal valley extending from the disc to the border, whereas in *Echinobrissus Hugii*, Ag., there is always an undepressed portion of test between the disc and upper limit of the anal valley. From all other Oolitic species it is so entirely distinct that it is not likely to be mistaken for either of them.

Locality and Stratigraphical position.—This is a true Cornbrash urchin. I have collected it from that "terrain" near Cirencester, and Miss Slatter found it in the same rock near Fairford. Professor Phillips obtained his original type specimens from the Cornbrash at Scarborough, and I have collected it out of the same bed. I am indebted to my friend, John Leckenby, Esq., for a good series from that locality; and Edward Wood, Esq., of Richmond, most liberally gave me the fine large specimen from Scarborough figured in Pl. XXV, fig. 2 *a*. I have received a very fine series of this species from my friend, the Rev. W. A. Griesbach, which he collected from the Cornbrash at Rushden, Northampton-shire; these specimens have supplied abundant materials for the curious structure of the disc and other interesting details.   One of these urchins (fig. 2 *f*, *g*) enabled me to exhibit a conoidal variety which sometimes occurs, but it would require many figures to do full justice to all the variations of outline observed in the Northamptonshire specimens; which likewise exhibit the sculpture on the plates much better than those from Yorkshire. The Gloucestershire specimens are rarely well preserved.   This species is exceedingly rare in the Wiltshire Cornbrash, which contains in such abundance *E. clunicularis*, Lhwyd.

M. Triger collected *E. orbicularis* from the Forest Marble, Ass. No. 4, department of the Sarthe, at Pécheseul, Noyen, route de Mamers à Montagne, where it is very rare. I know of no other foreign locality in which it has been found.

ECHINOBRISSUS QUADRATUS, *Wright*, nov. sp.   Pl. XXVI, fig. 1 *a, b, c, d.*

Test quadrate, elongated, depressed, narrow before, wide behind; posterior border deeply sulcated; anal valley short, wide, and with sloping sides, extending to the apical disc, which is excentral and nearer the posterior border; dorsal surface much inclined from the apical disc to the anterior border; sides narrow; base very concave; plates of the test closely covered with small tubercles.

Dimensions.—Height, seven tenths of an inch; antero-posterior diameter, one inch and nine twentieths; transverse diameter, one inch and three tenths of an inch.

Description.—This nucleolite has by some been considered to represent *Nucleolites major*, Agass., in the English Oolitic rocks, but a careful examination of M. Agassiz's figure, given in the 'Échinodermes Fossiles de la Suisse,' will convince the inquirer that this urchin is very distinct from the Swiss form; in our nucleolite the apical disc is excentral, and situated nearer the posterior border, whereas in *E. major* the excentral disc is nearer the anterior border; the mouth-opening in our nucleolite is small and nearly circular, whilst in *E. major* it is pentagonal, and much nearer the anterior border.   Besides these organic distinctions, they belong to widely different stages of the Oolitic rocks, *E. quadratus*, Wright, being found in the Cornbrash, whereas *E. major*, Agas., was collected from the "terrain Portlandien de la Vallée de la Birse."

E. quadratus is a large, elongated nucleolite, with a well-defined quadrate outline, much wider behind than before; the upper surface is flat, and the slope from the disc to the anterior border is long and gently declined; the ambulacral areas are nearly all of the same width, and terminate at a short distance from the apical disc, (fig. 1 *a*); they pass entirely straight from the border to the disc, without forming curves as in *E. clunicularis*, Lhwyd.

The poriferous zones are narrow; the pores forming the external row are only a little larger than those of the inner row (fig. 1 *d*), and there are seven pairs of pores opposite each large plate; the inter-ambulacral areas are of unequal width, the anterior pair are the narrowest, and the posterior pair and single inter-ambulacrum are the same diameter; their upper surface is uniformly convex, and the plates are covered with tubercles, set closely together (fig. 1 *d*). There are five or six rows on each plate, the areolas of which are so nearly approximated that there are much fewer granules between them than in the nearly allied *E. clunicularis*; the tubercles likewise are larger than in that species. (Compare Pl. XXIV, 1 *e*, with Pl. XXXVI, fig. 1 *d*.)

The upper surface of the single inter-ambulacrum is very much depressed to form the anal valley, the sides of which slope obliquely inwards, and form an angle of 65° with the base (fig. 1 *a*); the posterior border is much indented by this depression, and forms quite a concave depression behind (fig. 1 *a, b*).

The apical disc is rather larger than in *E. clunicularis*; it occupies the vertex of the test, and is placed nearer the posterior than the anterior border. The four genital holes are large and distinct, and the madreporiform body is not very prominent.

The base is concave, and the small excentral mouth-opening (fig. 1 *b*) is situated nearer the anterior than the posterior border; the peristome is circular or subpentagonal, with five rudimentary oval lobes. The poriferous zones lie in depressions, and distinctly radiate from its circumference (fig. 1 *b*). The basal tubercles are larger than the dorsal; they are placed so close together that the areolas are separated only by single rows of granules (fig. 1 *b*).

Affinities and differences.—I have already pointed out the diagnostic characters by which this species is distinguished from *E. major*, Agas., the nucleolite which most nearly resembles it in form, size, and general outline. Its next nearest affinity is with *E. clunicularis*, Lhwyd; from this species it is distinguished, however, by its quadrate shape and depressed dorsal surface, by the wide anal valley, with its oblique sloping sides and concave posterior border. The apical disc is likewise situated behind the centre of the test, and the tubercles are larger and more closely crowded together. The structure of the anal valley and the excavated character of the single inter-ambulacrum serve at a glance to distinguish it from *E. scutatus*, Lamk.

Locality and Stratigraphical position.—All the specimens that I know of this species were collected by Mr. William Buy from the Cornbrash near Sutton-Benger, Wilts, where it is extremely rare; it occurs with the small gray variety of *E. clunicularis*, associated with *Holectypus depressus*, Lamk., *Acrosalenia Wiltoni*, Wright, *Acrosalenia hemicidaroides*, Wright, *Stomechinus intermedius*, Agas., *Acrosalenia spinosa*, Agas., *Avicula echinata*, Sow., *Terebratula obovata*, Sow., *Terebratula lagenalis*, Schloth., and other well-known Cornbrash forms.

C.—*Species from the Coralline Oolite.*

ECHINOBRISSUS SCUTATUS, *Lamarck.*   Pl. XXXVI, fig. 2 *a, b, c, d, e, f.*

ECHINITES CORDATUS,	Lang, Lapid. Figur. Helvetiæ, p. 119, tab. xxxv, fig. 1, 1708.
ECHINOBRISSUS ELATIOR.	Breynius, Schediasma de Echinis, p. 63, tab. vi, fig. 3, 1732.
SPATANGUS DEPRESSUS.	Leske, Additamenta ad Kleinii Dispositionem Echinodermatum, p. 238, tab. li, figs. 1, 2, 1778.
NUCLEOLITES SCUTATUS.	Lamarck, Syst. Animaux sans vertèbres, tome iii, p. 36, 1816.
CLYPEUS CORDATUS.	Smith, Strata Identified, plate of Coral Rag Fossils, 1817.
ECHINUS DEPRESSUS.	Schlotheim, Petrefactenkunde, p. 313, 1822.
NUCLEOLITES SCUTATUS.	Defrance, Dict. des Sciences Naturelles, tome xxxv, p. 213, 1825.
— SCUTATA.	Blainville, Dict. des Sciences Naturelles, tome xxxv, p. 213, 1825.
— SCUTATUS.	Deslongchamps, Encycloped. Method., ii, p. 570, No. 1.
— —	Goldfuss, Petrefacta Germaniæ, Band. i, tab. xliii, fig. 6 *a—c,* p. 140, 1826.
— —	Desmoulins, Tableaux Synop. Échinides, p. 356, No. 5, 1836.
— —	Agassiz, Echinodermes Foss. de la Suisse, p. 45, tab. vii, figs. 12—21, 1839.
— CLUNICULARIS, var. *a major.*	Forbes, Memoirs of the Geological Survey, decade 1st, pl. ix, 1849.
— SCUTATUS.	D'Orbigny, Prodrome de Paléontologie, tome i, p. 379, 1850.
— —	Bronn, Lethæa Geognostica, 3d Aufl. zweit, Band., p. 151, tab. xvii, fig. 13, 1851.
— —	Wright, Annals and Magazine of Nat. Hist., 2d series, vol. 13, p. 185, 1851.
— DIMIDIATUS (pars)	Wright, Annals and Magazine of Nat. Hist., 2d series, vol. ix, p. 301, 1851.
— SCUTATUS.	Forbes, in Morris's Catalogue of British Fossils, 2d ed., p. 84, 1854.
ECHINOBRISSUS SCUTATUS.	Desor, Synopsis des Échinides Foss., p. 277.

Test elliptical, sub-quadrate, rounded before, enlarged, expanded, and bilobed behind; upper surface convex, more or less depressed; sides tumid; base concave; apical disc small, excentral, nearer the anterior border; dorsal valley wide, with perpendicular walls; apex separated from the disc by an undepressed portion of test; valley extending about two thirds the distance between the vertex and border; vent large and elliptical; base concave, much depressed at the excentral mouth-opening; peristome slightly pentagonal.

Dimensions.—Height, seven tenths of an inch; antero-posterior diameter, one inch and three tenths; transverse diameter, one inch and a quarter.

Description.—It is impossible to decide whether the urchin figured by Plot* in tab.

* 'History of Oxfordshire.'

ii, fig. 12, and by Lister* (tab. 7, fig. 26), refers to this species or *E. clunicularis*, in consequence of the omission of the anal valley in these figures, on which the true specific character strictly depends. Down to a very recent date, the two forms have been confounded with each other, although, when critically examined, the differences are very evident. The stratigraphical distribution of the two species is moreover well defined, *E. clunicularis* ranging from the superior zone of the Inferior Oolite to the Cornbrash, whilst *E. scutatus* is limited to the Calcareous Grit and other subdivisions of the Coral Rag. The abundance of this nucleolite in the Calcareous Grit near Oxford makes it highly probable that this was the form the older authors above cited had in view in their respective works. Lang's† " Echinites Cordatus quaternis radiis è duplici serie transversarum lineorarum conflatis" (tab. 35, fig. 1), probably represents a bad specimen of this urchin, in which the single ambulacrum had been obliterated. The specimen is described as being very rare, and was found in the hills around Bætstein and Luggeren. Leske's‡ figure of *Spatangus depressus* (tab. 51, fig. 1) apparently represents a quadrate depressed variety of this species, but in consequence of the anal valley being filled up with matrix, the true specific character is concealed. Many authors are of opinion that the figure of Goldfuss§ does not represent the true Lamarckian *Scutatus*, and Desmoulins has proposed to separate it, under the name *Goldfussii;* but after having studied a large number of individuals collected at Trouville, I have found many specimens with which the figure of Goldfuss entirely agrees. A sufficient margin has not been allowed for the varieties which the same species exhibits when obtained from different localities. M. Agassiz's ‖ figure of this nucleolite is very good, and represents, I think, the true type form of the species. This author participates in the opinion expressed by Desmoulins, in reference to which he says—

" En revanche je pense avec M. Desmoulins que Goldfuss a identifié à tort l'espèce qu'il a décrite sous le nom de *N. scutatus*, avec le *N. scutatus* de Lamarck, dont nous nous occupons en ce moment. Ce dernier en effet n'affecte nullement cette dépression de la face supérieure postérieure qui est trés saillante dans les figures de Goldfuss; c'est au contrare à la face antérieure qui est la plus inclinèe." " Afin de destinguer le *N. scutatus*, Lam., du *N. scutatus*, Goldf. (qui n'a point encore éte trouvé en Suisse), M. Desmoulins a donné à ce dernier le nom de *N. Goldfussii*."¶

Professor Edward Forbes** considered this species as a variety of *N. clunicularis*, and described it as " *Var. a major, sub-depressa, lata, lateribus, tumidiusculis. Spatangus depressus,* Leske, ap. Klein, p. 238, t. 51, fig. 12 (copied in 'Enc. Meth.,' pl. 157, figs.

---

* 'De Lapidibus Turbinates,' cap. ii, titulus xxvi.

† 'Historia Lapidum Figuratorum Helvetiæ,' tab. 35, p. 119.

‡ 'Additamenta ad Kleinii Echinodermata,' p. 238, tab. 51, figs. 1, 2.

§ 'Petrefacta Germaniæ,' tab. 43, fig. 6.

‖ 'Échinodermes Fossiles de la Suisse,' tab. vii, figs. 19—21.

¶ Ibid., partie première, p. 46.

** 'Memoirs of the Geological Survey, decade 1, pl. ix.

5, 6). *Nucleolites scutata*, Lamarck, ' An. s. Vert.,' iii, p. 35 ; Defrance, ' Dict. Sc. Nat.,' vol. xxxv, p. 213. *N. scutatus*, Agassiz, ' Echin. Suiss.,' p. 45, pl. 7, figs. 19—21." This quotation shows my lamented friend's opinion when the description of his Pl. IX was written. At that time I believe he had not seen my Trouville specimens of *E. scutatus*, for on making with him a comparison of some type-forms of *N. scutatus*, Lamk., he readily admitted, after that examination, the specific differences existing between the *N. clunicularis* and the true Lamarckian species.

In my first memoir on the Cassidulidæ of the Oolites,* I grouped several individuals of this species with *E. dimidiatus*, Phil. At that time nearly all the English Coral Rag nucleolites were referred to Phillips's species ; and I was only convinced of my error after I had examined a series of good type-specimens from Trouville, the original locality of Lamarck's species.

Echinobrissus scutatus, Lamk., when fully developed, is uniformly convex on the upper surface; it is rather narrower before than behind; its length nearly equals its breadth, when measured about the middle of the test; the flanks are rounded and tumid (fig. 2 *c*), and the posterior border is truncated (fig. 2 *b*); the vertex is situated nearer the anterior than the posterior border, and in the centre thereof is placed the apical disc ; from this point the test slopes gently towards the posterior border, but more abruptly to the anterior side (fig. 2 *c*); the ambulacral areas are narrow, and nearly uniform in width, the posterior pair being a little broader than the anterior areas. The poriferous zones are petaloidal only on two thirds of the upper surface (fig. 2 *a*, *c*). The pores of the inner row are round, those of the outer row form oblique slits ; on the flanks they are both round, more distant, and placed obliquely, thus ·.·..·.·; at the border they become very small, and at the base indistinct ; near the peristome they are again larger and more numerous ; in the more crowded portions of the zones there are from six to seven pairs of pores opposite one of the large plates, and where they are more distinct on the sides there are four small pairs opposite one plate.

The two anterior inter-ambulacral areas are narrower than the posterior pair (fig. 2 *a*). There are about sixteen plates in a column, each plate forming a double inclined plane, and having its surface crowded with small equal-sized tubercles, arranged close together in three or four rows (fig. 2 *f*) ; the postero-lateral pair are wider and longer, and contain more plates in each column ; in other respects they have a similar structure to the anterior areas. The single inter-ambulacrum is about as wide as the postero-lateral pair ; it is truncated behind, and its border is grooved by the anal valley ; this depression has a uniform width, is concave at its base and upper part, and there is always an undepressed portion of test between its termination and the apical disc (fig. 2 *a*); the anal opening is round, and is seen at the end of the valley (fig. 2 *a*, *d*).

The apical disc is small and excentral ; its elements are so intimately soldered together, that the sutures in all my specimens are obliterated ; the madriporiform tubercle is large

* ' Annals and Magazine of Natural History,' 2d series, vol. ix, p. 300.

and occupies the centre of the disc; it appears to cover the genital plates (fig. 2 *e*), and the four genital holes are large oblique slits, which extend into the inter-ambulacral areas (fig. 2 *e*).

The base is nearly flat at the sides, and slightly concave towards the mouth-opening (fig. 2 *b*). In large, well-developed specimens, the inter-ambulacral areas are a little prominent, and the course of the ambulacra is marked by corresponding depressions in the test; the poriferous zones are so feebly shown, that the pores can only be seen with a lens.

The mouth-opening is moderately large for a nucleolite; it is situated in a depression opposite the apical disc, and is nearer the anterior than the posterior border (fig. 2 *b*); the peristome is pentagonal, and each angle of the pentagon corresponds to an ambulacral area; the areas form inconsiderable petaloidal expansions as they radiate from the peristome, and the pores are crowded close together in the vicinity of the mouth-opening. The anal valley forms one of the distinctive characters of this species (fig. 2 *a*); it is of an ovate or lanceolate form, with a blunt apex; in some specimens it appears as if a portion of the inter-ambulacrum had been drilled out for the passage of the intestine; in some individuals it extends only half the distance between the margin and the vertex, whilst in others it reaches two thirds the length; in all the specimens I have examined, an undepressed portion of test separates the apical disc from the upper border of the anal valley; inferiorly, the valley forms a considerable sulcus, grooves the centre of the area, dividing its posterior border, and producing the cordate or bilobed form this species assumes (fig. 2 *a, b*).

Affinities and differences.—I have already stated that *E. scutatus* was formerly considered by English naturalists to be a variety of *E. clunicularis*, Lhwyd, and as such its history is more or less connected with that species. Were a student, therefore, to endeavour to unravel its synonyms from the books alone, he certainly would be puzzled in his search, as the critical remarks on *E. scutatus* have more frequently been made on book-statements than from an examination of specimens. If, however, a comparison be made between a series of *E. clunicularis*, Lhwyd, from the Cornbrash, with a corresponding series of *E. scutatus*, Lamk., from the Calcareous Grit, all doubts will be removed from the mind of the observer as to the specific differences existing between these species; and in default of such specimens, a careful study of Mr. Bone's most excellent figures in our Plates XXIV and XXVI, with the ample magnified details he has given therein, will afford sufficient evidence for our conclusion.

E. scutatus, Lamk., so nearly resembles *E. dimidiatus*, Phil., that the latter has by many been considered to be a variety of the former; this point, however, will be more properly discussed in the section devoted to the description of *E. dimidiatus*, Phil.

These are the only English nucleolites which at all resemble *E. scutatus*. Between all the other older species and this Corallian form the distinctions are numerous and self-evident. Between the newer Portland species, *E. Brodeii*, Wr., and *E. scutatus*, a detailed analysis will be given in the section on that species.

Locality and Stratigraphical position.—I have collected *E. scutatus*, Lamk., from the Lower Calcareous Grit at Bullington-Green, near Oxford, where it was associated with *Cidaris Smithii*, Wright. I have discovered it was from this same quarry that the large type-specimen of the Cidarite figured in our Pl. II was obtained. I have gathered it from the Lower Calcareous Grit at Filey Brig, Gristhorpe Bay, and Scarborough Castle Hill, Yorkshire, and from the same formation at Marcham, and Faringdon, Berks. The Yorkshire specimens I chiseled out of blocks containing *Pygurus pentagonalis*, Phil., and other Calcareous Grit shells, and the Berkshire specimens were associated with *Cidaris florigemma*, Phil., and *Hemipedina Marchamensis*, Wright. I have collected it from the Coralline Oolite at Calne, Wilts, where it is very abundant in some beds. In one slab, about nine inches square, obtained from a large quarry near the town, there are about fifty specimens, more or less imperfect, closely laid together. Like other urchins, it appears to have been gregarious, and would be found in great numbers if its head zone was exposed. *Cidaris florigemma*, Phil., *Hemicidaris intermedia*, Flem., *Acosalenia decorata*, Haime, *Pseudo-diadema versipora*, Phil., lie with it in the same slabs.

My friend, Charles Pierson, Esq., collected this species from a Pisolitic Oolitic rock—the Lower Calcareous Grit of English geologists—about one mile from Trouville, Calvados, where it is very abundant, and in a good state of preservation. M. Desor states that it is found in the Oxfordien of Trouville, and Vaches-Noires, Calvados, Lannois, Ardennes, and Chamsol, Doubs.

History.—The history of this species has been already so fully detailed in my analysis of its synonyms, that it is unnecessary to enter into any further details on the subject.

ECHINOBRISSUS DIMIDIATUS, *Phillips*.   Pl. XXVI, fig. 3 *a, b, c. d.*

CLYPEUS DIMIDIATUS.		Phillips, Geology of Yorkshire, pl. 3, fig. 16, p. 127, 1829.
NUCLEOLITES DIMIDIATA.		Desmoulins, Études sur les Échinides, Synonyme Générale, No. 25, p. 362, 1836.
—	DIMIDIATUS.	Agassiz, Prodrome d'une Monogr. des Radiares, p. 9, 1837.
—	PARAPLESIUS.	Agassiz, Catalogus Systematicus, p. 4, 1840.
—	DIMIDIATUS.	Morris, Catalogue of British Fossils, p. 55, 1843.
—	—	Agassiz and Desor, Catalog. Raisonné des Échinides Annales des Sciences Naturelles, 3ᵉ série, tome vii, p. 154, 1847.
—	DIMIDIATA.	Bronn, Index Palæontologicus, p. 818, 1848.
—	DIMIDIATUS.	Forbes, Memoirs of the Geological Survey, decade 3, description of pl. ix, 1849.
—	—	D'Orbigny, Prodrome de Paléontologie, tome 1, p. 379, 1850.

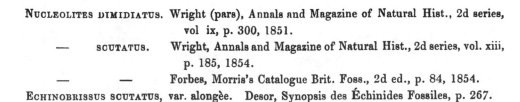

NUCLEOLITES DIMIDIATUS. Wright (pars), Annals and Magazine of Natural Hist., 2d series, vol ix, p. 300, 1851.
— SCUTATUS. Wright, Annals and Magazine of Natural Hist., 2d series, vol. xiii, p. 185, 1854.
— — Forbes, Morris's Catalogue Brit. Foss., 2d ed., p. 84, 1854.
ECHINOBRISSUS SCUTATUS, var. alongèe. Desor, Synopsis des Échinides Fossiles, p. 267.

Test elongated, narrower before than behind ; convex above, concave below ; sides tumid ; apical disc excentral, nearer the anterior border ; anal valley short and narrow, reaching two thirds the distance between the posterior border and summit, with a triangular undepressed space between its upper margin and disc ; posterior lobes obsolete, posterior border only slightly grooved.

Dimensions.—Antero-posterior diameter, one inch and three tenths ; transverse diameter, one inch and three twentieths ; height, seven tenths of an inch.

Description.—This urchin is very abundant in the Coralline Oolite of Malton, from which rock it was first figured by Professor Phillips. Unfortunately he gave no description of the species, and his figure represented only one view of the nucleolite. Professor Edward Forbes, who carefully studied this form, gave the following diagnosis of its characters : " *N. ambitu ovato, antice rotundato, postice bilobato ; dorso convexo, apice centrali, vertice sub-centrali, postice tumido ; ambulacris anguste lanceolatis ; sulco anali profundo, ovato obtuso, superne abbreviata, lobis posterioribus tumidis ; ventre plus minusve concavo.*

" This species rarely exceeds one inch in length, and varies greatly in the convexity of its upper surface. The ovate anal sulcus, reaching about two thirds of the distance between the posterior margin and the true summit, conspicuously distinguishes it from *clunicularis,* with which it was confounded before being distinguished by Phillips."

When Professor Forbes published the above 'Note on British Nucleolites' appended to the description of pl. ix, decade 1, of the 'Memoirs of the Geological Survey,' he was not aware of the existence, in our Oolites, of *N. scutatus ;* it is therefore probable that his diagnosis was framed to include some forms of that species.

E. dimidiatus has an oval outline, the upper surface of the test is uniformly convex ; the sides and anterior border are very tumid in some specimens, and moderately so in others ; the posterior border is rounded and only slightly grooved by the sulcus.

The ambulacral areas are narrow and lanceolate ; the single area and anterior pair are narrower than the posterior pair, which are much longer and better developed than the others. The poriferous zones lie in slight depressions of the test, their petaloidal portions are wider than the homologous part of this species in *E. scutatus,* and there are five to six pairs of pores opposite each large plate.

The inter-ambulacral areas are of unequal width, the anterior pair are the narrowest,

46

the posterior pair are one third wider than the anterior pair, and the inter-ambulacrum is the widest. There are about sixteen plates in each postero-lateral column, and the surface of each is covered with numerous small tubercles (fig. 3 *d*), arranged in five or six horizontal rows.

The anal valley extends about two thirds of the distance from the posterior border to the apical disc; it is a narrower and more shallow sulcus than the corresponding valley in *E. scutatus*, and the vent opens nearer the surface. Between its upper border and the disc there is a well-marked triangular space, in length about one third the distance between the disc and border; the base of which is formed by the arch of the sulcus and the lateral portions of undepressed test, and the sides, by the long, oblique, posterior poriferous zones (fig. 3 *a*); this space is undepressed, and the plates composing it differ in no respect from the plates occupying the same region in other areas. When viewed laterally, the test of *E. dimidiatus* forms a long gentle slope from the vertex to the posterior border (fig. 3 *c*), and a short abrupt slope towards the anterior border; this arises from the excentral position of the apical disc, and the greater height of the anterior portion of the test.

The base is concave towards the centre, and tumid at the sides; the small mouth-opening is situated opposite the apical disc; the peristome is pentagonal, and its angles correspond to each of the ambulacral areas; the tubercles at the base are larger and more closely crowded together than those on the upper surface; the poriferous zones are so indistinct at the base that they appear only as faint lines radiating from the peristome.

The apical disc is very small, the madreporiform body makes a slight pyriform prominence, and the pairs of perforated genital plates extend outwards between the ambulacral areas.

Affinities and differences.—In its general outline this species resembles *E. Goldfussii*, Desor, from the Kelloway ferrugineux of the Sarthe; in that species, however, the vent is nearer the border, and the test proportionately higher to its length. It so closely resembles *E. scutatus*, Lamk., that I formerly considered it a variety of the latter. It is, however, distinguished from *E. scutatus* by the following characters: the test is more elongated; the sides and base are more tumid; the apical disc is smaller and more excentral; the anal valley is smaller and shallower, and rarely extends so high up the area; the poriferous zones are wider, and their petaloid portions lie in slight depressions on the upper surface; the sides are very tumid, and they are frequently more irregular in outline and unsymmetrical in proportions than in *E. scutatus*; the base is more concave, and its sides more tumid; the slope of the dorsal surface likewise, from the disc to the posterior border, is longer and more inclined than in its nearly allied *E. scutatus*.

Locality and Stratigraphical position.—*E. dimidiatus*, Phil., is collected in considerable abundance from the Coralline Oolite at Malton, Yorkshire, where it is associated with *E. scutatus*, Lamk., *Clypeus subulatus*, Young, and *Pygurus giganteus*, Wr. The finest specimens are those which occur in the large Freestone quarry at the east of the town. C. W. Strickland, Esq., has obtained it in his quarry at Hildenley, near Malton, where it is associated with *Hemicidaris intermedia*, Flem., *Cidaris florigemma*, Phil., *Pseudo-diadema hemisphæricum*, Agas., *Collyrites bicordata*, Leske, *Pygurus pentagonalis*, Phil., and *Nucleolites scutatus*, Lamk.

D.—*Species from the Portland Oolite.*

ECHINOBRISSUS BRODIEI, *Wright*, nov. sp. Pl. XXXV, fig. 1 *a, b, c, d, e.*

Test elongated, much depressed, nearly as broad before as behind; anterior border rounded; posterior border angular; anal valley wide, short, limited to the lower half of the inter-ambulacrum; apical disc small, nearly central; poriferous zones slightly petaloidal; base undulated, from the convexity of the inter-ambulacra; mouth-opening large, excentral, pentagonal, oblique, situated near the anterior border; inter-ambulacrum produced, recurved, and truncated posteriorly.

Dimensions.—Antero-posterior diameter, one inch; transverse diameter, nine tenths of an inch; height, four tenths of an inch.

Description.—Urchins are so extremely rare in the Portland Oolite of England, that the discovery of this specimen in that formation at Brill enables me to give a figure of a nucleolite, of which I had formerly observed fragments in the same formation at Portland. The test is unfortunately not well preserved in the Brill urchins, the matrix having adhered so firmly to the test of the largest specimen that the surface could only be exposed by the use of dilute acid, a process which at all times destroys the fine sculpture of the shell, and ought never to be employed in cleaning Echinoderms but as a "dernier resort" for disclosing structure in doubtful cases. The shell is elongated, a little wider behind than before, the anterior border is rounded, and the posterior angular. The upper surface is much depressed and uniformly convex, forming regular curves in the length and width (fig 1 *c, d*).

The small apical disc occupies the vertex, which is anteriorly excentral; its elements unfortunately are concealed by some closely adherent matrix, which cannot be removed.

The ambulacral areas are nearly all of the same width; the poriferous zones are narrowly petaloid, and both rows of holes are about the same size.

The inter-ambulacral areas are of unequal width; the anterior pair are the narrowest, the posterior pair the widest, and the single area of intermediate width; they are all uniformly convex, and covered with very small tubercles; the sides of the test are much depressed, and form an angle of about 25° with the base; the margin is rounded and flattened before, and gradually expands to within the eighth of an inch of the postero-lateral ambulacra ; at this point it becomes rather abruptly truncated obliquely inwards and backwards, and is again transversely truncated at the posterior border (fig. 1 *b*) ; the upper half of the single inter-ambulacrum is smooth and undepressed (fig. 1 *a*) ; its lower half is occupied by the wide anal valley, with its perpendicular sides and large vent, which opens near the surface (fig. 1, *d*) ; from the sides of the valley two prominent ridges descend to the border, where they form obtuse prominences, the intermediate space being truncated (fig. 1 *a*).

The base is undulated from the convexity of the inter-ambulacra (fig. 1 *b*), and the depressions formed by the ambulacra as they radiate from the peristome ; the large mouth-opening is situated nearer the anterior than the posterior border, in a slight depression of the test ; the peristome is pentagonal, and its longest diameter is in the transverse direction (fig. 1 *b*). The tubercles at the base are large, and disposed with considerable regularity ; at the anterior margin the granules surrounding the areolas form hexagons, as shown in (fig. 1 *e*).

Affinities and differences.—This species is so entirely unlike any of its English congeners that it cannot be mistaken for either of them, the flat, depressed upper surface, the length of the test, which is rounded before and angular behind, the wide, short anal valley, and large, transversely elongated mouth-opening, form a group of characters by which it is separated from them. The length of the test and position of the anal valley, groups it naturally with *E. Goldfussii*, Desor, and *E. pulvinatus*, Cotteau, both from the " Kelloway ferrugineux " of the department of the Sarthe. It is distinguished from the first by the flatness of the upper surface, and the size of the anal valley, and from the latter by the absence of the tumid sides, flat dorsum, and marginal valley, which characterise the French urchins. The like absence of tumidity in the sides of *E. Brodiei* distinguishes this species from *E. dimidiatus*, Phil., of the Coralline Oolite.

Locality and Stratigraphical position.—This nucleolite was collected from the Portland Oolite at Brill, Buckinghamshire, by my friend, the Rev. P. B. Brodie, who has kindly supplied the following note on a section of the Portland beds at that locality to accompany my description of this beautiful new form :

" The occurrence of any of the *Echinodermata* in the Portland Oolite is so rare, that it is desirable to give a short notice of the strata at the locality whence the specimen described by my friend, Dr. Wright, was obtained. The sections at Brill, in Buckingham-shire, are well known from Dr. Fitton's able memoir, ' On the Strata below the Chalk,' and therefore I shall content myself by a very brief account of those which came more immediately under my own inspection, which are in fact identical with those given by that geologist. The summit of the hill is capped by the Lower Green Sand, as stated by Dr. Fitton, but as this seems to be identical with the beds above the Portland at Great Hazeley, whence I procured several small Paludina similar to a species described by Professor

Phillips, in his interesting paper on the 'Iron Sands of Shotover.' * It seems probable that those at Brill and Hazeley are of the same age, and must be classed as estuarine deposits, belonging rather to the Wealden than the Lower Green Sand. It is as well, perhaps, to mention these two localities, because they are not referred to in Professor Phillips's paper. Many years ago I found nodules of iron sand containing Paludinæ in the Vale of Wardour, in Wiltshire, and possibly a more careful examination might serve to identify them with the estuarine sands of Shotover above mentioned.

"The top of the Portland Rock consists of a white or gray calcareous stone, with *Perna mytiloides*, Lamk., *Trigonia gibbosa*, Sow., and *Trigonia incurva*, Sow., which is underlaid by beds of hard grit divided by clay. This is succeeded by a white limestone, seen also in the Vale of Wardour and other places; it contains casts of *Trigonias* and other shells. It is in many respects a remarkable stratum, being in parts of a soft, marly texture, which readily crumbles to pieces when struck with the hammer. A seam of clay with broken shells divides this from a hard, rough, calcareous stone, of a brown colour, used for building purposes, and from this some good specimens of *Perna mytiloides*, Lamk., with the shell attached, may be procured. Between this and the Portland sand there are several coarse bands of stone, more or less calcareous and sandy, with a large preponderance of green particles of silicate of iron; the nodules are full of the characteristic fossils of this formation, viz., *Pecten lamellosus*, Sow., *Trigonia gibbosa*, Sow., *Astarte cuneata*, Sow., *Cardium dissimile*, Sow., a large *Spondylus* with spines, *Panopœa*, *Exogyra*, and *Serpula* (which usually occur together in great perfection on the edges of the stone), a large species of *Mytilus*, with *Modiola*, *Natica elegans*, Sow., *Buccinum naticoides*, Sow. (both of which retain the shell in some specimens), and *Cerithium Portlandicum*, Sow. The shells in this lower division are numerous, and often better preserved than is usually the case in the Portland series; this locality therefore well deserves a careful search. I have little doubt that the *Echinobrissus* came from one of these beds, overlying the Portland sand, as the stone from whence I extracted it agrees exactly in lithological structure therewith. The inferior shelly strata are largely used round Brill for road-mending, and I may add that I found the two specimens of this new urchin in a heap of stone which was placed there for this purpose, at the foot of the hill close to the quarries. The total thickness of the strata exposed above the sand may be somewhere about twenty-five feet.

"I am not aware whether any *Bryozoa* have been previously noticed in the Portland Oolite; but it seems worth while to mention their occurrence at Swindon, near the Reservoir, where I obtained a few specimens in ferruginous sand attached to single valves of *Trigonia gibbosa*, Sow.

---

* 'Journal of the Geological Society,' vol. xiv, part 3, No. 55, for August, 1858.

NOTES

ON FOREIGN JURASSIC SPECIES OF THE GENUS ECHINOBRISSUS NEARLY ALLIED TO BRITISH FORMS, BUT WHICH HAVE NOT YET BEEN FOUND IN THE ENGLISH OOLITES.

A.—*Species in which the anal valley extends to the apical disc.*

ECHINOBRISSUS ELONGATUS, *Agass.* Syn. *Echinobrissus elongatus,* Cotteau and Triger, Échinides de la Sarthe, pl. x, figs. 8—11.

A narrow, oblong, elongated nucleolite, thin at the border, convex before, truncated, thin, and recurved behind; apical disc central; anal valley wide and reaching the disc; ambulacra narrow; zones slightly petaloidal; base concave; mouth excentral, small, and situated in a depression.

Dimensions.—Height, half an inch; length, one inch and one tenth of an inch; breadth, seven tenths of an inch.

Formation.—Calcaire à polypiers (Bathonien) Normandy. Forest marble (Sarthe).

Collections.—M. Deslongchamps. My cabinet; the specimens sent by Professor Deslongchamps.

ECHINOBRISSUS CREPIDULA, *Desor.* Syn. *Nucleolites crepidula,* Cotteau, Échinides Fossiles, pl. v, figs. 4—6, p. 68.

A small elongated urchin, very flat, rounded, and narrow anteriorly, dilated and sub-rostrated posteriorly; anal valley elongated, deep, and broad, extending from the disc to the border; mouth pentagonal and sub-medial.

Dimensions.—Height, five millimètres and a half; length, fourteen millimètres; breadth, eleven millimètres.

Formation.—Forest marble of Châtel-censoir (Yonne), where it is abundant only as siliceous moulds of the interior.

Collections.—M. Cotteau. Museum of Paris.

ECHINOBRISSUS AMPLUS, *Agass.*   Syn. *Nucleolites amplus*, Agassiz, Catal. raisonné, p. 96.

" A large species, as broad as it is long, convex above, and nearly square ; posterior border declined, thin, and emarginate ; anal valley extending to the ambulacral summit, which is central ; base concave ; mouth-opening excentral."   Desor.

Formation.—" Marnes à Discoidées (Vésulien) de Wolfliswyl (Argovie) ; Val de Laufen·

Collections.—Moesch, Mus. Bâle.   Coll., Gressly.

ECHINOBRISSUS PLANULATUS, *Roemer.*   Syn. *Nucleolites planulatus*, Roemer, Oolit. Gebirges, pl. i, fig. 19, p. 28.

A small elongated nucleolite, remarkable for its extremely depressed dorsal surface ; it is rounded before, truncated behind, and has the posterior border slightly emarginated.

Formation.—Upper Coral Rag of Lindner Berges (Hanover).

Collections.—M. Roemer.   My cabinet, specimen sent by Professor Roemer.

ECHINOBRISSUS MAJOR, *Agass.*   Syn. *Nucleolites major*, Agassiz, Échinoderm. Foss. Suisse, part i, pl. vii, figs. 22—24, p. 46.

A large, quadrate, elongated nucleolite, rounded before, enlarged behind, having the posterior border truncated, and strongly emarginated ; base concave ; mouth pentagonal, and near the anterior margin.

Dimensions.—Height, seven tenths of an inch ; length, one inch and three tenths ; breadth, one inch and three twentieths.

Formation.—" Portland inférieur (Astartien Oolitique) de Laufon avec *Pygurus Hartmanni*, Delémont (Jura Bernois)."   Desor.

Collection.—M. Gressley.

ECHINOBRISSUS GRACILIS, *Agass.*   Syn. *Nucleolites gracilis*, Agassiz, Échinoderm. Foss.
Suisse, part i, pl. vii, figs. 10—12, p. 44.

A small, beautiful nucleolite, rounded before, abruptly enlarged behind the disc, emarginated and truncated posteriorly; ambulacra very narrow; anal valley wide and deep; mouth-opening very excentral near the anterior border; summit slightly excentral.

Dimensions.—Height, half an inch; length, seven tenths of an inch; breadth at the widest part, nearly equal to the length.

Formation.—Portlandien inférieur (Astartien) de Rædersdorf Haut Rhin, Porrentruy.

Collection.—M. Gressley.   Very rare.

B.—*Species in which the anal valley does not extend to the apical disc.*

ECHINOBRISSUS GOLDFUSSII, *Desmoulins.*   Syn. *Echinobrissus Goldfussi*, Cotteau and
Triger, Échinides de la Sarthe, pl. xix,
figs. 1, 2, p. 86.

A small nucleolite, rounded before, sub-truncated and slightly dilated behind; upper surface convex and inflated; much sloped from the summit to the posterior border; base flat, vertex excentral and anterior; anal valley short, wide, and deep, arched above, expanded below, extending one third of the distance between the border and disc; vent elliptical near the surface; mouth-opening anteriorly excentral; peristome pentagonal.

Dimensions.—Height, half an inch; length, nineteen twentieths of an inch; breadth, nearly eight tenths of an inch.

Formation.—Kelloway ferrugineux, Montbizot, Department of the Sarthe, Largues (Haut Rhin); Launoy (Ardennes); Étage Oxfordien.

Collections.—MM. Cotteau and Triger.

ECHINOBRISSUS PULVINATUS, *Cotteau,* Échinides de la Sarthe, pl. xix, figs. 3, 4, p. 87.

A moderate sized, oblong nucleolite, round before, sub-truncated behind, thick and inflated at the borders, depressed at the upper surface; base sub-concave and slightly undulated; apical disc sub-central, nearer the anterior border; anal valley far removed from the summit, and forming a sulcus in the posterior border ; mouth-opening small and excentral.

Dimensions.—Height, eleven twentieths of an inch ; length, one inch and one twentieth ; breadth, nine tenths of an inch.

Formation.—Kelloway ferrugineux, near Mamers, Sarthe ; very rare.

Collections.—MM. Michelin and Cotteau ; my cabinet.

ECHINOBRISSUS ICAUNENSIS, *Cotteau.* Échinides Foss. de l'Yonne, pl. xlv, figs. 6—8, p. 326.

Test elongated, depressed, narrow before, wider behind ; apical disc excentral, nearer the anterior border ; anal valley short, wide, extending only half way between the border and disc ; vent large and elliptical ; base concave ; mouth-opening excentral, nearer the anterior border ; peristome pentagonal.

Dimensions.—Height, four tenths of an inch ; length, one inch and one tenth ; breadth, nine tenths of an inch.

Formation.—Kimmeridgien (" Calcaire des environs de Tonnerre et de Chablis), Gray (Haute Marne)," Cotteau ; very rare.

Collections.—M. Cotteau, M. Rathier.

ECHINOBRISSUS TRUNCATUS, *Desor.* Synopsis des Échinides Fossiles, p. 268.

A new species with a very elongated form, nearly uniform in width throughout, and slightly enlarged behind ; anal valley, supra-marginal ; the entire length of the sulcus seen only when the posterior border is examined.

Formation.—" Portlandien supérieur (Virgulien) d'All près Porrentruy.   Très rare." Desor.
Collection.—Mus. Bâle.

47

Genus—CLYPEUS, *Klein,* 1734.

This genus includes all the large discoidal urchins, with petaloid ambulacra, in which the vent opens at the upper surface into an anal valley; the dorsal portions of the poriferous zones are widely petaloidal; the ambulacral summit is central, or excentral, and when this is the case, the excentricity is always towards the posterior border; the elements of the apical disc are closely soldered together, there are two pairs of perforated genital plates, one single imperforate plate, and five ocular plates, the orbits are in general marginal and visible, and the spongy madreporiform body occupies the centre of the disc.

The base is flat, concave, or undulated, in different species; the mouth-opening is small and lodged in a depression, which is either central or sub-central; the peristome is surrounded by five oral lobes, formed by the terminations of the inter-ambulacra.

This group differs so slightly in all its essential characters from the genus *Echinobrissus*, that I formerly agreed with my late friend, Professor Forbes, to include all the species of Clypeus in the Lamarckian genus *Nucleolites*. After a more detailed study of the comparative anatomy of their tests, taking into consideration the magnitude and development of the long wide petaloidal poriferous zones, and the relation these parts had to the internal organs, I have determined to retain the genus.

Like *Echinobrissus*, the genus *Clypeus* includes two types of structure; in the first, the anal valley extends from the apical disc to the posterior border, this section includes the majority of the species, having for its type *Clypeus Plotii*, Klein. In the second, the anal valley does not extend to the disc, but is separated therefrom by a greater or less undepressed portion of test. The types of the second section are *Clypeus Hugii*, Ag., and *Clypeus subulatus*, Young.

All the species are limited to the Oolitic rocks, and they had their greatest development in the seas which deposited the lower division of the Jurassic series; as the Inferior Oolite, zone of Ammonites Parkinsoni; the Bath Oolite group, and the Cornbrash contain the most of the species. I know of none in the Oxford Clay, and the Coral Rag contains only one.

SECTION A. *Anal valley extends from the disc to the border.*

A. *Species from the Inferior Oolite.*

CLYPEUS PLOTII, *Klein.* Pl. XXVIII and XXIX.

POLAR STONE.	Plot, History of Oxfordshire, tab. ii, figs. 9 and 10, 1677.
ECHINITES.	Lister, Lapidibus Turbinatis, p. 224, pl. vii, fig. 27, 1678.
— CLYPEATUS.	Llhwyd, Lithophylacii Britannici Ichinographia, tab. xiii, p. 48, No. 971, 1698.
ECHINUS DISCOIDES.	Morton, Natural History of Northamptonshire, p. 233, 1712.
CLYPEUS PLOTII.	Klein, Natural. Disposit. Echinodermat., tab. xii, p. 22, 1734.
— —	Leske, Additamenta ad Kleinii Echinodermatum, p. 157, 1768.
ECHINUS SINUATUS.	Linnæus, Systema Naturæ, by Turton, vol. iv, p. 144, 1801.
CLYPEUS SINUATUS.	Parkinson, Organic Remains, vol. iii, p. 24, pl. ii, fig. 1, 1811.
GALERITES UMBRELLA.	Lamarck, Animaux sans Vertèbres, tom. iii, p. 23, No. 15, 1816.
— PATELLA.	Lamarck, Animaux sans Vertèbres, tom. iii, p. 23, No. 14, 1816.
ECHINITES SINUATUS.	Schlotheim, Petrefacktenkunde, i, p. 310, 1820.
NUCLEOLITES PATELLA.	Defrance, Dict. des Sciences Naturelles, tom. xxxv, p. 213, 1825.
ECHINOCLYPEUS UMBRELLA.	De Blainville, Diction. des Sciences Naturelles, tom. lx, p. 189, 1830.
— PATELLA.	De Blainville, Dict. des Sciences Naturelles, tom. lx, p. 189, 1830.
GALERITES PATELLA.	Deslongchamps, Encyclopéd. Méthodique, pl. cxliii, figs. 1, 2.
— —	Id., Enc., tom. ii, p. 434, No. 14.
ECHINOCLYPEUS UMBRELLA.	De Blainville, Manuel d'Actinologie, p. 208, 1834.
CLYPEUS PATELLA.	Agassiz, Prodrome, Mém. Soc. d'Hist. Nat. Neuchâtel, tom. i, p. 186, 1835.
— —	Agassiz, Échinodermes Fossiles de la Suisse, 1st part, p. 56, tab. v, figs. 4—6, 1839.
— —	Morris, Catalogue of British Fossils, p. 50, 1843.
— SINUATUS.	Morris, Ibid.
— ANGUSTIPORUS.	Agassiz, Cat. raisonnée, An. Sc. Nat., tom. vii, p. 156, 1847.
— EXCENTRICUS.	M'Coy, Annals and Magazine of Natural History, p. 417, 1848.
NUCLEOLITES SINUATUS.	Forbes, Memoirs of the Geol. Survey, decade i, 1849.
— —	Wright, Annals and Mag. of Nat. Hist., vol. ix, p. 306, 1851.
CLYPEUS PATELLA.	Bronn, Lethæa Geognostica, Band. ii, p. 152, tab. xv, fig. 9, 1852.
NUCLEOLITES SINUATUS.	Forbes, in Morris's Catalogue of British Fossils, 2d ed., p. 84, 1854.
CLYPEUS PLOTII.	Salter, in Hull's Memoirs of the Geological Survey, 1857.
CLYPEUS SINUATUS.	Desor, Synopsis des Échinides Fossiles, Pl. XXXV, p. 276, 1858.

Dimensions.—Height, one inch and seven tenths; antero-posterior and transverse diameters nearly equal, measuring four inches and four tenths.

Description.—This well-known and widely distributed urchin was first figured by Dr. Plot in his 'History of Oxfordshire,' tab. ii, figs. 9, 10, and was accompanied with the following quaint remarks:

" Of *Brontiæ*, therefore, or *Ombriæ* (call them which you will), we have several sorts in Oxfordshire, which yet all agree in this, that they are a sort of solid irregular hemispheres; some of them oblong, and having somewhat of an oval; others either more elevated, or depressed on their bases. All of them divided into five parts, most times inequal, rarely equal, by five rays issuant from an umbilicus or center, descending from it down the sides of the body, and terminating again somewhere in the base. They are never found in beds together, like some other formed stones, nor that I have yet heard of (says the ingenious Mr. Ray*) in great numbers in one place: but in the latter I must take leave to inform him, that though I think it in the main to be true, yet that at Tangley Fulbrook, and all about Burford, they are found in such plenty, that I believe it were easy in a little time to procure a cart-load of the first sort of them, carefully exhibited in tab. ii, figs. 9, 10, whose innermost texture, though it seem to be nothing more than a coarse rubble-stone, yet is thinly cased over with a fine laminated substance (the plates lying obliquely) much like *Lapis Judaicus*. In form they are flat, depressed upon the basis, in colour generally yellow, their rays made of a double rank of transverse lines, with void spaces between the ranks, visible enough on the top of the stone (fig. 9), but not so distinguishable on the bottom (fig. 10); the whole body of the stone, as well as the spaces included within the rays, being elsewhere filled with annulets much more curiously wrought by nature than by the tool of the graver.

" The center of these rays, by Pliny called modiolus, by Aristotle umbilicus, is never placed on the top of the stone, but always inclining to one side, as that at the bottom does to the other; the axis lying obliquely to the horizon of the stone. Which gave occasion to a learned Society of Virtuosi, that during the late usurpation lived obscurely at Tangley, and had then time to think of so mean a subject, by consent to term it the *Polar-stone*, having ingeniously found out by clapping two of them together, as suppose the figs. 9 and 10, that they made up a globe, with meridians descending to the horizon, and the pole elevated, very nearly corresponding to the real elevation of the pole of the place where the stones are found."†

Klein‡ gave the following diagnosis of this urchin: " Species 1, Plotii; maximus discum referens; Burfordinensis, Hist. Oxon., tab. ii, figs. 9, 10. Luydii prope Fulbrock in agro Oxon. Integra testa intra demissos circulos stellata, superficiem quinque, tæniæ profundæ duplicatæ ac crenatæ et unus altus sulcus lævis in undecim, basin vero quinque sulci angustiores in totidem segmenta dividunt." In tab. xii, this author figured a large

---

* 'Observations Topograph.,' p. 116.

† Plot's 'History of Oxfordshire,' pp. 90, 91.

‡ 'Naturalis Dispositio Echinodermatum,' p. 22.

Clypeus which was communicated to him by Dr. Heucher from the Royal Cabinet of Dresden ; this urchin was described by Leske* in his additions to Klein's Monogragh as a distinct species, under the name *Clypeus sinuatus ;* by subsequent authors, *C. Plotii,* Klein, with the author's reference to Plot's figure as the type of the same, is omitted. Although Leske described both species and pointed out the diagnostic differences which he supposed to exist between them, nevertheless Klein's reference to the type of the English species has been overlooked, and Leske's name given to this urchin. Lamarck† introduced still further confusion into the subject by describing the flattened varieties of *Clypeus Plotii,* Klein, under the name *Galerites patella,* and this new specific synonym was adopted by Defrance,‡ Deslongchamps,§ De Blainville,|| and Agassiz,¶ and is retained by most Continental palæontologists at the present time. As it has been one of my objects to trace the true history of every species described in this work, even at the risk of disturbing a name which has passed unchallenged for nearly a century, justice to Klein renders it imperative that his name should be retained to the urchin which was first figured by Dr. Plot, and that of *Clypeus sinuatus,* Leske, to the specimen contained in the Dresden collection ; should a further examination of that urchin show that it is only a tumid variety of *C. Plotii,* Klein, the priority of the latter will still entitle it to be retained as the name of this species.

Although this large buckler-shaped urchin has been well known to naturalists for nearly two centuries, nevertheless no good figure of the test, with details of its structure, has been given until now ; this is the more remarkable, when the beauty and abundance of the species is considered, together with its importance in Oolitic geology.

The test exhibits many varieties of form and outline ; some of these have received specific names and descriptions by different authors, of which *Clypeus angustiporus,* Agass., and *Cl. excentricus,* M'Coy, are examples. This urchin attained its most typical form and best development in the seas which deposited the beds constituting the zone of Ammonites Parkinsoni of the Inferior Oolite ; the specimens found in the Stonesfield Slate, Great Oolite, and Cornbrash are in general smaller, and deviate more or less from the Inferior Oolite forms.

The ambulacral areas on the dorsal surface are narrowly lanceolate ; the anterior pair, and single area, are about the same width ; but the posterior pair are wider ; they are all more or less slightly flexed (Pl. XXVIII, fig. 1 *a*) ; at the base they form narrow depressed bands, which radiate from the mouth to the circumference (fig. 1 *b*), and give a stellate character to the base of the test.

* 'Additamenta ad Kleinii Dispositionem Echinodermat
† 'Animaux sans Vertèbres,' tom. iii, p. 23, No. 14.
‡ 'Diction. des Sciences Naturelles.'
§ 'Ecyclopéd. Méthodique.'
|| 'Diction. des Sciences Naturelles.'
¶ 'Echinodermes Fossiles de la Suisse.'

The poriferous zones are broadly petaloidal on the dorsal surface; at the border the pores are as closely approximated as they are at the base (Pl. XXIX, fig. 1 *a, b*); about two lines up the sides they become abruptly apart, and continue so throughout the zones; they again gradually approximate in the vicinity of the disc (fig. 1 *a, b*); the holes of the inner row are round, those of the outer row are slit-like, and the outer and inner holes of each pair are united by fine fissures (Pl. XXIX, fig. 1 *c*); the intermediate septa having a single row of small granules on their surface; at the base the pores forming a pair are small, and closely approximated; the zones are very narrow, and form wavy lines, in which the bigeminal pores are placed at intervals apart, sometimes they are disposed in rectilineal order, and sometimes arranged in triple oblique pairs; at Pl. XXIX, fig. 1 *g*, I have represented this arrangement of the pores in one of the basal zones, where each small ambulacral plate is seen to be perforated by a pair of pores. As the ambulacral plates are much narrower on the upper surface than they are at the base, the pores are more numerous, and more closely approximated on the dorsal surface, where there are from eight to ten pores opposite each large plate (Pl. XXIX, fig. 1 *c*).

The inter-ambulacral areas are of unequal width, the anterior pair are the narrowest, the single area is the widest, and the posterior pair of intermediate dimensions; there is often a slight depression down the centre of each area in the line of the median suture, which, with the flat petaloidal depression formed by the poriferous zones, produces a series of undulations on the upper surface of the areas (Pl. XXIX, fig. 1 *b*). The single inter-ambulacrum is deeply cleft by the anal valley, which commences narrow at the disc, expands towards the middle, and still more so at the border (Pl. XXIX, fig. 1 *a*). The sides are vertical, and the vent is seen at the extremity of the channel, and is figured in shadow in fig. 1 *a*. The border of the test forms a series of graceful undulations, from the prominence and convexity of the inter-ambulacral and the narrowness of the ambulacral areas. (Pl. XXIX, fig. 1 *a* and fig. 1 *b*, shows the undulation of the border.)

The tubercles on the upper surface are small, and nearly equal in size throughout both areas; those in the ambulacra are arranged in oblique V-shaped rows, and those on the inter-ambulacral plates in nearly horizontal rows, of which there are five or six on each large plate (Pl. XXIX, fig. 1 *c*). At the base, they are more unequal in size, and larger than on the upper surface; in the middle of the area they are smaller; at the sides they are larger, and sometimes the large tubercles are perforated, as shown at fig. 1 *c* and fig. 1 *f*. They are surrounded by deep areolas, and between these depressions the surface of the plates is covered with fine granulations (fig. 1 *c*). It is right to remark, that in the specimens I have hitherto examined the perforation of the large basal tubercles is an exceptional, rather than a general character.

The apical disc is large, and lies in a depression behind the vertex. Its plate-elements are rarely seen separate, except in specimens which have been decomposed, as in this urchin, the genital and ocular plates are all firmly anchylosed together from early age;

the four ovarial holes project into the inter-ambulacral spaces, and the five ocular holes are seen at the apices of the ambulacra (Pl. XXIX, fig. 1 *d*). The madreporiform body occupies the centre of the disc, and presents a fine spongy structure on its surface.

The base is flat, or slightly concave; the basal portions of the inter-ambulacral areas are convex, and separated from each other by narrow, depressed ambulacral valleys (Pl. XXVIII, fig. 1 *b*), which radiate in straight lines from the mouth to the border, the margin of the ambulacra being defined by the narrow poriferous zones.

The mouth is excentral, and situated nearer the anterior than the posterior border (Pl. XXVIII, fig. 1 *b*). The peristome is surrounded by five prominent oral lobes, which are separated from each other by the depressed ambulacra, as they radiate from the mouth. At Pl. XXIX, fig. 1 *h*, another view is given of this quinque-lobed opening.

There is a very marked distinction in the varieties of *Clypeus Plotii* collected from the Inferior Oolite and the Great Oolite; so much so, that they have been considered, by good local observers, to belong to two distinct species. On this point my excellent friend, the Rev. A. W. Griesbach, remarks—

" *Clypeus Plotii.*—I have two urchins on the table, both said to be this species. One of them is a very good specimen from the *Great Oolite*, Kingsthorp, the other an *Inferior Oolite* specimen from Rodborough Hill. These urchins appear to belong to two distinct species, for I think their differences can hardly be less than specific. I have no doubt but you have specimens both from the Great and Inferior Oolite. Pray compare them. I will just say, meanwhile, that in the Inferior Oolite specimen the apical disc is *all but central*, while in the Great Oolite one it is *much nearer* the *posterior margin*. In the Great Oolite specimen the apices of the ambulacra are *deeply sunk below the plane of the test,* while in the Inferior Oolite specimen they are *in the same plane as the test*. And, in the Inferior Oolite specimen, the region of the shell which contains the anal furrow is *very tumid,* and towards the disc *nearly as high as the highest part of the shell;* in the Great Oolite specimen the posterior part of the test corresponding is *extremely and abruptly depressed and flattened*. In general form the Inferior Oolite specimen is high and spherical, the Great Oolite specimen low and flattened. May not these two forms eventually prove to be *C. sinuatus*, Leske, *C. Plotii*, Klein, respectively ?"

Affinities and differences.—This large discoidal urchin is so distinct from all other congeneric forms that it can scarcely be mistaken for any of them. Its nearest affinities are with *Clypeus Michelini*, Wr., and *Clypeus Mülleri*, Wr.

It is distinguished from *Clypeus Michelini* by its greater convexity, the wideness of the poriferous zones, the size of the tubercles, and the width of the anal valley. From *Clypeus Mülleri*, Wr., by its orbicular outline, and the absence of the produced, deflected, and truncated posterior border, so characteristic of that species. From *Clypeus Hugii*, Ag., it is distinguished by the extension of the anal valley from the disc to the border; whereas in that species a portion of undepressed test always separates the disc from the valley.

From *Clypeus subulatus*, Young, it is distinguished by its orbicular outline and anal valley; in that urchin the outline is oblong, and the anal valley quite short and marginal. The height and convexity of *Clypeus Agassizii*, Wr., with its shallow anal valley and inflated test, prevent the possibility of mistaking *Clypeus Plotii*, Kl., for that urchin, which replaces it in some of those regions where the latter is absent.

Locality and Stratigraphical position.—The metropolis of *Clypeus Plotii* is the Trigonia grit of the Inferior Oolite, in the zone of Ammonites Parkinsoni; it is extremely abundant in some localities, but rare in others; over the central parts of the Cotteswold Hills it is found in great numbers; where its bed crops out at the surface, as near Naunton Inn, a cartload of weathered Clypei might sometimes be collected. My best specimens were obtained from near Stow-in-the-Wold. I have likewise collected beautiful examples at Rodborough Hill, Scar Hill, Shurdington Hill, Leckhampton Hill, Cleeve Hill, and at Cubberley, Cowley Wood, Pen Hill, Little Rissington, Adlestrop, Northleach, and Hampen, Gloucestershire; in the same zone near Burford, and Sarsden, and in the Stonesfield Slate at Stonesfield, Oxon. Mr. Lycett obtained one specimen from the Fuller's-earth, at Minchinhampton Common; it is found likewise in the Great Oolite at Minchinhampton, Gloucestershire; Kiddington, Oxon; and Kingsthorp, Northampton. Mr. Macneil, after many years' collecting, found only two specimens in the Cornbrash at Trowbridge, Wilts, both of which are now in my cabinet.

The Foreign distribution of this species, according to M. Desor, is "Oolite vesulienne du Kornberg près Frick et Buren près Gensingen (Argovie), Muttenz (Bâle), Porrentruy, Plasne près Poligny, St. André près Salins.

"Grande Oolite de Boulogne-sur-Mer, Chayal (Ardennes), Montanville, Flincy (Meuse), Noviant, Besançon."

CLYPEUS ALTUS, *M'Coy.* Pl. XXVII, fig. 1 *a, b, c, d, e, f.*

CLYPEUS ALTUS.		M'Coy, Ann. and Mag. of Nat. Hist. 2d ser., vol. ii, p. 417, 1848.
NUCLEOLITES ALTUS.		Forbes, in Morris's Catalog. of Brit. Foss., 2d ed., p. 83, 1854.
CLYPEUS DAVOUSTIANUS.		Desor, Synopsis des Échinides Fossiles, p. 55, 1855.
—	ALTUS.	Wright, Stratigraph. Distrib. Ool. Echinoder., British Association Reports, 1856.
—	DAVOUSTIANUS.	Cotteau, in Davoust, Note sur les foss. spec. de la Sarthe, p. 7, 1856.
—	—	Cotteau, Note sur quelques Oursins de la Sarthe, Bull. de Geol. Soc. de France, 2ᵉ sér., t. xiii, p. 650, 1856.
—	—	Cotteau and Triger, Échinides du départ. de la Sarthe, pl. xii, figs. 1—7, p. 62, 1858.

Test sub-circular, broader than long, upper surface evenly convex, high, and sometimes sub-conical; base concave, very much undulated, from the extreme tumidity of the basal portions of the inter-ambulacra; apical disc central; poriferous zones petaloid on the upper two thirds of the dorsal surface; lower third, and basal portion, with parallel rows of unconnected pores; anal valley long, narrow, deep, extending from the disc to the border, fissure-like above, and slightly expanded near the margin; single inter-ambulacrum much deflected and beak-like, truncated at its extremity; mouth-opening small, sub-central, lodged in a deep depression, peristome surrounded by five oral lobes. Greatest diameter of the shell across the middle of the postero-lateral inter-ambulacra.

Dimensions.—Height, one inch and one fifth; antero-posterior diameter, two inches and six tenths; transverse diameter, two inches and nine tenths.

Description.—This beautiful Clypeus attains considerable dimensions, and preserves through all the phases of its development the specific characters enumerated in our diagnosis. It is remarkable for having the upper surface evenly convex, much elevated, or even sub-conical; it is likewise broader across the postero-lateral ambulacra than it is long, its base is more concave, the inter-ambulacral segments of this region are more tumid, and the outline of the basal margin more undulated (fig. 1 *c, d*), than in any other species.

The ambulacral areas are narrow at the margin, slightly enlarged at the upper third, and lanceolate at the apex. The poriferous zones in the upper two thirds of the dorsal surface are slightly petaloid; in the lower third the pores are small, parallel, equal, and unconnected (fig. 1 *a, c*); in the petaloid portion, the pores of the inner row are round, those of the outer row are slit-like and oblique (fig. 1 *e*); the ambulacral plates are very narrow, there being seven, with a corresponding number of pairs of holes, opposite each large inter-ambulacral plate (fig. 1 *e*).

The inter-ambulacral areas are of unequal width, the anterior pair are the narrowest, the posterior pair the widest, and the single area of intermediate dimensions; they are almost uniformly convex on the upper surface, and present very great inequalities at the base, more so, in fact, than in any other species of the ECHINOBRISSIDÆ. The single inter-ambulacrum is short, curved, much deflected, and truncated; it is deeply cleft by the anal valley, which extends from the apical disc to the margin; this sulcus is in the form of a deep fissure, with vertical parallel walls in the upper half of its length, but moderately expanded in the lower half (fig. 1 *a* and *d*).

The small, narrow, and elongated apical disc is well preserved in most of my specimens, and is situated near the centre of the test, immediately behind the vertex; it is composed of two small, anterior, perforated genitals (fig. 1 *f*), and two larger posterior perforated genitals, with a single, long, imperforate genital plate, which descends into the narrow anal valley; the spongy madreporiform body rises from the surface of the right anterior genital, and extends into the centre of the disc; the five small ocular plates

48

have marginal eyeholes, and the surface of all the discal elements is covered with fine granulations.

This urchin, when viewed in profile, presents a considerable undulation around the border (fig. 1 c and d), occasioned by the narrowness of the ambulacra, and the extreme tumidity of the marginal and basal portions of the inter-ambulacra. It is as remarkable for the concavity of its base, and the cushion-like structure of its basal inter-ambulacra, as for the evenness and convexity of the dorsal portions of the same segments : the basal portions of the antero-lateral inter-ambulacra are small, and moderately convex (fig. 1 c and d), the postero-lateral pair are very prominent and tumid, and the single inter-ambulacrum is angular, deflected, and truncated at its extremity.

The small mouth-opening lies in a deep sub-central depression (fig. 1 b), and the peristome is surrounded by five small oral lobes; the basal portions of the ambulacra are narrow, and the poriferous zones so small and indistinct that the pores, even with a lens, are seen with difficulty ; as they approach the mouth they increase in size and number, and form a series of triple oblique rows in the five petal-like expansions which radiate from the mouth (fig. 1 b).

The small tubercles are in general arranged in four horizontal rows on each plate (fig. 1 e) ; they are very uniform in size, and surrounded by sunken areolas, the tubercles at the base and border are larger than those of the upper surface ; all the intermediate surface of the plates is covered with fine homogeneous granules, microscopic in size, but regular in their arrangement.

Affinities and differences.—This species resembles some of the smaller varieties of *Clypeus Plotii*, Klein, but is readily distinguished from these, and from all congeners, by its greater proportional breadth across the postero-lateral inter-ambulacra, the narrowness and fissure-like character of the anal valley, the remarkable undulations of the border, the tumidity of the basal inter-ambulacral cushions, and the concavity of the base. It resembles *Clypeus Mülleri*, Wr., in the shortness of the petaloid portion of the dorsal ambulacra, and the narrowness of the anal valley ; but the oblong form of *Clypeus Mülleri*, the flatness of its base, and the depression of its dorsal surface, form diagnostic distinctions between them. In the narrowness of its ambulacral petals it resembles *Clypeus Michelini*, Wr., but the flat dorsal and basal surfaces of that species form distinctive characters by which the two urchins are readily distinguished from each other.

Locality and Stratigraphical position.—I have collected this urchin from the Inferior Oolite of Dorsetshire only, in the upper ragstones of that formation, appertaining to the zone of Ammonites Parkinsoni, as defined in the chapter on the Stratigraphical distribution of Oolitic Echinodermata. I have found it at Burton-Bradstock, and Walditch Hill, near Bridport, associated with *Ammonites Parkinsoni*, Sow., *Ammonites subradiatus*, Sow., *Stomechinus bigranularis*, Lamk., *Holectypus hemisphæricus*, Desor, *Collyrites ovalis*, Leske, and *Collyrites ringens*, Desml. It appears to take the place of *Clypeus Plotii* in the Parkinsoni zone of Dorsetshire, just as *Holectypus hemisphæricus*, Desor, replaces in the same rock in Dorsetshire *Holectypus depressus*, Lamk., which is so abundant a fossil in this zone of the Inferior Oolite in the Midland counties of England and so rare in the South.

History.—This urchin was first described by Professor M'Coy, in his paper on 'New Mesozoic Radiata;' the species was for a time overlooked, as no figure was given with the description; Professor M'Coy subsequently kindly sent me pen-and-ink outlines of all the species described in that memoir, by which I was enabled to identify the forms supposed to be new. This Clypeus has recently been beautifully figured, and well described as *Clypeus Davoustianus*, by my esteemed friend M. Cotteau, in his valuable monograph on the 'Échinides of the Sarthe.' It was not included in my memoir on the 'Cassidulidæ of the Oolites,' as at the time I could not find a specimen to study. Since the publication of that Prodrome I have collected a very fine series of this species; a comparison of the figures given by M. Cotteau, and the one now published in our Pl. XXVII, fig. 1, leaves no doubt as to the identity of the Sarthe *Clypeus Davoustianus*, Cot., and the Dorsetshire *Clypeus altus*, M'Coy.

CLYPEUS MICHELINI, *Wright.* Pl. XXX, fig. 2 *a, b, c, d.*

NUCLEOLITES MICHELINI. Wright, Ann. and Mag. of Nat. Hist., 2d series, vol. xiii, p. 161, pl. xii, fig. 6, 1854.
— — Forbes, in Morris's Catalogue of Brit. Fossils, 2d ed., additional sp. of Echinodermata, 1854.
— — Wright, Report Oolitic Echinod., Brit. Assoc. Reports, 1857.
CLYPEUS MICHELINI. Desor, Synopsis des Échinides Fossiles, p. 277, 1858.

Test circular or oblong, discoidal and much depressed, posterior border produced, truncated and slightly deflected in old individuals; ambulacral areas narrowly lanceolate; poriferous zones narrow, only slightly petaloid on the dorsal surface, vertex and apical disc nearly central; anterior half of the upper surface convex, posterior half much declined; anal valley narrow above, diverging below, extending from the apical disc to the border; base flat, slightly concave; mouth excentral, peristome with five small lobes; postero-lateral inter-ambulacral areas slightly tumid at the base.

Dimensions.—Height, nine tenths of an inch; antero-posterior diameter, three inches and a quarter; transverse diameter, two inches and one fifth.

Description.—The outline of this urchin varies in different individuals, and likewise in the same individual at different periods of life; its most typical form is oblong, convex anteriorly, produced and truncated posteriorly, and enlarged in the region of the postero-lateral inter-ambulacra; in others the circumference is nearly circular, and in some few transversely oval; the first form is, probably, characteristic of adult life, as the elongation and truncation of the single inter-ambulacrum are markedly shown in the only two large specimens I have seen of this rare species. In all the test is very flat; the anterior half is gently and nearly equally convex, and the posterior half much declined towards the posterior border. The ambulacral areas are narrow, the anterior one most so; the antero-lateral and postero-laterals are about the same width; they have a lanceolate form, and are composed of very narrow plates; about three tenths of an inch above the margin, the pores slightly diverge, and continue about the same width apart until they approach the apical disc; the distance between the rows of pores in this species is less than in any other known Clypeus, and forms one of its diagnostic characters; in all the specimens I have examined the ambulacral areas are likewise slightly elevated above the general surface of the test. The inter-ambulacral areas are of unequal width, the antero-lateral pair are the narrowest, they are, however, about nine times the width of the anterior single ambulacral area; the postero-lateral pair are three tenths of an inch wider than the antero-laterals, and the single inter-ambulacrum is about the same width as the latter. The anal valley extends from the apical disc to the posterior border; it is very narrow, with deep perpendicular sides above, which become shallow and expanded below; the postero-lateral inter-ambulacra are enlarged at the margin; the single inter-ambulacrum is considerably produced, its posterior border is broadly truncated and slightly deflected, within which the expanded sides of the anal valley are excavated.

The base is nearly flat; the elevations are produced by the prominence of the postero-lateral inter-ambulacra, and the deflection of the single inter-ambulacrum. The small mouth-opening is excentral, and placed nearer the anterior than the posterior border; the peristome is surrounded by five small oral lobes, which make inconsiderable prominences at the base. The apical disc is small, and absent in most of my specimens; in one only is it preserved. The genital plates are nearly equal sized; the anterior and posterior pair are perforated, and the single plate is imperforate. The madreporiform tubercle rests on the right anterior plate, and extends into the middle of the disc. The ocular plates are small, and firmly wedged between the disc and summits of ambulacra; the eyeholes are large, and, with the four genital holes, form a circle of perforations around the circumference of the disc.

Affinities and differences.—*Clypeus Michelini*, in its oblong form, truncated posterior border, and narrow anal valley, resembles *C. Mülleri*, Wr.; but is readily distinguished from the latter by the form, narrowness, and structure of the ambulacral areas; in *C. Mülleri* they are expanded and petaloid, and in *C. Michelini* they are narrow and lanceolate; the pores at no point are at any great distance apart; the anal valley in both species extends from the apical disc to the margin, but in *Clypeus Michelini* it is more expanded below and deeper above than in *Clypeus Mülleri*.

I have now before me *Clypeus angustiporus*, Ag., from a coarse Oolitic rock (Bradfordien?) near Metz, collected by M. Terquem, and kindly sent me by M. De Lorière; this urchin appears to be a variety of *Clypeus Plotii*; from which *C. Michelini* differs in many particulars. In the French urchin the apical disc is excentral, the anal valley wide above and not much expanded below, the ambulacral areas are narrow, and the test declines gradually from the vertex to the anterior border, which forms a rather acute angle; the base is undulated, and the mouth-opening nearly central; these characters distinguish it from our urchin.

Clypeus Michelini differs so widely from all the varieties of *Clypeus Plotii* with which I am acquainted, that it cannot possibly be mistaken for either of them, if proper care be taken when a comparison is made between them.

Locality and Stratigraphical position.—I have collected this species chiefly from the Freestone beds of the Inferior Oolite at Wallsquarry and Nailsworth, Gloucestershire; the specimens figured in Pl. XXX, were cut out of the centre of a block of building stone; the Oolitic grains were imbedded in the plates of the test, and have in some measure injured their surface. I found two small specimens in the zone of Ammonites Humphriesianus, in a sandy bed of this middle division of the Inferior Oolite at Cleeve Hill, near Cheltenham, where it was associated with *Ammonites Brongniarti*, Sow., *Am. Humphriesianus*, Sow., *Am. Brocchii*, Sow. Mr. Reed, of York, collected several specimens *germinans* from the Inferior Oolite at Whitwell, Yorkshire; one of these, kindly given me by that gentleman, I have figured in Pl. XXVI, fig. 2. It was associated with *Stomechinus germinans*, Phil., *Trigonia costata*, Sow., *Gervillia Hartmanni*, Münst., and other Inferior Oolite shells.

B. *Species from the Great Oolite.*

CLYPEUS MÜLLERI, *Wright.* Pl. XXXIII, figs. 1, 2, 3, 4, 5, 6.

> NUCLEOLITES SOLODURINUS. Wright, Ann. and Mag. of Nat. Hist., 2d ser., vol. ix, p. 305, 1851.
> — — Forbes, in Morris's Cat. of Brit. Foss., 2d ed., p. 84, 1854.
> — — Wright, Report on Brit. Ool. Echinodermata, Brit. Assoc. Report, 1857.

Test oblong, posterior border much produced, deflected, and truncated; ambulacral areas largely petaloidal, with their apices closely approximated; apical disc and vertex excentral posteriorly, anal valley narrow, acutely lanceolate, with vertical walls, extending from the disc to the border; base concave, mouth-opening sub-central, peristome surrounded by five prominent oral lobes.

Dimensions.—Large specimen from the Forest Marble. Antero-posterior diameter, three inches and a quarter; breadth, three inches and three tenths.

Average-sized specimen from the Great Oolite. Height, eight tenths of an inch; antero-posterior diameter, two inches and a quarter; breadth, two inches and three twentieths.

Description.—This beautiful Clypeus was first found by my esteemed friend S. P. Woodward, Esq., in the Great Oolite near Cirencester, and was referred by him to *Clypeus Solodurinus*, Ag. In this opinion I formerly concurred, and described it under that name in my memoir on the ' Cassidulidæ of the Oolites,' already referred to in the synonyms of this species. Having discovered a marly vein in the Great Oolite which contained a number of specimens, I had an opportunity of studying this urchin in different stages of development; from this examination I ascertained that our original determination could not be maintained, as the three specific characters insisted upon by M. Agassiz, namely, the angular and truncated form of the posterior border, the nearer approximation of the vent to the summit, and the possession of a very thick test (" Le test est assez épais et recouvert d'une granulation assez uniforme sur toutes les parties intactes"*), were characters which did not hold good in the suite of specimens collected. I therefore determined to describe it as a distinct species, associating with it a name most justly esteemed by all physiologists.

Clypeus Mülleri has an oval form (fig. 2 *a*); rounded before, and slightly truncated behind; the test is very thin, and on that account is not often well preserved, the upper surface is flat and much depressed, it slopes more towards the posterior than the anterior border; the marginal fold is rounded, and the sides are tumid in proportion to the height of the test (fig. 2 *c*).

The ambulacral areas are narrowly lanceolate, nearly of equal width, and closely approximated around the disc; the dorsal portions of the poriferous zones are widely petaloid, each petal has an elegant leaf-like form, being narrow below, expanded in the middle, and lanceolate above; for a short distance above the border, the pores forming a pair lie close together (figs. 1 and 3), then gradually become wider apart, until they attain their maximum separation in the middle of the dorsal surface; from this point they again gradually approximate until they lie close together at the apex. The form of the ambulacral petals (fig. 2 *c*) is one of the diagnostic characters of this species when compared with *Clypeus Plotii;* in the latter the separation of the pores takes place much nearer the border than in *Clypeus Mülleri;* they likewise taper more towards the apex, and form a much more graceful figure on the upper surface of the shell than they do in that species.

* ' Échinodermes Fossilles de la Suisse,' prem. partie, p. 35.

The holes of the inner row are round, those of the outer row in the form of long slits (figs. 5 and 6); there is a narrow space external to the inner row (fig. 5), beyond that the slit commences and passes transversely across the zones (fig. 6), where it glides into the outer hole; one row of small granules occupies the space external to the inner row of holes (fig. 6), and on the upper surface of the septa dividing the pairs of pores a row of granules is disposed with great regularity (fig. 6); the ambulacral plates are narrow above and broader below; on the upper surface there are seven pairs of pores opposite one large inter-ambulacral plate, and at the base there are four pairs of pores opposite one large plate. The basal portions of the poriferous zones are narrow, and lie in shallow depressions of the surface; for about three fourths of the distance between the border and mouth-opening, the pairs of pores are placed wide apart, at the inner fourth they become more numerous and are disposed in close-set, triple, oblique pairs, which form a penta-phylloid floscelle around the mouth (fig. 2 b, fig. 3 b).

The inter-ambulacral areas are of unequal width; the anterior pair are the narrowest (fig. 2 a), the posterior pair the widest (fig. 2 a), and the single inter-ambulacrum about the width of the latter; this area is slightly produced, deflected, and abruptly truncated posteriorly.

The anal valley extends from the disc to the border; it is a narrow, lanceolate depression, with vertical sides (fig. 2 a, fig. 1), which gradually expands from the apex to the border; the vent opens at the extreme end of the valley beneath the disc, and the sides of the channel bulge slightly outwards to give increased space to the intestinal aperture: the base is nearly flat, and the inter-ambulacral segments form only inconsiderable elevations (fig. 2 b); the surface of the plates is covered with tolerably regular rows of very small tubercles, encircled by microscopic granules, the fineness and minuteness of the sculpture on the plates is therefore another character (fig. 5) by which it is distinguished from *Clypeus Plotii*; like other Clypei, the tubercles of the base are larger than those on the dorsal surface.

The vertex is excentral and posterior (fig. 2 c), and immediately behind it is placed the apical disc; which is small and closely wedged in between the apices of the ambulacra; the disc is composed of two pairs of perforated genital plates, the posterior being larger than the anterior pair; the large, spongy, madreporiform body occupies the center of the disc, and extends as far as the ocular plates (fig. 4), which are small and scarcely visible without the aid of a lens; their minute marginal orbits are seen opposite the apices of the ambulacra.

The mouth-opening is sub-central, nearer the anterior than the posterior border (fig. 2 b); the peristome is surrounded by five small prominent lobes, formed by the terminations of the inter-ambulacra; between the oral lobes the poriferous zones assume a depressed, leaf-like figure, freely perforated in this region for the passage of tubular organs, which, in this species, appear to have been very numerous around the mouth.

Affinities and differences.—*Clypeus Mülleri* more closely resembles *Clypeus Plotii* than any other English species ; having affinities with it in the form and structure of the ambulacra, the extent and narrowness of the anal valley, and the depression of its dorsal surface. It is distinguished from *C. Plotii*, however, by its oblong figure, truncated posterior border, shorter and more graceful petaloidal ambulacra, finer and more minute sculpture on the plates, a flatter base, with smaller tubercles thereon. The large specimen of *Clypeus Mülleri* (fig. 1) very much resembles *Clypeus Michelini*, but the widely petaloidal character of the ambulacral areas in the former species present a great contrast to the structure of the homologous portion of the test in the latter urchin, and serve to distinguish them from each other; whilst the depression of the dorsal surface, the narrowness of the anal valley, the smallness of the apical disc, and the microscopic sculpture on the test, assimilate the two forms closely together.

It is distinguished from *Clypeus Solodurinus*, Ag., by having a much flatter under surface, with inconsiderable undulations of the inter-ambulacra ; whilst, according to Agassiz, in the Swiss urchin, " La face inférieure est régulièrement ondulée par suite de la dépression des ambulacres." The test in *C. Mülleri* is likewise extremely thin, the sculpture fine, and almost microscopic ; whilst in *C. Solodurinus*, " Le test est assez épais, et recouvert d'une granulation assez uniforme sur toutes les parties intactes."

Locality and Stratigraphical position.—I collected this Clypeus from the white marly vein which traverses the upper region of the Great Oolite in some parts of Gloucestershire, as near Cirencester, near Northleach, at Salperton tunnel, Great Western Railway, near Minchinhampton, and near Cowley Wood; in all these localities it was associated more or less abundantly with *Echinobrissus Woodwardi*, Wr. Mr. Frederick Bravender collected the fine large specimen figured in Pl. XXXIII, fig. 1, from the Forest Marble near Cirencester. The Rev. A. W. Griesbach found one specimen in the Cornbrash of Rushden, Northamptonshire, where it is extremely rare, as the specimen which my kind friend has communicated is the only one he has seen in that locality. M. Bouchard-Chantereaux sent me a specimen which he collected from the Great Oolite near Boulogne-sur-Mer.

History.—This urchin was first discovered by Mr. S. P. Woodward, in the Great Oolite near Cirencester ; it was subsequently described in my memoir on the ' Cassidulidæ of the Oolites ' as *Nucleolites Solodurinus*, and is now figured for the first time as *Clypeus Mülleri*. I dedicate the species to the memory of Jöhannes Müller, late Professor of Physiology in the University of Berlin, whose profound observations on the anatomy, physiology, and metamorphoses of the Echinodermata have thrown so much new and important light on the natural history of this class of the Animal Kingdom.

SECTION B. *Anal valley does not extend from the border to the disc.*

A. *Species from the Inferior Oolite.*

CLYPEUS HUGII, *Agassiz.* Pl. XXX, fig. 1 *a, b, c, d, e, f.*

CLYPEUS HUGII.	Agassiz, Échinodermes Foss. de la Suisse, 1ᵉ Partie, tab. x, fig. 2—4, p. 34, 1839.
— —	Agassiz and Desor, Cat. rais. des Éch. Ann. Sc. Nat., 3 série, tom. vii, p. 156, 1847.
— —	D'Orbigny, Prod. de Pal. Strata, t. i. p. 290, No. 496, 1850.
NUCLEOLITES HUGII.	Forbes, Mem. of the Geol. Surv. of Great Britain, Decade i, description of Pl. ix, 1850.
— —	Wright, Ann. and Mag. of Nat. Hist. 2d series, 1851, vol. ix, p. 303.
— —	Forbes, in Morris's Catalogue of Brit. Fossils, p. 84, 1854.
CLYPEOPYGUS HUGII.	Desor, Synopsis des Échinides Fossiles, p. 274, 1857.
ECHINOBRISSUS HUGII.	Cotteau and Triger, Échinides du département de la Sarthe, pl. vi, fig. 10—12, p. 58, 1858.

Test sub-orbicular, dorsal surface convex; apical disc central; ambulacral areas narrowly petaloid in the upper two thirds of the dorsal surface; anal valley short, and wide, occupying the lower half of the area; a considerable portion of undepressed test between the disc and valley, single inter-ambulacrum produced and deflected; base nearly flat, mouth-opening sub-central, nearer the anterior than the posterior border, peristome pentagonal, surrounded by five oral lobes; a penta-phylloid floscule around the mouth.

Dimensions.—Height, one inch; length and breadth nearly equal, two inches and one eighth.

Description.—All the specimens of this urchin I examined before the fine example figured in Pl. XXX, were small, and resembled *Echinobrissus*, but in this urchin the characters of *Clypeus* are well marked; the test is circular, being nearly as long as it is broad; it is rounded before (fig. 1 *a*), and slightly rostrated behind, by the prominence and deflection of the single inter-ambulacrum (fig. 1 *b*); the upper surface is uniformly convex (fig. 1 *c*), in the figured specimen it is rather conical, rising high at the vertex, and declining rapidly on all sides, more especially towards the posterior border; the base is flat or slightly concave, and the basal portions of the inter-ambulacra form

49

prominent undulations between the narrow ambulacra (fig. 1 *b, c*). The ambulacral areas are narrowly lanceolate, the anterior pair curve gently upwards and outwards, and the posterior pair upwards and inwards; the poriferous zones are narrowly petaloid three fourths of their length between the border and disc; at the lower fourth the pores forming a pair are closely approximated, and the rows are very narrow; above this point the pores of the inner row are round, those of the outer row in the form of oblong slits (fig. 1 *e*), which are connected with the inner row by fine sutures; the transverse sulci disappear some distance above the border, the pores then become simple, oblique, and wider apart; in the basal portion of the zones the pores are minute and far apart, and their track is only traced by the depression formed by the ambulacra near the mouth-opening; around the peristome (fig. 1 *b*) the pores become more numerous, and form five leaf-like expansions. The inter-ambulacral areas are of unequal width, the anterior pair are the narrowest, the single area the widest, and the posterior pair of unequal dimensions (fig. 1 *a*); the anterior border is obtusely rounded, the sides swell gradually outwards to the junction of the posterior pair with the single area, the widest part of the test is in the direction of a line passing transversely across the vertex; the single inter-ambulacrum is slightly produced, deflected, and truncated (fig. 1 *a, d*); the anal valley is short and wide, and occupies the lower half of the inter-ambulacrum (fig. 1 *a, c, d*); between the upper portion of the valley-arch and disc the test is undepressed (fig. 1 *a*); the vent occupies the extreme termination of the valley, and its perpendicular sides are scooped out (fig. 1 *d*) to afford greater space for the passage of the intestinal tube and the closure of the aperture by its circle of anal plates (Pl. XLI, fig. 1).

The apical disc is small, and its elements so intimately soldered together that their separate study is impossible in the specimen figured; the disc, moreover, is so much covered over by the madreporiform body that the sutures are all concealed (fig. 1 *f*). In another small specimen the disc is composed of two anterior and two posterior perforated ovarial plates, and a single imperforate ovarial; the five ocular plates are very small and their orbits marginal.

The test is moderately thick, and the surface of the plates is covered with several horizontal rows of small tubercles (fig. 1 *e*); the base of each is encircled by a sunken areola, and the intermediate portion delicately sculptured with microscopic granules (fig. 1 *e*); the tubercles at the base are only a little larger than those on the upper surface.

The base is flat, slightly concave, or undulated, the ambulacra form straight valleys, and the inter-ambulacra moderately convex elevations; the prominence of these undulations and the deflection of the inter-ambulacrum are greater in proportion in small individuals. The mouth-opening is excentral, being situated near the junction of the anterior with the middle third of the antero-posterior basal diameter; the pentagonal peristome is surrounded by five lobes, and the terminations of the ambulacra form five leaf-like expansions, or a penta-phylloid floscule in which the pores are arranged in triple oblique ranks.

Affinities and differences.—The general outline of some of the smaller specimens of this urchin resembles *Echinobrissus orbicularis*, but the structure of the ambulacra, and of the apical disc, the size and position of the anal valley, together with the undepressed portion of test between the apex of the valley and disc, readily distinguish it. The larger form (fig. 1 *a*) differs so entirely from other Clypei that it cannot be mistaken for any of its congeners, whilst its orbicular outline, petaloidal ambulacra, small solidified apical disc, and mouth-opening provided with oral lobes, justify its position among the Clypei rather than with *Echinobrissus*, among which I formerly placed it.

Locality and Stratigraphical position.—I have collected *Clypeus Hugii* in the Lower Trigonia Grit, zone of Ammonites Parkinsoni Inferior Oolite, at Rodborough Hill, Shurdington Hill, Leckhampton Hill, and Ravensgate Hill; the large figured specimen was found at Shurdington Hill. I have collected many specimens from the Trigonia Grit at Hampen; many years ago it was found in considerable abundance in the upper ragstones of the Inferior Oolite with *Trigonia costata* at Charlcombe, near Bath. The Hampen specimens are associated with *Pedina rotata*, Wr., *Hyboclypus ovalis*, Wr., *Holectypus depressus*, Lamk., and *Clypeus Plotii*, Kl., with several species of Conchifera characteristic of that zone of life, as *Trigonia costata*, Sow., *Pecten symmetricus*, Mor., *Tancredia donaciformis*, Lyc., and *Quenstedtia oblita*, Phil., together with Anthozoa, as *Anabacia orbulites*, Lamx.

The foreign distribution of *Clypeus Hugii*, is, according to M. Desor, "Marnes à Discoïdes (Vesulien), de Hornussen et de Bozen (Argovie), des environs d'Olten, du Mont Terrible."

In Switzerland, he adds, this species is the faithful companion of *Holectypus depressus*, *Collyrites ovalis*, and *Echinobrissus clunicularis*, and one of the most characteristic fossils of the Marnes à Discoïdes, the zone of Ammonites Parkinsoni Inferior Oolite, so that its statigraphical position and palæontological associates is the same in the Alpine regions of Switzerland, as in the Cotteswold Hills.

The true zone of this species in the department of the Sarthe is yet uncertain, the only specimen found was collected near Mamers. I have no doubt its bed will be found to be the Inferior Oolite.

CLYPEUS AGASSIZII, *Wright.*   Pl. XXXI and XXXII.

NUCLEOLITES AGASSIZII.   Wright, Ann. and Mag. of Nat. Hist., 2d series, vol. ix, p. 308,
                         Pl. iii, fig. 3 *a—c*, 1851.
   —   —   Forbes, in Morris's Catalogue of Brit. Foss., 2d ed. p. 84, 1854.
CLYPEUS AGASSIZII.   Desor, Synopsis des Échinides Fossiles, p. 278, 1857.
   —   —   Davoust. Note sur les Fossiles spéciaux á la Sarthe, p. 25, 1856.
   —   —   Cotteau and Triger, Échinides du département de la Sarthe, pp.
                         16 and 61, pl. iii, fig. 1; pl. ix, fig. 9; pl. ix, fig. 1—3.

Test large, hemispherical, or sub-conoidal; margin sub-circular; a little longer in the antero-posterior diameter; rounded before, slightly rostrated, and truncated behind; vertex central, apical disc excentral, and inclined backwards; ambulacral areas very narrow; poriferous zones widely petaloid and conjugate from above the margin to the disc; at the margin and base, pores simple, non-conjugate, and scarcely visible; near the mouth more numerous and apparent, and arranged in triple oblique pairs; vent oblong, and near the surface; anal valley very shallow, commencing in the middle of the single inter-ambulacrum; an undepressed portion of test between the vent and disc, sometimes sulcated in the middle; base flat, or slightly undulated; mouth-opening large, subcentral; peristome surrounded by five prominent lobes; dorsal tubercles nearly microscopic; basal, a little more conspicuous.

Dimensions.—Height, two inches and one quarter; antero-posterior diameter, four inches and one fifth; transverse diameter, four inches.

Description.—The large size, the hemispherical or sub-conoidal form of the upper surface, the flat base, and superficial anal valley, readily distinguish this magnificent species from all other Clypei. The outline of the margin is nearly circular, the antero-posterior being a little more than the transverse diameter; the test is rounded before, and slightly rostrated, and truncated behind; the upper surface assumes a conoidal figure (Pl. XXXII, fig. 1, *a*); the vertex is central, and the test declines more towards the posterior than the anterior border.

The ambulacra are long, narrow, and lanceolate; the anterior area is straight; the antero-lateral pair describe an *f*-shaped curve on the sides of the test (Pl. XXXII, fig. 1, *a*); and the postero-lateral curve forwards upwards and inwards at the posterior surface (Pl. XXXII, fig. 1, *b*).

The poriferous zones are very large, and widely petaloid on the upper surface (Pl. XXXI, fig. 1, *a*); they are formed of an inner row of small, simple, round, or oval pores, which extend equidistant from the margin to the disc, and an external row, of long, narrow, transverse pores, connected with the inner row by conjugate sulci; the septa between the transverse pores support, on their upper surface, a regular row of small tubercles (Pl. XXXII, fig. 1, c). Near the marginal fold the pores of the outer row are round, like those of the inner row, and form narrow zones of close-set pores, only apparent by faint depressed lines, which mark the track of the zones from the border to the mouth; near the peristome the pores become more numerous, and here form a series of triple oblique rows of pairs. Pl. XXXII, fig. 1, d, represents one of the basal ambulacral areas, highly magnified (Pl. XXXI, fig. 1, b).

The inter-ambulacral areas are of unequal width at the margin; the anterior pair are the narrowest, the posterior pair are wider, and the single area is the widest; they are uniformly convex on the upper surface, and are formed of very large pentagonal plates, bent in the middle; each plate supports four horizontal rows of small tubercles, surrounded by sunken areolas, and having the inter-tubercular surface covered with fine, microscopic, homogeneous granules. Pl. XXXII, fig. 1 c, exhibits the ambulacral plates, poriferous zones, and two inter-ambulacral plates magnified three diameters, and shows that there are seven pairs of pores opposite each large plate.

The under surface is flat, and the basal portions of the inter-ambulacra are only slightly cushioned; the tubercles in this region are larger, and not so regularly arranged in rows as on the upper surface (Pl. XXXI, fig. 1 b), but are scattered more promiscuously over the surface.

The height of the test, and the excentral position of the apical disc, inclined backwards behind the vertex, occasions the curvature in the lateral ambulacra already described (Pl. XXXII, fig. 1 a). The disc is small, in proportion to the size of the test (Pl. XXXI, fig. 1 a), and is composed of two anterior, and two posterior perforated ovarial plates, and a single imperforate plate. The madreporiform body covers all the centre of the disc, and the five ocular plates are only indicated by their marginal orbits at the apices of the ambulacra; the elements of the disc are closely soldered together, and only occasionally seen distinct in some decomposed specimens.

The dorsal portion of the single inter-ambulacrum, when viewed posteriorly (Pl. XXXII, fig. 1 b), presents a triangular figure; the border forms the base, and the two postero-lateral-ambulacra the sides of the triangle; near the middle of this space is placed the oval vent, which opened near the surface (fig. 1 b); the anal valley, is a very shallow depression, from its sides, two nodulated ridges descend downwards and outwards towards the border, the plane of the intermediate space being a little beneath the general plane of the test (fig. 1 b). In some smaller specimens there is sometimes a narrow depression which passes upwards from the summit of the arch above the anal valley towards the disc, which is effaced in the large type specimen (fig. 1 b); in this urchin the space between the upper border of the opening and the disc is occupied to the extent of one inch by a smooth, slightly depressed portion of test; which, added to the oval vent, shallow valley, and superficial depression beneath, form a group of characters which well characterise this noble urchin; in the fine large specimen figured by my friend M. Cotteau from the Inferior Oolite of the Sarthe, a narrow sulcus extends from the disc to the summit of the anal arch.

The mouth-opening is large and sub-central (Pl. XXXI, fig. 1 *b*); it has a pentagonal form, and the peristome is surrounded by five large, prominent, oral lobes, formed by the terminal folds of the inter-ambulacra; alternating with the lobes, the five ambulacra form a depressed penta-phylloid floscule around the mouth opening, which imparts a marked stellate character to the flat base of this Clypeus.

Affinities and differences.—This fine species is distinguished from its congeners by its elevated, sub-conoidal, dorsal surface, its flat base, its excentral apical disc, declining towards the posterior border, its short, shallow, anal valley, and oblong vent. In some specimens the anal valley is more developed than in our large fine type urchin; this is apparent in two other individuals I possess, and M. Cotteau has made the same remark on some of the specimens which he has examined from the Sarthe: " Suivant, M. Wright, cette ouverture est à fleur du test, presque superficielle (*nearly superficial*). Ce caractère n'existe pas toujours, et dans plusieurs de nos exemplaires, notamment dans celui que nous avons fait représenter, l'anus s'ouvre dans un sillon aigu, au milieu d'une depression très-apparente de l'inter-ambulacre-postérieur; mais cette différence ne nous a pas empêché de le réunir au *Clypeus Agassizii.* Les échantillons d'Angleterre présentent eux-mêmes quelques traces de sillon et appartiennent certainement au même type que les nôtres." *

Locality and Stratigraphical position.—This urchin has hitherto been found only in the upper ragstones of the Inferior Oolite, near Bridport, in the zone of Ammonites Parkinsoni, at Barton-Bradstock, and Walditch Hill; its associates in the same bed are *Clypeus altus*, M'Coy; *Holectypus hemisphæricus*, Desor; *Stomechinus bigranularis*, Lamk.; *Collyrites ringens*, Desml.; *Collyrites ovalis*, Leske; *Hyboclypus gibberulus*, Ag.; *Ammonites Parkinsoni*, Sow.; *Ammonites subradiatus*, Sow.; *Trigonia costata*, Sow.

History.—First figured and described in my memoir on the Cassidulidæ of the Oolites ' Ann. and Mag. of Nat. Hist.', for 1851. It has recently been beautifully figured and well described by MM. Cotteau and Triger, in their fine monograph ' on the Échinides' of department of the Sarthe.

* Cotteau and Triger, Échinides du département de la Sarthe, p. 17, 1858.

CLYPEUS RIMOSUS, *Agassiz.*

CLYPEUS RIMOSUS.   Agassiz and Desor, Catalogue raisonnè des Échinides des Sciences
Naturalles, 3ᵉ serie, tome vii, p. 156.
—   —   Desor, Synopsis Échinides Fossiles, p. 277.

This urchin was entered in the catalogue raisonnè as, "Espece plate, discoide, à ambulacres costulès, Terr., Jurass. du Gloucestershire, Deluc;" in the 'Synopsis des Échinides Fossiles,' M. Desor gave the following detailed diagnosis of it: "Species discoidal, slightly convex, subrostrated posteriorly; ambulacral summit central; anal valley very narrow, extending to the apical disc; ambulacral petals convex and prominent, with very large poriferous zones, which equal in width the inter-poriferous spaces; base undulated, peristome excentrally forwards; the pores disposed in double ranks in the phyllodes; oral lobes small and not approximated." M. Desor, in a note, adds, "by its general form, as well as by its costulated petals, this species approaches much to *C. Michelini*, Wr., but the poriferous zones, instead of being distinguished by their narrowness, are remarkable for their breadth. Should it happen to be demonstrated by a series of examples that this character is not constant, it would be possible to unite these two species."

After a careful examination of all our Oolitic Clypei, with the view to identify M. Deluc's specimen, which was said to have been collected from the Inferior Oolite of Gloucestershire, I have been unsuccessful.

Long before the publication of M. Desor's Synopsis, it occurred to me that *C. rimosus* might be a variety of *C. Michelini*, one character, however, pointed out in the diagnosis, "poriferous zones very large, equalling the inter-poriferous space," is sufficient to prove that this is not the case, seeing that *C. Michelini* is as remarkable for the narrowness of the petaloid portion of the poriferous zones, as *C. rimosus* is for their breadth, and this character is constant in the large series of specimens I have examined. The breadth of the poriferous zones establishes an affinity with *C. Mülleri*, but in that species the test is oval, the ambulacra are on the same plane with the inter-ambulacra and not costulated as in *C. rimosus*. It is probable that this urchin may prove to be a depressed variety of *C. Plotii*.

CLYPEUS SUBULATUS, *Young and Bird.*   Pl. XXXIV, fig. 1 *a, b, c, d, e, f, g.*

ECHINITES SUBULATUS.	Young and Bird, Geol. Surv. of the Yorkshire Coast, pl. vi, fig. 11, p. 214, 1827.
CLYPEUS EMARGINATUS.	Phillips, Geology of Yorkshire, pl. iii, fig. 18, p. 127, 1829.
— —	Morris's Catalogue of British Fossils, p. 50, 1843.
NUCLEOLITES EMARGINATUS.	Forbes, Memoirs of the Geological Survey, Decade 1, descrip. pl. i.
— —	Wright, Ann. and Mag. of Nat. Hist., 2d ser. vol. ix, p. 310, 1851.
PYGURUS EMARGINATUS.	Desor, Synopses des Échinides Fossiles, p. 316, 1857.

Test large, oblong, and much depressed; sides equally declining; vertex and apical disc nearly central; ambulacra narrowly lanceolate; poriferous zones petalloid, on the upper three fourths of the dorsal surface; simple, in the lower fourth; anal valley short and deep, far removed from the disc, and occupying the lower third of the area; anterior border rounded; posterior border produced, rostrated, and deflected; base concave; mouth subcentral; basal inter-ambulacra much cushioned; margin gently undulated; tubercles small, and disposed in rows.

Dimensions.—Large specimen, fig. 1 *a, b.* Height, one inch; antero-posterior diameter, three inches and four tenths; transverse diameter, three inches and one fifth. Smaller specimen, fig. 1 *c, d.* Height, nine tenths of an inch; antero-posterior diameter, two inches and six tenths; transverse diameter, two inches and six tenths.

Description.—This beautiful Clypeus has hitherto been found only in the Coralline Oolite of Yorkshire, and was first described from that formation by the Rev. George Young, in his ' Survey of the Yorkshire Coast.' The figure given in that work is very bad, but the description is sufficiently accurate to identify the species: "The dorsal surface has the same elegant markings as No. 5 (*Pygurus pentagonalis*), but the petals are rather awl-shaped than lanceolate, from which peculiarity we name it *Echinites subulatus*. The middle part of each petal forms a slight ridge; on the contrary the five corresponding marks on the base, meeting in the mouth, are depressed. The base is concave, and the mouth is situated immediately under the vertex. The vent is in a short groove on the edge, but more towards the upper surface, as in some of the Spatangus family."*

When I published my memoir on the Cassidulidæ of the Oolites † I could not obtain a specimen of this urchin; most of those contained in the Yorkshire collections are either

---

* Geological Survey of the Yorkshire Coast, 2d ed. p. 214.
† Annals and Magazine of Natural History, 2d series, vol. ix, p. 310, 1851.

crushed or have been much injured in cleaning; fortunately I have obtained two good examples, which have been beautifully figured in Pl. XXXIV, one of these specimens was obtained from a band of clay and has the form and sculpture finely preserved.

The thin test has an oblong form, rounded before, dilated in the middle, and a little rostrated behind (fig. 1 a); the upper surface is uniformly convex (fig. 1 c); the sides decline equally towards the border (fig. 1 d), which is rather attenuated; when the shell is viewed in profile, the margin is seen to be much undulated (fig. 1 c, d).

The ambulacral areas taper gradually from the border to the disc (fig. 1 a), and lie on the same plane with the general surface of the test; the poriferous zones on the dorsal surface are petalloid for three fourths of the space between the border and disc, they lie rather beneath the general plane of the test, and form very uniform graceful petals on the upper surface, narrow below, gradually swelling out in the middle, and tapering to narrow lanceolate terminations at the disc (fig. 1 a); the pores of the inner row are round (fig. 1 e), those of the outer row in the form of long, narrow, transverse slits; the pores of both rows are conjugate through fine sulci; there are eight pairs of pores opposite one of the large plates, and a septum between each transverse slit, the outer surface of which supports a regular row of fine granules (fig. 1 e).

At the lower fourth of the areas the pores lie close together, the rows are here very narrow, and continue so across the base, the pairs of pores in this region, being placed at wider distances apart (fig. 1 b), near the mouth-opening they lie closer together in triple oblique rows in the penta-phylloid floscelles of the ambulacra, which radiate outwards between the lobes in depressions of the test (fig 1 b).

The inter-ambulacral areas are of unequal width; the antero-lateral pair are the narrowest, and the postero-lateral pair, and single area, are of the same width, but broader than the anterior pair; their surface is marked by two slight ridges, which radiate from the disc to the border, and subdivide each area into three segments. In specimens which have been scraped or filed, unfortunately almost universally the case with this Clypeus, these ridges are not seen; but in the small, nearly perfect specimen (fig. 1 c, d), they form a very prominent character in its upper surface, and which is faintly represented in fig. 1 a in the left antero-lateral segment. The inter-ambulacral ridges form distinct prominences at the border, and on the surface of the basal cushions; they likewise form two small carinæ on the deflected basal surface of the small specimen (fig. 1 d). The small tubercles are arranged in rows with great regularity on the plates, of these there are in general four on each; the tubercles are surrounded by sunken areolas, and the intermediate surface is covered with rather large granules (fig. 1 f), which give a highly ornamented sculptured surface to the test, when examined with an inch lens. The tubercles at the base are very much larger than those on the upper surface; some of them likewise are perforated, especially those situated in the concave depression in the vicinity of the mouth (fig. 1 f).

The single inter-ambulacrum forms one of the most remarkable features in the anatomy of the test; the anal valley is short, deep, and limited to the lower third of the dorsal portion of this area (fig. 1 *a*); the posterior part of this segment is rostrated, and much deflected (fig. 1 *d*), and its extremity abruptly truncated (fig. 1 *a, d*); the arch over the vent is narrow above, its sides slope a little inwards, and the opening expands towards the border (fig. 1, *a, d*); the vent opens near the surface beneath the arch (fig. 1 *d*); it has an oval form with its long diameter towards the vertex.

The base is concave (fig. 1 *b*), and the inter-ambulacra form prominent cushions between the narrow ambulacra, the convexity of these segments produces a considerable undulation in the border, which is well represented in fig. 1 *c, d*; the mouth-opening is sub-central, nearer the anterior than the posterior border, it has a pentagonal form, and the peristome is surrounded by five prominent oral lobes (fig. 1 *b*).

The apical disc is very small, it occupies the centre of the upper surface, and forms the vertex of the test; I have only seen one specimen in which this part is preserved. The separate plates are so intimately soldered together, that I can only recognise the four genital holes which perforate the test obliquely (fig. 1 *g*), and the five ocular holes at the summits of the ambulacra; the surface of the discal plates is covered by a spongy madreporiform body which forms the most conspicuous element in the small central disc (fig. 1 *g*).

Affinities and differences.—*Clypeus subulatus* is frequently mistaken for *Pygurus pentagonalis*, Phil., and lies with this name attached, in several Yorkshire collections. In fact, the oblong figure of the test, the apparent absence of an anal valley, which is always filled up with matrix, and seldom cleared out, the form of the ambulacral petals, and the central position of the small apical disc, produce an assemblage of characters which have misled local observers. M. Desor, in his synopsis, has placed this Clypeus in the genus Pygurus, and has added this note to his diagnosis—"Cette position exceptionnelle du périprocte n'est pas une raison suffisante pour éloigner cette espèce au *Pygurus* auxquels elle correspond par tours ses autres caractères." I can readily understand how this mistake has been committed, if M. Desor's opinion was formed from Phillips's figure, the only one up to the present time worthy of the name of a sketch which has been published. I trust, however, that Mr. Bones's beautiful plate, with its ample details will set the question at rest, for there can be no doubt that this urchin is a true Clypeus. It has no affinity with any other English congener; its concave under surface, and strongly-cushioned basal inter-ambulacra, resemble the base of *Clypeus altus*; but that species has an orbicular outline, and a long, narrow, anal valley, which widely separate it from *Clypeus subulatus*. It more nearly resembles *Clypeus Rathieri*, Cott., from the Forest marble of Châtel-Gérard (Yonne), than any other species; but the narrowness of the anterior border, the flatness of the base, and the excentral position of the disc, prove how specifically distinct they are from each other. I may add, that the small specimen from the clay band (fig. 1), is the only example I have yet seen which shows the basal surface, now published in fig. 1 *b*, for the first time.

Locality and Stratigraphical position.—This urchin has hitherto been collected only from the Coralline Oolite of Malton and Scarborough, although I have seen fragments of it in the Ayton quarries. It is usually imbedded in a white Oolitic limestone, firmly attached by the base in consequence of its inequalities, and, having the upper surface exposed, the matrix is usually removed by scraping; but the Oolitic grains frequently indent the surface of the plates. This species was formerly more abundant, than now, probably because the vein containing the fossil was worked more in former years. Almost all the specimens are fractured and crushed, and it is rare to find one which preserves the true type-form of the species. For the same reason we seldom observe the ridges which radiate from the disc to the mouth-opening, on the surface of each inter-ambulacral segment. I have never seen them distinctly, but in the two specimens I have figured, where they undoubtedly exist, and form a good diagnostic character for the species.

History.—It was first figured and described as *Echinites subulatus*, by Young and Bird, in their ' Geological Survey of the Yorkshire Coast,' and afterwards by Professor Phillips, as *Clypeus emarginatus*, in his ' Geology of Yorkshire;' the first name must therefore be retained. Professor Forbes gave a diagnosis of the species in his ' Notes on British Oolitic Nucleolites,' in the first decade of his ' Memoirs of the Geological Survey.' It was imperfectly described, for want of specimens, in my ' Memoir on the Cassidulidæ of the Oolites;' but is now figured, with full details, from two fine specimens, for the first time.

NOTES

On Foreign Jurassic species of the genus CLYPEUS nearly allied to British forms, but which have not yet been found in the English Oolites.

Clypeus solodurinus, *Agassiz.* Syn. *Échinoderm, Foss, Suisse* I, p. 35, tab. v, figs. 1—3.

Test very thick, narrow, and elongated, posterior border angular, produced, and truncated; ambulacra narrow, poriferous zones petaloidal, contracted at the lower fourth, apical disc central; anal valley lanceolate; vent in its upper part, near the apical disc, which is small and nearly central; border undulated, base nearly flat, with prominent inter-ambulacral cushions; mouth-opening large, peristome surrounded by five lobes.

Dimensions.—Antero-posterior diameter, two inches and nine tenths; transverse diameter, two inches and eight tenths; height, eight tenths of an inch.

Formation.—" Oolite Vesulienne, d'Obergoesschen (Jura Soleurois), Egg (Argovie); Plasne près Poligny (Jura)." Desor.

Collections.—MM. Strohmeyer, Bronn, Marcou.

Clypeus Boblayi, *Michelin.* Cotteau and Triger. Échinides du depart. de la Sarthe, pl. xi, figs. 4, 5; p. 64.

Test moderate size, sub-circular, round before, slightly truncated behind; dorsal surface much depressed, almost flat; under surface sub-concave. Summit very excentrical posteriorly; anterior ambulacral areas much longer than the posterior. Poriferous zones widely petaloid on the upper surface, and becoming abruptly contracted near the border, where the pores are simple and non-conjugate; anal valley narrowly lanceolate, and becoming regularly wider from the disc to the border.

Dimensions.—Height, three quarters of an inch; antero-posterior diameter, three inches and nine tenths; transverse diameter, four inches and one tenth.

Formation.—Great Oolite, near Mamers Sarthe.

Collection.—M. Michelin ; very rare.

CLYPEUS RATHIERI, *Cotteau.* Échinides Foss., de l'Yonne, pl. vi, p. 71.

Test elongated, very flat on the dorsal surface, and concave at the base; anterior border rounded, and sub-truncated; posterior half of the test much wider and thicker; posterior border rostrated. Ambulacral areas widely petaloid in the middle of the dorsal surface, narrowly lanceolate near the disc, and contracted near the border; anal valley short, narrow, and marginal; occupying the lower third of the inter-ambulacrum, and forming a groove in the posterior border; mouth-opening small, sub-central; peristome surrounded by five lobes.

> *Dimensions.*—Height, six tenths of an inch; antero-posterior diameter, two inches and a half; transverse diameter, two inches and a quarter.

> *Formation.*—Siliceous beds of the Forest marble (Great Oolite) at Châtel-Gérard, Yonne.

> *Collection.*—M. Rathier ; very rare.

CLYPEUS OSTERWALDI, *Desor.* Synopsis des Échinides Foss. p. 277.

Test large, depressed, enlarged, and rostrated posteriorly; anal valley very narrow, extending to the apical disc, which is central. Poriferous zones, in width about one half the ambulacral area; base, much undulated; mouth-opening excentral; peristome, with five lobes and a penta-phylloid rosette, in which additional pores are arranged in double and triple oblique ranks.

> *Formation.*—" Great Oolite (Bathonien) de Noiraigue (Canton de Neuchâtel) au dessous des marnes à Discoidées. Abondant." Desor.

> *Collections.*—M. Gresley, Mus. de Neuchâtel.

CLYPEUS ROSTRATUS, *Desor.* Synopsis des Echinides Foss. p. 278.

Test elevated, sub-conical; posterior border much rostrated; anal valley very much inclined, and almost vertical; ambulacral petals less elongated than in the preceding species; base concave, much undulated; mouth-opening excentral; peristome with three small lobes.   Type of the species, T. 4.

Formation.—" Marnes à Discoidèes (Vesulien) du Kornberg près Frick et de Hornussen (Argovie)." Desor.

Collections.—Mus. Bâle, MM. Moesch, Schmidlin.

Family 10. ECHINOLAMPIDÆ, *Wright*, 1855.

NUCLEOLIDÉES (pars), *Albin Gras.*, 1848.
ECHINOBRISSIDÆ (pars), *D'Orbigny*, 1855.
ECHINOLAMPASIDÆ (pars), *Grey*, 1855.
CASSIDULIDES (pars), *Desor*, 1858.

The family ECHINOLAMPIDÆ includes all the urchins which have petaloid or sub-petaloid ambulacra; the vent, supra-marginal, marginal, or infra-marginal, opening at the surface of the test, and not into an anal valley; the mouth edentulous; the peristome surrounded by five oral lobes, with which petaloidal expansions of the basal ambulacra alternate

The thin test has in general an oval, oblong, sub-pentagonal, or orbicular form, the upper surface is convex, and depressed, elevated, or conoidal; the vertex is usually excentral, and situated nearer the anterior border.

The ambulacral areas and poriferous zones form elegant leaf-like expansions on the dorsal surface, and miniature petals at the base, where they develope an "oral rosette," or a penta-phylloid floscelle around the mouth; the leaves alternate with the prominent peristomal lobes, in which the pores are arranged in crowded oblique ranks.

The small apical disc is composed of a single imperforate, and four small perforated ovarial plates; the madreporiform body is proportionally large, extending over the other discal elements; the five ocular plates are very small, with marginal orbits.

The vent always opens at the surface of the test, and never into a valley, as in the ECHINOBRISSIDÆ. The opening has an oblong form; its long diameter corresponding with the transverse diameter of the test in some genera, and with the longitudinal axis in others; it occupies a marginal, supra-marginal, or infra-marginal position. In some existing species this aperture is closed by three thin, shelly valves, covered with tubercles; the lateral valves are larger and triangular, the central one linear, erect.*

The tubercles are often perforated, and surrounded by sunken areolas; in *Echinolampas* and *Pygurus*, the inter-tubercular surface is covered with a close set granulation, and the tubercles at the base are much larger than those on the upper surface.

The ECHINOLAMPIDÆ form a natural group, nearly equivalent to the *Echinanthi*, of Breynius,† and which that author thus defined:

" ECHINANTHUS *est Echinus cujus apertura pro ore est prope centrum, pro ano in, vel ad marginen, longissime ab ore distantem.*"

Figuram omnes hujus species habent *ovalem*, cujus altera extremitas angustior, altera latior, in qua semper apertura pro ano observatur. Ceterum pori in vertice *schema* efformant *floris* cujusdam *penta-petali*, quasi acu artificiose delineatum; et hæc ratio est cur huic Generi *Echinanthi* nomen imposuerim.

---

* 'Gray's Catalogue of the Recent Echinida of the British Museum,' p. 35, 1855.
† 'De Echinus et Echinites,' p. 59.

This definition unfortunately embraces forms which appertain to several genera. Even the three species figured as *Echinanthus*, in Tab. iv, of Breynius's work, from the position of the vent, represent two distinct types.

The ECHINOLAMPIDÆ resemble the ECHINOBRISSIDÆ in the general organization of their test; but the species are distinguished from those of that group by the position of the vent, the development of the peristomal lobes, the basal ambulacral rosette, and by the tubercles, in some genera at least, being perforated.

In the following table I have given a short diagnosis of the genera, and include in the family ECHINOLAMPIDÆ, together with their stratigraphical distribution in time.

A TABLE SHOWING THE CLASSIFICATION AND DISTRIBUTION OF THE ECHINOLAMPIDÆ.

FAMILY.	DIAGNOSIS.	GENERA.	FORMATION.
ECHINOLAMPIDÆ.	Test, oblong, inflated; petals long, narrow, nearly reaching the border; vent transverse, supra-marginal.	PYGORHYNCHUS, *Agassiz.*	Tertiary.
	Test oblong, depressed; petals short, narrow, limited to the dorsum; vent oval, marginal, or supra-marginal.	ECHINANTHUS, *Breynius.*	Tertiary and Upper Cretaceous.
	Test ovoid or discoid; petals wide, long, nearly reaching the border; vent, transverse, infra-marginal.	ECHINOLAMPAS, *Grey.*	Living and Tertiary.
	Test large, discoidal, or elevated; petals large, long, reaching the border; vent infra-marginal, longitudinal, surrounded by a distinct area.	PYGURUS, *Agassiz.*	Cretaceous and Oolitic.
	Test oblong, depressed; petals long, narrow, reaching the border; vent longitudinal, marginal.	BOTRIOPYGUS, *D'Orbigny.*	Cretaceous.
	Test elevated, or conical; petals narrow, lanceolate, short, limited to the dorsum; vent small, transverse, without area; base flat, without elevation.	FAUJASIA, *D'Orbigny.*	Upper Cretaceous.
	Test large, oval or circular, much elevated, or conoidal; petals long, straight, wide, equal, not contracted below; vent infra marginal, triangular; base flat.	CONOCLYPUS, *Agassiz.*	Tertiary and Cretaceous.

*Genus—*PYGURUS, *d'Orbigny,* 1855.

ECHINANTHITES, *Leske,* 1778.
CLYPEASTER (pars), *Lamarck,* 1801.
ECHINOLAMPAS (pars), *Agassiz,* 1836.
PYGURUS (pars), *Agassiz* 1840.
PYGURUS, *d'Orbigny,* 1855.
PYGURUS, *Desor,* 1858.

The genus *Pygurus,* as now limited, is composed of large, discoidal, or clypeiform urchins, in which the test in general is more or less enlarged at the sides, and rostrated posteriorly; its upper surface is usually depressed, and rarely elevated.

The ambulacral areas and poriferous zones in the upper surface form petaloidal expansions, which have an elegant form, being in general contracted at the border, enlarged in the middle, and attenuated at the apex. The anterior single area is narrower than the antero- and postero-lateral areas; the summit is in general central, or slightly excentral, the inclination being always forwards.

The base is concave and much undulated, the wide basal inter-ambulacra swell into prominent cushions, and the narrow ambulacra form narrow valleys between them.

The mouth-opening is pentagonal, and always excentral; the peristome is surrounded by five prominent lobes, with which five expanded ambulacral petals alternate; in the poriferous zones near the mouth the pores are closely crowded in triple oblique ranks; these perforated petals form an oral rosette or a penta-phylloid floscule (Pl. XXXVII, fig. 1 *b, e*).

The vent is infra-marginal; it is in general oval, and surrounded by a distinct area, which occupies the rostrated portion of the single inter-ambulacrum; the long diameter of the opening in general corresponds with the longitudinal axis of the test, although it is sometimes transverse (Pl. XXX, fig. 2 *b*).

The apical disc is very small, and occupies the summit; it is composed of two pairs of narrow, perforated, and a single rudimentary imperforate, ovarial plate; five minute ocular plates, with central eyeholes, are interposed between the ovarials (Pl. XXXV, fig. 3). The small madreporiform body is attached to the surface of the right anterior ovarial, and forms thereon a spongy eminence, which extends over the other discal elements (fig. 3 *g*).

The tubercles are very small on the upper surface, but larger at the base; they are surrounded by sunken areolas, have their summits perforated, and the inter-tubercular space covered with close-set miliary granules (Pl. XXXV, fig. 2 *e, g*).

51

PYGURUS MICHELINI, *Cotteau.* Pl. XXXV, fig. 2 *a, b, c, d, e, f, g.*

PYGURUS MICHELINI.		Cotteau, Études sur les Échinides Foss. de l'Yonne, p. 70, pl. v, fig. 7, 1849.
—	PENTAGONALIS.	Wright, Ann. and Mag. of Nat. Hist., 2d ser., vol. ix, pl. iv, fig. 3, p. 313, 1851.
—	—	Forbes, in Morris's Catalogue of British Fossils, 2d ed., p. 88, 1854.
—	MICHELINI.	D'Orbigny, Pal. Franç. ter. Crétacés, t. vi, p. 301, 1855.
—	DAVOUSTIANUS.	Davoust., Note sur les Foss. spéciaux à la Sarthe, p. 6, 1856.
—	MICHELINI.	Desor, Synopsis des Échinides Fossiles, p. 315, 1857.
—	—	Cotteau et Triger, Échinides du département de la Sarthe, pl. xiii, figs. 1—5, p. 65, 1858.

Test oval, or sub-pentagonal, very slightly indented before and rostrated behind ; upper surface convex, under surface concave, with prominent, cushioned, basal inter-ambulacra ; apical disc nearly central ; ambulacral areas and poriferous zones widely petalloid on the upper surface, contracted at the lower fifth, and lanceolate at the apex ; inter-ambulacra with two flat ridges, which in each area extend from the disc to the mouth ; vent elliptical, infra-marginal, lodged in a deep anal depression, with inclining sides ; mouth-opening large, pentagonal, excentral ; peristome surrounded by five prominent oral lobes and five depressed phylloidal floscules.

Dimensions.—Height, one inch and one tenth ; antero-posterior diameter, three inches and one tenth ; transverse diameter, three inches and one tenth.

Description.—The first specimen of this *Pygurus* I obtained was an elongated subpentagonal variety, resembling in outline some of the Yorkshire specimens, and which I erroneously referred, in my 'Memoir on the Cassidulidæ of the Oolites,' to *P. pentagonalis,* Phil. ; since that time I have collected a very fine series of this urchin, which I have carefully compared with M. Cotteau's beautiful figures, and have no hesitation in referring them to *Pygurus Michelini,* Cott. Our English examples are larger than those from the Sarthe, but in all the details of their anatomy they are identical with that form.

The test is nearly orbicular ; in some specimens it is longer than broad, flattened, or slightly concave before, and produced or rostrated behind ; the upper surface is convex, and rises to a prominent vertex (fig. 2 *c, d*), which is sub-central, and from which the sides decline unequally ; in consequence of the prominence of the single ambulacrum, the posterior side forms a more regular inclined plane than the anterior side (fig. 2 *d*) ; the border is very much undulated, and the base concave.

The ambulacral areas are widely petalloid on the upper surface, narrow at the border and base, and again expanded near the mouth ; they are sharply lanceolate at the apex, and closely approximated at the disc (fig. 2 *a*) ; the poriferous zones on the upper surface are formed of an inner row of round holes and an outer row of oblique, slit-like apertures, the length of which gradually diminish from the middle of the petal upwards to the apex, and downwards to about the lower fifth of the area, where the oblong pores are reduced to simple pores, like those of the inner row (fig. 2 *d*); at the border the pairs of pores lie close together, at the base they are much wider apart (fig. 2 *b*), and about half an inch from the mouth they again greatly increase in number, and in the depressed phylloidal floscule form a regular series of triple oblique pairs (fig. 2 *b*), with twelve rows in each zone.

The inter-ambulacral areas are uniformly convex above, the postero-lateral and single inter-ambulacrum are of the same width, and the anterior pair are narrower at the under surface; these segments form prominent cushions, which are very tumid at the border, but less so near the mouth; the most prominent part of each cushion is flattened at the interspace between the ridges which radiate from the disc to the mouth.

The single inter-ambulacrum is produced, rostrated, and deflected; at its infra-marginal border is a deep anal depression, with prominent and inclined sides, at the bottom of which the elliptical anus opens (fig. 2 *b*); the anal valley indents the border, and forms a conspicuous notch in the margin when the test is viewed from behind forwards, as is well shown in fig. 2 *c*, which likewise exhibits the undulations of the base.

The apical disc is remarkable for the size of the madreporiform body and the smallness of the genital plates (fig. 2 *f*); the anterior pair are less than the posterior pair, and the four are perforated; the single plate is posterior, and imperforate (fig. 3); the small rhomboidal ocular plates alternate with the genitals, the very minute eyeholes are perforated in the centre of the plates; the madreporiform body rises from the surface of the right antero-lateral genital, extends over the surface of all the others, and occupies the centre of the disc (fig. 3).

The large mouth-opening is sub-central and forward; the peristome is pentagonal, and surrounded by five prominent oral lobes (fig. 2 *b*); alternating with them are the five depressed phylloidal terminations of the ambulacra, in which numerous pores are arranged in triple oblique rows; this crowding together of the pores in regular order, imparts an ornamental character to the five oral ambulacra.

The tubercles on the upper surface are very small, on each plate they are arranged in four or five tolerably regular horizontal rows, and surrounded by sunken areolas; the inter-tubercular surface is covered with microscopic granules, placed so close together that all the intermediate portion of the plates, when examined with a low magnifying power, is seen to have a finely sculptured appearance. At the base the tubercles are larger, and disposed with less regularity; they increase in size, and are set closer together at the border and on the convex surface of the five basal cushions, whilst they are still larger and placed wider apart near the mouth-opening and on the sides of the ambulacra.

Affinities and differences.—This species has been frequently confused with *Pygurus depressus*, Ag., which it very much resembles; according to M. Cotteau, it is distinguished from that species by the following characters, " Par sa forme plus oblongue, plus allongée, par ses ambulacres relativement plus larges, se rétrécissant moins brusquement et logés, aux approches du péristôme, dans des sillons plus droits et plus prononcés, par sa face inférieure moins déprimée. Ces deux espèces caractérisent d'ailleurs un horizon différent ; le *Pygurus Michélini*, propre aux couches de la Grande Oolite, se rencontre associé à *l'Echinobrissus clunicularis*, au *Collyrites ovalis*, à *l'Hyboclypus gibberulus*, tandis que le *Pygurus depressus* se trouve dans le Kelloway ferrugineux avec le *Collyrites elliptica*, *l'Echinobrissus Gold-fussii*, le *Pseudodiadema Calloviense*, etc."*

Pygurus Michelini, Cott., resembles *Pygurus pentagonalis*, Phil., in its general outline ; the former, however, is more depressed on the upper surface and more undulated at the base, the anterior border is more concave, and the single inter-ambulacrum more rostrated and deflected.

Locality and Stratigraphical position.—I found one small example of this species in the Great Oolite at Minchinhampton, where it is excessively rare. I have collected beautiful specimens in the Cornbrash near Trowbridge, Wilts, and my kind friend, the Rev. A. W. Griesbach, presented me with the magnificent specimen figured at Pl. XXXV, which he obtained from the Cornbrash at Rushden (Northamptonshire). I have two inferior specimens from the Cornbrash near Yeovil. The specimen of this species I first figured was said to have been found in the Lower Trigonia Grit (zone of *Ammonites Parkinsoni*) at Shurdington Hill, near Cheltenham. This urchin has been collected from a marly bed of Cornbrash near Fairford ; it was found by my friend, J. Lowe, Esq., in the Cornbrash at Wincanton (Somerset), and I have seen specimens which were obtained from the Bradford clay at Bradford, Wilts. It seems therefore that the range of this species, like many other Echinodermata which first appeared in the zone of *Ammonites Parkinsoni*, Inferior Oolite, extended onwards through Fuller's earth, Great Oolite, Bradford clay, and Forest marble, into the Cornbrash, in which formation they all became extinct.

The foreign distribution of this *Pygurus* is the same as in our English Oolites. M. Bouchard-Chantereaux kindly sent me a specimen which he collected from the Great Oolite near Boulogne-sur-Mer.

It has been collected by M. Triger at " Monné (carrière de Bernay), La Jaunelière, Hyéré, Noyen, Pécheseul, route de Coutilly, route de Suré à Mortagne.

" Tabl. de M. Triger, Bradford Clay, Ass. No 1, et Forest marble, Ass. No. 4."

It has been found by M. Cotteau in the " Grande Oolite de Asnières, Châtel-Gérard (Yonne)," and by M. Desor in the " Marnes Vésuliennes du Jura Soleurois et Argovien."†

* Cotteau et Triger, ' Échinides du département de la Sarthe,' p. 67.
† Ibid.   Ibid.   p. 67.

PYGURUS PENTAGONALIS, *Phil.* Pl. XXXVI, fig. 1 *a, b, c, d ;* fig. 2 *a, b.*

ECHINANTHITES ORBICULARIS.	Young and Bird, Geol. Surv. of Yorksh. Coast, pl. vi, fig. 5, p. 213, 1822.
CLYPEASTER PENTAGONALIS. .	Phillips, Geology of Yorkshire, pl. iv, fig. 24, 1829.
ECHINOLAMPAS PENTAGONALIS.	Morris, Catalogue of British Fossils, p. 52, 1843.
PYGURUS PENTAGONALIS.	Forbes, in Morris's Catalogue of Brit. Fossils, 2d ed. p. 8, 1854.
— —	Desor, Synopsis des Échinides Fossiles, p. 314, 1858.

Test sub-pentagonal, emarginate, and concave anteriorly, wide in the middle, produced, and deflected posteriorly ; upper surface convex, with a conical vertex, ambulacral areas and poriferous zones widely petalloid in the upper two thirds of the dorsal surface, and very narrow in the lower third ; apical disc small, central, forming the vertex of the test ; under surface concave ; basal inter-ambulacral cushions moderately prominent ; mouth-opening small, sub-central, forwards ; peristome pentagonal, with five mammillated oral lobes and five narrow ambulacral phylloidal floscules ; vent elliptical, infra-marginal, situated in a deep anal valley.

Dimensions.—A. Height, one inch and four tenths ; transverse diameter, three inches and three quarters ; antero-posterior diameter, three inches and nine tenths.

B. Transverse diameter, three inches and a half ; antero-posterior diameter, three inches and four tenths.

C. Height, one inch and one fifth ; transverse diameter, three inches ; antero-posterior diameter, three inches.

Description.—I have given the measurements of three different Yorkshire specimens of this *Pygurus.* The specimen A was collected from the Coralline Oolite at Hildenley, near Malton, by C. W. Strickland, Esq. ; the specimen B was obtained near Scarborough, and belongs to the Scarborough Museum ; and the specimen C was collected from the Lower Calcareous Grit near Scarborough. The examples from the Coralline Oolite are in general much larger than those found in the Lower Calcareous Grit.

The test has an orbicular or sub-pentagonal form ; the anterior border is emarginate and concave ; the antero-lateral border expands outwards to the middle of the test, where its greatest diameter is attained. The postero-lateral border slopes inwards and backwards, and forms a rostrated and deflected termination. This is the form of the urchin figured in fig. 1 *a, b,* from the Calcareous Grit, which may be taken as a type of the species. In the one from the Coralline Oolite (fig. 2 *a*) the test is altogether orbicular, and more convex.

The ambulacral areas are remarkably petalloid (fig. 1 *a*) on the upper surface, the lower third of the area is very narrow, the middle third much expanded, and the upper third lanceolate. The poriferous zones on the dorsal surface are formed of an inner row of oblong holes, and an outer row of oblique, slit-like apertures, both rows being conjugate through very fine sulci (fig. 2 *b*); at the lower fourth of the zone the slit-like outer pores contract into simple pores like those of the inner row; at the border the pairs of pores lie close together, whilst at the base they are wide apart; near the mouth-opening the ambulacra form phylloidal expansions, in which the pores lie crowded in triple, oblique pairs; the mammillated character of the oral lobes occasions a considerable contraction of the areas around the peristome (fig. 1 *b*).

The inter-ambulacral areas on the upper surface are uniformly convex; the anterior pair are narrower than the posterior pair; the single area, which is about the same width of the latter, is produced, rostrated, and deflected (fig. 1 *c*), and in some varieties forms a caudal prolongation; at the under surface, which is concave, the basal inter-ambulacra are moderately convex; the postero-lateral pair and single inter-ambulacrum are more so than the anterior pair; the anal valley is deep, with inclined sides, and the vent is large and elliptical (fig. 1 *b*).

The mouth-opening is small, sub-central, and forwards; it lies at the most concave part of the base; the peristome is pentagonal, and the five oral lobes, formed by the terminations of the inter-ambulacra, have a prominent, mammillated character, which occasions a contraction of the ambulacra at the point where they join the peristome; the protrusion of the lobes removes the mouth itself far from the surface; the emargination of the anterior border is more conspicuous at the base, and the anteal sulcus occasions a considerable depression in the anterior border.

The tubercles are very small, and arranged in close-set rows, of which there are five or six on each plate (fig. 2 *b*); the areolas are well defined; the tubercles at the base are much larger than those on the upper surface; they are more developed at the border, on the declining sides of the cushions, and near the mouth, than on the convex surface of these prominences, where they are small and closely aggregated together; the tubercles here are distinctly perforated (Pl. XXXV, fig. 2 *g*), whilst the intermediate surface is covered with close-set microscopic granules. Few portions of the specimens I have seen are sufficiently well preserved to show the sculpture; in those from the Calcareous Grit it is almost always effaced, whilst in those from the Coralline Oolite it is concealed by the Oolitic Coralline mud in which the most of them is enveloped.

The apical disc is small and central (fig. 1 *d*); the madreporiform body, which is round, convex, and prominent, forms the vertex, the large ambulacral petals covering nearly all the other portion of the upper surface. In fig. 1 *c* the dorsum forms a conical eminence through the development of the petals and disc (fig. 1 *a*).

Affinities and differences.—This species so closely resembles *Pygurus Michelini*, Cott., that it is only after a careful analysis of its specific characters the differences between them are discovered. *Pygurus pentagonalis* is more convex on the upper surface, and more concave at the base ; the ambulacral petals on the dorsal surface are petalloid only in the upper two thirds of their length, and very narrow and graceful in the lower third of the areas (fig. 1 *a, c* ; fig. 2 *a*) ; the anterior border is emarginate and concave, and the under side is impressed by the anteal sulcus ; the posterior inter-ambulacrum is rostrated and deflected ; the mouth-opening is small, and the oral lobes large and prominent. When these characters, which are permanent in all the examples I have examined, are compared with the homologous parts in *Pygurus Michelini*, which have been already described, the distinction between the species will be readily determined. (Compare Pl. XXXV and XXXVI.)

Locality and Stratigraphical position.—I have collected this urchin from the Lower Calcareous Grit at Bullington-green, near Oxford, at Farringdon, Berks, from the same rock at Scarborough, Castle Hill, and Gristhorpe Bay, on the Yorkshire coast. It is likewise obtained from the Coralline Oolite at Scarborough and Ayton, and at Malton and Hildenley ; from the latter locality I have been enabled to study a fine large specimen, kindly presented to me by C. W. Strickland, Esq., and beautifully developed by him. My kind friend, John Leckenby, Esq., communicated the beautiful specimen figured in Pl. XXXVI, fig. 2 *a*, which came from the Coralline Oolite of Malton, and belongs to the Scarborough Museum. The small specimen (fig. 1 *a, b*) was collected by my friend, Dr. Murray, of Scarborough, from the Lower Calcareous Grit, near that town, and generously given to me by him for this work ; *Pygurus pentagonalis* is, therefore, a true Corallian form, and a most characteristic urchin of this formation, both in the Midland Counties as well as in Yorkshire.

PYGURUS COSTATUS, *Wright*, nov. sp. Pl. XXXVII, fig. 1 *a, b, c, d, e, f.*

Test sub-pentagonal, discoidal, much depressed at the upper surface ; ambulacral petals large, costated, extending over four fifths of the upper surface ; anterior border flat, emarginate ; posterior border rostrated ; postero-lateral border very thin ; apical disc excentral, forwards ; base flat ; mouth-opening large, sub-central ; peristome surrounded by five large oral lobes and five spoon-shaped phylloidal floscules.

Dimensions.—One large specimen.—Height, one inch and one tenth ; transverse diameter and antero-posterior diameters, equal, four inches.

Specimen, Pl. XXXVII.—Height, nineteen twentieths of an inch ; transverse diameter, three inches and three tenths ; antero-posterior diameter, three inches and three tenths.

Description.—This urchin is remarkable for its discoidal form and for the prominent, costated character of its dorsal ambulacra ; the anterior border is flat, the sides form obtuse angles near the middle of the disc, and the posterior border is rostrated, but not deflected ;

the upper surface is very much depressed, and remarkable for the costated character of the ambulacral areas, which are widely petalloid four fifths of the distance between the disc and border (fig. 1 *a*).   The poriferous zones consist of an inner row of oblong pores, and an outer row of oblique, slit-like pores, conjugate throughout by fine sulci (fig. 1 *d*); the transverse pores are separated by thin septa, on the surface of which a series of minute granules are arranged; there are from seven to nine pair of pores opposite each large inter-ambulacral plate (fig. 1 *d*).   Near the border the pores are simple, and set close together; they continue so round the marginal fold; at the base they are placed wide apart, from their ambulacral plates being large and rhomboidal; on the outer two thirds of these segments, near the mouth, the areas suddenly expand (fig. 1 *b*), and between the oral lobes they again as suddenly contract, forming thereby spoon-shaped depressions around the mouth (fig. 1 *e*).   On the sides of these depressions the poriferous zones present a remarkable development (fig. 1 *e*); the pores form three crescentic rows on each side of the depression, between the single pair and the marginal contraction (fig. 1 *e*).   The lobes are very large and mammillated, and form considerable eminences around the mouth, their lip-like forms extending over the border of the peristome (fig. 1 *f*).

The inter-ambulacral areas are of unequal width; the antero-lateral pair are the narrowest, the single posterior rostrated area the widest, and the postero-lateral pair of intermediate dimensions; their upper surface is almost uniformly sloped, and their basal portions extremely flat; in most Pyguri the prominence of the inter-ambulacral cushions might be considered a generic character; in this species, however, the convexity of these basal segments is very inconsiderable (fig. 1 *c*).

The mouth-opening is large and sub-central, nearer the anterior border; the peristome is pentagonal, and surrounded by five mammillated prominent lobes, which project, with lip-like processes, over the oral opening (fig. 1 *b*, *f*); between the lobes the ambulacra are much contracted, and beyond the lobes they form wide phylloidal expansions, on the sides of which the poriferous zones consist of three concentric rows of holes (fig. 1 *e*); the structure of the oral lobes, phylloidal ambulacra, and trigeminal pores, form the most remarkable features in this disciform species.

The apical disc is small and sub-central, it forms the vertex of the test (fig. 1 *c*), and the centre of a conoidal elevation, occasioned by the costated character of the ambulacral areas; the disc consists of four small perforated genital, a single imperforate genital, and five very small ocular, plates; the small madreporiform body rises from the surface of the right antero-lateral genital, covers the surface of the other plates, and forms a round spongy prominence in the centre of the disc (fig. 1 *a*).

The dorsal tubercles are very small, and arranged in four or five rows on each plate; they are surrounded by well-defined sunken areolas, and all the intermediate surface is occupied by microscopic miliary granules, which are likewise encircled by sunken areolas; when viewed with an inch object-glass, the plates are seen to possess a delicately sculptured surface (fig. 1 *a*); the basal tubercles on the sides of the inter-ambulacra, and the convex part of the lobes are fewer in number, and larger in size (fig. 1 *b*), than those of the dorsum, and their surface is perforated; the tubercles on the convex portions of the inter-ambulacra are very small, and placed so close together that the borders of the areolas form hexagonal figures; the basal ambulacra are destitute of tubercles, and the plates are covered only with miliary granules; the nakedness of the surface clearly displays the size of the plates, and their peculiar figure, with the distant pores in the narrow zones in this region of these segments.

The oblong vent is infra-marginal, and the anal valley is shallow, with sloping sides, on which some larger tubercles are disposed.

Affinities and differences.—This species resembles *Pygurus orbiculatus*, Leske, so beautifully figured by M. Cotteau; *Pygurus costatus*, however, has a much larger test, it is more depressed at the upper surface, and has more prominent, costated, ambulacra; the pentagonal border, is more rostrated behind, and more emarginate before; the larger mouth-opening has more prominent lobes and larger phylloidal ambulacra, and the test presents an ensemble of characters by which these two allied forms may readily be distinguished.

Pygurus costatus resembles *Pygurus Marmonti*, Beaud., from the Kelloway ferrugineux of the Sarthe, in the general disciform shape of the test and flatness of the base, but in the specimen of *P. Marmonti* before me the poriferous zones are petaloid to the border, the vent is removed inwards some distance from the margin, and the test has, moreover, an orbicular circumference.

The depression of the dorsum, thinness of the border, angularity of the sides, flatness of the base, and prominence of the dorsal ambulacra, clearly distinguish *Pygurus costatus* from its other Oolitic congeners, and I am unacquainted with any other foreign form beside those enumerated with which to compare our urchin.

Locality and Stratigraphical position.—This species was collected from the Lower Calcareous Grit of Oxfordshire and Wiltshire; the specimen I have figured was found near Oxford, and I have another from the same rock near Calne. It is not a common species, as I have rarely seen it in collections of Calcareous Grit fossils. The specimen I have figured was most kindly given to me for this work by my friend, the Rev. P. B. Brodie.

PYGURUS BLUMENBACHII, *Koch* and *Dunker*.   Pl. XXXVIII, figs. 1 and 2.

CLYPEASTER BLUMENBACHII.	Koch and Dunker, Norddeutschen Oolithgebildes, pl. iv, fig. *a, b, c,* p. 37, 1837.
PYGURUS BLUMENBACHII.	Agassiz and Desor, Catalog. raisonné des Échinides, Annales des Sciences Naturelles, 3ᵉ série, t. vii, p. 162, 1847.
— —	D'Orbigny, Prodrome de Paléont. Stratigr., t. i, p. 26, étage 14ᵉ, 1850.
— —	Wright, Ann. and Mag. of Nat. Hist., 2d series, vol. ix, p. 312, 1851.
— —	Forbes, in Morris's Catalog. of Brit. Fossils, 2d ed., p. 88, 1854.
— —	Cotteau, Études sur les Échinides Fossiles (Yonne), pls. xxxiii and xxxvi, p. 233, 1856.
— —	Desor, Synopsis des Échinides Fossiles, p. 313, 1858.

Test thin, sub-quadrate, with a sinuous border; upper surface elevated anteriorly, gradually declining posteriorly; apical disc excentral forwards, forming the vertex; ambulacral areas and poriferous zones broadly petaloid on two thirds of the dorsal surface; anterior border emarginate and concave; sides crescentic; posterior border produced, rostrated, and much deflected; under surface concave, with prominent basal inter-ambulacra.   Mouth-opening large, sub-central; peristome with five very prominent oral lobes, and five phylloid ambulacral floscules.   Tubercles in general small, but larger on the anterior part and at the base.

Dimensions.—Height, one inch and a quarter; antero-posterior diameter, two inches and one fifth; transverse diameter, two inches and three tenths.

Description.—The Oolitic Pyguri, in general, have a remarkable similarity in their external form; so much so, that it frequently requires a careful examination of their characters to distinguish allied species from each other.   It is, however, altogether different with *Pygurus Blumenbachii*, which forms a remarkable exception to the general rule.   In this singular urchin the outline is sub-quadrate, the anterior border is emarginate and concave, and deeply indented by the central sulcus (fig. 1 *a*, fig. 2 *a*); the lateral parts of the margin are convex; the posterior border consists of a double sinuous line, in the centre of which is the single inter-ambulacrum, this forms a convex, rostrated prominence, slightly deflected downwards (fig. 1 *a*, fig. 2 *a*).   The upper surface presents a most singular profile (fig. 1 *d*); the anterior half is relatively much elevated and turgid, and the posterior half slopes gently downwards to the border.

The dorsal ambulacra are broadly petaloid at the upper half and extremely narrow at the lower half, and their wide, lanceolate apices are closely approximated around the disc (fig. 1 *a*, fig. 2 *a*).

The poriferous zones consist of an inner row of round holes and an outer row of oblique, slit-like apertures, which are limited to the upper half of the rows (fig. 2 *b*); in the lower half, the pores are simple, like those of the inner row; at the border, the holes are so minute they cannot be distinguished; at the base they are placed wide apart (fig. 1 *b*), and near the mouth the ambulacra expand into phylloid expansions, which, near the peristome, are contracted by large oral lobes.

The inter-ambulacral areas are very unequal in width and development; the anterior pair are narrow, convex, and prominent; they rise nearly perpendicular, forming with the base an angle of 80°, and near the vertex curve backwards; the plates on the inner sides of the areas, as well as on the single ambulacrum, carry much larger tubercles than the other dorsal plates (fig. 1 *a*, fig. 2 *a*); the posterior pair and the single area incline to an angle of about 35°. The upper surface thus acquires the remarkable anterior elevation which gives so marked a character to this species, and allies it with a Neocomian form— *Pygurus Montmollini*, Ag. The single inter-ambulacrum possesses a central elevated portion on its upper surface, made more apparent by two lateral depressions commencing at the inner zone of the postero-lateral ambulacra, which gradually rise and blend with the central elevation (fig. 1 *c*); this is continued downwards and backwards, and forms the rostrated portion of the single area, which is slightly deflected at its termination (fig. *c*, *d*).

The apical disc is small and excentral, and nearer the anterior border; it consists of four small, perforated, ovarial plates (fig. 3), a single smaller, imperforate plate, and five very small ocular plates, perforated near their centre; the spongy, madreporiform body rises from the surface of the right antero-lateral plate, extends into the centre of the disc covering the inner portions of the ovarial plates, and having the ocular plates disposed around its circumference.

The under surface is concave, and very much undulated, the ambulacra forming narrow, depressed valleys from the border to the mouth, and the basal inter-ambulacra extremely convex eminences between them (fig. 1 *b*, *c*); near the mouth-opening their terminal portions are developed into five tumid lobes.

The large sub-central mouth-opening is directly beneath the apical disc; it is consequently nearer the anterior than the posterior border; the peristome is surrounded by five oral lobes, which alternate with five phylloid ambulacra, filled with several longitudinal rows of pores (fig. 1 *b*).

The anal valley is a slight depression, formed out of a prominent portion of the basal inter-ambulacrum; it has declining sides, covered with large tubercles, and is quite infra-marginal; the vent is oval, and elongated in the antero-posterior diameter (fig. 1 *b*, fig. 1 *c*).

The tubercles on the dorsal surface are very small, and arranged in five concentric rows on each plate (fig. 2 *b*); they are encircled by sunken areolas, and the intermediate space is covered with close-set miliary granules (fig 2 *d*). On the antero-lateral inter-ambulacra and single ambulacrum the tubercles are considerably larger (fig. 1 *a*, fig. 2 *a*); at the base they are still larger, and their deep areolas form hexagonal cells on different portions of the base.

Affinities and differences.—In its general characters, but more especially in the oblique, tumid, conoidal elevation of the anterior half of its upper surface, *Pygurus Blumenbachii*, Koch, resembles three other congeneric forms—*Pygurus Montmollini*, Ag., *P. Orbignianus*, Cott., and *P. Rogerianus*, Cott., from each of which it is distinguished, however, by specific characters; the first and second are Neocomian, and the latter Kimmeridge species. It differs, according to M. Cotteau, from *Pygurus Montmollini*, Ag., in its greater size, less elevated upper surface, and more rostrated posterior border. It differs from *Pygurus Orbignianus*, Cott.,* equally by its size, by its less conical upper surface and more tumid anterior border, by its petalloid ambulacra being more slender, by its inter-ambulacral tubercles being closer together and more irregularly disposed. *Pygurus Rogerianus*, Cott.,† more closely resembles *P. Blumenbachii*, but it appears to M. Cotteau, who has carefully compared these two species, that *P. Rogerianus* is distinguished from the latter by the test being much longer than it is wide, by the upper surface being more depressed, its tubercles being less numerous, and its intermediate granules disposed in regular and concentric series.‡

The only two English specimens of this urchin which I know are those figured in our plate; the largest belongs to the Museum of Practical Geology, and was collected by the officers of the Geological Survey; the other is in my cabinet. These Pyguri are much smaller than the very fine specimens which my friend M. Cotteau has so well figured and described in his work, hence the comparison which he has made was between these fine large specimens and the other species above enumerated, and which all belong to the secondary rocks of France. M. Cotteau's specimen measures in height 34 millimètres = one inch and nine twentieths; antero-posterior diameter, 87 millimètres = nearly three inches and a half; and transverse diameter, 86 millimètres = three inches and four tenths.

Locality and Stratigraphical position.—The specimen collected by the officers of the Geological Survey was obtained from the Coral Rag at Abbotsbury, Dorsetshire, where it is extremely rare. My specimen was said to have been procured from the Inferior Oolite, near Yeovil, but this I have discovered to be a mistake. I have reason to believe that it was collected from the Lower Calcareous Grit at Bullington Green, near Oxford, associated with *Cidaris Smithii*, Wr., and *Echinobrissus scutatus*, Lamk.

The foreign distribution of this species is as follows: In France it characterises the inferior and superior stages of the Corallien. M. Cotteau collected it in " Calcaire blancs

* 'Catalogue raisonné des Échinides du Terrain Néocomien,' p. 12.

† "Note sur les Échinides de l'étage Kimmeridgien de l'Aube," 'Bull., de Géol. Soc. de France,' 2ᵉ série, t. xi, p. 356.

‡ 'Études sur les Échinides Fossiles du department de l'Yonne,' p. 238.

et Pisolitique" of Châtel-Censoir, and of Coulanges-sur-Yonne, where it is very rare; it is found more frequently in the " Couches Coralliennes supérieures" of Baily, of Thury, and of Tonnerre; the specimens collected from the latter locality by M. Rathier were in fine preservation, and were nearly as large as those found at Thury.

This species has likewise been found by MM. Cotteau and Royer in the " Calcaires à Astartes de l'Aube et de la Haute-Marnes."

The original German specimen was found, according to Koch and Dunker, in the "krystallinischen Dolomitquadern des oberen Korallenkalkes am Waltersberge bei Eschershausen."*

Professor Roemer kindly sent me a specimen of *Pygurus Blumanbachii*, Koch, which was collected from the so-called Portland-Kalk, *zone of Pterocera Oceani*, at Hildesheim, Hanover. This rock Dr. Oppel † considers to be the equivalent of our Kimmeridge clay, and not of the true Portland stage. The occurrence of this urchin in the zone of *Pterocera Oceani* strengthens my learned friend's opinion, as MM. Cotteau and Rathier have already found it in the " Calcaires à Astartes," in l'Aube, and Haute-Marne, which is the true equivalent of the Astartekalke of Lindener Berg, and of the environs of Hildesheim.

PYGURUS PHILLIPSII, *Wright*, nov. sp. Pl. XXXIX, fig. 1, *a, b, c, d.*

PYGURUS PHILLIPSII. Wright, Report on British Oolitic Echinodermata, British Association Reports, p. 402, vol. for 1856.

Test nearly orbicular, rather longer antero-posteriorly than transversely; rounded before, slightly produced behind; upper surface very much depressed; sides rounded; anal valley very near the border; ambulacral areas narrow; poriferous zones petaloid on five sixths of the dorsal surface; apical disc small, nearly central, four rows of tubercles on the large plates; inter-ambulacra with slight central triangular elevations, which occasion corresponding tumidities at the border; anal valley wide and deep, causing an emargination of the posterior border.

Dimensions —Height, one inch; antero-posterior diameter, nearly four inches; transverse diameter, three inches and three quarters.

* Norddeutsch. Oolithgebildes, p. 38.
† Die Juraformation, p. 763.

Description.—The beautiful specimen figured in Pl. XXXIX is the only one of this form I have seen. The upper surface, sides, and outer part of the base, are in fine preservation, but the greater portion of the under side is concealed by the matrix.

The test is thin, and has a sub-orbicular circumference; it is rounded before, and slightly produced behind, the difference between the length and width being only one quarter of an inch.

The dorsal ambulacra are narrow, only slightly expanded in their upper half, and terminating in sharp, lanceolate apices around the disc (fig. 1 *a*); they have six rows of tubercles, disposed alternately on the plates, so that they form double oblique rows, with three tubercles in each (fig. 1 *d*).

The poriferous zones, of moderate width, are petalloid five sixths of the distance between the border and disc; as in all other Pyguri, the holes of the inner row are nearly round, those of the outer row are oblique or nearly transverse slits, which about equal in length the width of one half of the area (fig. 1 *c*); between each slit-like aperture there is a partition of the test, on the surface of which a series of ten granules are very regularly arranged (fig. 1 *d*) in a single row; at the borders the pores lie close together, in single pairs, but they are wider asunder at the base.

The inter-ambulacral areas are of unequal width; the anterior pair are about one sixth narrower than the posterior pair; they are formed of long, narrow plates, which are bent to an obtuse angle in the middle; along the line of these angles the surface of the test is slightly elevated, producing in the middle of each area a triangular elevation, the base of which is at the border, and the apex towards the disc (fig. 1 *a*); the margin, in like manner, exhibits a fulness corresponding with the bases of these elevations.

The apical disc is absent, and the space for its reception is small in comparison with the size of the test.

The tubercles are beautifully preserved in this species (fig. 1, *c*, *d*); on the upper surface each plate carries four horizontal rows, which are arranged in zigzag order above one another; the tubercles are all perforated, and crenulated, and raised on small mammillary eminences; the areolas which encircle them are wide and well defined; a circle of miliary granules surrounds the areolas, and other granules fill up all the intermediate spaces; the granules are surrounded by narrow areolas, which impart a highly sculptured character to the surface of the test.

The anal valley lies so near the posterior border that it produces an emargination thereof (fig. 1 *a*); when viewed from behind, the vent is seen quite in the border of the rostrated portion of the single inter-ambulacrum (fig. 1 *b*).

The tubercles at the border and base are much larger and more prominent than those on the upper surface; and the areolas present a regular hexagonal disposition around the margin and at the base.

I have only seen one specimen of this urchin, embedded on a portion of Coralline Oolite limestone; the test is very thin, and has been fractured, the joint having been closed again by crystallization in the rock; the base is nearly entirely concealed by adhering matrix.

Affinities and differences.—The flatness of the upper surface and the form of the ambulacral areas in this species closely resemble *Pygurus Hausmanni*, but the test is narrower before, wider near the middle, and more rostrated posteriorly, than the usual specimens of that large species. I have, therefore, described it under a distinct name, not, however, without misgivings of its propriety, as a series of specimens might exhibit intermediate links, by which the two forms would blend into one type. This is one of the many difficulties to be encountered in describing new species from single examples, which in the present instance is unavoidable; time and additional specimens, however, will prove how far my doubts are well-founded or otherwise.

Locality and Stratigraphical position.—This species was collected from the Coralline Oolite at Malton, Yorkshire, where it is extremely rare; the specimen I have figured is the only one I have seen.

History.—This urchin was first recorded in my memoir ' On the Stratigraphical Distribution of the Oolitic Echinodermata,' afterwards published as one of the ' Reports of the British Association for the Advancement of Science,' for the year 1856. It is now figured for the first time, and dedicated to my learned friend, Professor John Phillips, of Oxford.

PYGURUS HAUSMANNI, *Koch* and *Dunker.*   Pl. XL ;   Pl. XXX, fig. 2.

CLYPEASTER HAUSMANNI.		Koch and Dunker, Versteinerungen des Ool. Gebirg., tab. iv, fig. 3, p. 38, 1837.
—	—	Leymerie, Stat. Géol. et Min. du dép. de l'Aube, p. 239, 1846.
PYGURUS HAUSMANNI.		Agassiz et Desor, Cat. raisonné des Échinides, An. des Sciences Naturelles, 3ᵉ serie, tom. vii, p. 162, 1847.
—	—	D'Orbigny, Prodrome de Paléontologie, tom. ii, p. 26, 14ᵉ étage, 1850.
—	—	Cotteau, Note sur les Échid. de l'étage Kimmeridg., Bull. Géol. Soc. de France, 2ᵉ serie, tom. xi, p. 317, 1853.
—	—	Forbes, in Morris's Catalogue of British Fossils, 2d ed., p. 83, 1854.
—	—	D'Orbigny, Paléontologie Française Ter. Cretacés, t. vi, p. 301, 1856.
—	—	Cotteau, Études sur les Échinides Fossiles, p. 328, 1856.
—	—	Desor, Synopsis des Échinides Fossiles, p. 314, 1858.
—	GIGANTEUS.	Wright, Oolitic Echinodermata, Report of the British Association for the Advancement of Science for 1856, p. 396.

Test large, sub-circular, sometimes oval, and slightly rostrated posteriorly; upper surface flattened, and much depressed; base sub-concave, rounded anteriorly, and slightly produced posteriorly; ambulacral areas on the upper surface nearly equal-sized and lanceolate; poriferous zones petalloid near to the border; apical disc small, nearly central; inter-ambulacral areas broad and flat, with a very distinct zigzag median suture; margin very thin; base sub-concave; mouth-opening small, situated nearer the anterior than the posterior border.

Dimensions.—A. Antero-posterior diameter, six inches and four tenths; transverse diameter, six inches; height indeterminable.

B. Antero-posterior diameter, five inches and one fourth; transverse diameter, four inches and nine tenths; height, one inch and three tenths.

Description.—This large discoidal urchin is remarkable for the great size it attains; nearly all the specimens I have seen are broken, and more or less imperfect, so that the identification of the species is extremely difficult. Last summer, however, I met with one which had retained the form of its circumference, as well as the shape of its upper surface, and this example enabled me to identify the species I had formerly named *Pygurus giganteus* with Koch and Dunker's *Clypeaster Hausmanni.* It is, therefore, extremely interesting to find this urchin in the same horizon of the Coralline Oolite of Malton, the zone of *Cidaris Blumenbachii,* the one it occupies in the Korallenkalk of northern Germany.

Pygurus Hausmanni has in general a sub-circular outline, rather inclining to an oval, its transverse diameter being always less than its antero-posterior measurement; the anterior border is rounded, and in specimen B the posterior border is a little produced; the upper surface is moderately convex in the smaller specimen, but is very much flattened in the larger ones, and the anterior half is more convex than the posterior half.

The ambulacral areas are narrow and lanceolate; they have six rows of small tubercles in their widest part, which are not all arranged in a horizontal series on the two corresponding plates of the area, but are disposed thereon so as to form oblique V-shaped rows. Plate XXXIX, fig. 2, exhibits this arrangement of the tubercles. The poriferous zones are moderately wide, the holes of the inner row are round, those of the outer row are slit-like, of which there are eight pairs opposite each large plate (fig. 2); the septum between each pair of holes supports on its upper surface a horizontal row of nine small granules. The ambulacral areas and poriferous zones form together a series of five elegant leaf-shaped petals, which are enlarged in the middle, become lanceolate near the disc, and are contracted at the circumference; the poriferous zones are petaloidal six sevenths of their length; and near the lower seventh the pores approximate; in their course round the margin, and across the base they remain close together in pairs.

The inter-ambulacral areas are of unequal width, the anterior pair are the narrowest, and the posterior pair and single area are the widest; the former in B measures two inches and four tenths, and the latter, which are about the same width, measure two inches and nine tenths across. The long plates forming these areas are bent in the middle (Pl. XL), and their surface is covered with four rows of small, regularly arranged, crenulated and perforated tubercles, raised on bosses, and surrounded by sunken areolas; the inter-tubercular portion of the plates is covered with close-set miliary granules. (Pl. XXXIX, fig. 2.)

The small apical disc is situated at the vertex, rather nearer the anterior than the posterior border; the discal elements are soldered together, and nothing but the four ovarial holes, and small central madreporiform body are visible in the specimens I have hitherto seen.

In all the specimens of this urchin I have examined in different collections, the under surface is covered with the Oolitic matrix, which adheres so firmly that it is impossible to remove it; the structure of the base is therefore unknown to me. M. Cotteau, however, states, " That at the inferior surface, in his specimen, the ambulacra converge in a straight line to the mouth; they are narrow, bordered with pores, set wide apart, and enclosed in very apparent depressions, which alternate with the elevations of the inter-ambulacral areas."

The anal opening is situated just below the posterior border, it has an oval form, its long diameter corresponding to the antero-posterior diameter of the test.

The mouth-opening, according to M. Cotteau, is excentral, nearer the anterior border, the peristome is pentagonal, and surrounded with five prominent oral lobes.

The test of this species is very thin and delicate, a circumstance which may account for the fractured condition in which it is so often found; in general it is met with in masses of Oolitic limestone, from which it has to be cut out with great care.

Affinities and differences.—This gigantic urchin so much resembles *Pygurus Phillipsii,* Wr., in all the leading points of its structure, that it is possible the latter may be only a young form of *Pygurus Hausmanni*; it requires, however, more specimens than I have hitherto had at my disposal to state this as a fact. *Pygurus Hausmanni* in its magnitude resembles *Pygurus Icaunensis,* Cotteau, but it differs from the latter in having its upper surface more depressed, its ambulacral areas narrower, and in preserving their petaloidal figure near to the border, whilst in *P. Icaunensis* they are wider in the upper half, and much narrower in the lower half; the base of this urchin is likewise more concave and the inter-ambulacra more prominent and cushioned. *P. Hausmanni* differs from *Pygurus pentagonalis* in having the dorsal surface more depressed, the general outline more sub-circular, the ambulacral areas narrower above and wider below, and the single inter-ambulacrum less rostrated than in the latter species. The prominence and elevation of the ambulacra in *Pygurus costatus* with its pentagonal form readily distinguish it from

Pygurus Hausmanni, although the great depression of the test in both these species produces a close resemblance between them.

Locality and Stratigraphical position.—This large species has hitherto been found in England, only in the Coralline Oolite, at Malton, Yorkshire, and always in the thick bedded limestones of that formation, associated with *Clypeus subulatus, Cidaris florigemma, Echinobrissus dimidiatus* and *Collyrites bicordata.*

In France it has been collected, according to M. Cotteau, in the "Calcaires à Astartes de l'Aube," at the environs of Longchamps, by M. Royer, and at Polisot by M. Leymerie; M. d'Orbigny states that it is found in the "étage corallien" of Tonnerre, and Thury, (Yonne).

In Germany it was collected by Koch and Dunker* in the "Oberen Korallenkalk," at Kleinenbremen, near Bückeburg, associated with *Astrea? helianthoides,* Goldf., *A. agaricites,* Goldf., *Terebratula lacunosa,* Schl., and *Cidaris Blumenbachii,* Münst.

History.—This large urchin was described and figured in 1837 for the first time by Koch and Dunker; it has been subsequently described by M. Cotteau, and mentioned in the different works enumerated in the synonyms of this species.

* Beiträge des Norddeutch Oolithgebirg, p. 38.

NOTES

ON FOREIGN JURASSIC SPECIES OF THE GENUS PYGURUS, NEARLY ALLIED TO BRITISH FORMS, BUT WHICH HAVE NOT YET BEEN FOUND IN THE ENGLISH OOLITES.

PYGURUS ACUTUS, *Agassiz*. Catalogue raisonné des Échinides, p. 104.

Test elongated, and depressed, sensibly enlarged before and behind, its form resembling *Pygurus productus* from the Neocomian, posterior border rostrated, vent oblong, and infra-marginal.

Formation.—Inferior Oolite of Nantua.

Collection.—M. d'Orbigny.

PYGURUS DEPRESSUS, *Agassiz*. Syn. *Pygurus depressus*, Cotteau and Triger. Échinides du départ. de la Sarthe, pl. xx, fig. 1—6; p. 90.

Test sub-orbicular, or elongated, slightly depressed before, and much rostrated behind; upper surface elevated and uniformly convex, inferior surface depressed; deeply concave in the middle, and having the basal inter-ambulacra much cushioned; discal summit nearly central; ambulacral areas widely petaloid; poriferous zones very large, contracted at the lower fourth of the dorsal surface, apical disc small, slightly prominent and excentrally forwards; vent elongated, opening near the posterior border in a deep depression; mouth-opening, excentral, nearer the anterior border; peristome pentagonal, surrounded by five oral lobes, with which five ambulacral petals alternate.

Dimensions.—Height eight tenths of an inch; antero-posterior diameter, one inch and four tenths; transverse diameter, two inches and one quarter.

Formation.—Kelloway ferrugineux, Ass. No. 2, M. Triger; Chauffour, Sarthe.

Collections.—MM. Triger, Cotteau, one specimen in my cabinet collected by M. Sœmann, at Chauffour, Sarthe, and kindly sent me for this work.

PYGURUS (ECHINANTHUS) ORBICULATUS, *Leske.*   Syn. *Pygurus orbiculatus*, Cotteau and
Triger, Échinides du la départ. de la Sarthe, pl. xix, fig. 6—7, p. 88.

Test large, sub-circular, longer in length than in width; upper surface a little
elevated, sub-conical, and depressed at the border; base almost flat; basal inter-
ambulacra not prominent; summit slightly excentrally forwards; ambulacral areas lanceolate; poriferous zones very wide, and petaloidal almost to the margin; apical disc small,
slightly excentral; vent oval, infra-marginal, situated in a deep depression.   Mouth
opening small, sub-central nearer the anterior border; peristome pentagonal, surrounded
by five small oral lobes, with which five wide ambulacral petals alternate.

Dimensions.—Height, nine tenths of an inch; antero-posterior diameter three inches;
transverse diameter, two inches and eight tenths.

Formation.—Kelloway ferrugineux, Ass., No. 2, M. Triger, Coulans, Chauffour, Montbizot, environs of Mamers, Sarthe.

Collections.—MM. Michelin, Triger, Guéranger.

PYGURUS MARMONTI, *Beaudouin.*   Syn. *Pygurus Marmonti*, Desor, Synopsis des Échinides
Fossiles (p. 316).

Test large, sub-circular, much depressed on the upper surface; ambulacral areas
lanceolate; poriferous zones wide, petaloidal to the margin; base flat; inter-ambulacral
basal cushions distinctly flattened on the surface; mouth-opening small, pentagonal, and
sub-central; peristome surrounded by five oral lobes, which alternate with five ambulacral
petals; vent small, oval, situated in a deep depression removed a short distance from the
border.   M. Cotteau suggests that it might be necessary to unite this species with *P.
orbiculatus*, from which it differs in having a more circular form, and likewise in the
vent being removed a little farther inwards from the border; it is found in the same zone
with that urchin, and may be only a variety of it.

Dimensions.—Height, nine tenths of an inch; antero-posterior diameter three inches
and one quarter; transverse diameter, three inches.

Formation.—"Kellovien de Chatillon sur Seine, Mamers Estrochey (Côte d'Or),
Grande Oolite de Normandie."   Desor.
One of the two specimens in my cabinet was collected by M. de
Loriere from the "Étage Kellovien Chauffour, Sarthe."   The other
was obtained from the Kelloway ferrugineux Estrochey (Côte d'Or).

Collections.—MM. Michelin, Deslongchamps, Cotteau, my cabinet.

PYGURUS ICAUNENSIS, *Cotteau*. Études sur les Échinides Foss., pl. xxxvii, xxxviii.

Test large, sub-circular, length and breadth nearly equal; upper surface elevated and conoidal; under surface concave; ambulacral areas petaloid, very narrow below, dilated in the middle, and lanceolate above; poriferous zones wide, and petaloidal two thirds the distance between the disc and border; base deeply concave; basal inter-ambulacral areas very convex and prominent; mouth-opening excentral, small, pentagonal, nearer the anterior border, surrounded by five oral lobes, and five wide ambulacral petals; anal opening large, oval, infra-marginal, situate in a deep depression. This large species resembles *P. Hausmanni*, of which M. Desor suggests it may probably be a variety; the plaster mould in my collection exhibits very decided specific differences between it and the discoidal urchin I have identified with the German species.

> *Dimensions*.—Height, one inch and a quarter; antero-posterior diameter four inches and a half; transverse diameter, four inches and three tenths.

> *Formation*.—Calcáreo-siliceous strata of the Inferior Coralline Oolite at Druyes, (Yonne). Rare.

> *Collections*.—M. Cotteau; only two specimens known. A plaster mould of the figured specimen in my collection.

PYGURUS TENUIS, *Desor*. Synopsis des Échinides Fossiles, p. 315.

"Test large, circular, much dilated; border thin, almost trenchant; summit central; ambulacral petals very long, petaloidal almost to the margin." Desor.

> *Formation*.—" Portlandien inférieur (Astartien) d'Oberbuchsitten (Canton de Soleure), Oolite Astartienne de Laufon (Jura Soleurois), Delémont." Desor.

> *Collections*.—Mus. de Neuchâtel; M. Michelin.

PYGURUS JURENSIS, *Marcou.*　　Mem. Soc. Geol. de France, 2de serie, tom. iii, p. 114.

"Test dilated, and rostrated; summit excentral; ambulacral areas large, extending petaloidal almost to the border; under surface much undulated; anal opening infra-marginal." Desor.

Formation.—" Portlandien supérieur (Virgulien) de Suziau près Salins, des environs de Morteau, Gray, Haute-Saône.
Portlandien moyen (Ptérocérien moyen) de Montbéliard." Desor.

Collections.—MM. Marcou, Thurmann, Jaccard, d'Orbigny.

ON

THE STRATIGRAPHICAL DISTRIBUTION

OF THE

OOLITIC ECHINODERMATA.*

ALL the classes of the animal kingdom, when viewed in relation to their stratigraphical distribution, are not of the same value to the palæontologist. Some Mollusca, as the Conchifera and Gasteropoda, have a much greater extension in time than the Cephalopoda, and among Radiata, the Echinodermata and Anthozoa may be adduced as examples of classes whose life was alike limited; in estimating the value of palæontological evidence, therefore, it is necessary to take into consideration this important fact, which has not received the attention it is so justly entitled to.

The Echinodermata, although occupying a low position in the animal series, in a zoological point of view, still afford the palæontologist most important data for discussing questions relative to the distribution of species in time and space; it is well known, for example, that the Silurian, Devonian, and Carboniferous rocks are all characterised by distinct forms of Crinoidea, most of which are limited in their range to the different stages of these great groups. It is the object of this chapter, however, to show that the species of Oolitic Echinodermata had a like limited range in time, and that the different stages of the Oolitic formations are characterised by species which are special to each.

Dr. William Smith was doubtless aware of the value of the Echinodermata in stratigraphical geology, for he carefully noted the different species known to him which characterised the different subdivisions of the secondary rocks; and it is a remarkable fact, in connection with this subject, that although our knowledge of the species of this class

---

* The stratigraphical distribution of the Echinodermata, originally written for this work, was communicated, in the form of a memoir, to the Geological Section of the British Association, at the meeting held at Cheltenham, in August, 1856. The Council did me the honour, to order the communication to be printed entire among their Reports; for this reason it appeared in the 'Report of the British Association for the Advancement of Science for 1856,' and is now corrected down to the present date.

has been nearly quadrupled since the publication of his works,* still the outlines sketched by the hand of our great master remain nearly the same as laid out by him.

I have already shown that the test of the Echinodermata constitutes an internal and integral part of the body of the animal, participating in its life, intimately connected with the organs of digestion, respiration, and generation, as well as with those of vision and locomotion, and consequently having many of the distinctive characters of the organism indelibly impressed on different parts of the skeleton. The individual plates composing the columns of the test of the ECHINOIDEA, and the ossicula forming the skeletons of the ASTEROIDEA, OPHIUROIDEA, and CRINOIDEA, are organized after distinct plans ; they are therefore of great value in determining the species, as the specific characters are often well preserved on even fragmentary portions of the skeleton ; for this reason the remains of these animals are of the highest value in stratigraphical geology, and second in importance to no other class of the animal kingdom.

In the ECHINOIDEA the body is spheroidal, oval, depressed or discoidal, and enclosed in a calcareous test or shell composed of ten columns of large plates constituting the *inter-ambulacral* areas ; and ten columns of small plates constituting the *ambulacral areas*, which segments are separated from each other by ten rows of holes constituting the *poriferous zones*. The external surface of the plates is studded with tubercles of different sizes, in the different families ; to these the spines are articulated, by a kind of ball-and-socket joint, which are of different sizes, forms, and dimensions in the different families, and serve to characterise the genera and species.

At the summit of the test is the apical disc, composed of five genital plates perforated for the passage of the ovarial and seminal canals ; and five ocular plates notched or per-forated for lodging the eyes : in one family, the SALENIADÆ, an additional or suranal plate, composed of one or many pieces, is introduced within the circle formed by the genital and ocular plates.

There are two great apertures in the shell, one for the mouth, which is always at the base ; the other for the anus, which occupies different positions on the test ; in one section it is in the centre of the upper surface, directly opposite to the mouth, and surrounded by the genital and ocular plates ; in a second section the vent is external to the circle of genital plates, and never opposite to the mouth, but situated in different positions in relation to that opening, being placed on the upper surface, on the sides, the border, the infra-border, or the base, in the different groups.

The mouth is often armed with a complicated apparatus of jaws and teeth, or it is sometimes edentulous, and provided with lobes formed of the plates of the test itself.

The ASTEROIDEA have a depressed stelliform body provided with five or more lobes or hollow arms, which are a continuation of the body, and contain prolongations of the

---

* 'Strata identified by Organized Fossils,' 4to, 1816.—'Stratigraphical System of Organized Fossils,' 4to, 1817.

viscera. The mouth is always below and central; two or four rows of tubular retractile suckers occupy the centre of the rays; and in two families an anal vent opens at the central or sub-central part of the dorsal surface. The complicated skeleton is composed of numerous solid calcareous ossicula, variable as to number, size, and arrangement in the different genera which they serve to characterise. Their coriaceous integument is often studded with pedicellariæ and calcareous spines of various forms; they have a spongy madreporiform body situated on the upper surface of the disc near the angle between two rays; and reptation is accomplished by retractile tubular ambulacral suckers.

The OPHIUROIDEA have a distinct depressed discoidal body surrounded by long slender rays, in which there is no excavation for any prolongation of the viscera; they are special organs of locomotion, independent of the visceral cavity, and provided with spines which are supported on their sides; they have no pedicellariæ; the mouth is basal and central, surrounded by membranous tentacula, and they have no anal vent. The skeleton is composed of a series of plates which form the disc or centrum, and the long slender rays are sustained by numerous elongated vertebra-like ossicula, having numerous plates or spines disposed along the borders of the rays to assist in reptation. The form, structure, arrangement, and covering of the discal plates, and of the ossicles of the rays, afford good characters for distinguishing the genera.

The CRINOIDEA have a distinct bursiform body formed of a calyx, composed of a definite number of plates, provided with five solid rays, independent of the visceral cavity, and adapted for prehension; they have a distinct mouth and vent, no retractile suckers, and the ovaries open into special apertures at the base of the arms. The skeleton is extremely complicated, being composed in some genera of many thousands of ossicula articulated together, the number, form, and arrangement of which are determinate in the different families, the multiples of five being the numbers which in general predominate; the central plate of the calyx is supported on a long jointed column composed of circular, pentagonal or stelliform plates, the articulating surfaces are sculptured with crenulations which interlock into each other; in many genera the stem was attached by a calcareous root to the bed of the sea, and supported the calyx and arms upwards like a plant; in others it appears to have been moveable, and was used as a point of suspension from submarine bodies, the calyx and arms having had a pendent position.

The mouth is central and prominent, and the vent opens near its side; the arms are mostly ramose and multiarticulate, and when extended form a net-like instrument of considerable dimensions.

The four orders of Echinodermata thus briefly described are the only ones found fossil in the oolitic rocks, and of these by far the largest number of species belong to the ECHINOIDEA; for this order I have proposed the following classification, which differs in many essential particulars from that of previous authors.

As the mouth is always basal, central, subcentral, or excentral, the excentricity being invariably towards the anterior border, this aperture does not afford a character of primary importance, although when taken in connexion with others it is valuable in the definition of families.

The position of the anal opening affords a good primary character ; in one section the vent opens *within* the centre of the apical disc, surrounded by the genital and ocular plates ; in another section the vent opens *without* the apical disc, and is external to, and at a greater or less distance from, the genital and ocular plates ; these two sections may be thus defined.

ECHINOIDEA ENDOCYCLICA.

A. Test circular, spheroidal, more or less depressed, rarely oblong ; mouth central and basal ; vent in the centre of the upper surface directly opposite the mouth, and surrounded by five perforated genital and five ocular plates. Mouth always armed with five powerful calcareous jaws, formed of many elements disposed in a vertical direction.

ECHINOIDEA EXOCYCLICA.

B. Test sometimes circular and hemispherical, oftener oblong, pentagonal, depressed, clypeiform or discoidal ; mouth central or excentral ; vent external to the circle of genital and ocular plates, never opposite the mouth, situated in different positions in relation to that opening : four of the genital plates are generally perforated ; the fifth is in general imperforate. Mouth sometimes armed with jaws, but oftener edentulous. Jaws disposed in a more or less horizontal direction.

The structure of the ambulacral areas and poriferous zones, the form, number, and arrangement of the tubercles and their spines, the presence or absence of fascioles or semitæ, the size and form of the elements of the apical disc, and the position of the anus, afford collectively good characters for defining the genera.

The minute details in the structure of the plates ; the size, form, and number of the tubercles on each ; the form and arrangement of the pores in the zones ; their proximity or remoteness from each other ; the general outline of the body, which has only certain limits of variation ; the character of the sculpture on the plates ; the form of the areolas ; the greater or less prominence of the base ; the size of the tubercles ; the presence or absence, the size and arrangement of the granules forming the areolar circle ; the completeness or incompleteness of the same ; the width of the miliary zone, the number and size of the rows of granules composing it ; the length of the spines ; the form of their stems ; the character of the sculpture thereon ; the size of the head, the prominence and milling of the ring,—are all details of structure which individually and collectively afford good specific characters, as they are persistent details which are more or less developed on every considerable fragment of the test and spines of ECHINOIDEA.

Taking these characters for our guidance, I have grouped the genera, already so numerous by the discovery of extinct forms, into the following natural families:

A TABLE, SHOWING THE SECTIONS AND FAMILIES OF THE ECHINOIDEA.

ORDER.	SECTIONS.	FAMILIES.
ORDER—ECHINOIDEA	**SECTION A.** *Echinoidea endocyclica.* Vent within the genital plates, always opposite the mouth.	CIDARIDÆ. HEMICIDARIDÆ. DIADEMADÆ. ECHINIDÆ. SALENIADÆ.
	SECTION B. *Echinoidea exocyclica.* Vent without the genital plates, never opposite the mouth.	ECHINOCONIDÆ. COLLYRITIDÆ. ECHINONIDÆ. ECHINANTHIDÆ. ECHINOLAMPIDÆ. CLYPEASTERIDÆ. ECHINOCORIDÆ. SPATANGIDÆ.

A Table showing the Stratigraphical distribution of

FAMILIES, GENERA, AND SPECIES.	LOWER DIVISION.									
	LIAS.			INFERIOR OOLITE.			GREAT OOLITE.			
	Lower Lias.	Middle Lias.	Upper Lias.	Murchisonæ zone.	Humphriesianus zone.	Parkinsoni zone.	Fuller's Earth.	Stonesfield Slate.	Great Oolite.	Bradford Clay.
Fam. CIDARIDÆ.										
Cidaris Edwardsii, *Wright* . .	*	*								
Ilminsterensis, *Wright* . .	...	...	*							
Mooreii, *Wright* . . .	...	...	*							
Fowleri, *Wright* . . .	...	..	...	*						
Bouchardii, *Wright* . .	...	...	...	*	...	*				
Wrightii, *Desor* . . .	...	...	...	*						
confluens, *Forbes* . . .	...	...	...	...	...	*				
Bradfordensis, *Wright* . .	...	...	...	...	...	...	...	...	...	*
florigemma, *Phillips* . .	...	...	...	...	...	...	...	...	...	...
Smithii, *Wright* . . .	..	...	...	...	...	...	...	...	...	...
spinosa, *Agassiz* . . .	...	...	...	...	...	...	...	...	...	...
Boloniensis, *Wright* . .	...	...	...	...	...	...	...	...	...	...
Rabdocidaris Moraldina, *Cotteau* .	...	*	...	...	...	...	...	...	...	...
maxima, *Münster* . . .	...	...	...	...	*	...	...	...	...	...
Diplocidaris Desori, *Wright* . .	...	...	...	*						
Wrightii, *Desor* . . .	...	...	...	*						
Cotteauana, *Wright* . .	...	...	...	...	*					
Fam. HEMICIDARIDÆ.										
Hemicidaris granulosa, *Wright* .	...	...	...	...	...	*	...	...	...	...
pustulosa, *Agassiz* . . .	...	...	...	...	...	*	...	...	...	...
Stokesii, *Wright* . . .	...	...	...	...	...	...	...	*		
Luciensis, *d' Orbigny* . .	...	...	...	...	...	...	...	...	*	...
minor, *Agassiz* . . .	...	...	...	...	...	...	...	...	*	...
Ramsayii, *Wright* . . .	...	...	...	...	...	...	...	...	*	

the genera and species of the Oolitic Echinodermata.

		MIDDLE DIVISION.					UPPER DIVISION.				OBSERVATIONS AND FOREIGN LOCALITIES.
Forest Marble.	Cornbrash.	Kelloway Rock.	Oxford Clay.	Lower Calc. Grit.	Coralline Oolite.	Upper Calc. Grit.	Kimmeridge Clay.	Portland Sand.	Portland Oolite.	Purbeck Beds.	
...	...	...	..	*	*	...	...	...	...	...	France, Germany, and Switzerland.
...	...	...	...	*	*		...	...	...		Switzerland.
...	...	...	...	...	...	...	*	...	...	...	Boulogne-sur-Mer.
...	...	...	...	...	...	..	*	...	...	..	France.
...	...	...	...	...	...	..	...	...	.	..	Germany.
...	...	...	...	...	...	...	...	...	.	...	
...	...	...	...	...	...	...	...	...	...	...	Ranville, France.
...	...	...	...	...	...	...	...	...	...	...	Ranville, France.
...	...	...	...	...	...	...	...	...	...	...	Luc, France.
...	...	...	...	...	...	...	...	...	...	...	Ranville, France.

FAMILIES, GENERA, AND SPECIES.	LOWER DIVISION.									
	LIAS.			INFERIOR OOLITE.			GREAT OOLITE.			
	Lower Lias.	Middle Lias.	Upper Lias.	Murchisonæ zone.	Humphriesianus zone.	Parkinsoni zone.	Fuller's Earth.	Stonesfield Slate.	Great Oolite.	Bradford Clay.
Hemicidaris Bravenderi, *Wright*	...	...	...	...	...	...	...	...	*	
Wrightii, *Desor*	...	...	...	...	...	...	...	...	...	*
Icaunensis, *Cotteau*	...	...	...	...	...	...	...	...	*	...
intermedia, *Fleming*	...	...	...	...	...	...	...	...		
Davidsonii, *Wright*	...	...	...	...	...	...	...	...		
Purbeckensis, *Forbes*	...	...	...	...	...	...	...	...	...	...
Fam. DIADEMADÆ.										
Pseudodiadema lobata, *Wright*	*									
Mooreii, *Wright*	...	...	*							
Wickense, *Wright*	...	...	*							
depressum, *Agassiz*	...	...	...	*	*	...	...	...	*	...
Parkinsoni, *Desor*	...	...	...	...	...	...	...	*		
pentagonum, *M'Coy*	...	...	...	...	...	...	...	...	*	
homostigma, *Agassiz*	...	...	...	...	...	...	...	...	...	*
Bailyi, *Wright*	...	...	...	...	...	...	...	...	...	...
vagans, *Phillips*	...	...	...	...	...	...	...	...	*	...
Bakeriæ, *Woodward*	...	...	...	...	...	...	...	...		...
versipora, *Phillips*	...	...	...	...	...	...	...	...		
hemisphæricum, *Agassiz*	...	...	...	...	...	...	...	...		...
radiatum, *Wright*	...	...	...	...	...	...	...	...	...	
mamillanum, *Roemer*	...	...	...	...	...	...	...	...	...	...
Hemipedina Bechei, *Broderip*	*									
Bowerbankii, *Wright*	*									
Tomesii, *Wright*	*									
Jardinii, *Wright*	...	*								
Etheridgii, *Wright*	...	...	*							

Forest Marble.	Cornbrash.	MIDDLE DIVISION.					UPPER DIVISION.				OBSERVATIONS AND FOREIGN LOCALITIES.
		Kelloway Rock.	Oxford Clay.	Lower Calc. Grit.	Coralline Oolite.	Upper Calc. Grit.	Kimmeridge Clay.	Portland Sand.	Portland Oolite.	Purbeck Beds.	
...	...	..	...	...	...	...	..	...	...	...	France.
...	...	...	...	...	*	...	...	...	...	...	France and Germany.
...	...	...	..	...	...	...	...	*	...	...	Boulogne-sur-Mer.
...	...	...	...	...	...	...	...	...	...	*	France.
...	...	...	...	...	...	...	...	...	...	...	France.
..	*	...	..	...	...	...	...	...	...	...	France and Switzerland.
...	*										
...	*	...	...	...	...	...	...	...	...	...	Switzerland.
...	*										
...	...	...	...	...	*						
..	...	...	...	...	*	...	...	...	...	...	France and Switzerland.
...	...	...	...	...	*						
...	...	...	...	...	*	...	..	...	...	...	France and Germany.

FAMILIES GENERA, AND SPECIES.	LOWER DIVISION.									
	LIAS.			INFERIOR OOLITE.			GREAT OOLITE.			
	Lower Lias.	Middle Lias.	Upper Lias.	Murchisonæ zone.	Humphriesianus zone.	Parkinsoni zone.	Fuller's Earth.	Stonesfield Slate.	Great Oolite.	Bradford Clay.
Hemipedina Bakeriæ, *Wright*	...	...	...	*	...	...	...	...	...	...
perforata, *Wright*	...	...	...	*						
tetragramma, *Wright*	...	...	...	*						
Waterhousei, *Wright*	...	...	...	*						
Bonei, *Wright*	...	...	...	*						
Davidsoni, *Wright*	...	...	...	...	...	...		...	*	
Woodwardii, *Wright*	...	...	..	...	...		...	...	...	
microgramma, *Wright*	...	...	...	...			...	...		..
Marchamensis, *Wright*	...	...	...	...			...	...		..
Corallina, *Wright*	...	...	...	...		..	...	...		
tuberculosa, *Wright*	...	...	...	...		...	...	...		...
Morrisii, *Wright*	...	...	...	...		..	...			
Cunningtoni, *Wright*	...	...	...							
Pedina rotata, *Wright*	...	...	...	...	...	*	...	...	...	...
Smithii, *Forbes*	...	...	...		...	*	...	..	*	
Fam. ECHINIDÆ.										
Glypticus hieroglyphicus, *Goldfuss*	...	...	...		...		...		...	...
Magnotia Forbesii, *Wright*	...	...	...		...	*				
Polycyphus Normannus, *Desor*	...	...	...	...	...	*	...	*	*	...
Deslongchampsii, *Wright*	...	...	...	*						
Stomechinus germinans, *Phillips*	...	...	...	*						
intermedius, *Agassiz*	...	...	...	...	*	*	...		...	...
bigranularis, *Lamarck*	...	...	...	...	...	*	...	...	...	...
microcyphus, *Wright*	...	...	...	...	...	...	...	...	*	
gyratus, *Agassiz*	...	...	...	...	...	...	...	...	...	...
nudus, *Wright*	...	...	...	...	...		...	...	...	...

Forest Marble.	Cornbrash.	MIDDLE DIVISION.					UPPER DIVISION.				Observations and Foreign Localities.
		Kelloway Rock.	Oxford Clay.	Lower Calc. Grit.	Coralline Oolite.	Upper Calc. Grit.	Kimme-ridge Clay.	Portland Sand.	Portland Oolite.	Purbeck Beds.	
...	...	...	...	...	...	...	...	...	...	...	France.
...	*										
...	*										
...	...	...	...	*							
...	...	...	...	...	*						
...	...	...	...	...	*						
...	...	...	...	...	...	...	*				
...	...	...	...	...	...	...	*				
...	*										
...	...	...	...	...	*	...	...	...	...	...	France, Germany, and Switzerland.
...	*	...	...	...	...	...	...	...	...	...	Ranville and Luc, France.
...	*	...	...	...	...	...	...	...	...	...	Ranville, France.
...	...	...	...	...	...	...	...	...	...	...	Ranville, Port-en-Bessin, France.
...	...	...	...	...	*	...	..	..	...	...	France and Switzerland.
...	...	...	...	...	*						

FAMILIES, GENERA, AND SPECIES.	LOWER DIVISION.									
	LIAS.			INFERIOR OOLITE.			GREAT OOLITE.			
	Lower Lias.	Middle Lias.	Upper Lias.	Murchisonæ zone.	Humphriesianus zone.	Parkinsoni zone.	Fuller's Earth.	Stonesfield Slate.	Great Oolite.	Bradford Clay.
Fam. SALENIADÆ.										
Acrosalenia minuta, *Buckman*	*	...	...	...	...	...	...	...	...	...
crinifera, *Quenstedt*	...	...	*	...	...	...	...	...	...	...
Lycetti, *Wright*	...	...	...	*	*	*	...	...	...	...
pustulata, *Forbes*	...	...	...	...	...	...	...	...	*	...
Wiltonii, *Wright*	...	...	...	...	...	...	...	...	·	...
Loweana, *Wright*	...	...	...	...	...	...	...	...	...	...
spinosa, *Agassiz*	...	...	...	...	...	*	...	··	*	...
hemicidaroides, *Wright*	...	...	...	...	...	...	...	...	*	...
decorata, *Haine*	...	...	...	...	...	...	...	...	...	...
Fam. ECHINOCONIDÆ.										
Holectypus depressus, *Leske*	...	...	...	...	...	*	...	*	*	...
hemisphæricus, *Desor*	...	...	...	·	...	*	...	...	...	...
oblongus, *Wright*	...	...	...	...	...	...	...	...	...	...
Pygaster semisulcatus, *Phillips*	...	...	...	*	...	...	...	...	*	...
conoideus, *Wright*	...	...	...	*	...	...	...	...	...	...
macrostomus, *Wright*	...	...	...	...	...	...	...	...	...	...
Morrisii, *Wright*	...	...	...	...	...	...	...	...	...	...
umbrella, *Lamarck*	...	...	...	...	...	...	...	...	...	...
Fam. ECHINOBRISSIDÆ.										
Echinobrissus clunicularis, *Llhwyd*	...	...	...	...	...	*	...	*	*	...
Woodwardii, *Wright*	...	...	...	...	...	...	...	...	*	...
orbicularis, *Phillips*	...	...	...	...	...	...	...	...	...	...
quadratus, *Wright*	...	...	...	...	...	...	...	...	...	...
Griesbachii, *Wright*	...	...	...	...	...	...	...	...	...	...

Forest Marble.	Cornbrash.	MIDDLE DIVISION.					UPPER DIVISION.				OBSERVATIONS AND FOREIGN LOCALITIES.
		Kelloway Rock.	Oxford Clay.	Lower Calc. Grit.	Coralline Oolite.	Upper Calc. Grit.	Kimme- ridge Clay.	Portland Sand.	Portland Oolite.	Purbeck Beds.	
...	...	...	...	...	...	...	...	...	...	...	Pliensbach bei Boll, Württemberg.
*											
...	*										
*											
*	*	...	...	...	...	...	...	...	...	..	Ranville, Châtel-Censoir, France; and Soleure, Switzerland.
*	*										
...	...	...	...	...	*						Bar-sur-Aube, Yonne, France.
...	*	...	...	...	...	...	...	...	...	...	France, Germany, Switzerland.
...	...	...	...	...	...	...	...	...	...	...	France, Germany, Switzerland.
...	...	...	...	...	*						
*											
...	*										
...	...	...	...	*	*	...	...	...	...	...	Druyes, Châtel-Censoir, Coulanges-sur-Yonne, St. Mihiel, France.
*	*	...	...	...	...	...	...	...	...	...	France, Germany, Switzerland.
...	*	...	...	...	...	...	...	...	...	...	Mamers, Sarthe, France.
...	*										
...	*										

FAMILIES, GENERA, AND SPECIES.	LOWER DIVISION.									
	LIAS.			INFERIOR OOLITE.			GREAT OOLITE.			
	Lower Lias.	Middle Lias.	Upper Lias.	Murchisonæ zone.	Humphriesianus zone.	Parkinsoni zone.	Fuller's Earth.	Stonesfield Slate.	Great Oolite.	Bradford Clay.
Echinobrissus dimidiatus, *Phillips*	...	...	...	...	...	...	...	...	...	...
scutatus, *Lamarck*	...	...	...	...	...	...	...	...	...	...
Brodiei, *Wright*	...	...	...	...	...	...	...	...	...	...
Clypeus Plotii, *Klein*	...	...	...	...	...	*	*	*	*	...
Agassizii, *Wright*	...	...	...	...	...	*	...	...	...	...
altus, *M'Coy*	...	...	...	...	...	*	...	...	...	...
Michelini, *Wright*	...	...	...	*	*					
Hugii, *Agassiz*	...	...	...	...	...	*	...	...	...	...
rimosus, *Agassiz*	...	...	...	...	...	*				
Mülleri, *Wright*	...	...	...	...	...	...	...	...	*	
subulatus, *Young and Bird*	...	...	...	...	...	...	...	...	...	...
Fam. COLLYRITIDÆ.										
Collyrites ringens, *Agassiz*	...	...	...	...	...	*	...	...	...	...
ovalis, *Leske*	...	...	...	...	...	*	...	...	...	...
bicordata, *Leske*	...	...	...	...	...	...	...	...	...	...
Hyboclypus agariciformis, *Forbes*	...	...	...	*	...	...	...	...	*	...
caudatus, *Wright*	...	...	...	...	...	*	...	...	...	...
gibberulus, *Agassiz*	...	...	...	...	...	*	...	...	...	...
ovalis, *Wright*	...	...	...	...	...	*				
Fam. ECHINANTHIDÆ.										
Pygurus Michelini, *Cotteau*	...	...	...	...	...	*	...	...	*	*
pentagonalis, *Phillips*	...	...	...	...	...	...	...	...	...	...
Blumenbachii, *Koch & Dunker*	...	...	...	...	...	...	...	...	...	...
Phillipsii, *Wright*	...	...	...	...	...	...	...	...	...	...
Hausmanni, *Koch & Dunker*	...	...	...	...	...	...	...	...	...	...
costatus, *Wright*	...	...	...	...	...	...	...	...	...	

Forest Marble.	Cornbrash.	MIDDLE DIVISION.					UPPER DIVISION.				Observations and Foreign Localities.
		Kelloway Rock.	Oxford Clay.	Lower Calc. Grit.	Coralline Oolite.	Upper Calc. Grit.	Kimmeridge Clay.	Portland Sand.	Portland Oolite.	Purbeck Beds.	
...	...	...	...	...	*						
...	...	...	...	*	*	...	...	...	...	...	France, Germany, Switzerland.
...	...	...	...	...	...	...	...	....	*		
...	*	...	...	...	...	...	...	...	...	...	France and Switzerland.
...	.	...	...	...	...	...	...	..	...	...	Conlie, Mamers, Sarthe, France.
...	...	...	...	...	..	...	...	...	...	...	Pécheseul, Sarthe, France.
...	...	...	...	...	...	...	..	...	...	...	France and Switzerland.
...	...	...	...	...	*						
...	*	...	...	..	...	...	...	...	...	...	France and Switzerland.
...	. ...	...	..	...	...	...	...	...	...	...	France and Switzerland.
...	...	...	..	*	*	...	...	...	...	...	France and Switzerland.
...	...	...	...	...	...	...	...	...	...	...	Bayeux, France.
...	...	...	...	...	...	...	...	...	...	...	France and Switzerland.
...	*	...	...	...	...	...	...	...	...	...	France and Switzerland.
...	..	..	...	*	*						
...	...	...	...	*	*	...	...	...	...	...	France and Germany.
...	...	...	..	...	*						
...	...	...	...	...	*	...	..	...	...	...	France and Germany.

FAMILIES, GENERA, AND SPECIES.	LOWER DIVISION.									
	LIAS.			INFERIOR OOLITE.			GREAT OOLITE.			
	Lower Lias.	Middle Lias.	Upper Lias.	Murchisonæ zone.	Humphriesianus zone.	Parkinsoni zone.	Fuller's Earth.	Stonesfield Slate.	Great Oolite.	Bradford Clay.
Order ASTEROIDEA.										
Fam. URASTERIDÆ.										
Uraster Gaveyi, *Forbes*	...	*								
carinatus, *Wright*	...	*								
Fam. SOLASTERIDÆ.										
Solaster Moretonis, *Forbes*	...	...	..	...	...	...	...	...	*	
Fam. GONIASTERIDÆ.										
Goniaster Hamptonensis, *Wright*	...	...	...	...	...	...	...	...	*	
obtusus, *Wright*	...	...	...	*						
Fam. ASTERIDÆ.										
Tropidaster pectinatus, *Forbes*	...	*								
Astropecten Hastingsiæ, *Forbes*	...	*								
Orion, *Forbes*	...	*								
Phillipsii, *Forbes*	...	*								
Leckenbyi, *Wright*	...	...	...	...	*					
Scarburgensis, *Wright*	...	...	...	...	*					
Cotteswoldiæ, *Buckman*	...	...	...	...	...	...	...	*		
Wittsii, *Wright*	...	...	...	...	...	...	...	*		
Forbesii, *Wright*	...	...	.	...	...	...	...	*		
clavæformis, *Wright*	...	...	.	...	...	...	...	...	...	...
rectus, *M'Coy*	...	...	.	...	...	...	...	...	...	...
Luidia Murchisonii, *Williamson*	...	*								
Plumaster Ophiuroides, *Wright*	*									

		MIDDLE DIVISION.					UPPER DIVISION.				OBSERVATIONS AND FOREIGN LOCALITIES.
Forest Marble.	Cornbrash.	Kelloway Rock.	Oxford Clay.	Lower Calc. Grit.	Coralline Oolite.	Upper Calc. Grit.	Kimme-ridge Clay.	Portland Sand.	Portland Oolite.	Purbeck Beds.	
...	...	...	...	*							
...	...		...	*		...					

FAMILIES, GENERA, AND SPECIES.	LIAS.			INFERIOR OOLITE.			GREAT OOLITE.			
	Lower Lias.	Middle Lias.	Upper Lias.	Murchisonæ zone.	Humphriesianus zone.	Parkinsoni zone.	Fuller's Earth.	Stonesfield Slate.	Great Oolite.	Bradford Clay.
Order OPHIUROIDEA.										
·Fam. OPHIURIDÆ.										
Palæocoma Gaveyi, *Wright*	…	*								
Milleri, *Phillips*	…	*								
Egertoni, *Broderip*	…	…	*							
tenuibrachiata, *Forbes*	…	…	*							
Brodiei, *Wright*	…	…	*							
Murravii, *Forbes*	…	…	…	…	*					
Griesbachii, *Wright*	…	…	…	…	…	…	…	…	…	…
Order CRINOIDEA.										
Fam. PENTACRINIDÆ.										
Pentacrinus tuberculatus, *Miller*	*	…	…	…	…	…	…	…	…	…
scalaris, *Goldfuss*	*	…	…	…	…	…	…	…	…	…
basaltiformis, *Miller*	…	*	…	…	…	…	…	…	…	…
Goldfussii, *M'Coy*	…	*								
robustus, *Wright*	…	*								
punctiferus, *Quenstedt*	…	*	…	…	…	…	…	…	…	…
Johnsonii, *Austin*	…	…	*	…	…	…	…	…	…	…
dichotomus, *M'Coy*	…	…	*							
Phillipsii, *Wright*	…	…	*							
Milleri, *Austin*	…	…	…	…	*					
Austenii, *Wright*	…	…	…	*						
subsulcatus, *Goldfuss*	…	…	…	…	…	…	…	…	…	*
subteres, *Goldfuss*	…	…	…	…	…	…	…	…	…	*
Extracrinus Briareus, *Miller*	*	…	…	…	…	…	…	…	…	…
subangularis, *Miller*	…	*	…	…	…	…	…	…	…	…

Forest Marble.	Cornbrash	MIDDLE DIVISION.					UPPER DIVISION.				OBSERVATIONS AND FOREIGN LOCALITIES.
		Kelloway Rock.	Oxford Clay.	Lower Calc. Grit.	Coralline Oolite.	Upper Calc. Grit.	Kimme-ridge Clay.	Portland Sand.	Portland Oolite.	Purbeck Beds.	
...	...	...	...	...	...	...	...	...	...	...	France, Germany.
...	...	...	...	...	...	...	...	...	...	...	France, Germany.
...	...	...	...	...	...	...	...	...	...	...	France, Germany.
...	...	...	...	...	...	...	...	...	...	...	Germany.
...	...	...	...	...	...	...	...	...	...	...	France.
...	...	...	...	...	...	...	...	...	...	...	Germany.
...	...	...	...	...	...	...	...	...	...	...	Germany.
...	...	...	...	...	...	...	...	...	...	...	France, Germany.
...	...	...	...	...	...	...	...	...	...	...	France, Germany.

FAMILIES, GENERA, AND SPECIES.	LOWER DIVISION.									
	LIAS.			INFERIOR OOLITE.			GREAT OOLITE.			
	Lower Lias.	Middle Lias.	Upper Lias.	Murchisonæ zone.	Humphriesianus zone.	Parkinsoni zone.	Fuller's Earth.	Stonesfield Slate.	Great Oolite.	Bradford Clay.
Fam. APIOCRINIDÆ.										
Apiocrinus Parkinsoni, *Schlotheim* .	...	...	...	...	...	...	...	...	...	*
elegans, *Defrance* . . .	...	...	...	...	...	...	...	...	...	*
exutus, *M'Coy* . . .	...	...	...	...	...	...	...	...	...	*
Millericrinus Prattii, *Gray* . .	...	...	...	...	...	...	...	...	*	
obconicus, *Goldfuss* . .	...	...	...	...	...	...	...	...	*	
Koninckii, *Wright* . .	...	...	...	...	...	...	...	...	...	...
echinatus, *Schlotheim* . .	...	...	...	...	...	...	...	...	...	...
	10	17	12	0	49	0	1	8	28	9

166 Species.

ECHINOIDEA ....

ASTEROIDEA ....

OPHIUROIDEA ....

CRINOIDEA ....

Forest Marble.	Cornbrash	MIDDLE DIVISION.					UPPER DIVISION.				Observations and Foreign Localities.
		Kelloway Rock.	Oxford Clay.	Lower Calc. Grit.	Coralline Oolite.	Upper Calc. Grit.	Kimmeridge Clay.	Portland Sand.	Portland Oolite.	Purbeck Beds.	
...	...	...	...	...	...	...	...	...	...	...	} Ranville, Mamers, France; Alsace, Germany.
...	...	...	...	...	...	...	...	...	...	...	
*											
...	...	...	...	...	*	...	...	...	...	...	France, Germany.
8	21	0	0	11	24	0	4	1	1	1	

.... 119
.... 18
.... 7
.... 22—166

From the above Tables, it appears that the English Oolitic rocks are known at present to contain 166 species of fossil Echinodermata, of which 119 species belong to the Order ECHINOIDEA; 18 species to the Order ASTEROIDEA; 7 species to the Order OPHIUROIDEA; and 22 to the Order CRINOIDEA. All the species belonging to the families CIDARIDÆ, HEMICIDARIDÆ, DIADEMADÆ, ECHINIDÆ, SALENIADÆ, ECHINOCONIDÆ, ECHINOBRISSIDÆ, COLLYRITIDÆ, ECHINANTHIDÆ, and ECHINOLAMPIDÆ, have been figured in this work.

The ASTEROIDEA, OPHIUROIDEA, and CRINOIDEA, will form the subject of a second Monograph. An analysis of the Tables gives the following distribution of the species in each stage:

Lower Lias	10	Species.
Middle Lias	17	,,
Upper Lias	12	,,
Inferior Oolite	49	,,
Fuller's Earth	1	,,
Stonesfield Slate	8	,,
Great Oolite	28	,,
Bradford Clay	9	,,
Forest Marble	8	,,
Cornbrash	21	,,
Oxford Clay and Kelloway	0	,,
Lower Calcareous Grit	11	,,
Coral Rag	24	,,
Upper Calcareous Grit	?	,,
Kimmeridge Clay	4	,,
Portland Sand	1	,,
Portland Oolite	1	,,
Marine Purbeck Beds	1	,,

The Lias forms appear to be special to the three subdivisions of that formation, so well characterised by the species of Ammonites which indicate these three zones of Liassic life. The Inferior Oolite contains forty-nine species, of which forty-three are ECHINOIDEA, three ASTEROIDEA, one OPHIUROIDEA, and two CRINOIDEA; of these, ten species extend into the Great Oolite, and nine species pass into the Cornbrash; the Inferior Oolite has therefore thirty species which up to this time have not been found in any other formation; all the species from the Lias to the Cornbrash inclusive became extinct before the deposition of the Kelloway rock and Oxford clay. The Fuller's earth has yielded one species, and the Stonesfield slate contains eight species, several of which are special to this fissile rock. The Great Oolite has yielded twenty-eight species, of which ten extend into the Cornbrash, fourteen are special to the Great Oolite stage, and four are common to the different stages of the lower division of the Oolites. The nine species of the Bradford clay are mostly common to this argillaceous bed, and the Great Oolite limestone on which it rests. The Forest Marble contains eight species, of which four are common to this rock and the Cornbrash, which contains twenty-one species, many of which are found in older formations; with the deposition of the Cornbrash the lower division of the Oolites terminate, and with that formation all the species of Echinodermata found in these rocks became extinct.

The middle division of the Oolites contains far fewer species than the lower. The Kelloway rock and Oxford clay, so rich in Cephalopoda, have not in England, as far as I can learn, yielded any remains of Echinodermata. The Lower Calcareous grit, Coral rag, and Upper Calcareous grit, have several species in common; of the eleven species of the Lower Calcareous grit, six are common to it and the Coral rag, which contains twenty-four species; I have not ascertained how many, if any, pass into the Upper Calcareous grit; in fact, these three stages represent in reality only one stratigraphical zone of life.

The Kimmeridge clay up to the present time is known to contain only four species, which are all special to it. There is one species in the Portland sand, one in the Portland Oolite, and one in the Marine Purbeck beds. The Portland Oolitic limestone is said to contain the remains of several Echinoderms, although I have been able to obtain only two specimens of the same species for examination from that formation.

BIBLIOGRAPHY OF THE ECHINODERMATA,

ARRANGED IN

CHRONOLOGICAL ORDER.

1551. BELON. P.............................Histoire naturelle des Poissons marins. Paris, 4to, fig.

1554. RONDELET. GULIELM, ...Libri de Piscibus marinis, in quibus veræ Piscium effigies expressæ
sunt. Lugd., folio, libri xviii, cap. xxix, "De Echinio," pp. 577—
583.

1558. GESNER. CONR. ....Historiæ Animalium, lib. iv, Tiguri, folio.

1601. LIBAVIUS. ANDR. ....Batrachiorum, libri II, cap. xxiii, p, 424, "Echinatas opinatur esse
lapides qui in Bufonibus reperiantur." Leake, p. xi. Francf., 8vo.

1622. BESLER. BASIL.....................Continuatio rariorum et aspectu digniorum varii generis, etc.
Norb., folio.

1642. BESLER. M. R. ....Gazophylacium rerum naturalium. Norimb. Folio, fig. Leipzig,
1733.

1642. ALDROVANDI. UL. ....De reliquis Animalibus exsanguibus, utpotè de Mollibus, Crustaceis,
Testaceis, et Zoophytis. Bonon., folio.

1647. BOOT. A. B. DE ....Gemmarum et Lapidium Historia. Leyden, 8vo.

1648. ALDROVANDI. U. ... .Museum Metallicum, in libros IV distributum, cum fig. ligno in-
cisis. Bonon., folio, fig.

1655. WORM. OL. ....Museum Wormianum, seu Historia rerum rariorum, &c. Amstel.,
folio, fig.

1656. MOSCARDO. LUD..................Note overo Memorie del Museo suo.   Padua, folio, fig.

1666. MERRETT. CHR. ....Pinax Rerum naturalium Britannicarum, continens Vegetabilia,
Animalia, et Fossilia in hâc Insulâ reperta. London, 1666, 1677,
1704, 8vo.

1668. CHARLETON. GUALT. ....Onomasticon Zoicon, plerorumque Animalium differentias, etc., 4to.

1669. LACHMUND. FR. ....Oryctographia Hildesheimensis, sive admirandorum Fossilium quæ
in Tractu Hildesheimensi reperiuntur Descriptio. Hildesh.,
4to, fig.

1670. Boccone. P. Silv. ....Recherches et Observations d'Histoire naturelle touchant le Corail, la Pierre étoilée, etc. Paris, 1670, 12mo, fig. Amst., 1674, 8vo, fig.

1672. Imperato. Ferr..................Historia naturale, nella quale ordinatamente si tratta della diversa Condizione de Minere, Pietre preziose, e altre curiosità. Folio, cum fig.

1676. Plot. Robert ....Natural History of Oxfordshire, being an essay toward the Natural History of England. Oxford, folio, tab.

1678. Lister. Martin ....Historiæ Animalium Angliæ tres tractatus. London, 4to, tab.

1680. Wagner. J. J.....................Historia naturalis Helvetiæ curiosa in VII Sectiones compendiosè digesta. Tiguri, 1680—1701, 12mo.

1681. Grew. Neh.......................Musæum Regalis Societatis, or a Catalogue of Natural and Artificial Rarities, &c., preserved in Gresham College. Folio.

1681. Bonnani. Phil. ....Ricreazione dell' occhio e della mente nell' osservazione delle Chiocciole.' Rom., 1681, 4to, fig.—'Recreatio mentis et oculi in observatione Animalium Testaceorum.' Romæ, 1684, 4to, fig.

1686. Plot. Robert ....Natural History of Staffordshire. Oxford, folio, tab.

1689. Wagner. J. J......................De Mineris Ferri sub diversis figuris, Lapidum, Testaceorum marinorum, &c. 'Misc. Nat. Cur.,' Dec. II, 1689, Obs. 149, fig. Norimb.

1690. Mentzel. Chr. ....De Generatione Lapidum vulgò Bufonum et Echinometris. 'Miscell. Nat. Cur.,' Dec. II, p. 118, fig.

1695. Petiver. J. ....Musæi Petiveriani Centuria prima, continens Animalia, fossilia, plantas ex variis mundi plagis advecta, ordine digesta et nominibus propriis signata. In 8vo, cum fig.

1698. Luidius. Edvardus ....Lithophylacii Brittannici Ichnographia. London, 1685. 8vo, fig.

1702. Scheuchzer. J. J. ....Specimen Lithographiæ Helveticæ curiosæ. Tiguri, 8vo, fig.

1705. Hooke. Robert ....The Posthumous Works published by Richard Waller. London, folio, the figures of Echini drawn by Dr. Hooke.

1707. Sloane. Hans....................A Voyage to the Islands of Madeira, Barbados, Nieves, St. Christopher's, and Jamaica. Lond., 1707—1725, 2 vols. folio, fig.

1707. Mylius. Gottl. Fr. ....Memorabilia Saxoniæ subterraneæ, &c. Lips., 1709—1718, 2 vols. 4to, fig.

1708. Langius. Car. Nic. ....Historia Lapidum figuratorum Helvetiæ ejusque viciniæ. Venetiis, 4to, cum tabl.

1708. Baier. J. J. ....Oryctographia Norica, sive Rerum fossilium et ad regnum minerale pertinentium, in territorio Norimbergensi ejusque viciniâ observatarum succincta descriptio. Norimb., 4to, fig.

1709. Mylius. G. F. ....Rerum naturalium Historia existentium in Museo Kircheriano. Rom., 1709, folio, fig.; 1773, 1782, folio, fig. (curâ J. Ant. Batarra).

1709. Scheuchzer. J. J. ....Herbarium Diluvianum. Tiguri, folio, fig.

1709. Gandolphus. ....In 'Histoire de l'Academie Royale des Sciences,' an. 1709, "De Echini incessu exponit."

1712. Morton. John ....A Natural History of Northamptonshire, with an account of the Antiquities. Lond., folio, plates.

1712. Réaumur. R. Ant. de...........Histoire de l'Academie Royale des Sciences. Par., an. 1712, p. 177, "Echini animalis historiam proposuit." Leske, xiv.

1713. Petiver. J. ....Aquatilium Animalium Amboinæ Icones et Nomina. Lond., folio, fig.

1714. BÜTTNER. DAN. SIG. ....Coralliographia subterranea, etc. Lips., 4to, fig.

DACOSTA. EM. MENDEZ ....Letter concerning two beautiful *Echinites*. 'Phil. Trans.,' XLVI, p. 143.

1715. MERCATI. MICH. ....Metallotheca Vaticana, opus posthumum, edit. à J. M. Lancisio, acc. Appendix cum Iconibus. Romæ, 1715-19, folio, fig.

1716. SCHEUCHZER. J. J. ....Museum diluvianum. Tiguri, 8vo.

1717. HELWING. G. A. ....Lithographia Angerburgica, etc. Regiomonti. 4to.

1717. MERCATI. MICHL...............Michaelis Mercati Samminiatensis metallotheca, opus posthumum, editum opera Ioannis Mariæ Lancisii. Folio, cum tabl., Romæ.

1718. MELLE. JAC. À...................De Echinites Wagricis Epistola ad Woodwardum. Lubecæ, 4to, fig.

1718. SCHEUCHZER. J. J. ....Meteorologia et Oryctographia Helvetica. Zurich, 4to, fig.

1718.   „   ....Naturgeschichte des Schweizerlandes. Band 3, in 4to, mit Kupfert.

1719. ROSINUS. M. R. ....Tentaminis de Lithozois ac Lithophytis, olim marinis, jam verò subterraneis, Prodromus ; sive de Stellis marinis quondam, nunc fossilibus Disquisitio. Hamb., in 4to.

1719. VALENTINI. M. B...............Stella marina rarissima petrefacta et in Hassia reperta. (*Academiæ Caesareo-Leopoldinæ naturæ curios. Ephem.*), Centur. VIII, obs. LI, p. 334.

1720. MELLE. JAC. À ....De Lapidibus figuratis agri littorisque Lubecensis Commentatio epistolica. Lubecæ, 4to, fig.

1720. HELWING. G. A. ....Lithographiæ Angerbergicæ. Pars II, in qua de Lapidibus figuratis aliisque fossilibus in Districtu Angerburgensi ejusque vicinia noviter detectis. Lips. In 4to, cum fig.

1720. VOLKMANN. G. ANT. ....Silesia subterranea, &c. Leipzig, 4to, plates.

1723. SCHEUCHZER. J. J. ....Herbarium diluvianum, editio novissima, duplo auctior. In fol., cum fig.

1724. WAGNER. P. C. ....Dessertatio inauguralis physico-medica de lapidibus judaicis. In 4to, fig.

1728. BRÜCKMANN. F. E. ....Thesaurus subterraneus Ducatus Brunsvigii. In 4to, fig.

1729. WOODWARD. J. ....Natural History of the Fossils of England. London, 2 vols. 8vo.

1730. GIMMA. GIAC. ....Della storia naturale delle Gemme. Nap., 4to, 2 vols. In tomo secundo, libro quinto, etiam de *Echinite* agitur. Cf. Gronovius, 'Bibliotheca Reg. Animal.,' p. 110.

1732. SIVERS. H. JAC. ....Curiosorum Nicudorpensium Specimena IV. Lubecæ, 1732-34, 8vo, fig.

1732. BREYNIUS. J. P. ....Dissertatio de polythalamiis, etc., cui aducitur Schediasma de Echinis methodice disponendis. Cum figuris. Gedani. 4to.

1733. LINCK. J. H....................De Stellis Marinis liber singularis. Folio, plates.

1733. RITTER. ALB. ....Epistolica historico-physica Oryctographiæ Goslariensis. Helmst., 4to, fig.

1734. KLEIN. J. T. ....Naturalis Dispositio Echinodermatum et de aculeis Echinorum Marinorum. Gedani. 4to, tabl. ' *Ordre naturel des Oursins de mer et fossiles, avec des Observations sur les Piquans des Oursins de mer*,' Gall. edit. *Desbois*. Paris, 1754, 8vo, fig.

1734. SEBA. ALBERTUS ....Locupletissimi rerum naturalium Thesauri accurata Descriptio et Iconibus artificiosissimis expressio, per universam Physices historiam. Amstel., 1734—1765, 4 vols. folio, fig., tom. iii, pp. 18—37, Pl. X—XV.

1735. BYTEMEISTER. H. J...............Bibliothecæ appendix sive catalogus curiosorum artificialium et naturalium, &c. Folio, plates.

BRANDER. GUSTAV. ....An Account of a remarkable *Echinus*. 'Phil. Trans.,' XLIX, p. 295, table viii, fig. 2, *Echinus atratus*, Lin.

1737. KUNDMANN. J. C ....Rariora naturæ et artis item in re medica. Leipzig, folio.

1738. SHAW. THOMAS...................Travels or Observations relating to several parts of Barbary and of the Levant. Oxf., 1738, folio, fig. ; Lond., 1757, 4to ; Supplement, Oxf., 1746, folio.

1739. PLANCUS. J. ....Liber de conchis minùs notis Littore Ariminensi. Venet., 4to, cum tab.

1740. RUMPHIUS. G. EV. ....Thesaurus Piscium, Testaceorum, et Cochlearum, quibus acc. Conchylia, Mineralia, Metalla, Lapides variis in locis reperta. Hagæ, folio, tabl.

1740. BROMELL. MAGN. V. ....Mineralogia et Lithographia Suecana. Stockholm, Leipzig, 8vo.

1741. FRISCH. J. LEOP.................Musei Hoffmanniani Petrificata et Lapides descripta. Halæ, 4to, fig.

1741. PLANCUS. J. ....Oryctographiæ Calenbergicæ. Sondersh., 1741-43, 4to, fig.

1742. BRÜCKMANN. FR. ERN. ....Epistolæ itinerariæ Cent. I—III. Wolfenb., 1742-50, 4to, fig.

1742. BOURGUET. L. ....Traité des Pétrifications. In 4to, with plates.

1743. HEBENSTREIT. J. ERN. ....Museum Richterianum. Lips., folio, fig.

1744. GUALTIERI. NIC. ....Index Testarum Conchyliorum quæ adservantur in Museo suo et methodicè distributæ exhibentur tabulis CX. Florent., fol., tab.

HACQUET. BALTH. ....Abhandlung von einem neuentdeckten Echiniten. Naturf., XI, p. 105, fig.

1747. SCILLA. A. ....De Corporibus marinis lapidescentibus quæ defossa reperiuntur. Romæ, fol., cum tab.

1748. BAKER. HENRY ....A Description of curious Echinites. 'Phil. Trans.,' XLIV, p. 432, fig.

1748. HILL. JOHN ....The History of Fossils. London, folio.

1750. HUGHES. GRIFFITH...............A Natural History of Barbadoes. Lond., folio, fig.

1751. PONTOPPIDAN. ERIC. ....Norviges Natural Historie, &c. Kiöb, 1751-53, 2 vols. 4to, fig. '(Ang.) Natural History of Norway.' London, 1755, folio.

1752. ALLIONE. CARL. ....Oryctographiæ Pedemontanæ specimen exhibens corpora fossilia terræ adventitia. Paris, 8vo, fig.

1752. MOHR. G. FR. ....Specimen Historiæ naturalis subterraneæ agri Giengensis ejusque viciniæ. 'Ephem. Nat. Cur. Norimb.,' vol. IX, p. 120.

1752. STOBÆUS. KIL.....................Opuscula, in quibus Petrefactorum, Numismatum et Antiquitatum Historia illustratur. Dantisci, 1752-53, 4to, fig.

1752. SCHEUCHZER. J. J. ....Physica sacra Iconibus æneis illustrata. Amst., 5 vols. folio, fig.

1753. TESSIN. CAR. GUST.............Museum Tessinianum. Holmiæ, folio.

1754. LINNÆUS. CAR. ....Museum S. R. M. Adolphi Frederici Regis Suecorum. Holmiæ, folio, fig.

1754. TORRUBIA. JOS. ....Aparato para la Historia naturel Española. Madr., 1754-1765, fol., fig. (Germ. Chr. G. v. *Murr*.): 'Vorbereitung zur Naturgeschichte von Spanien.' Halle, 1773, 4to, fig. Many Spanish Fossils are figured in this work.

1757. GEHLER. J. C. ....Dissert. de Characteribus Fossilium externis. Lips., 4to.

1757. D'ARGENVILLE. A. J. D. ....L'Histoire Naturelle éclaircie dans deux de ses parties principales, la Lithologie et la Conchyliologie. Paris, in 4to, avec figures.

1783-93. LINNÆUS. CAR. ....Systema Naturæ, seu Regna tria Naturæ systematicè proposita per classes, ordines, genera, et species. Ed. J. Fr. Gmelin, 9 vols. 8vo, Leipzig, 1783-93.

1758. SCHREBER. J. CHR. D. ....Lithographia Halensis: Dissertatio. Halle, 4to.

1758. BORLASE. WILLIAM.............The Natural History of Cornwall. Oxford, 1758, folio, fig.

1758. GESNER. J. ....Tractatus physicus de Petrificatis in duos partes distinctus, &c. Lugd. Bat., 8vo.

1759. ABILGAARD. SÖREN.............Beskrivelse over Stevens Klint. Copenh., 1759, 4to, fig. 'Beschreibung von Stevens Klint und dessen Merkwürdigkeiten, mit mineralogischen und chemischen Betrachtungen erläutert, und mit Kupferstichen versehen. Aus dem Dänischen übersetzt.' Copenh. und Leipz., 1764, 8vo, fig.

1760. D'ANNONE. J. J. ....Acta Helvetica physico-mathematico anatomico - botanico - medica, figuris aeneis illustrata. Basil, vol. iv, 4to.

1760. HOFER. J. FIL. ....Tentaminis lithologici de Polyporitis et Zoophytis petrefactis Missus. 'Act. Helvet.,' iv, p. 169, fig.

1760. SCHULZE. C. F. ....Betrachtung der versteinerten Seesterne und ihrer Theile. In 4to, mit Kupfert.

1761. GUETTARD. J. ET. ....Mémoire sur les Encrinites et les pierres étoilées dans lequel on traitera aussi des Entroques, &c. (*Mémoires de l'Académie Royale des Sciences de Paris*, an. 1755, pp. 224 et 318.)

1761. ELLIS. JOHN....................An account of an Encrinus or Star-fish, with a jointed stem. 'Phil. Transac.,' vol. lii, part 1, p. 357.

1762. STRÖM. HANS ....Physisk og œkonomisk Beskrivelse over Fögderiet Söndmör beliggende i Bergens stift i Norge. Soröe, 4to, fig.

1762. WALCH. J. E. J. ....Das Steinreich systematisch entworfen. Halle, 1762-64, 2 vols. 8vo, fig.

1763. ANDREÆ. J. G. R. ....Briefe aus der Schweiz nach Hannover geschrieben in dem Jahre 1763. Zür. u. Winterth., 1776, 4to, fig.

1763. BERTRAND. E. ....Dictionnaire universel des Fossiles propres et des Fossiles accidentels. La Haye, 2 vols. 8vo.

1763. DE LUC........................Mémoires de Mathématique et de Physique, presentés à l'Académie Royale des Sciences par divers Savans, à Paris, 1763. 4to, tome IV, p. 467. 'Mémoire sur un Echinite singulier.' See Leske, p. 134.

1763. WALLERIUS. J. G. ....Mineralogie oder Mineralreich, von ihm eingetheilt und beschreiben; ins Deutsche übersetzt. Von J. D. Denso. Zweite Auflage, mit Kupfertaf., in 8vo.

1766. MÜLLER. PH. L. ST. ....*Deliciæ Naturæ Selectæ*, oder auserlesenes Naturalien-Kabinet, etc.; ehemals herausgegeben von G. W. Knorr und fortgesetzt von dessen Erben. Nürnb., folio, fig.

1767. PALLAS. P. S. ....Spicilegia Zoologica. Tome i, Berlin, 4to, 1767-74.

PARSONS. JAMES ....Remarks on a Petrified Echinus (*Cidaris papillata*, Leske). 'Philosophical Trans.,' XLIX, p. 155, fig.

1767. DAVILA. DON P. FR. ....Catalogue systèmatique et raisonné des Curiosités de la nature et de l'art qui composent son Cabinet. Paris, 3 vols. 8vo.

1768. WALCH. J. E. J. ....Die Naturgeschichte der Versteinerungen, zur Erläuterung der Knorrischen Sammlung von Merkwürdigkeiten der Natur. Nürnb., 1768-73, 4 vols. folio, fig.

1771. BECKMANN. JOH.................Commentatio de reductione rerum fossilium ad genera naturalia Protyporum, pars secunda. N. Comm. Gœtt., 1771, II, p. 68; III, p. 95.

1771. FAVANNE DE MONTCERVELLE ...Conchyliologie de *d'Argenville*. Édit. fort augmentée, 4to, 3 vols., 1771-80.

1774. PHELSUM (MARK VAN) ....Brieef aan *Cornelius Nozemann* over de Gewebslekken of Zee-egeln. Rotterdam, 8vo, fig.

1775. WALCOT. J. ....Descriptions and Figures of Petrifactions found in the Quarries, Gravel-pits, &c., near Bath. 8vo, plates.

1776. WALCH. J. E. J. ....Lithologische Beobachtungen ; I. Ein Versteinerter Ostracion, und III. Beytrag zur Naturgeschichte der Enkriniten und Pentacriniten Naturforscher, VIII, Stück, pp. 259—272.

1776. PENNANT. THOMAS ....British Zoology. Lond., 4 vols. 8vo, plates.

1776. SCHRÖTER. J. S. ....Journal für die Liebhaber des Steinreiches und der Konchyliologie, mit Kupfertal. 8vo. In the third volume is described *Cidaris papillatæ*, Leske, with spines.

1776. MÜLLER. O. FR. ....Zoologii Daniciæ Prodromus, seu Animalium Daniæ et Norwegiæ indigenorum characteres, nomina, etc. Hafn., 8vo.

1777. „ ....Zoologia Danica, seu Animalium Daniæ et Norwegiæ rariorum Descriptiones et Historia. Hafn. et Lips., 1779—1784, 2 vols. 8vo. Lips., 1781, folio. '*Splendidarum et diligentissime sculptarum harum iconum uberior explicatio seu historia animalium nondum produit, at vehementer a plurimis exoptatur.*' Leske, xiii.

1778. LESKE. N. G. ....Additamenta ad J. T. Klein naturalem dispositionem Echinodermatum. Lips., folio, plates.

1786. ELLIS. J. ....The Natural History of many curious Zoophytes, systematically arranged and described, by Dr. Solander. 4to, with 63 plates.

1791. BRUGUIÈRE. J. G..................Encyclopéd. Méthodique partie des Vers. 4to. Plates.

1798. CUVIER. G. ....Tableau Élémentaire de l'Histoire naturelle des Animaux. Paris, 8vo, fig.

1799. FAUJAS DE ST.-FOND. B..........Histoire naturelle de la Montagne de St.-Pierre de Maestricht. Par., folio, fig. (pp. 168—175, *Oursins fossiles ou Échinites.* Pl. 29, 30.)

1801. LAMARCK. J. B. DE..............Système des Animaux sans vertèbres, ou tableau général des classes, des ordres, et des genres de ces Animaux. 8vo.

1804. PARKINSON. JAMES ....Organic Remains of a former World. Lond., 1804-11, 3 vols. 4to, plates.

1806. ABILGAARD. P. C. ....Zoologia Danica de Müller. Copenhague, 4to, cahier, avec figures.

1809. SAVIGNY. J. CÈS.................Description de l'Egypte, ou recueil des recherches qui ont étés faites en Egypte, pendent l'expédition de l'armée Française. *Les Échinodermes*, par C. Savigny. Folio, plates.

1812. LAMARCK. J. B. DE ....Extrait du Cours de Zoologie du Muséum d'Histoire naturelle sur les Animaux sans vertèbres. Paris, 8vo.

1812. BRONGNIART. ALEX...............Geographie physique des environs de Paris, avec M. G. Cuvier. Paris, 4to, 2° edit., 1822.

1813. SCHLOTHEIM. E. F. VON.........Beiträge zur Naturgeschichte der Versteinerungen in geognostischer Hinsicht. Taschenb. für Miner. von Leonhard, vol. VII.

1814. LEACH. WILL. ELFORD ....Zoological Miscellany, being a description of new and interesting Animals, illustrated with coloured figures. Lond., 1814-17, 3 vols. 8vo, plates.

1815. LAMARCK. J. B. DE ....Histoire des Animaux sans vertèbres. Paris, 1815-22. 7 vols. 8vo.

1816. SCHLOTHEIM. E. F. VON ....Beiträge zur Naturgeschichte der Versteinerungen in geognosticher Hinsicht. (*Denkschr. der kon Akad. der Wissensch. zu München Jahn* 1816-17.)

1816. SMITH. WILLIAM ....Strata identified by Organized Fossils. Containing prints on coloured paper of the most characteristic specimens in each stratum. 4to, plates.

1817.   „   ....Stratigraphical System of Organized Fossils, with reference to the specimens of the original collection in the British Museum. Lond., 4to.

1817. CUVIER. GEORGE ....Règne Animal distribué d'après son organisation. Paris, 4 vols. 8vo, fig.

1818. BOSC. L. A. G.....................Nouveau Dictionnaire d'Histoire Naturelle (Déterville). Art. *Oursin*, par M. Bosc, t. 24.

1820. SCHWEIGGER. A. FR. ....Handbuch der Naturgeschichte der skelettlosen ungegliederten Thiere. Leiptz., 8vo.

1821. MILLER. J. S. ....A Natural History of the Crinoidea or Lily-shaped Animals; with observations on the Genera Asteria, Euryale, Comatula, and Marsupites. Bristol, 4to, with 50 plates.

1822. MANTELL. G. ....The Fossils of the South Downs; or, Illustrations of the Geology of Sussex. Lond., 4to, plates.

1822. PARKINSON. JAMES ....Outlines of Oryctology. Lond., 8vo, fig.

1822. SCHLOTHEIM. E. F. VON ....Nachträge zur Petrefaktenkunde. 2 vols., in 8vo, 37 plates, 1822-23.

1822. YOUNG & BIRD ....A Geological Survey of the Yorkshire Coast. Lond., 4to, plates.

1822-30. LAMOUROUX. J. V. F.........Dictionnaire classique d'Histoire Naturelle. Art. 'On different Genera of *Echinidæ*,' by Lamouroux.

1823. DELLE CHIAJE. STEF. ....Memorie sulla storia e notomia degli animali senza vertebre del regno di Napoli. Naples, 4 vols. 4to, plates, 1823-29.

1816-45. DE BLAINVILLE et DEFRANCE..Dictionnaire des Sciences Naturelles. Levrault. Paris, 1816-45, 61 vols. 8vo, plates. 'The Echinidæ, Recent and Fossil,' by M. De Blainville and M. Defrance.

1824. DEFRANCE. M.....................Tableau des corps organisés Fossiles. Paris, 8vo.

1824. DESLONGCHAMPS. EUD. ....Encyclopédie Méthodique. 'Histoire Naturelle des Zoophytes, ou Animaux rayonnés, faisant suite à l'Histoire naturelle des Vèrs de Bruguière.' 2 vols. 4to, plates. "The Echinodermata," by Professor Deslongchamps.

1824-31. MECKEL. T. F. ....System der vergleichenden Anatomie. 5 Band 8vo. Halle, 1824-31.

1824. STOKES. CHARLES ....On Three Specimens of Fossil Echini. 'Trans. Geol. Soc.,' ser. 2, II, p. 406.

1825. GRAY. JOHN EDWARD ....An attempt to divide the Echinidæ, or Sea-eggs, into natural families. 'Ann. of Phil.,' 2d ser., vol. X, p. 423.

1825. KŒNIG. CHARLES ....Icones Fossilium Sectiles. Lond., 4to, plates.

1826. GOLDFUSS. G. A. ....Petrifacta Germaniæ, tam ea, quæ in Museo Universitatis regiæ Borussicæ Fridericiæ Wilhelmiæ Rhenanæ servantur, etc., iconibus et descriptionibus illustrata. 3 vols. folio, cum tab., Düsseld., 1826-33.

1826. RISSO. A. ....Histoire naturelle des principales Productions de l'Europe meridionale, particulièrement de celles des environs de Nice et des Alpes maritimes. 5 vols. 8vo, fig.

1827. Griffith & Pidgeon ....Cuvier's Animal Kingdom, with supplementary additions to each Order, by Edw. Griffith and Edw. Pidgeon. 16 vols. 8vo, and 814 plates, 1827-35.

1827. Thompson. John V. ....Memoir on the Star-fish of the Genus *Comatula*. 'Jameson's Edin. New Phil. Jour.,' vol. XX, p. 295, fig.

1828. Fleming. Rev. John ....British Animals. Lond., 8vo.

1828. Münster ....Ueber Versteinerungen, etc. (On the Petrefactions of the Ferruginous Clay and Green Sands of Kressenberg.) *Keferstein's Teutschland*, vol. VI, p. 93.

1829. Cuvier et Latreille ....Le Règne Animal distribué d'après son organisation. 2d ed., 5 vols. 8vo plates.

1829. Phillips. John ....Illustrations of the Geology of Yorkshire. 2 vols. 4to, with plates and sections.

1829. Serres. Marcel de ....Géognosie des Terrains tertiaires du Midi de la France. Montp., 8vo, fig.

1829. Guérin-Méneville. F. E. ......Iconographie du Règne Animal de *M. le Baron Cuvier*, ou Représentation, d'après nature, des espèces les plus remarquables et souvent non encore figurées de tous les genres d'Animaux. Paris, 1829-39, 7 vols. 8vo of beautiful plates.

1830. Pidgeon. Edward ....The Fossil Remains of the Animal Kingdom. Lond., 8vo, plates.

1830. Woodward. Samuel ....A Synoptical Table of British Organic Remains. Norwich, 1830, in 8vo.

1831. Deshayes. G. P. ....Description de Coquilles Caractéristiques des Terrains. 8vo, plates.

1832. Dechen. H. von ....Handbuch der Geognosie von Sir H. T. De la Bèche. In 8vo.

1833. Woodward. Samuel ....Outline of the Geology of Norfolk. Norwich, 8vo, plates. Figures of several new species of Echinodermata.

1833. de la Beche. Sir H. T. ....Geological Manual. 3d ed., 8vo, fig.

        „   ....On the Geological Distribution of Organic Remains contained in the Oolitic Series of the Great London and Paris Basins, and in the same Series of the South of France. 'Phil. Mag.,' ser. 2d, VII, pp. 81, 202, 250, 334; VIII, pp. 35, 208.

        „   ....On Echinoneus lampas. 'Trans. Geol. Soc.,' ser. 1, I, table 3, figs. 3—5.

1834. De Blainville. H. M. D.........Manuel d'Actinologie ou de Zoophytologie. Paris, 2 vols. 8vo, plates.

1834. Agassiz. Louis ....Ueber die äussere Organisation der Echinodermen. Isis, 1834.

        „   ....On the Growth, and on the Bilateral symmetry of Echinodermata. 'Phil. Mag.,' ser. 3, V, p. 369.

1835. Desmoulins. Charles ....Études sur les Échinides. Premier Mémoire, Prodrome d'une nouvelle classification de ces Animaux. Second Mémoire, Généralités: étude analytique des parties solides de ces Animaux. Troisième Mémoire, Synonymie Générale. Bordeaux, 1835-37, 8vo.

1835. Cuvier et Brongniart ....Description Géologique des Environs de Paris. 8vo and 4to, atlas of plates, 3d edition.

1835. Agassiz. Louis ....Prodrome d'une Monographie des Radiares ou Échinodermes. (*Mémoires de la Soc. des Sc. Nat. de Neufchatel*, tom. I, p. 168.)

1835. Bronn. H. G. ....Lethaea Geognostica, oder Abbildungen und Beschreibungen der für die Gebirgs-formationen bezeichnendsten Versteinerungen. 2 Band. 8vo, atlas 47, plates 4to.

1835. BRODERIP. W. J..................On some Fossil Crustacea and Radiata found at Lyme Regis, in Dorsetshire. 'Trans. Geol. Soc.,' ser. 2, v, p. 171.

1835-45. LAMARCK. J. B. DE............Histoire Naturelle des Animaux sans vertèbres. 2d edit. Revue et augmentée par MM. G. P. Deshayes, Milne Edwards, et Dujardin. 11 vols. 8vo.

1836. GRATELOUP. DR. ....Mémoire de Géo-Zoologie sur les Oursins Fossiles (Échinides), qui se rencontrent dans les Terrains calcaires des environs de Dax. 'Actes de la Soc. Linn. de Bordeaux,' tome viii.

1836. ROEMER. FR. ADOLPH ....Die Versteinerungen des Norddeutschen Oolithen-Gebirges. Hanover, 4to, plates.

1836. BUCKLAND. REV. W. ....Geology and Mineralogy considered with reference to Natural Theology. 2 vols. 8vo, plates.

1837. PHILIPPI. TH. ....Beschreibung zweier missgebildeter See-Igel, nebst Bemerkungen über die Echiniden überhaupt. 'Wiegm. Arch.,' 1837, I, p. 241.

1837. KOCH F. und DUNKER ....Beiträge zur Kenntniss des Norddeutschen Oolithgebildes und dessen Versteinerungen. Brausnchweig. 4to, with plates.

1837. PUSCH. G. G. ....Polens Paléontologie. Stuttg., 4to, with plates.

1838. AGASSIZ. LOUIS ....Monographies d'Échinodermes, vivans et Fossiles. Première Monographie, des Salénies. Neuchâtel, folio, plates.

1839. LEYMERIE, A. ....Mémoire sur la partie inférieure au Système secondaire du département du Rhône. Mem. de la Société Géologique de France, tom. iii, p. 113, "Lias Échinodermes." Plates.

1839-42. DUBOIS DE MONTPEREUX. F. Voyage autour du Caucase. 6 vols. 8vo, atlas, plates, Paris.

1839. AGASSIZ. LOUIS ....Description des Échinodermes Fossiles de la Suisse. 2 parts, 4to, plates, 1839-40.

1839. SHARPEY. W. ....Cyclopædia of Anatomy, Article 'Echinodermata,' vol. ii.

1839-43. MUNSTER. G. VON ....Beiträge zur Petrefactenkunde. Heft 6. Bayreuth, 4to.

1840. SOWERBY. J. DE CARL...........Description of the Fossils from the Province of Cutch. 'Trans. Geol. Soc. Lond.,' 2d series, vol. iv.

1840. HAGENOW. F. VON ....Monographie der Rügenschen Kreide-Versteinerungen. IIte Abtheilung. Radiarien und Annulaten. ('Neues Jahrbuch für Min. Geo. und Petrefact. von Leonhard und Bronn.')

1841. DUNKER und VON MEYER ......Palæontographica: Beiträge zur Naturgeschichte der Vorwelt. 4to, plates, 1841-51.

1841. VALENTIN. G. ....Anatomie des Échinodermes. Première Monographie, Anatomie du genre Echinus. Folio, plates.

1841. SISMONDA. EUG. ....Monographia degli Echinidi fossili del Piemonte Torino. 4to, fig. 'Mem. Accad. Tor.,' IV, 2 ser., p. 3.

1841. KROHN. A. ....Ueber die Anordnung des Nervensystems der Echiniden und Holothurien im allgemeinen. 'Müller's Archiv,' and in Ann. des Sc. Nat.,' t. xvi, 1841.

1841. JONES. T. RYMER................A General Outline of the Animal Kingdom. 1 vol. 8vo, with woodcuts.

1841.   „   ....Lectures on the Natural History of Animals. 2 vols. 12mo.

1841. FORBES. EDWARD ....A History of British Star-fishes and other Animals of the Class Echinodermata. In 8vo, woodcuts of all the species.

1841. AGASSIZ. LOUIS ....Monographies d'Échinodermes, vivans et Fossiles. Seconde Monographie, des Scutelles. Neuchâtel, folio, plates.

1842. AUSTIN. T............Proposed Arrangement of the *Echinodermata*, particularly as regards the *Crinoidea* and a subdivision of the Class *Adelostella* (*Echinidæ*). 'Ann. and Mag. N. H.,' X, p. 106.

1842-51. KONINCK. L. DE............Description des Animaux Fossiles qui se trouvent dans le terrain Carbonifère de Belgique. 2 vols. 4to, plates and Supplement.

1842. DESOR. ED. ....Monographie d'Échinodermes. Troisième Monographie, des Galerites. Neuchâtel, folio, plates.

1842.   „   ....Monographie d'Échinodermes. Quatrième Monographie, des Dysaster. Neuchâtel, folio, plates.

1842. PHILIPPI. TH............Ueber *Clypeaster altus*, Scill., und einege Verwandte. 'L. & Br. N. Jahrb.,' 1842, p. 52, fig.

1842. MÜLLER und TROSCHEL ....System der Asteriden mit Zwölf Kupfertafeln. 1842.

1843. PORTLOCK. JOSEPH ELLISON ...Report on the Geology of the County of Londonderry. 8vo, plates.

1843. MORRIS. JOHN............A Catalogue of British Fossils. 8vo.

1843. SISMONDA. EUG. ....Memoria geo-zoologica sugli Echinidi del Contado di Nizza. Tor. 8vo, fig.

1843. KLIPSTEIN. A. ....Kentniss der Ostlichen Alpen. Giessen, 4to, 21 plates.

1844. PICTET. F. T. ....Traité élémentaire de Paléontologie. 4 vols. 8vo, plates, 1844-46.

1845. MURCHISON. SIR RODERICK ...Outline of the Geology of the Neighbourhood of Cheltenham. 2d ed., by J. Buckman and H. E. Strickland, 8vo, plates.

1845. GEINITZ. H. B............Grundriss der Versteinerungskunde. 8vo, mit 26 Steindrucktafeln, 1845-46.

1845-50 DUJARDIN. FEL. ....Dictionnaire universel d'Histoire Naturelle (d'Orbigny). 'The Genera of the Echinoderms,' by Dujardin.

1846. AGASSIZ et DESOR ....Catalogue raisonné des Familles, des genres, et des espèces de la classe des Échinodermes. 'Annales des Sciences Naturelles,' 3me série, vols. VI, VII, 1846-47.

1846. D'ARCHIAC. LE VICOMTE ......Description des Fossiles recueillis par M. Thorent dans les Falaises de Biaritz. 'Mem. Soc. Géol. France,' t. II, 2me série, p. 189.

1846.   „   ....Description des Fossiles recueillis dans les couches à Nummulites des environs de Bayonne. *Mém. de la Soc. Géol. de France*, 2e sér., vol. II, p. 189.

1846. SCHMIDT. F. A. ....Petrefacten-Buch order allgemeine und besondere Versteinerungskunde, mit Berücksichtigung der Lagerungsverhältnisse besonders in Deutschland. In 4to, mit Taf.

1846. LEYMERIE. A. ....Mém. sur le terrain à Nummulites des Corbières et de la montagne Noire. *Mém. de la Soc. Géol. de France*, 2e sér., vol. i, 337, 1846.

1846. MÜLLER. JOHANN ....Ueber die Larven und die Metamorphose der Ophiuren. 'Vorgetragen in der Königl. Akademie der Wissenschaften zu Berlin.'

1847. TALLAVIGNES ....Résumé d'un Mém. sur le terrain à Nummulites au département de l'Aude, et des Pyrénées. *Bull.*, 2e série, vol. iv, p. 1140.

1847. VOGT. C. ....Lehrbuch der Geologie und Petrefactenkunde. 2 Band. 8vo, plates und Illustrationen in Holzstich.

1847-52. HALL. J. ....Palæontology of New York. 2 vols. 4to, plates; vol. i, 1847, vol. ii, 1852.

1847-57. D'ARCHIAC. LE VICOMTE ...Histoire des Progrès de la Géologie de 1834 à 1845. Tome viii, 8vo.

1848. M'COY. FRED. ....On some New Species of Mesozoic Radiata. 'Ann. and Mag. Nat. Hist.,' ser. 2, ii, p. 397.

1848. GRAS. ALBIN ....Description des Oursins Fossiles du département de l'Isère. Grenoble, 8vo, plates.

1848. MÜLLER. JOHANN.................Ueber die Larven und die Metamorphose der Echinodermen. 'Vorgetragen in der Königl. Akademie der Wissenschaften zu Berlin.'

1848. DUVERNOY. G. L.................Mémoire sur l'analogie de composition, et sur quelques points de l'organisation des Échinodermes. Extrait du tome xx des 'Mém. de l'Acad. des Sciences,' pl. iii, Paris, 1848.

1848. FORBES. EDWARD ....Memoirs of the Geological Survey of Great Britain. 'On the ASTERIADÆ found fossil in British Strata,' p. 457. 'On the CYSTIDEÆ of the Silurian rocks of the British Islands,' p. 483, vol. II, part II.

1849.   „   ....Memoirs of the Geological Survey of the United Kingdom. Figures and descriptions illustrative of British organic remains. Decades I, III, IV. Asteriadæ and Echinidæ.

1849. BRONN. H. G. ....Index Palæontologicus. 3 Band. 8vo, 1849.

1849. AGASSIZ. LOUIS ....Lectures on Comparative Embryology. Boston, 8vo, woodcuts.

      „   ....Additions to Mr. S. Wood's Catalogue of Crag Radiaria. 'Ann. and Mag. Nat. Hist.,' vi, p. 343.

1849. D'ORBIGNY. ALCIDE ....Cours Élémentaire de Paléontologie et de Géologie. 2 vols., plates.

1849. HAIME. JULES ....On Milnia decorata. 'Ann. des Sc. Nat.,' 3me série, tome xii, p. 217, pl. 2.

1849-56. COTTEAU. GUSTAVE ....Études sur les Échinides Fossiles du département de l'Yonne. Tome I, Terrain Jurassique. Plates. Tome II, Terrain Crétacé.

1849-50. MÜLLER. JOHANN ....Ueber die Larven und die Metamorphose der Holothurien und Asterien. 'Vorgetragen in der Königl. Akademie der Wissenschaften zu Berlin.'

1850.   „   ....Anatomische Studien über die Echinodermen. 'Müller's Archiv,' Heft ii, 1850.

1850.   „   ....Berichtigung und Nachtrag zu den Anatomischen Studien über die Echinodermen. 'Müller's Archiv,' Heft iii.

1850.   „   ....Fortsetzung der Untersuchungen über die Metamorphose der Echinodermen. Ibid., Heft v.

1850. D'ORBIGNY. ALCIDE.............Prodrome de Paléontologie. 3 vols. 12mo.

1850. D'ARCHIAC. LE VICOMTE ......Description des Fossiles du groupe Nummulitique des environs de Dax et de Bayonne. Bull., 2e série, vol. iv, p. 1006, 1847; Mém. de la Soc. Géol. de France, 2e série, vol. iii, p. 397, 1850.

1850. BELLARDI. L. ....Liste des Fossiles de la formation Nummulitique du comté de Nice. 'Bull.,' 2e série, vol. vii, séance 17 Juin, 1850.

1851. VOGT. C. ....Zoologische Briefe Naturgeschichte. 2 Band 8vo, fig. Frankfurt.

1851. COTTEAU. GUSTAVE ....Catalogue Méthod. des Échinides Néocomiens. 'Bull. Soc. des Sc. Nat. de l'Yonne,' t. v, p, 285.

1851. MÜLLER. JOHANN ....Ueber die Ophiuren-larven des Adriatischen Meeres. 'Müller's Archiv,' Heft i, 1851.

1851. HUXLEY. THOS. H. ....Report upon the Researches of Professor Müller into the Anatomy and Development of the Echinoderms. 'Ann. and Mag. of Nat. Hist.,' 2d series, vol. viii, p. 1.

1851. GRAY. JOHN EDWARD ....Description of some new genera and species of Spantangidæ in the British Museum. 'Ann. and Mag. N. H.,' 1851, vol. vii.

1851. GRAY. JOHN EDWARD ....Catalogue of Echinidæ, or Sea-eggs, in the collection of the British Museum. Part I, *Echinida irregularia.* 12mo, 1851.

1851. WRIGHT. THOMAS ....On the CIDARIDÆ of the Oolites, with a description of some new Species of that Family. 'Ann. and Mag. of Nat. Hist.,' 2d series, vol. viii, p. 241.

1851. „ ....On the CASSIDULIDÆ of the Oolites, with descriptions of some new Species of that Family. 'Ann. and Mag. of Nat. Hist.,' 2d series, vol. ix, p. 81.

1852. BUVIGNIER ....Statistique Géologique. Mineral. et paléontologique du département de la Meuse. Paris, 1 vol. 8vo, atlas folio.

1852. QUENSTEDT. FR. AUG...........Handbuch der Petrefaktenkunde. 8vo, plates.

1852. WRIGHT. THOMAS ....Contributions to the Palæontology of the Isle of Wight. 'Ann. and Mag. of Nat. Hist.,' 2d series, vol. x, p. 87.

1852. „ ....Contributions to the Palæontology of Gloucestershire. A description and figures of some new species of Echinodermata from the Lias and Oolites. 'Ann. and Mag. of Nat. Hist.,' 2d series, vol. xiii, p. 161.

1853. MÜLLER. JOHANN.................Ueber den Bau der Echinodermen. 'Vorgetragen in der Königl. Akademie der Wissenschaften zu Berlin.' Plates.

1853-55. D'ORBIGNY. ALCIDE ....Paléontologie Française. Terrains Crétacés, tome Sixième, contenant les Échinodermes. Plates, 8vo.

1854. DE KONINCK et LE HON.........Recherches sur les Crinoides du Terrain Carbonifère de la Belgique, suivis d'une notice sur le genre Woodocrinus. Brux., 4to, plates.

1854. HUXLEY. THOS. H. ....On the Structure of the Echinoderms. By Johann Müller, translated by Thomas Huxley, F.R.S. 'Ann. and Mag. of Nat. Hist.,' 2d series, vol. xiii, pp. 1, 112, 241 ; 1854.

1854. STROMBECK. A. VON ....On the Echinodermata of the Hilo-conglomerate in north-western Germany. 'Quart. Jour. of the Geol. Soc. ; Geol. Memoirs,' p. 13, vol. ii.

1854-58. PICTET et RENEVIER ....Fossiles de Terrain Aptien de la Perte du Rhône. 4to, plates.

1855. GRAY. J. E. ....Catalogue of the recent Echinidæ, or Sea-eggs, in the Collection of the British Museum. Part I, *Echinida irregularia.'* 12mo, with plates.

1855. WRIGHT. THOMAS ....On Fossil Echinoderms from the Island of Malta, with notes on the stratigraphical distribution of the Fossil Organisms in the Maltese beds. 'Ann. and Mag. of Nat. Hist.,' 2d series, vol. xv, plates vii.

1855. PETERS. WILHELM ....Ueber die au der Küste von Mossambique beobachteten Seeigel, und inbesondere über die Gruppe der Diademin. 'Vertrag. Königl. Akademie Wissenschaften zu Berlin,' 1853, plate.

1855. COTTEAU. G........................Note sur un nouveau genre d'Échinide fossile, Genre *Desorella.* 'Bull.,' 2e sér., t. xii, p. 710.

1855. TUOMEY and HOLMES ....Fossils of South Carolina, Echinidæ, 4to, plates.

1855. WRIGHT. THOMAS ....On a new genus of Fossil CIDARIDÆ, with a Synopsis of the species included therein. 'Ann. and Mag. of Nat. Hist.,' 2d series, vol. xvi, p. 96.

„ ....On some new species of *Hemipedina* from the Oolites. 'Ann. and Mag. of Nat. Hist.,' 2d series, vol. xvi, p. 98.

1855-60. WRIGHT. THOMAS ....Monograph on the British Fossil Echinodermata from the Oolitic Formations. Published by the Palæontographical Society of London. Vol. i, 4to. The Echinoidea, with 44 plates.

1856. DESOR. ED. ....Classification des Cidarides. 'Bullet. de la Soc. des Sciences Naturelles de Neuchâtel,' tom. iv, 8vo, plate.

1856. LEYMERIE et COTTEAU ....Catalogue des Échinides Fossiles des Pyrénées. 'Bull. de la Soc. Géol. de France,' t. xiii, p. 319.

1856. COTTEAU. G......................Sur quelques Oursins du département de la Sarthe. 'Bull. de la Soc. Géol. de France,' t. xiii, p. 646.

1856. SALTER and WOODWARD.........Memoirs of the Geological Survey of the United Kingdom. Figures and descriptions illustrative of British organic remains. Decade V. Echinidæ.

1857. ETALLON. M. A. ....Ésquisse d'une description Géologique du Haut-Jura. Paris, 8vo.

1857. COTTEAU et TRIGER.............Échinides du département de la Sarthe. Liv. 1—5, with beautiful plates.

1858-60. COTTEAU. G. ....Échinides nouveaux ou peu connus. 'Revue et Mag. de Zoologie,' No. 5, Paris.

1858. „ ....Note sur l'appareil apicial du Genre *Goniopygus*. 'Bull. Soc. Géol. Fr.,' t. xvi, p. 162.

1858. ETALLON. M. A. ....Études Paléontologiques sur le Haut-Jura Rayonnés du Corallien. Ext. des 'Mém. de la Soc. d'Emulation du départ. du Doubs,' séance Mai 8, 1858.

1858. EBRAY. TH. ....Études Paléontologiques sur le départ. de la Nièvre; famille des COLLYRITIDÆ.

1858. QUENSTEDT. FR. AUG............Der Jura. 8vo, plates, Tübingen.

1858-59. BILLINGS and SALTER ......Geological Survey of Canada. Figures and descriptions of Canadian organic remains. Decade III. CYSTIDEÆ and ASTERIDEÆ, by E. Billings. Genus *Cyclocystoides*, by E. Billings and J. W. Salter. Decade IV. CRINOIDEÆ from the Lower Silurian, by E. Billings.

1859. MICHELIN. HARDOUIN...........Notice descriptive de quelques espèces nouvelles d'Échinides, famille des Clypéastroides, tribu des Laganides. 'Revue et Mag. de Zoologie,' No. 9, 1859.

1859. COTTEAU. G......................Note sur le Genre *Galeropygus*. 'Bull. de la Soc. Géol. de Fr.,' t. xvi, p. 289, Paris.

1859. „ ....Notice Bibliographique. Synopsis des Échinides Fossiles, par Ed. Desor. 'Revue et Mag. de Zoologie,' No. 2, Paris.

1860. „ ....Note sur le genre *Heterocidaris*, nouveau type de la famille des Cidaridées. 'Bull. de la Soc. Géol de France,' 2e série, t. xvii, p. 378.

1860. „ ....Note sur les Echinides recueillis en Espagne par MM. de Verneuil, Triger, et Collomb. 'Bull. de la Soc. Géol. de France,' 2e série, t. xvii, p. 372.

1860. MICHELIN. HARDOUIN...........Monograph on the genus *Clypeaster*, with figures of all the species, Living and Fossil. Folio, plates. As this work is not yet published, I am unable to give its correct title. The author has kindly sent me some of its magnificent plates.

ADDITIONAL NOTES ON THE ECHINOIDEA.

SINCE the commencement of this Monograph in 1855, I have obtained additional information on many of the species described therein, and have lately discovered some new forms. I purpose, therefore, giving notes on the described species under their respective names, and in the same order as they were originally figured in the body of the work. The descriptions of the new species which are lithographed in the supplementary plates will be found among the genera to which they belong.

CIDARIDÆ.

CIDARIS FOWLERI, *Wright.* Supplement, Pl. XLII, fig. 1, *a, b, c, d, e, f.*

The fine specimen figured in this plate was found in a mass of pea-grit on Leckhampton Hill; the block containing the urchin had been long exposed to the atmosphere, and the test, in consequence, is a little weathered in parts. It has seven primary spines, more or less perfect, with several secondary spines attached *in situ* to the test. The primary spines (Pl. XLII, fig. 1 *e*) are long, slender, and nearly of a uniform diameter throughout, apparently tapering very little towards their free extremity. The head is strongly crenulated, the milled rim prominent, the neck smooth and of the same thickness as the body; the surface of the long stem is covered with short, thorn-like tubercles, which have their points directed forwards. As all the spines are more or less fractured, their proportionate length to the diameter of the test cannot be ascertained. The secondary spines (fig. 1 *f*) are small and spatulate; many of them are still adherent to the plates of the test. Fig. 1 *b* represents an inter-ambulacral plate, with its primary tubercle and circle of areolar granules; fig. 1 *c*, a profile of one of the large tubercles, showing the prominence of the crenulations; fig. 1 *d*, one of the jaws of the lantern, magnified twice.

CIDARIS SMITHII, *Wright.* Supplement to pages 50—52.

When I figured the very fine specimens of Pl. II, considerable doubts existed relative to the locality whence they were collected. My lamented colleague, Professor Forbes, was impressed with the idea that they were foreign fossils, and was averse to their being drawn as British Echinidæ; the history of the specimens, however, convinced me that they were English, although their locality was then unknown. Some time after the publication of the first part of this work, I went to Oxford for the purpose of examining the late Dr. Buckland's collection, and in one of the drawers of his cabinet I saw a *Cidaris Smithii*, from the same rock as that in which my doubtful specimens were imbedded. I lost no time in visiting the locality, Bullington Green, near Oxford, whence it was obtained, where I found several plates of tests and fragments of spines of the species. From this locality Mr. Whiteaves, of Oxford, lately collected a very large specimen of *Cidaris Smithii*, which, through that gentleman's kindness, is now in my cabinet. This test measures three inches and three quarters in diameter, one third more than the largest specimen previously known, and has the jaws and teeth *in situ*. In the same stratum of Coral Rag at Bullington, I found *Echinobrissus scutatus*, Lamk., *Pygaster umbrella*, Agas., and *Pygurus pentagonalis*, Phil. I have lately obtained *Cidaris Smithii*, Wr., from the Coral Rag at Hillmarton, Wilts, where it is associated with *Cidaris florigemma*, Phil., *Pseudodiadema versipora*, Woodward, and *Pygaster umbrella*, Agass. This is the original locality whence Dr. William Smith obtained the specimen now in the British Museum.

DIPLOCIDARIS WRIGHTII, *Desor*.   Pl. XLI, fig. 6, 7, Supplement to page 58.

A very fine, large specimen of this gigantic Cidaris was discovered by my friend, the Rev. T. W. Norwood, in the Inferior Oolite at Shurdington Hill, near Cheltenham. Unfortunately, I have only been able to figure one of its largest spines. This was undoubtedly, one of our largest Oolitic Cidaridæ, and, from Mr. Norwood's description, must have attained a gigantic size, as appears by the following note, which that gentleman has kindly supplied.

"The urchin, of which I sent you the fragments about a year ago, was found by me in the Pisolite of Shurdington Hill, under the following circumstances. A very thin, sandy, band divided two compact and indurated rock-masses forming a plane of easy and natural separation between them. In this band the urchin had been locked up, apparently in a state of wonderful preservation, and in the posture of life, with its equator evenly parallel to the divisional plane of the strata, and its magnificent spines (such as I have nowhere else seen) radiating regularly around it. The ground had chanced to be broken at this point for the purpose of quarrying stone for wall-making; the upper rock-bed had been removed down to the sandy band; and, in its removal, had torn the urchin in two at the equator, and carried away half the test and a corresponding number of spines. Therefore, when I came to the place and discovered the specimen, it was lying on the surface of the lower rock-bed, showing five or six large spines, which appeared to diverge from a circular space, about equal in diameter to the equatorial section of *Cidaris Fowleri* or *C. florigemma*. On the spur of the moment and with insufficient tools I rashly attempted to detach my treasure; and as it turned out that the lower portion of the test was fast imbedded in the hard rock beneath, I had the sad mortification to fracture and destroy it. As I soon afterwards handed over to you the fragments that remained, I now leave the description of them in your hands. This urchin was associated, in the same rock-surface, with *Pygaster semisulcatus, Terebratula simplex,* and other equally characteristic fossils of the Cheltenham Pisolite."

HEMICIDARIS BRILLENSIS, *Wright*, nov. sp. Supplement, Pl. XLIII, fig. 2 *a, b, c, d.*

Test sub-globose, ambulacral areas wide, with two rows of primary tubercles, which extend over three fourths of the area; inter-ambulacral areas narrow, with two rows of small, nearly equal-sized, tubercles, ten in each row; apical disc large, plates very narrow, in consequence of the wideness of the vent; mouth-opening large, peristome divided into ten nearly equal lobes; poriferous zones narrow and much undulated; pores very much crowded at the base.

Dimensions.—Height, one inch; transverse diameter, one inch and a half.

Description.—This remarkable urchin, at first glance, resembles an *Acrocidaris*, from the size, number, and development of the tubercles on the ambulacral areas. In most of the other forms of *Hemicidaris* the semi-tubercles are limited to the basal region of the ambulacra, but in this species they extend through nearly three fourths of the area; this region of the test is likewise much wider than in other congeneric forms, in which the semi-tubercles are limited to the base of the area, and the margins thereof are occupied by rows of small granules. There are about ten tubercles, very regularly arranged, in each row; those at the base are very small, whilst the upper six pair are nearly of a uniform size, although rather smaller than the primary tubercles of the inter-ambulacra; the upper part of the area has only a few, small granules on its margin. The poriferous zones are very narrow, and extremely flexuous, winding round the border of the large semi-tubercles, and only becoming straight at the upper fourth, where they cease (fig. 2 *c*). The pores are separated by a thick septum, and there are six pairs of holes opposite each inter-ambulacral plate; they are much crowded together, in oblique rows, in the wide spaces left by the small semi-tubercles at the base of the areas.

The inter-ambulacral areas are narrow, scarcely twice the width of the ambulacra; they are occupied by two rows of primary tubercles, about ten in each row, of a moderate size, and nearly uniform magnitude throughout (figs. 2 *a, b*); they are raised on prominent bosses, with deeply crenulated summits (fig. 2 *c*); the areolas are transversely oblong, and confluent above and below; a double row of small granules descends in a zigzag line down the middle of the area, and the zonal border of the plates has a single row of the same sized granules, which separates the areolas from the poriferous zones, and forms a series of crescents throughout the area (fig. 2 c).

The apical disc is large (fig. 2 a), and placed rather behind the vertex of the test; the ovarial plates are narrow (fig. 2 d), and the duct-holes perforated near the apices of the plates; the madreporiform body occupies, as usual, the right antero-lateral plate, which is the largest; the three anterior ocular plates rest upon the ovarial plates, with which they alternate, whilst the two posterior oculars are placed between the two postero-lateral and single ovarial plates (fig. 2 d); the vent-opening is very large.

The base is flat, and the mouth-opening wide; the peristome is divided into ten nearly equal-sized lobes; all the tubercles in this region are small, and the poriferous zones at the base of the ambulacral areas are very much crowded, the pores being arranged in oblique rows.

Affinities and differences.—This species differs so much in its general physiognomy from the typical forms of its congeners that it requires a careful examination to be satisfied that it is a *Hemicidaris*, the size and number of the semi-tubercles, extending as they do so high up the area, and the small and uniform magnitude of the primary tubercles, produce so many rows of tubercles on the flanks of this urchin (fig. 2 b, d), that it might readily be referred to the genus *Acrocidaris* rather than to the group to which it belongs. In *Acrocidaris*, however, the tubercles are very unequal in magnitude on the sides of the test, and each ovarial plate supports on its centre a small, primary, perforated and crenulated tubercle, a character which is quite diagnostic of this genus. The greater width of the ambulacra, and the presence of a double row of semi-tubercles, extending three parts up the sides, distinguish this species from its congeners; the smallness of the primary tubercles in the inter-ambulacra, and the increased number and nearly uniform size of the same throughout the rows, distinguish it likewise from *Hemicidaris Davidsoni*, Wr., another Portland species, with which it has many affinities.

Locality and Stratigraphical position.—This urchin was discovered in the Portland Oolite, at Brill, and was purchased from the person who collected it therefrom by my friend, the Rev. P. B. Brodie, who kindly communicated the specimen for this work; the test is rather distorted, and much concealed by a small, encrusting oyster. At the same locality my friend collected the large *Echinobrissus Brodiei*, Wr., of which I have given a figure of the natural size in Pl. XLIII, fig. 3.

DIADEMADÆ.

Genus—HETEROCIDARIS, *Cotteau*, 1860.*

Test large, circular, depressed, inflated at the sides, sub-convex above, almost flat below; the inter-ambulacral areas very wide, and provided with from six to eight rows of large, nearly equal-sized, perforated tubercles, raised on prominent bosses, with crenulated summits; the areolæ are narrow, and their circumference surrounded by a circle of small, equidistant granules, a few only of which are distributed on the intermediate surface of the plates.

The ambulacral areas are straight, very narrow, and slightly flexuous above; they are furnished with two rows of small, distinct, perforated tubercles, uniform in size, and raised on small bosses, which are placed in regular rows on the margin of the area; three tubercles occupying the depth of each inter-ambulacral plate.

The poriferous zones are narrow; the pores are small, simple, non-conjugate, and superimposed, having a slight disposition to a trigeminal arrangement near the mouth.

The mouth-opening is large and pentagonal, about one third the diameter of the test; from the narrowness of the ambulacra, the lobes of this portion of the peristome are much smaller than those of the inter-ambulacra.

The spines are long and cylindrical; their surface is covered with fine, longitudinal lines, having small, indistinct tubercles interspersed amongst them.

I refer this genus to the family DIADEMADÆ, as I have defined it.† The size of the test, the narrowness of the ambulacra, the width of the inter-ambulacra, and the numerous rows of primary tubercles thereon, indicate that *Heterocidaris* has certain affinities with *Astropyga*, although it possesses many characters by which it is readily distinguished from that genus. *Heterocidaris* resembles some of the large forms of *Hemipedina*, as *H. Marchamensis*, Wr., but the deep crenulations on all the bosses shows it to be distinct from that form. M. Cotteau observes,‡ that the genus *Heterocidaris*, notwithstanding its resemblance to the DIADEMADÆ, is separated from that family by a character of the first order, namely, the structure of the peristome, which is pentagonal, and furnished with ambulacral lips much more narrow than those which correspond to the inter-ambulacra, whilst in the DIADEMADÆ the peristome is always decagonal, and notched by ten incisions more or less deep. He therefore places this genus in the family CIDARIDÆ of Desor, which forms, according to that author, however, a much larger group than the family CIDARIDÆ of this Monograph.§

* Extrait du 'Bulletin de la Société Géologique de France,' 2me série, tom. xvii, p. 378, pl. iv.
† See p. 18 of this **Monograph.**
‡ Extrait du 'Bulletin de la Société Géologique de France,' 2me série, tom. xvii, p. 380, pl. iv.
§ See p. 18.

HETEROCIDARIS WICKENSE, *Wright*, nov. sp.   Supplement, Pl. XLIII, fig. 5 *a, b, c*.

The only portions of this urchin I have seen were some fragments I found in the collection of my friend, Mr. Leckenby, of Scarborough, and which were collected from a sandy bed of Inferior Oolite at Blue Wick, near Robin Hood's Bay, on the Yorkshire coast. The fragment figured consists of four plates (fig. 5 *a*), representing one half of an inter-ambulacral area. There are three rows of large, equal-sized tubercles on each plate, which are perforated; the bosses are prominent, and their summits deeply crenulated (fig. 5 *c*); the narrow areolæ are surrounded by a circle of small granules; other granules are likewise sparsely scattered over the intermediate surface of the plates. This is the urchin which Professor Phillips refers to in his work on 'The Geology of Yorkshire,' of which he figures a single tubercle (pl. xi, fig. 2), and catalogues, at page 155, as a Cidaris from the Dogger of Blue Wick.

The only *Heterocidaris* known to M. Cotteau was obtained by M. Triger from the Inferior Oolite of Chevain (Sarthe). This magnificent specimen (*Heterocidaris Trigeri*, Cot.) has been figured in the 'Bulletin de la Société Géologique de France,' 2me série, t. xvii, p. 378, pl. iv, and likewise in the 'Échinides du département de la Sarthe,' pl. lvi. M. Babeau has collected a fragment of another test from the Inferior Oolite of the environs of the Langres (Haute-Marne), from a rock which contained *Cidaris spinosa* and *C. Courtandina*.

PSEUDODIADEMA LOBATUM, *Wright*, nov. sp.   Pl. XLI, fig. 3 *a, b*, Supplement.

Test depressed; ambulacral areas narrow; inter-ambulacra wide, with two prominent rows of primary tubercles; spines long, smooth, and pin-shaped; neck, ring, and head covered with fine, longitudinal lines; stem smooth and uniform in thickness, tapering only near the point.

Dimensions.—Indeterminate.

Description.—This small urchin was found at Pinhay Bay, near Lyme Regis, in a thin band of marl appertaining to the zone of *Ammonites planorbis*. All the specimens I have hitherto seen are so imperfect that they are insufficient for drawing up a complete diagnosis of the species.

The ambulacral areas are narrow, but as all the specimens I have seen are unfortunately fractured through this region, the number and size of the tubercles thereon cannot be examined.

The inter-ambulacral areas are wide and well-developed, possessing two rows of large, primary, perforated tubercles, raised on prominent bosses, with deeply crenulated summits; they are surrounded by wide areolas, bounded by a defined margin, and encircled with microscopic tubercles, which impart an ornamented character to the test.

The spines are long and slender, the head is stout, the margin of the acetabulum deeply crenulated (fig. 3 *b*); the milled ring prominent, the neck short, and both are sculptured with fine, longitudinal lines; the long, slender stem is nearly of the same thickness throughout, tapering to a point near the extremit. The proportionate length of the spine to the diameter of the test cannot be ascertained, for, although entire individual spines are abundant in the marl, those attached to the test are nearly all fractured.

Affinities and differences.—As *P. lobatum* is the oldest representative of the genus Pseudodiadema, in the Oolitic rocks, it is unfortunate, from the crushed state in which the test is found, that a critical comparison cannot be made between this species and its other Oolitic congeners. The very narrow ambulacra have few tubercles thereon, but as all the tests I have examined are fractured across this part, the details of its structure cannot be seen. The affinities between this urchin and some of the depressed *Acrosalenias* is considerable, and the length of the spines in proportion to the size of the test renders that relation still more remarkable.

Locality and Stratigraphical position.—This urchin was recently discovered at Pinhay Bay, near Lyme Regis, in a bed of mottled clay, on the shore at low-water mark; many tests were found together, with numerous long, slender spines strewed in abundance amongst them; some of the spines were attached to their respective tubercles, so that the identity of the spines is satisfactorily proved. This bed of clay appertains to the lower division of the Lower Lias, and may probably correspond to a similar urchin vein found in Warwickshire and Gloucestershire, at the base of the zone of *Ammonites planorbis*.

HEMIPEDINA TOMESII, *Wright*, nov. sp.

Hemipedina Tomesii, Wr.   *Hemipedina Tomesii*, Wr.

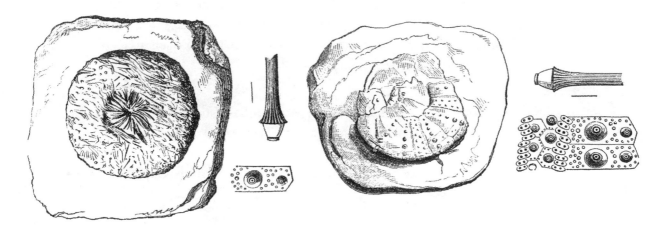

Test circular, depressed; ambulacral areas wide, with two rows of small tubercles on the margin of the area, set moderately distant apart; poriferous zones narrow; pairs of pores superimposed in groups of threes; inter-ambulacral areas with two rows of primary tubercles on the centre of the plates, and two rows of secondary tubercles internal to the primaries, which extend from the base and sides above the equator; areolæ wide, encircled by granules which likewise cover the surface of the plates. Spines long, slender, and needle-shaped; surface covered with fine, longitudinal lines.

Dimensions.—Height unknown; transverse diameter, one inch and two tenths.

Description.—The specimens of this urchin hitherto found are so much crushed and broken that it is impossible to make an accurate description of the species. The test is circular and depressed; the ambulacra are one third the width of the inter-ambulacra, and provided with two rows of small tubercles, which occupy the margin of the area; they are placed at a distance equal to the diameter of their areolæ apart from each other, and a delicate, zigzag line of small granules descends down the centre of the area; the poriferous zones are narrow, and in their upper part the pairs of holes manifest a disposition to a trigeminal arrangement, the inclination of the rows being upwards and outwards, the reverse of the direction in the genus *Pedina*. There are four or five pairs of holes opposite one large plate. The inter-ambulacral areas are three times as wide as the ambulacra at the circumference; one complete row of primary tubercles occupies the centre of the plates, and one incomplete row the inner portion thereof; the latter extend from the base and sides to two plates above the circumference. The primary tubercles at the equator have prominent bosses, which diminish in size on the upper surface; the areolæ are wide, smooth, well defined, and confluent above and below, laterally they are bounded by semicircles of small granules; at the zonal side of the plates there are three or four rows of the same-sized granules, and they likewise form circles around the incomplete rows on the centre of the area. The spines are slender and needle-shaped; the milled ring is prominent, and the surface of the stem covered with well-marked, longitudinal lines. All the specimens I have seen lie on their base on the matrix, and in none of them is the apical disc preserved.

Affinities and differences.—This urchin is much larger than *Hemipedina Bowerbankii*, Wr., which was collected from the same zone of the Lower Lias, near Lyme Regis. It resembles *Hemipedina seriale*, Leym. (Pl. IX, fig. 3 *a*), from the Lower Lias of France, but the inner row of tubercles in the inter-ambulacra are more developed and have a greater extension in that species.

Locality and Stratigraphical position.—This urchin was discovered by Mr. R. Tomes on a slab of Lower Lias, at Binton, in Warwickshire,* and in the White Lias at Stoney-thorpe, in the same county. The rock at Binton which contained this urchin comes from the base of the zone of *Ammonites planorbis*, and is known to the workmen as the Guinea Bed. It contains the bones of Saurian reptiles, &c., with the shells of *Avicula longicostata*, Stutch, *Lima punctata*, Sow., *Ostrea liassica*, Strick., and a small Coral. It may be justly considered as one of the basement beds of the Lower Lias, and this Hemipedina one of the earliest forms of the DIADEMADÆ in the Liassic rocks.

PEDINA SMITHII, *Forbes*.   Supplement, Pl. XLI, fig. 2 *a, b, c;* Pl. XLIII, fig. 1 *a, b, c, d.*
See pages 176-178, 'Monograph.'

PSEUDOPEDINA NODOTI. Cotteau, Revue et Magasin de Zoologie, No. 5, 1858, pl. ii, figs. 4—7.

Since I figured the original specimen of this species (Pl. XIII, fig. 2), which was collected by Dr. William Smith from the Inferior Oolite at Tucking Mill, I have met with two specimens from the Inferior Oolite, near Cheltenham; one from the Great Oolite near Cirencester, and one from the Cornbrash at Islip, near Oxford. One of the Inferior Oolite specimens was obtained from the Oolite marl near the Seven Springs, and is figured in Pl. XLI, fig. 2. The inter-ambulacral areas are very wide, and the plates com-posing them large; on the sides and upper surface there is only one row of primary tubercles situated very near the poriferous zones (fig. 2 *b*), and all the inter-tubercular space is covered with very small granules. The ambulacral areas are narrow, and taper much; the tubercles are few in number, very small, and sparsely distributed on the upper part of the area, but are larger and more numerous below. The proximity of the primary tubercles to the poriferous zones, the narrowness of the ambulacra, and the sparse distribution of tubercles thereon, with the wide space down the middle of the inter-ambulacra, which is occupied entirely with small granules, produce a remarkable physiognomy in this urchin.

The specimen from Oxfordshire was discovered by Mr. Whiteaves in the Cornbrash at Islip, and was presented by him to the Oxford Museum. I have figured this beautiful fossil in Pl. XLIII, fig. 1; as it is much more depressed than the Inferior Oolite varieties, although it evidently belongs to the same species; the base is flat, the mouth-opening large, and the peristome deeply divided by notches (fig. 1 *b*); the tubercles are much more abundant at the base (fig. 1 *b*), a second row occupying the middle of the inter-am-

* The reader is referred for a detailed description of this section to the author's memoir, 'On the Zone of *Avicula contorta*, and the Lower Lias in the South of England.' "Quart. Journ. of the Geol. Soc.," vol. xvi, p. 394.

bulacra in this region of the test; these, however, are limited to the base, for on the sides (fig. 1 c) and upper surface (fig. 1 a) there is only a single row of tubercles; at the base of the ambulacral areas the tubercles are larger, and the pores, closely packed together, lie obliquely across the zone in groups of threes (fig. 1 d).

Stratigraphical distribution.—I know this urchin from the Pea Grit and Oolite Marl, zone of *Ammonites Murchisonæ*, Inferior Oolite, from Crickley Hill, and the Seven Springs, near Cheltenham.

The specimen from the Great Oolite near Cirencester was almost entirely denuded of its test, but the position of the tubercles near the poriferous zones served to identify the species. This specimen was collected by Mr. Bravender, and kindly communicated for this work.

The Cornbrash specimen is circular, and much more depressed than any of the other varieties. It has enabled us to describe and figure correctly the entire external structure of this singular form.

Dr. Smith's specimen had a very marked pentagonal base, and although this character is absent in most of the specimens I have examined, still I have found the pentagonal outline to characterise one specimen from the Pea Grit and one from the Oolite Marl.

M. Cotteau recognised the resemblance which exists between this urchin and *Pedina Bakeri*, Wr., but the absence of a good figure of *Pedina Smithii*, Forbes, rendered it impossible for that learned author to discover the identity of his *Pseudopedina Nodoti* with *Pedina Smithii*. The excellent figures which I have now given will show that the French and English forms belong to the same species. M. Cotteau found the specimen figured by him in the Museum of Dijon; it was obtained from the Étage Bathonien, route de Fauge (Côte-d'Or), where it is very rare.

SALENIADÆ.

Acrosalenia pustulata, *Forbes.* Supplemental to pages 242–245.

I am indebted to Frederick Bravender, Esq., for the following notes on the discovery, in December, 1858, of a bed of marl in the Great Oolite near Cirencester, which contained immense numbers of *Acrosalenia pustulata*. He remarks—"We have discovered an extraordinary urchin-bed in a quarry near the town, but unfortunately the urchins are nearly all of the same sort. They occur in a marly bed in the Great Oolite, about four inches above the clay bed. I have now as many as 500 specimens, and might have got 1000 if I wanted them, as they were as thick as bees in a hive. If the bed extends any further, which will be ascertained when the quarrymen proceed, any quantity might be obtained.

The following section will afford an idea of the relative position of this marly vein with its *Acrosolenia*.

Section of the Urchin Quarry near Cirencester.

No. 1.	BRADFORD CLAY REPRESENTATIVE, with *Terebratula digona*, Sow.	Six feet.
2.	GREAT OOLITE.	Six feet.
3.	MARLY VEIN, with *Acrosalenia pustulata*, Forb.	Two inches.
4.	BAND OF STONE.	Four inches.
5.	CLAY BED.	Two feet.
6.	GREAT OOLITE.	

" The bed No. 3 is the one where the urchins *Acrosalenia pustulata*, Forbes, occur. The only other specimen of a different kind is *Holectypus depressus*, Lamk. The beds Nos. 2, 3, 4, and 5 have been called Forest Marble, which we do not approve of, as bed No. 2 is as decidedly freestone as the bed No. 6."

Thomas C. Brown, Esq., of Cirencester, has likewise kindly furnished me with the following note on this remarkable urchin-bed. He says—" In January, 1859, a great number of the *Acrosalenia pustulata* were found at Cirencester in the Great Oolite, eight or ten feet below the top of that stratum. A space of four or five yards square, two inches thick, was filled with this urchin, about 1000 in every superficial yard. They were found one upon another, about three deep, in a bed of white, marly clay. The tests were filled with this clay, and were found in a high state of preservation, with their spines recumbent upon them. It is presumed that this species is gregarious, and that a shoal of them were choked in a stream of mud; that they fell down together with the mud upon the Oolitic Rock then in course of formation, and were covered up with subsequent deposits of Oolitic matter. This species is not numerous in this district. The tests vary in size and shape, probably from a difference in age and sex, and there is great diversity in the form and number of the plates forming the apical disc. After repeated washings of the clay in water, fragments of the test were found, with broken spines of the larger and smaller ones; some of the latter are of a purple colour, many loose teeth, and one perfect set, together with Oolitic grains, but scarcely any other fossil."

ACROSALENIA PARVA, *Wright*, nov. sp.

Ambulacral areas with two rows of small marginal tubercles; inter-ambulacral tubercles large at the equator, and small on the upper and under surfaces; mouth opening wide, indistinctly decagonal, spines long and hair-like.

Dimensions.—Transverse diameter of the largest test two lines; height unknown.

Description.—The ambulacral areas of this Acrosalenia are moderately wide with two rows of small, perforated tubercles on the margins of the areas, a smaller tubercle alternating with a larger one throughout the row. The poriferous zones are narrow, and the holes large and distant from each other. The inter-ambulacral areas are wide, the plates have a single row of primary tubercles near their zonal sides, the tubercles near the equator are large, and raised on prominent bosses with crenulated summits; the areolæ are narrow, and confluent above and below, a semicircle of microscopic tubercles encircles the boss on its zonal side, and a zig-zag line of tubercles occupies the middle of the area which forms similar crescents on the sutural side of the areolæ; all these microscopic tubercles are perforated, a fact which can only be ascertained by the aid of a microscope with a half-inch object glass. The crenulations on the summit of the large bosses, when seen in profile with the microscope, resemble a circle of beads around that prominence.

The mouth-opening is wide, and indistinctly decagonal. The long, fine, and hair-like spines are scattered in profusion over the surface of the slab. When examined with a half-inch object glass, their surface is seen to be covered with sharp longitudinal lines, having an indistinctly undulated edge.

Affinities and differences.—This species differs from *Acrosalenia minuta* of the Lower Lias in having the ambulacral areas much better defined, and the tubercles of the inter-ambulacra larger and more prominent. Although a very small urchin, its generic characters are well marked; in the general neatness of its test it resembles some of the young forms of *Acrosalenia spinosa*, Ag., from the Cornbrash; this urchin affords another of those examples, so numerous among the Echinodermata, that the earliest forms of genera are, in general, those in which the typical characters of the group are best developed.

Locality and Stratigraphical position.—This small urchin was found by Mr. Tomes, who has kindly communicated it for description in the Lower Lias of Warwickshire, in the zone of *Ammonites obtusus*, it was associated with *Ammonites Birchii*, Sow., and has numerous small Gasteropoda and Conchifera imbedded with it on the same slab. It is the oldest *Acrosalenia* that has yet been found in the Lower Lias.

ECHINOCONIDÆ.

PYGASTER SEMISULCATUS, *Phillips.*   Pl. XLIII, fig. 6, Supplement to pages 275-78.

Professor Phillips found in the Inferior Oolite of Whitwell the original specimen of *Pygaster semisulcatus;* from the outline of the vent of that figure, it is still, however, doubtful whether the urchin he figured in Pl. III, fig. 17, of his 'Geology of Yorkshire,' was a Whitwell specimen.   I am of opinion that it was *Pygaster umbrella,* Ag., from the Coralline Oolite of Ayton or Hildenley that formed the type, and not a Whitwell specimen at all.   At that time, and for long afterwards, both urchins were considered to belong to one species, and it is very probable that, as much finer specimens of *Pygaster* were collected from the Coralline Oolite than had been obtained from Whitwell, a specimen from the Coralline Oolite was preferred for the drawing.   I have shown in my articles on these two species how perfectly distinct *Pygaster semisulcatus* is from *Pygaster umbrella,* and any one carefully comparing our description with Professor Phillips's figure will at once discover that the vent-opening in his drawing is the keyhole-shaped vent of *P. umbrella,* and not the wide opening of *P. semisulcatus.*

Having lately found a very good specimen of the true *Pygaster semisulcatus* from the Inferior Oolite of Whitwell in the collection of my friend, C. W. Strickland, Esq., of Hildenley, I have figured a portion of the posterior view of this urchin, with the view to exhibit the form of the vent-opening.   It will be observed that this aperture is much smaller than the vent-opening of *P. semisulcatus* from the Inferior Oolite of Gloucestershire (Pl. XIX, fig. 1), and does not extend so far down the single inter-ambulacrum as in that specimen; the tubercles are likewise more sparse upon the Yorkshire urchin, and the mouth-opening is relatively smaller.   It is important to note these characters, as they belong more to varieties of a given type than to a new specific form, and serve to teach us that, before the history of a species can be fully written, it is necessary to collect different individuals of the same species from localities widely apart, in order that we may estimate the degree of variation which changes of physical conditions were capable of exercising on the secondary characters of specific forms.

PYGASTER MACROSTOMA, *Wright.*   Supplement, Pl. XLI, fig. 4 *a, b, c;* fig. 5 *a, b.*

Test depressed, pentagonal; anal opening large, wide, occupying nearly two thirds of the single inter-ambulacrum; sides tumid, base convex from the peristome to the border, mouth-opening large, one fourth the diameter of the test.

Dimensions.—Height, seven tenths of an inch; transverse and antero-posterior diameters of the test nearly equal, two inches.

Description.—This urchin is remarkable for the length and width of the anal aperture, and for the great size of its mouth-opening; it is likewise much depressed and pentagonal, and covered with very small tubercles, sparsely distributed on the plates (fig. 5 *a*); in these respects it presents an assemblage of characters which, taken together, produce a form very different to any of the many varieties of *Pygaster semisulcatus* which have hitherto passed through my hands; for these reasons I have separated it from them under a distinct name. Knowing, however, the wide variations which many species exhibit in different individuals, and how necessary it is to possess examples of a series of these forms for comparison. I am most reluctant, in the absence of such materials, to multiply specific names. Still, for the sake of clearness, the provisional name *macrostoma* is proposed for this form.

Having only seen three or four examples of *Pygaster macrostoma*, the evidence, to my mind, is not sufficient to write positively on the subject, although all these specimens were remarkable for the great size of the two openings in the test.

Should a number of specimens of this urchin be hereafter gathered, and carefully compared with each other, it will then be seen whether the characters I have pointed out are persistent in the group, or shade off into forms, which may blend with other varieties of *Pygaster semisulcatus*. In the mean time it is right to register this urchin under a provisional name, and wait for the future discovery of more specimens for determination. The one proposed indicates its characters.

Part of the apical disc is preserved in a smaller specimen (fig. 5 *b*); it consists of four ovarial plates, the right antero-lateral supporting the madreporiform body being the largest, the single ovarial plate is absent in this specimen; the five small heart-shaped ocular plates are wedged in the interspaces between the ovarials (fig. 5 *b*), forming a crescent around the sub-compact disc; the posterior margin of the plates is free (fig. 5 *b*); it does not appear, however, in what manner the anal membrane and plates were connected therewith.

Locality and Stratigraphical position.—This urchin was collected in a bed of sandy Oolite, near Hampen, but whether it belongs to the Inferior Oolite or Cornbrash, I have, at present, no means of determining.

GALEROPYGUS AGARICIFORMIS, *Forbes.* Supplement, Pl. XLII, fig. 2 *a* and fig. 3, pages 292-95.

GALEROPYGUS AGARICIFORMIS. Cotteau, Bulletin Soc. Géol. de France, 2me série, tom. xvi, p. 289.

Although many hundreds of specimens of this urchin had passed through my hands when I described the species, still I had not then seen any traces of the apical disc; I was, therefore, unable to give any opinion upon M. Cotteau's proposal to separate into a distinct genus, under the name *Galeropygus*, those *Hyboclypi* which possessed a sub-compact and not an elongated disc. Very lately, however, I met with two specimens of this urchin which possessed portions of the disc *in situ*, and these form the subjects of figs. 2 and 3 of the Supplemental Pl. XLII.

The four ovarial plates, which are of a rhomboidal figure, are arranged in a crescentic form around the concave, anterior opening of the round, discal aperture; the right antero-lateral plate is the largest, and supports the madreporiform body; the plates are small, and externally present acute angles, which are inserted into the **V**-shaped notches of the inter-ambulacral segments of the discal opening; the foramina for the ovarial tubes are at the extreme point of the angle, and in some almost marginal. Four of the ocular plates are very small, and intercalated between the angles of the ovarial plates; the left postero-lateral is larger than the others, and wedged between the two left lateral ovarials, and all the orbits are distinctly marginal (fig. 3). The posterior part of this singular structure is absent, and it does not appear in what manner the single ovarial was articulated with the others, nor how the membrane of the vent, with its anal plates, was united to the test.

M. Cotteau first proposed the separation of *Hyboclypus disculus*, Ag., into the genus *Galeropygus*, from observing the difference which the disc of that urchin presented when compared with the true type form of the genus, *Hyboclypus gibberulus*, Ag.

The apical discs of the ECHINOIDEA EXOCYCLICA may be arranged, as M. Cotteau observes, into three groups—1st, *compact;* 2d, *sub-compact;* and 3d, *elongated.*

The disc is said to be *compact* when the ovarial plates form a circle around the madreporiform body, and when the five small ocular plates are intercalated between the angles formed by the ovarial plates, as in *Holectypus, Clypeus, Galerites,* and *Echinoconus.*

The disc is *sub-compact* when the three anterior ocular plates are intercalated between the angles of the ovarial plates, whilst the two posterior ocular plates are longitudinally on the same line as the postero-lateral ovarials; sometimes the single plate is altogether wanting; but it is oftener represented by two or three small, complementary, imperforate pieces, which reach the madreporiform body. This disposition of the plates gives the disc a sub-circular form, such as is seen in *Pyrina* and *Galeropygus.*

The disc is *elongated* when the four ocular plates, the anterior, lateral, and posterior, are longitudinally on the same line with the ovarials. The single ovarial plate is sometimes absent, as in the sub-compact disc; it is oftener, however, represented by one or many small, irregular, and imperforate complementary pieces, as in *Hyboclypus* and *Collyrites*.

The discs which I have discovered in *Galeropygus agariciformis* and *G. caudatus* undoubtedly belong to the sub-compact group, and justifies M. Cotteau in removing them into the genus he has established for their reception.*

ECHINOBRISSIDÆ.

Clypeus Plotii, *Klein.*   Pl. XLIII, fig. 4 *a*, *b*, Supplement to page 364.

Previous to the publication of the figures and description of *Clypeus Plotii* in this Monograph, no notice had been taken by former authors that the small tubercles of this urchin were perforated; the specimen I figured was supposed by some to have been exceptional rather than typical, as several accurate observers had failed to verify Mr. Bone's figures. Accordingly I exposed several specimens of *Clypeus Plotii* on my garden-wall during two winters, and effectually weathered the surface of their tests; by this process I have ascertained, that all the tubercles on the inter-ambulacra, and likewise on the ambulacra, including even the minute granules ranged on the edge of the zonal septa, are perforated. I have in fig. 4 *a* represented a portion of the upper surface of this urchin, of the natural size, and in fig. 4 *b* given a magnified view of one of the plates thereof; the bosses of the tubercles, with their deep-encircling areolæ, and the perforation of the summits, are well represented in this drawing, together with the miliary granules which are freely scattered over the surface of the test. The form and structure of the apical disc are likewise well seen in fig. 4 *a*; the elements of this compact disc are covered by the madreporiform body, which in this species extends over the surface of all the genital and ocular plates, the only indication of these bodies being the five orbits at the summits of the ambulacra and the four openings of the genital ducts opposite the inter-ambulacra; the spongy structure of the madreporiform body is likewise beautifully exemplified in this weathered specimen.

* For ample details on the genus *Galeropygus* the reader is referred to M. Cotteau's excellent memoir on that new genus, in the ' Bulletin Soc. Géol. de France,' 2me série, t. xvi, p. 289.

ECHINOLAMPIDÆ.

PYGURUS HAUSMANNI, *Koch and Dunker.*   Pl. XXXIX, XL to Supplement, p. 405.

Pygurus Hausmanni.

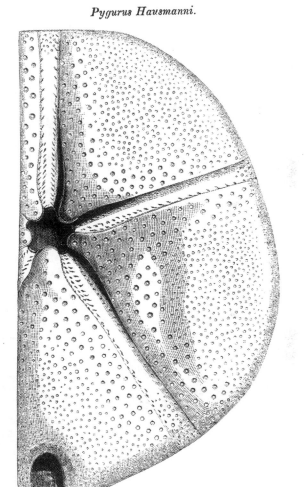

In all the specimens of this large urchin hitherto found it is the upper surface alone that is exposed. My friend, W. C. Strickland, Esq., having met with an uninjured test, he determined to remove the matrix from the base, with the intention of showing the mouth- and vent-openings; this he has succeeded in doing, and I am now, through his kindness, enabled to give a figure of the under surface and complete the description of this remarkable species.

The base is concave, inclining on all sides towards the mouth-opening, which is nearly central, being only three tenths of an inch nearer the anterior border; the peristome is pentagonal, and surrounded by five lobes, which are moderately prominent; the ambulacral valleys are slightly depressed, and converge in straight lines to the mouth; the wide inter-ambulacra are convex near the margin, and, with the depressed ambulacra, present a series of gentle undulations throughout the under surface; the anterior border is rounded, without any trace of anteal sulcus; the anal-opening lies in a deep, oblong depression near the posterior border; the test is very thin, and the tubercles in this region are very small; the apical disc is well preserved, and is only two lines in diameter; the small ovarial plates are closely soldered together, and covered by the madreporiform body which entirely envelopes the disc, and appears like a central spongy button at the vertex; the four oviductal holes are visible, and alone indicate the extent of the plates; the ocular plates are not visible, mere depressions only showing their position.

This specimen was collected from the Coralline Oolite at Settrington, near Malton, whence several good specimens have been obtained; the oolitic rock is here very fine, and cuts almost as white as chalk. Coral banks are likewise very abundant in this locality, and the fineness of the Oolite is probably due to the coralline mud which abounded in the vicinity of these *Anthozoa*.

ADDITIONAL NOTES

ON THE

BIBLIOGRAPHY OF THE ECHINODERMATA.

1850. FORBES, EDWARD ...."Description of Fossil Echinidæ from Portugal." 'Quart. Jour. Geol. Soc.,' vol. vi, p. 195, pl. xxv.

1850. SORIGNET........ ...."Oursins Fossiles du départment de l'Eure." 8vo.

1852.   ,,   "Note on Eocene Echinoderms procured by Sir Charles Lyell in Belgium." 'Quart. Jour. Geol. Soc.,' vol. viii, p, 340, pl. vi.

1852.   ,,   'Monograph of the Echinodermata of the British Tertiaries. Palæontographical Society, vol. 1852.

1854. D'ARCHIAC et HAIME........'Description des Animaux Fossiles du Groupe Nummulitique de l'Inde.'

1855. ROLLE, FREDERICH..........."Die Echinoiden der oberen Jura-Schichten von Nikolsburg in Mähren." 'Sitzungs-berichte der Akad. der Wissenschaften,' Vienna, B. xv, p. 521.

1855. HUXLEY, THOMAS..... ...."Lectures on the Structure of the Echinodermata." London, 'Medical Times.'

1856. COTTEAU et LEYMERIE........."Catalogue des Échinides Fossiles de Pyrenèes." 'Bull. de Soc. Géol. de France,' 2e série, t. xiii, p. 319.

1857. SALTER, J. W. ...."On some New Palæozoic Star-fishes." 'Ann. and Mag. of Nat. Hist.,' 2d series, vol. xx, pl. ix.

1857. HÈBERT, ED. ...."Les Mers anciennes et leur ravages dans le bassin de Paris Terrain Jurassique." 8vo.

1857. HELLER, CAMEL ...." Ueber neue fossile Stelleriden." 'Sitzungs-berichte K. Akad. Wissenschaft.,' Vienna, Bd. xxviii, p. 155.

1858. HALL, JAMES ....'Geological Survey of the State of Iowa,' 2 vols. 4to, 1 vol. of beautiful plates.

1859. WINKLER, J. J. ....'Die Schichten der *Avicula contorta* inner-und ausserhalb der Alpen.' 8vo, pl., München.

1860. COQUAND ...."Synops. des fossiles de la formation Cretacèe du Sud-ouest de la France." 'Bull. Soc. Géol. de France,' 3e série, t. xvi, p. 1013.

1860. COTTEAU, G. ...."Sur le genre Metaporhinus et la famille des Collyritides." 'Bull. Soc. Hist. et Naturelles,' Yonne.

PLATE I.

Cidaris from the Lias.

Fig.

1 *a.* CIDARIS EDWARDSII, *Wright*, p. 26.   Natural size, and restored to its globular form.

  b. One inter-ambulacral plate, and a portion of an ambulacral area, with the poriferous zones; magnified three diameters.

  c. A view in profile of one of the primary tubercles, magnified three times.

  d. A secondary spine, from an ambulacral area.

  e. The same, magnified five times.

  f. One of the primary spines, magnified three times.

  g. Lateral view of a jaw and tooth, magnified one and a half times.   This specimen shows a portion of the buccal membrane, with the spines which clothed the same.

Cidaris from the Inferior Oolite.

2 *a.* CIDARIS BOUCHARDII, *Wright*, p. 36.   Shell natural size, showing the base.

  b. Ditto, a side view of the same.

  c. One inter-ambulacral plate, and a portion of an ambulacral area, with the poriferous zones; magnified four diameters.

3 *a.* CIDARIS WRIGHTII, *Desor*, p. 39.   Test the natural size, showing the upper surface.

  b. Test the natural size, showing a lateral view.

  c. One inter-ambulacral plate, and a portion of an ambulacral area, with the poriferous zones; magnified four times.

  d. A primary spine, supposed to belong to *C. Wrightii.*

  e. Another spine, referred to the same species.

  f. A portion of the same, magnified four diameters.

4 *a.* CIDARIS FOWLERI, *Wright*, p. 32.   Natural size, showing the base of the test, with the jaws *in situ.*

  b. Lateral view of the same, showing the projection of the jaws.

  c. One inter-ambulacral plate, and a portion of an ambulacral area, with the poriferous zones; magnified three diameters.

  d. Portion of a spine attached to the test, this specimen belongs to the Museum of Geology, Jermyn Street.

5 *a.* DIPLOCIDARIS WRIGHTII, *Desor*, p. 58.   Fragment of a spine.

  b. The same, magnified two and a half diameters.

  c. nov. sp.

W.H.Baily del. et lith.

Printed by Hullmandel & Walton.

PLATE II.

Cidaris from the Coralline Oolite.

Fig.

1. *a.* CIDARIS SMITHII, *Wright,* p. 50.   Under surface of the test, natural size, showing the jaws *in situ*.

   b. Lateral view of another specimen, natural size.

   c. One inter-ambulacral plate, and a portion of an ambulacral area, with the poriferous zones, magnified two diameters.

   d. Inter-ambulacral plate, with its primary tubercle seen in profile, and magnified two diameters.

   e. Fragment of a spine attached to the test " *a*," magnified two diameters.

2. *a.* CIDARIS FLORIGEMMA, *Phillips,* p. 44.   Test with spines attached, on a slab of Coralline Oolite.

   b. A large test, natural size, showing the base and mouth opening.

   c. Lateral view of the same test, natural size.

   d. Primary spines of *Cidaris florigemma,* natural size.

   e. Head, neck, and acetabulum of the same, magnified two diameters.

   f. Small, spatulate, secondary spines, magnified three and a half times.

   g. One inter-ambulacral plate, and a portion of an ambulacral area with the poriferous zones ; magnified two diameters.

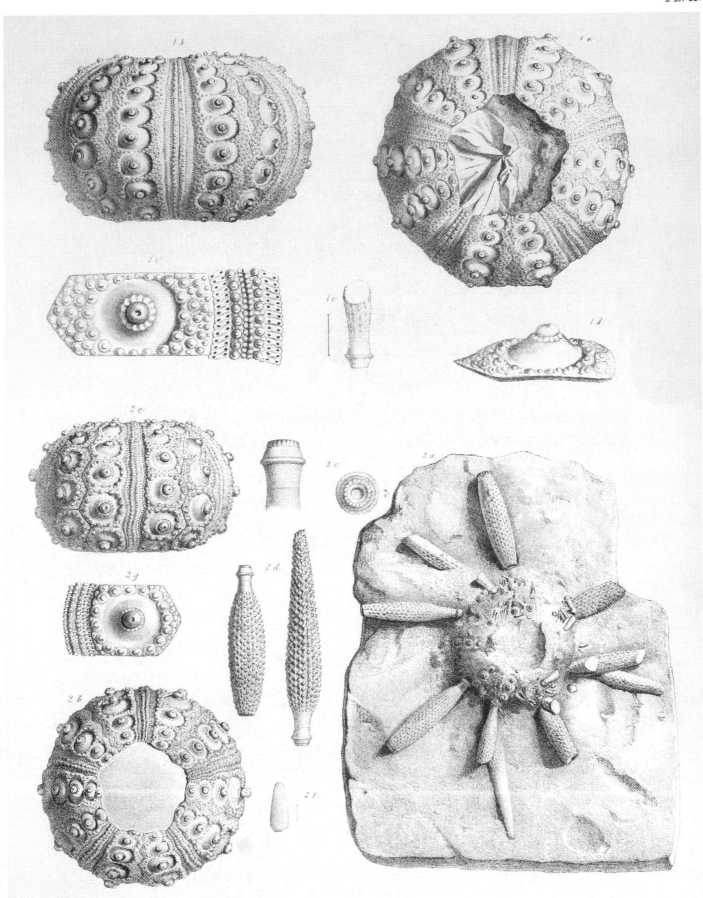

C. R. Bone, del. et lith.

Printed by Hullmandel & Walton.

PLATE III.

Hemicidaris from the Inferior Oolite.

Fig.

1 *a.* HEMICIDARIS PUSTULOSA, *Agassiz*, p. 73.   Upper surface of the test, natural size.

   b. Lateral view of the same, natural size.

   c. Two inter-ambulacral plates, and a portion of an ambulacral area, showing the two upper pairs of semi-tubercles and poriferous zones; magnified three diameters.

   d. A primary tubercle seen in profile, magnified two diameters.

   e. Apical disc, magnified two diameters.

2 *a.* HEMICIDARIS GRANULOSA, *Wright*, p. 71.   Upper surface of the test of a small urchin, natural size.

   b. Base of a large specimen, showing the mouth opening.

   c. Lateral view of the same test, both the natural size.

   d, One inter-ambulacral plate, and a portion of an ambulacral area with the poriferous zones; magnified three diameters.

   e. Inter-ambulacral plate and tubercle seen in profile, magnified three times.

Hemicidaris from the Stonesfield Slate and Great Oolite.

3 *a.* HEMICIDARIS STOKESII, *Wright*, p. 75.   Test natural size, imbedded in Stonesfield Slate.

   b. One inter-ambulacral plate, and a portion of an ambulacral area with the poriferous zones; magnified three diameters.

   c. Apical disc, magnified two diameters.

4 *a.* HEMICIDARIS ICAUNENSIS, *Cotteau*, p. 90.   Lateral view of the test, an interior mould.

   b. Upper surface of the same, both the natural size.

5 *a.* HEMICIDARIS MINOR, *Agassiz*, p. 80.   Test natural size, showing a lateral view.

   b. Upper surface, magnified two diameters.

   c. One inter-ambulacral plate, and a portion of an ambulacral area, with the poriferous zones; magnified three diameters.

   d. Inter-ambulacral plate and tubercle, seen in profile, and magnified three diameters.

6 *a.* HEMICIDARIS LUCIENSIS, *d' Orb.*, p. 78.   Test natural size, showing the upper surface.

   b. Base of the same, showing the mouth opening, both the natural size.

   c. Lateral view of the same test.

   d. One inter-ambulacral plate, and a portion of an ambulacral area, with the poriferous zones; magnified four diameters.

   e. Inter-ambulacral plate and tubercle, seen in profile, magnified four times.

   f. Apical disc, magnified three diameters.

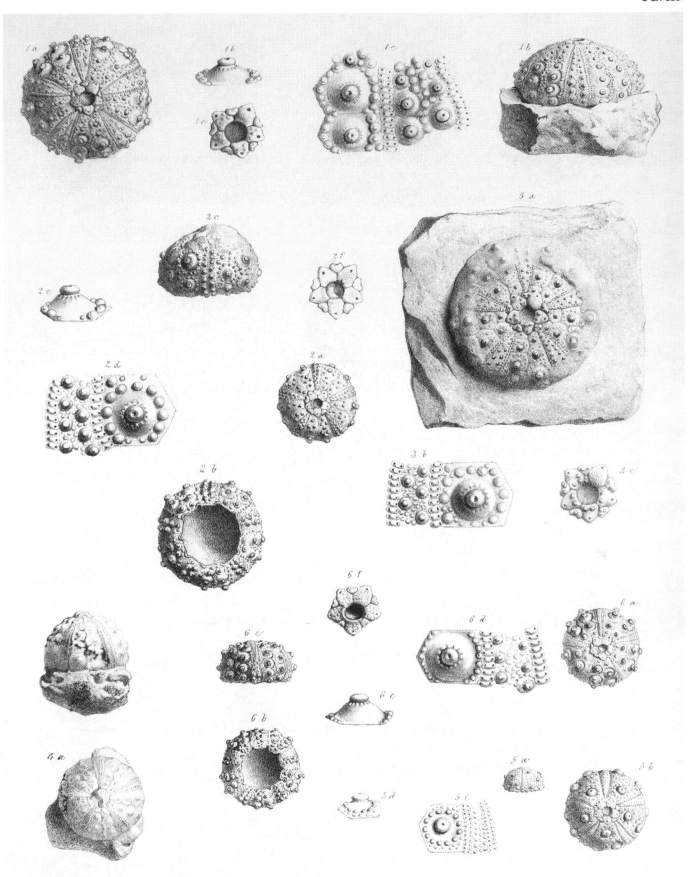

C. R. Bone, del. et lith.

Printed by Hullmandel & Walton.

PLATE IV.

Hemicidaris from the Great Oolite.

Fig.

1 *a*. HEMICIDARIS BRAVENDERI, *Wright*, p. 84.   Upper surface of the test, natural size.

 b. Lateral view of the same, both the natural size.

 c. One inter-ambulacral plate, and a portion of an ambulacral area, **magnified** three diameters.

 d. An inter-ambulacral plate and tubercle, seen in profile, and magnified three times.

 e. Mouth opening and peristome.

 f. Apical disc, magnified two diameters.

2 *a*. HEMICIDARIS WRIGHTII, *Desor*, p. 88.   Upper surface of the test, natural size.

 b. Lateral view of the same, natural size.

 c. Apical disc, magnified two diameters.

 d. One inter-ambulacral plate, and a portion of an ambulacral area, with the poriferous zones; magnified three times.

 e. Base of an ambulacral area, showing the semi-tubercles.

 f. An inter-ambulacral plate and tubercle, seen in profile, and magnified twice.

3 *a*. HEMICIDARIS SMITHII, *Woodward*.   Under surface, showing the base, natural size.

 b. Lateral view of the same, natural size.

Hemicidaris from the Purbeck Beds.

4 *a* HEMICIDARIS PURBECKENSIS, *Forbes*, p. 98.   Lateral view, showing likewise the upper surface, natural size.

 b. Two inter-ambulacral areas, and a portion of an ambulacral area, with the poriferous zones; magnified three times.

 c. Primary spine, natural size, showing the head, ring, and neck, magnified four times.

 d. A portion of the stem, with the same, magnified.

5 *a*. CIDARIS SMITHII, *Wright*, p. 50.   Primary spine, with the surface of the stem, magnified three times.

 b, c. Head, milled ring, and neck of the same, magnified four times.

 d. The acetabulum.

 e. A transverse section.

6 *a*. CIDARIS ILMINSTERENSIS, *Wright*, p. 31.   Inter-ambulacral plate, and a portion of an ambulacral area, natural size.

 b. The same, magnified three diameters.   The only fragment of the species I know.

7 *a*. CIDARIS BRADFORDENSIS, *Wright*, p. 42.   Inter-ambulacral plates, natural size.

 b. The same, magnified two and a half diameters.

 c. Primary spine, found in the same bed of Bradford Clay.

 d. The same, magnified three diameters.

C. R. Bone, del. et lith.

Printed by Hullmandel & Walton.

PLATE V.*

Hemicidaris from the Coralline Oolite.

Fig.

1 *a.* HEMICIDARIS INTERMEDIA, *Fleming*, p. 92.   Upper surface of the test, natural size.

  b. Under surface, showing the mouth opening, with the jaws and teeth *in situ.*

  c. Lateral view of the test "*a.*"

  e. Inter-ambulacral plate and tubercle, seen in profile, and magnified three diameters.

  d. Two inter-ambulacral plates, and a portion of an ambulacral area, showing the upper pair of semi-tubercles and poriferous zones; magnified three diameters.

  f. One of the jaws, magnified two diameters.

  g. Apical disc magnified two diameters.

  h. Test with spines attached, imbedded in a slab of Coralline Oolite.

  i. Outline of a conical variety of test in my collection.

  j. A small spatulate tertiary spine, natural size.

  k. The same, magnified five times.

  l. A small secondary spine, natural size.

  m. The same, magnified five times.

  n. A primary spine, natural size.

  o. The same, magnified twice.

Hemicidaris from the Portland Oolite.

2 *a.* HEMICIDARIS DAVIDSONI, *Wright*, p. 96.   Upper surface of the test, natural size.

  b. Lateral view of the same, showing the single row of semi-tubercles.

  c. Two inter-ambulacral plates, and a portion of an ambulacral area, with the poriferous zones; magnified three diameters.

  d. An inter-ambulacral plate and tubercle seen in profile, and magnified twice.

  e. Mouth opening and peristome, natural size.

---

* In consequence of the artist having by mistake lettered Plate IV Plate V, and Plate V Plate IV, the reference in the text does not agree with the numbers now unavoidably adopted, as the text and plates were both printed before the error was discovered. This, however, is the less to be regretted, as any confusion which might have arisen will now be prevented when the reason for the discrepancy is explained.

T. W.

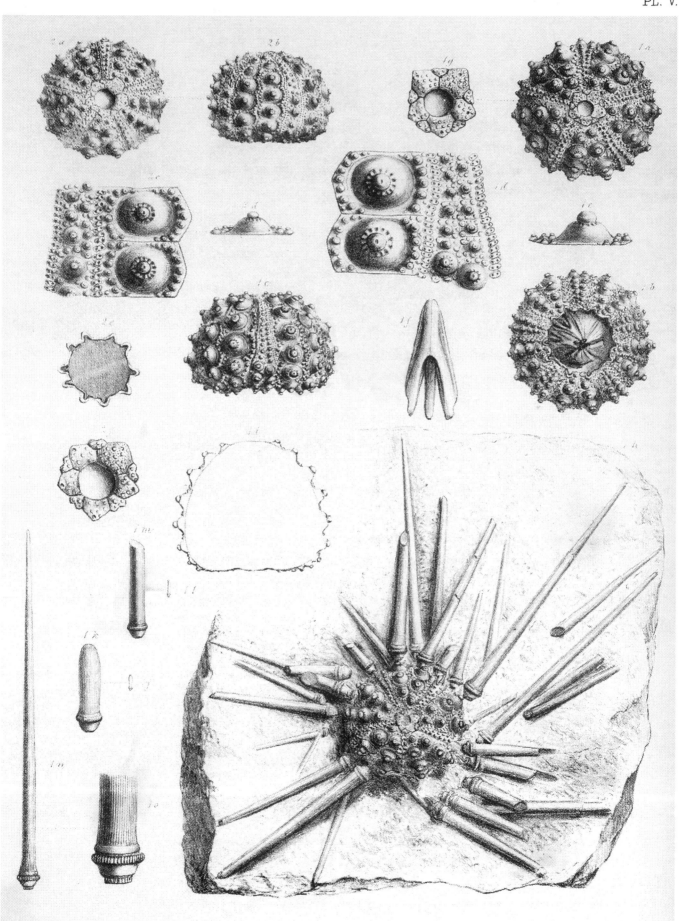

W.H.Baily del. et lith.

Printed by Hullmandel & Walton.

PLATE VI.

Pseudodiademas from the Lias.

Fig.

1 *a*. Pseudodiadema Mooreii, *Wright*, p. 110. Test, the natural size.

  b. Upper surface of the same, magnified two diameters.

  c. Base and mouth opening of the same, magnified two diameters.

  d. Two inter-ambulacral plates, and a portion of an ambulacral area, with the poriferous zones; magnified four times.

Pseudodiademas from the Inferior Oolite.

2 *a*. Pseudodiadema depressum, *Agassiz*, p. 112. Upper surface of the test, natural size.

  b. Base and mouth opening of the same, natural size.

  c. Lateral view of the same, natural size.

  d. Two inter-ambulacral plates, a portion of an ambulacral area, with the poriferous zones; magnified three diameters.

  e. Fragment of a spine, magnified four diameters.

  f. An entire spine, magnified four diameters.

  g. Base of an ambulacral area, showing the tubercles and trigeminal arrangement of the pores in this region, magnified three times.

  h. Inter-ambulacral plate and tubercle, seen in profile, magnified three times.

  i. A small specimen with its spines attached, lying on a block of Pea Grit.

Pseudodiademas from the Great Oolite.

3 *a*. Pseudodiadema pentagonum, *M'Coy*, p. 115. Upper surface of the test, natural size.

  b. Lateral view of the same, natural size.

  c. A portion of an ambulacral area, with the poriferous zones, showing the bigeminal arrangement of the pores, magnified four times.

  d. One inter-ambulacral plate, and a portion of an ambulacral area with the poriferous zones; magnified four times.

4. Pseudodiadema Parkinsoni, *Desor*, p. 114. Test and spines, copied from Parkinson's 'Organic Remains.'

Pseudodiademas from the Cornbrash.

5 *a*. Pseudodiadema homostigma, *Agassiz*, p. 118. Test, the natural size.

  b. Upper surface of the same, magnified two diameters.

  c. The base and mouth opening, magnified two diameters.

  d. Lateral view of the same, magnified two diameters.

  e. Two inter-ambulacral plates, and a portion of an ambulacral area, with the poriferous zones; magnified five times.

  f. An inter-ambulacral plate and tubercle, magnified five times.

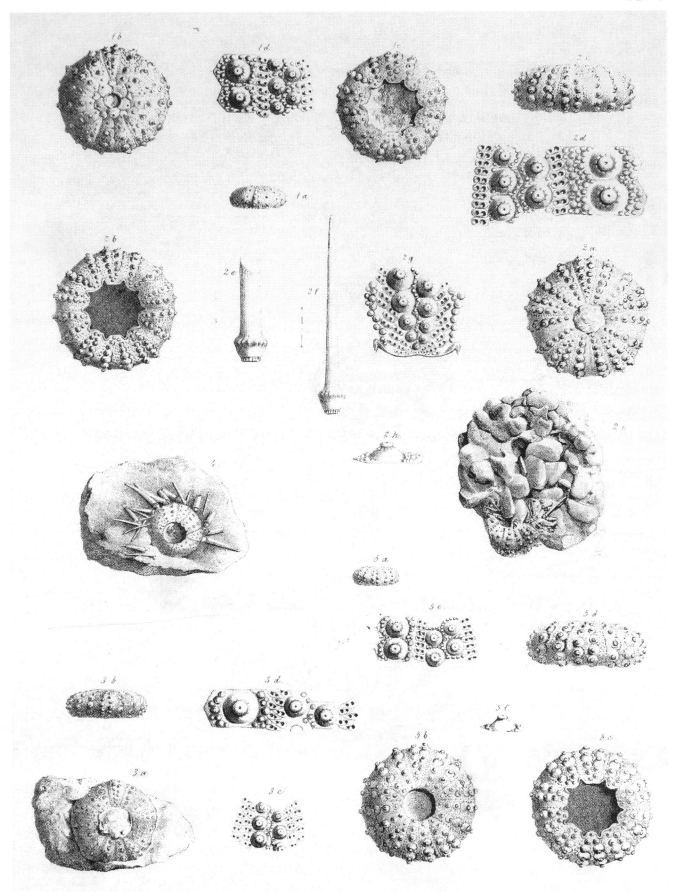

C. R. Bone, del. et lith.

Printed by Hullmandel & Walton.

PLATE VII.

Pseudodiademas from the Cornbrash.

Fig.

1 *a*. PSEUDODIADEMA BAILYI, *Wright*, p. 120.   Upper surface of the test, natural size.

   b. Under surface and mouth opening of the same, natural size.

   c. Lateral view of the same, natural size.

   d Half of the base, showing the peristome, &c., magnified three diameters.

   e. Half of the upper surface, magnified three diameters.

   f. One inter-ambulacral plate, and a portion of an ambulacral area, with the poriferous zones; magnified six diameters.

   g. A plate and tubercle, seen in profile, magnified six diameters.

2 *a*. PSEUDODIADEMA BAKERIÆ, *Woodward*, p. 121.   Upper surface of the test, natural size.

   b. Lateral view of the same, natural size.

   c. One inter-ambulacral plate, and a portion of an ambulacral area, with the poriferous zones; magnified four diameters.

Pseudodiademas from the Coralline Oolite.

3 *a*. PSEUDODIADEMA RADIATA, *Wright*, p. 131.   Upper surface of the test, natural size.

   b. Under surface of the test, natural size.

   c. Lateral view of the test, natural size.

   d. The half of the upper surface, magnified three diameters.

   e. The half of the under surface, magnified three diameters.

   f. One inter-ambulacral plate, and a portion of an ambulacral area, with the poriferous zones ; magnified six diameters.

   g. An inter-ambulacral plate and tubercle, seen in profile, magnified six times.

4 *a*. PSEUDODIADEMA VERSIPORA, *Phillips*, p. 124.   Upper surface of the test, natural size.

   b. Base and mouth opening of the same, natural size.

   c. Lateral view of the same, natural size.

   d. Half of the upper surface, magnified two diameters.

   e. Half of the under surface, magnified two diameters.

   f. One inter-ambulacral plate, and a portion of an ambulacral area, with the poriferous zones ; magnified four diameters.

   g. An inter-ambulacral plate and tubercle, seen in profile, and magnified four times.

W.H.Baily del. et lith.

Printed by Hullmandel & Walton.

PLATE VIII.

Pseudodiademas from the Coralline Oolite.

Fig.

1 *a*. PSEUDODIADEMA HEMISPHÆRICUM, *Agassiz*, p. 127. Upper surface of the test, natural size.

   b. The half of the under surface of the same, natural size.

   c. A lateral view of the same, the natural size.

   d. The half of a portion of an inter-ambulacral area, and a portion of an ambulacral area, with the poriferous zones ; magnified twice.

   e. Apical disc, magnified two diameters.

   f. Base of an ambulacral area, showing the trigeminal pores in this region.

2 *a*. PSEUDODIADEMA MAMILLANUM, *Roemer*, p. 132. Upper surface of the test, natural size.

   b. Under surface of the test, natural size.

   c. Lateral view of the test, natural size.

   d. The half of a portion of an inter-ambulacral area, and a portion of the ambulacral area, with the poriferous zones ; magnified three diameters.

3 *a*. CIDARIS BOUCHARDII, *Wright*, p. 36. Lateral view of a very fine test of this species, from the Inferior Oolite.

   b. One inter-ambulacral plate, and a portion of an ambulacral area of *Cidaris Bouchardii*, with a small secondary spine found on the test, magnified four times.

4 *a*. CIDARIS FLORIGEMMA, *Phillips*, p. 44. Under surface of a young test, natural size.

   b. Upper surface of the same, natural size.

   c. Lateral view of the same, natural size.

   d. One inter-ambulacral plate, and a portion of an ambulacral area, with the poriferous zones, magnified three diameters.

   e. Inter-ambulacral plate and tubercle, seen in profile, and magnified three times.

5. DIPLOCIDARIS DESORI, *Wright*, p. 56. Inter-ambulacral plate, ambulacral area, and poriferous zones ; natural size ; the only fragment I know.

6 *a*. HEMICIDARIS RAMSAYII, *Wright*, p. 83. Lateral view of the test, natural size.

   b. Upper surface, magnified three diameters.

   c. Under surface, magnified three diameters.

   d. Two inter-ambulacral plates, and a portion of an ambulacral area, with the poriferous zones ; magnified five times.

   e. The apical disc, magnified once and a half.

C. R. Bone, del. et lith.

Printed by Hullmandel & Walton.

PLATE IX.

Hemipedinas from the Lias.

Fig.

1 *a*. HEMIPEDINA BECHEI, *Brodrip*, p. 144.   Test with spines *in situ*, natural size.
  b. A primary spine, magnified three times.

2 *a*. HEMIPEDINA BOWERBANKII, *Wright*, p. 145.   Test with spines, natural size.
  b. One inter-ambulacral plate, and a portion of an ambulacral area, with the poriferous zones ; magnified five diameters.
  c. A primary spine, magnified three times.

3 *a*. HEMIPEDINA (DIADEMA) SERIALE, *Leymerie*, p. 146.   From the Lias of France copied from the 'Mem. de la Société Géologique de France,' t. ii, pl. 24, fig. 1. To show the form and structure of the test of a very rare allied species.
  b. One of the inter-ambulacral plates and poriferous zones, magnified.

4 *a*. HEMIPEDINA JARDINII, *Wright*, p. 146.   Upper surface of the test, natural size.
  b. Under surface of the test, natural size.
  c. Lateral view of the test, natural size.
  d. Upper surface of the test, magnified three diameters.
  e. Under surface of the test, magnified three diameters.
  f. Lateral view of the test, magnified three diameters.
  g. One inter-ambulacral plate, and a portion of an ambulacral area, with the poriferous zones, magnified nine times.

5 *a*. HEMIPEDINA ETHERIDGII, *Wright*, p. 148. Upper surface of the test, natural size.
  b. Under surface of the test, natural size.
  c. Lateral view of the test, natural size.
  d. Upper surface of the test, magnified three diameters.
  e. Under surface of the test, magnified three diameters.
  f. Lateral view of the test, magnified three diameters.
  g. One inter-ambulacral plate, and a portion of an ambulacral area, with the poriferous zones, magnified nine times.

PL. IX.

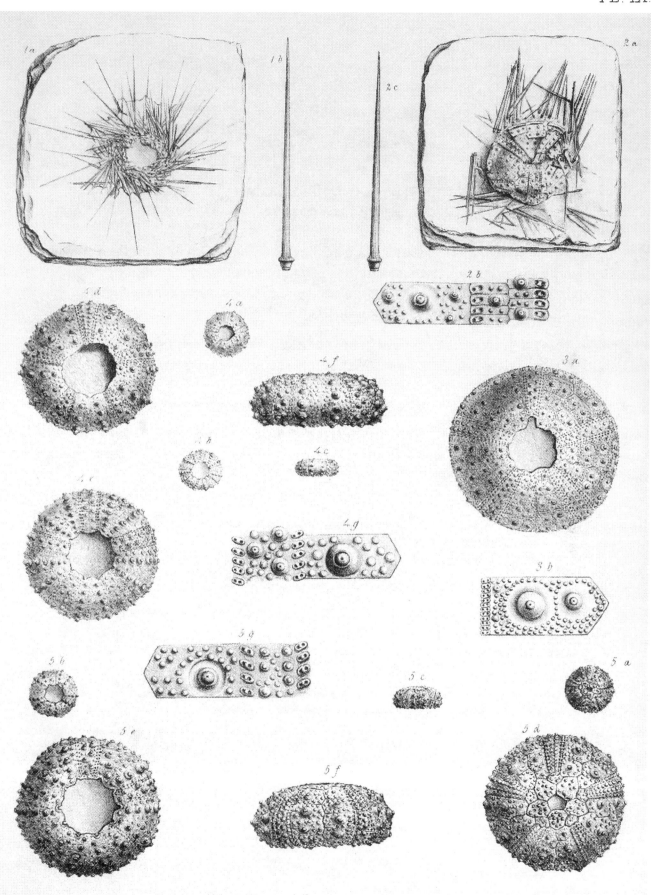

W. H. Baily del. et lith.

Printed by Hullmandel & Walton.

PLATE X.

Hemipedinas from the Inferior Oolite.

Fig.

1 *a.* HEMIPEDINA BAKERI, *Wright*, p. 149. Under surface of the test, natural size.
  b. Upper surface of the test, magnified two diameters.
  c. Lateral view of the test, magnified two diameters.
  d. Two inter-ambulacral plates, and a portion of an ambulacral area, with the poriferous rows ; magnified two diameters.
  e. Inter-ambulacral plate and tubercle, seen in profile, and magnified three times.
  f. The apical disc, magnified three diameters.

2 *a.* HEMIPEDINA PERFORATA, *Wright*, p. 151. Lateral view of a test, natural size.
  b. Upper surface, magnified two diameters.
  c. Under surface, magnified two diameters.
  d. Two inter-ambulacral plates, and a portion of an ambulacral area, with the poriferous rows ; magnified six times.
  e. Apical disc, magnified two diameters.
  f. Base of an ambulacral area, showing the trigeminal arrangement of the poriferous zones in this region.

3 *a.* HEMIPEDINA TETRAGRAMMA, *Wright*, p. 152. Lateral view of the test, natural size.
  b. The upper surface, magnified two diameters.
  c. Two inter-ambulacral plates, and a portion of an ambulacral area, with the poriferous zones ; magnified five diameters.

4 *a.* HEMIPEDINA WATERHOUSEI, *Wright*, p. 154. Lateral view of the test, natural size.
  b. Upper surface, magnified two diameters.
  c. Lateral view, magnified two diameters.
  d. Apical disc, magnified four diameters.
  e. Two inter-ambulacral plates, and a portion of an ambulacral area, with the poriferous zones ; magnified three diameters.

5 *a.* HEMIPEDINA BONEI, *Wright*, p. 156. Lateral view of the test, natural size.
  b. Upper surface of the test, magnified two diameters.
  c. Under surface of the test, magnified two diameters.
  d. Three inter-ambulacral plates, and a portion of an ambulacral area, with the poriferous zones ; magnified six diameters.

W.H. Baily del. et lith.

Printed by Hullmandel & Walton.

PLATE XI.

Hemipedinas from the Coral Rag.

Fig.

1 *a*. HEMIPEDINA MARCHAMENSIS, *Wright*, p. 161.   Base of the test, natural size.

  b. One inter-ambulacral plate, a portion of the ambulacral area, and poriferous zones, magnified twice.

  c. Lateral view of a primary tubercle, showing its prominent boss, with smooth summit.

2 *a*. HEMIPEDINA TUBERCULOSA, *Wright*, p. 164.   Upper surface of the test, natural size.

  b. Base of the test, natural size.

  c. Lateral view of the test, natural size.

  d. Two inter-ambulacral plates, a portion of the ambulacral area, and poriferous zones, magnified three times.

  e. Lateral view of a primary tubercle, showing the prominent boss, with its smooth summit, and the scrobicular granules encircling the areola.

  f. A primary spine, magnified three times.

Hemicidaris from the Great Oolite.

3 *a*. HEMICIDARIS BRAVENDERI, *Wright*, p. 84. Test and spines *in situ*, on a slab of Great Oolite from Stratton, near Cirencester.

  b. A primary spine, magnified three times.

  c. Base of a spine, showing the crenulated rim of its acetabulum and the milled ring.

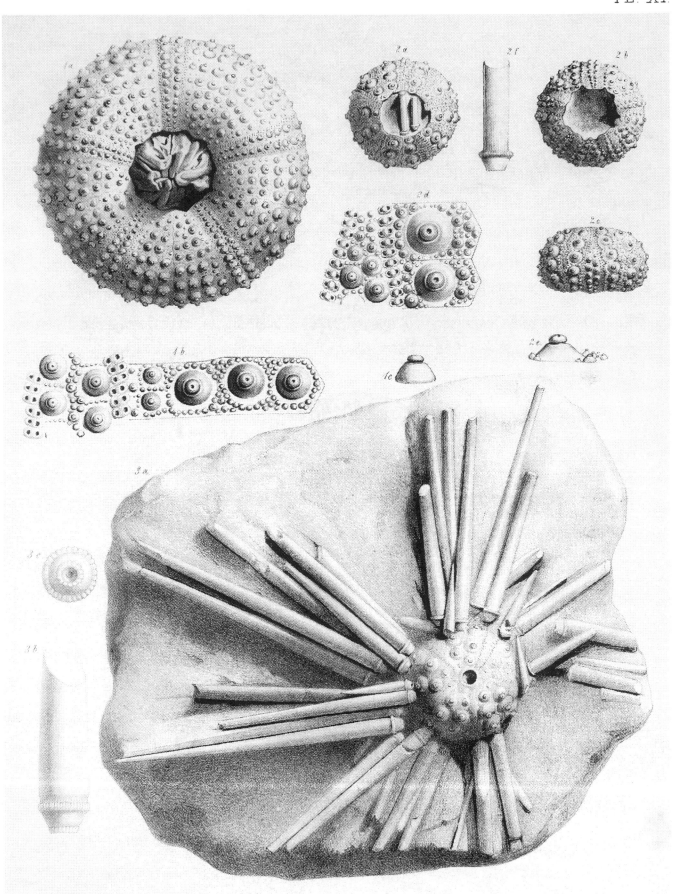

C. R. Bone, del. et lith.

Printed by Hullmandel & Walton.

PLATE XII.

Hemipedinas from the Great Oolite, Coral Rag, and Kimmeridge Clay.

Fig.

1 *a.* HEMIPEDINA CORALLINA, *Wright*, p. 163.   Portion of the base and jaws *in situ*.
  b. Fragment, with plates and spines, from the Coral Rag, Wilts.
  c. One inter-ambulacral plate, magnified three times.
  d. A primary spine, from fragment *b*, magnified three times.

2 *a.* HEMIPEDINA MORRISII, *Wright*, p. 166.   A fragment of the test, natural size.
  b. Ambulacral area, poriferous zones, and inter-ambulacral plates, magnified four times.
  c. A primary spine, magnified four times.

3 *a.* HEMIPEDINA CUNNINGTONI, *Wright*, p. 167.   A fragment of the test, natural size.
  b. The same, magnified four times.

4 *a.* HEMIPEDINA MICROGRAMMA, *Wright*, p. 159. Upper surface of the test, natural size.
  b. A lateral view of the test, natural size.
  c. One inter-ambulacral plate, a portion of the ambulacra, and poriferous zones, magnified four diameters.

4.* CIDARIS SPINOSA, *Agassiz*, p. 53.   Spine, natural size.

5 *a.* CIDARIS BOLONIENSIS, *Wright*, p. 53.   A flattened spine with prickly ridges.
  b. A spine with thorny prickles, resembling *C. spinosa*.

6 *a.* HEMIPEDINA DAVIDSONI, *Wright*, p. 156.   A lateral view of the test, natural size.
  b. The upper surface of the test, natural size.
  c. The under surface of the test, natural size.
  d. One inter-ambulacral plate, a portion of the ambulacra, and poriferous zones; magnified six diameters.

7 *a.* HEMIPEDINA WOODWARDI, *Wright*, p. 158.   A lateral view of the test, magnified twice.
  b. Upper surface of the test, magnified twice.
  c. Under surface of the test, magnified twice.
  d. One inter-ambulacral plate, ambulacra, and poriferous zones, magnified six times.

8. PSEUDODIADEMA MAMMILLANUM, *Roemer*, p. 132.   Primary spine, magnified three times.

9. PSEUDODIADEMA VERSIPORA, *Phillips*, p. 124.   Primary spine, magnified three times.

10 RABDOCIDARIS MAXIMA, *Münster*, p. 65.   Primary spine, natural size.

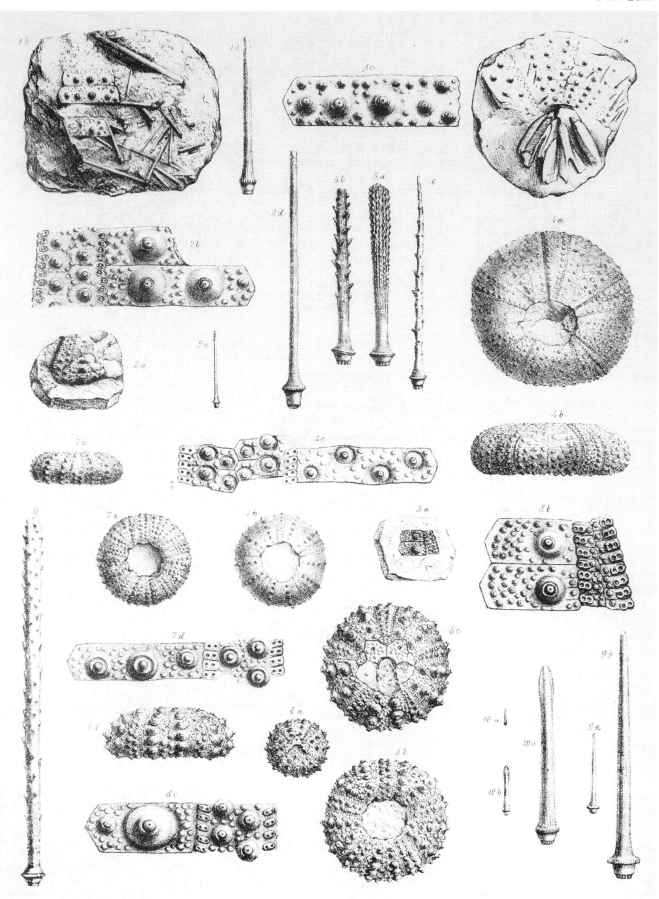

W.H.Baily del.et lith.

Printed by Hullmandel & Walton.

PLATE XIII.

Pedinas from the Inferior Oolite.

Glypticus from the Coral Rag.

Polycyphus from the Inferior Oolite.

PLATE XIII (*continued*).

Magnotia from the Inferior Oolite.

Fig.

6 *a.* MAGNOTIA FORBESII, *Wright*, p. 191.   Upper surface of the test, natural size.

  b. Upper surface of the test, magnified once and a half.

  c. Under surface of the test, magnified once and a half.

  d. Lateral view of the same, magnified once and a half.

  e. Apical disc, magnified four times.

  f. Inter-ambulacral and ambulacral plates, with the poriferous zones, showing the arrangement of the tubercles and the unigeminal disposition of the pores.

PL. XIII.

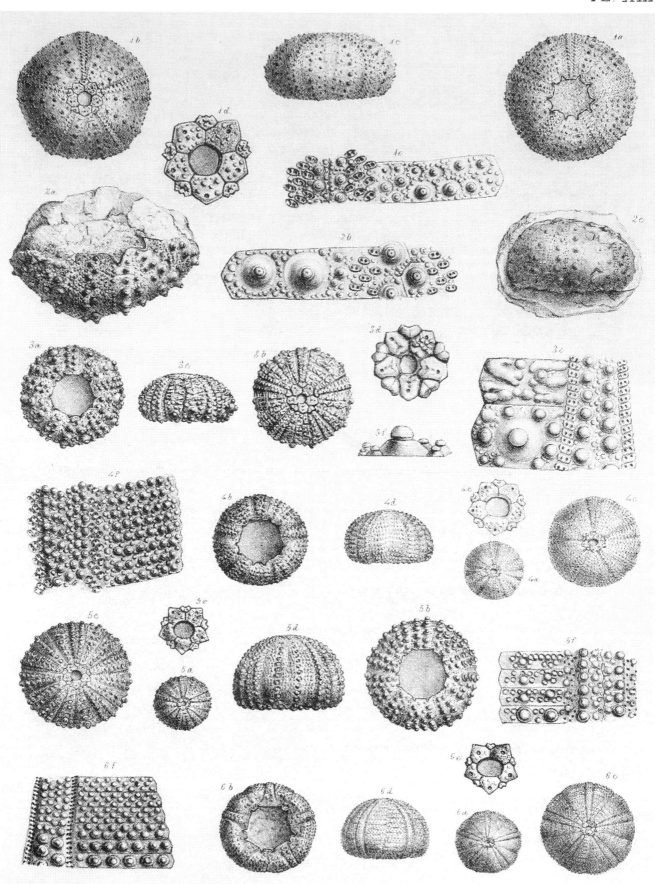

W.H.Baily del.et lith.

Printed by Hullmandel & Walton.

PLATE XIV.

Stomechini from the Inferior Oolite.

Stomechinus from the Coral Rag.

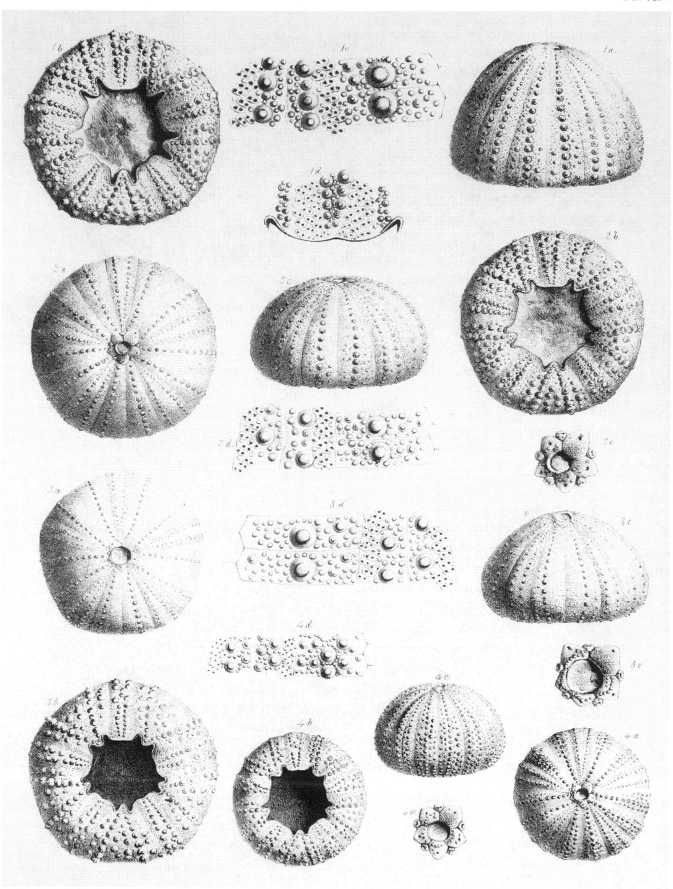

C. R. Bone, del. et lith.

Printed by Hullmandel & Walton.

PLATE XV.

Stomechini from the Great Oolite and Coral Rag.

Acrosalenias from the Lias and Cornbrash.

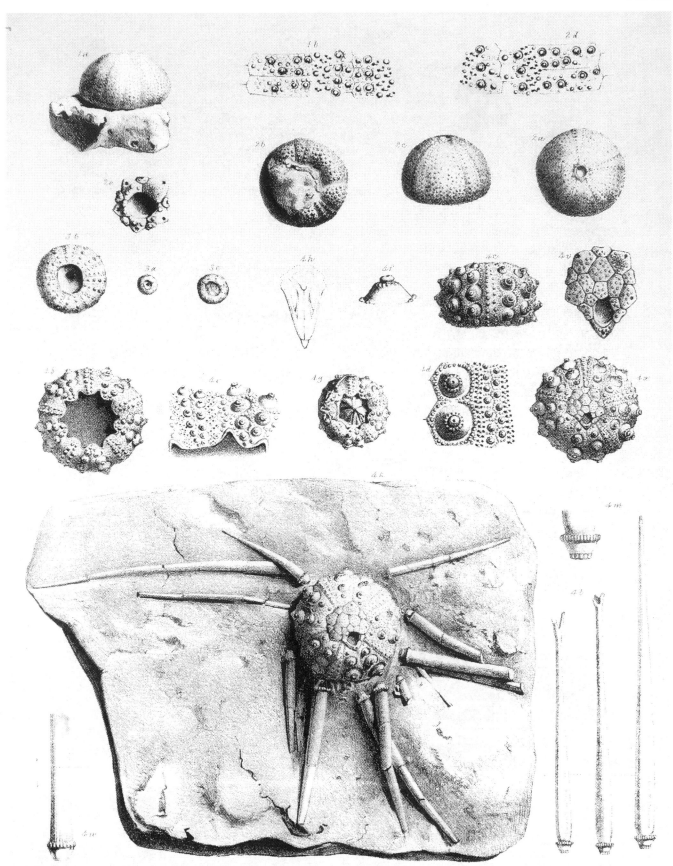

C. R. Bone, del. et lith.

Printed by Hullmandel & Walton.

PLATE XVI.

Acrosalenias from the Inferior and Great Oolite.

Fig.

1 *a.* ACROSALENIA LYCETTII, *Wright*, p. 232.   Upper surface of the test, natural size.
  b. Under surface of the same, showing the wide decagonal peristome, natural size.
  c. Lateral view of the same, showing both areas.
  d. Two inter-ambulacral plates, a portion of an ambulacral area, with the poriferous zones, magnified three times.
  e. Base of an ambulacral area, showing the trigeminal ranks of pores in this region, magnified three times.
  f. A primary tubercle, with its prominent conical boss, highly magnified.

2 *a.* ACROSALENIA PUSTULATA, *Forbes*, p. 242.   Upper surface of the test, natural size.
  b. Under surface of the same, showing the wide decagonal peristome, natural size.
  c. Lateral view of the same, showing the small ambulacral tubercles, natural size.
  d. Two inter-ambulacral plates, a portion of an ambulacral area, with the poriferous zones, magnified three times.
  e. A primary spine, natural size, and the base of the same, magnified three diameters.
  f. The body and spines *in situ*, on a slab of Great Oolite, natural size.

3 *a.* ACROSALENIA WILTONII, *Wright*, p. 246.   Upper surface of the test, natural size.
  b. Under surface of the same, natural size, showing the narrow mouth opening.
  c. Lateral view of the same, natural size, showing the wide miliary zone.
  d. Two inter-ambulacral plates, a portion of an ambulacral area, and the poriferous zones, magnified three times.
  e. A primary tubercle on its large prominent boss, magnified three times.

4 *a.* ACROSALENIA LOWEANA, *Wright*, p. 240.   Test and spines *in situ*, natural size, on a slab of Forest Marble.
  b. One of the primary spines, magnified five times, with transverse section of the same.

5.   Spine of CIDARIS YEOVILENSIS, *Wright*. ⎫ These new species will be described in
6.   CIDARIS MOOREI, *Wright*.   ⎭ the Appendix.

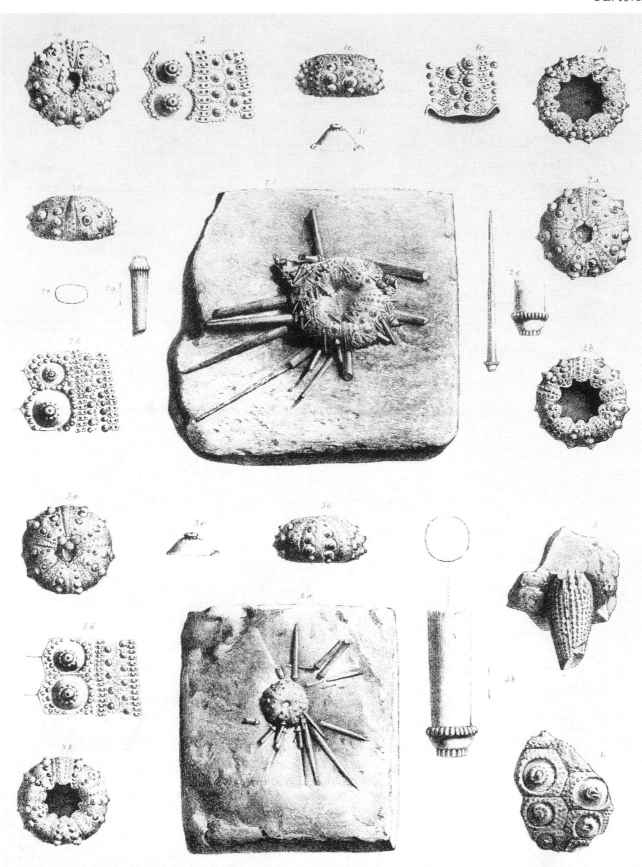

C. R. Bone, del. et lith.

Printed by Hullmandel & Walton.

PLATE XVII.

Acrosalenias from the Lias, Cornbrash, and Coral Rag.

Fig.

1 *a.* ACROSALENIA DECORATA, *Haime*, p. 247. Upper surface of the test, natural size.

 b. Under surface of the test, natural size, showing the concave base and equal-lobed peristome.

 c. Lateral view of the same, natural size, showing the magnitude of the equatorial tubercles.

 d. Upper surface of the inter-ambulacral and ambulacral areas, magnified four times.

 e. Two equatorial inter-ambulacral plates and a portion of the ambulacral area, magnified four diameters.

 f. The apical disc, with all the elements of the sur-anal plate *in situ*, magnified three times.

 g. The apical disc of another specimen, showing the single crescentic genital plate, magnified four times.

 h. Base of the ambulacral and inter-ambulacral areas, with two lobes of the peristome, magnified four times.

 i. Lateral view of an equatorial primary tubercle, with its highly crenulated boss.

 k. Slab from the Coralline Oolite of Malton, with tests and spines *in situ*, natural size.

 l. Primary spine, natural size, with base and transverse section, magnified four times.

 m. Secondary spine, magnified, showing the longitudinal lines on its surface.

2 *a.* ACROSALENIA MINUTA, *Buckman*, p. 230. Test, the natural size.

 b. Base of the same, magnified three diameters.

 c. Upper surface of the same, magnified three diameters.

 d. Lateral view of the same, magnified three diameters.

 e. Two inter-ambulacral plates, a portion of an ambulacral area, and poriferous zones, magnified ten times.

3 *a.* ACROSALENIA SPINOSA, *Agassiz*, p. 238. Upper surface of the test, natural size.

 b. Base of the same, natural size.

 c. Lateral view of the same, natural size.

 d. Two inter-ambulacral plates, a portion of the ambulacral area, and poriferous zones, magnified four times.

 e. The conical boss, with its crenulated summit, and perforated tubercle, magnified.

 f. The apical disc, with its single sur-anal plate, magnified four diameters.

PLATE XVII (*continued*).

Fig.

4 *a*. ACROSALENIA LOWEANA, *Wright*, p. 240.   Upper surface of the test, natural size.

  b. Lateral view of the test, natural size.

  c. Two inter-ambulacral plates, a portion of the ambulacral area, and poriferous zones, magnified three diameters.

5. ACROSALENIA WILTONII, *Wright*, p. 246.   The apical disc, magnified three times.

Spines of unknown Species.

6. Globular spine from the Great Oolite near Bath, magnified three times.

7. Spine from the Forest Marble of Upper Cubberly, near Cheltenham, natural size, and magnified three times.

8 Spine from the Great Oolite of Bath.

  Spine   ditto   ditto.

9. Spine from the Stonesfield slate of Eyeford, Gloucestershire, and surface magnified five times.

10. Spine from the Stonesfield slate of Eyeford, Gloucestershire, and surface magnified four times.

11. Spine from the Great Oolite of Minchinhampton, natural size.

12. Spine from the Pea-grit, near Cheltenham, fragment natural size.

13. Spine from the Lower Lias of Bushley, near Tewkesbury, and a portion magnified five times.

14. Spine from the Lower Lias of Lyme Regis, and portion magnified four times.

15. Spine   ditto   ditto,   and portion magnified five times.

16. Spine   ditto   ditto,   and portion magnified three times.

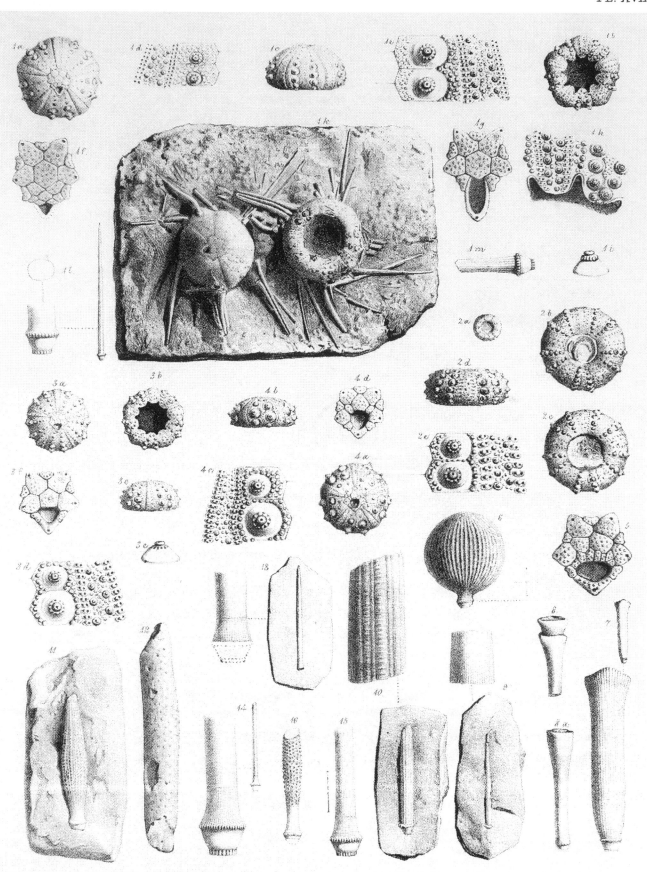

C. R. Bone, del. et lith.

Printed by Hullmandel & Walton.

PLATE XVIII.

Holectypi from the Inferior Oolite and Cornbrash.

Fig.

1 *a*. HOLECTYPUS DEPRESSUS, *Leske*, p. 260. Upper surface, natural size.

  b. The base, showing the mouth and anal openings, natural size.

  c. A side view of the same test, natural size.

  d. Ambulacral areas, poriferous zones, and inter-ambulacral plates, magnified three diameters.

  e. Primary tubercle from the upper surface, with its circle of areolar granules, magnified seven times.

  f. Primary tubercle from the base, magnified seven times.

  g. Under surface of another specimen, showing the jaws *in situ*, natural size.

  h. Portion of a primary spine, magnified eight times.

  i. The apical disc, magnified three and a half times.

2 *a*. HOLECTYPUS HEMISPHÆRICUS, *Agassiz*, p. 264. Upper surface, natural size.

  b. Under surface, natural size.

  c. Posterior view, showing the marginal anal opening.

  d. Ambulacral area, poriferous zones, and three inter-ambulacral plates, magnified three diameters.

  e. The apical disc, magnified three times.

  f. Lateral view, showing the greater length of the posterior half.

  g. A conical variety, showing the height of the pyriform marginal anal opening.

  h. The base of a large specimen.

Holectypus from the Coral Rag.

3 *a*. HOLECTYPUS OBLONGUS, *Wright*, p. 267. Upper surface, natural size.

  b. Base of a small specimen, natural size.

  c. Lateral view of *a*, natural size.

  d. Ambulacral area, poriferous zones, and two inter-ambulacral plates, magnified four diameters.

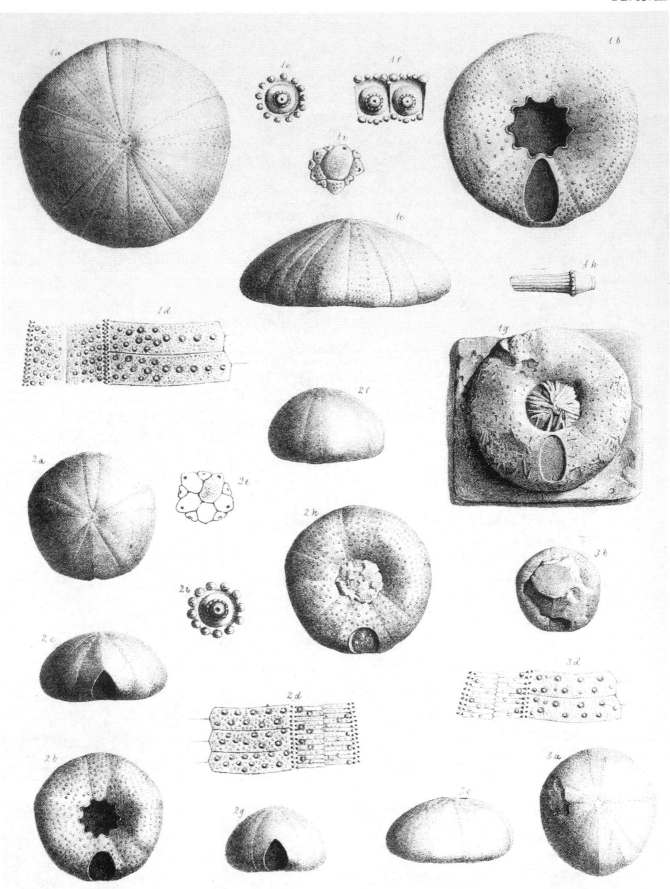

C. R. Bone, del. et lith.

Printed by Hullmandel & Walton.

PLATE XIX.

Pygasters from the Inferior Oolite.

Fig.

1 *a*. PYGASTER SEMISULCATUS, *Phillips*, p. 275.   Upper surface, natural size.

  b. Under surface, natural size.

  c. Lateral view of the same test.

  d. Ambulacral area, poriferous zones, and two inter-ambulacral plates, magnified three diameters.

  e. Primary tubercle, with the areolar circle of granules, magnified eight times.

  f. Primary tubercle of the ambulacral area, magnified eight times.

  g. Primary tubercle from the base, magnified eight times.

  h. Primary tubercle from the ambulacral area, magnified eight times.

2 *a*. PYGASTER CONOIDEUS, *Wright*, p. 278.   Upper surface, natural size.

  b. Under surface, natural size.

  c. Lateral view, natural size.

  d. Ambulacral area, poriferous zones, and two inter-ambulacral plates, magnified two diameters.

  e. Primary tubercles from the upper surface, magnified eight times.

  f. Primary tubercles from the base, magnified eight times.

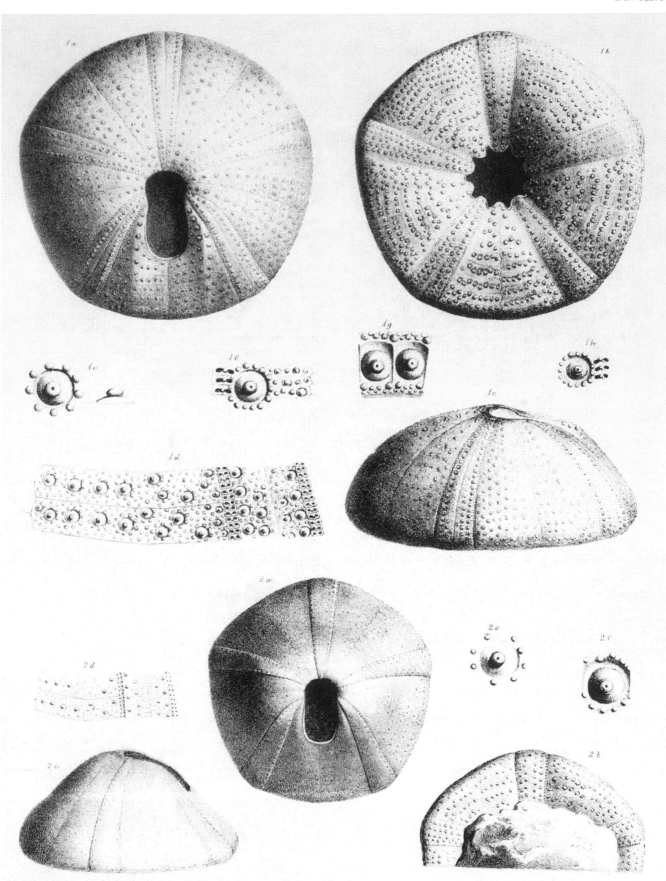

C. R. Bone, del. et lith.

Printed by Hullmandel & Walton.

PLATE XX.

Pygaster from the Cornbrash.

Fig.

1 *a.* Pygaster Morrisii, *Wright,* p. 280.   Upper surface, natural size.

  b. Under surface, natural size.

  c. Lateral view, natural size.

  d. Ambulacral area, poriferous zones, and inter-ambulacral plates, magnified four diameters.

  e. Primary tubercle from the upper surface, with areolar circle of granules, magnified eight times.

  f. Primary tubercle from the base, magnified eight times.

Pygaster from the Coral Rag.

2 *a.* Pygaster umbrella, *Agassiz,* p. 282.   Upper surface, natural size.

  b. Under surface of another specimen, natural size.

  c. Lateral view of *a.*

  d. Ambulacral area, poriferous zones, and inter-ambulacral plates, magnified three diameters.

  e. A portion of the apical disc, showing the four perforated ovarial plates and madreporiform body, magnified two and a half times.

  f. A primary tubercle, with its areolar circle of granules, magnified eight times.

  g. A primary tubercle from the base, magnified eight times.

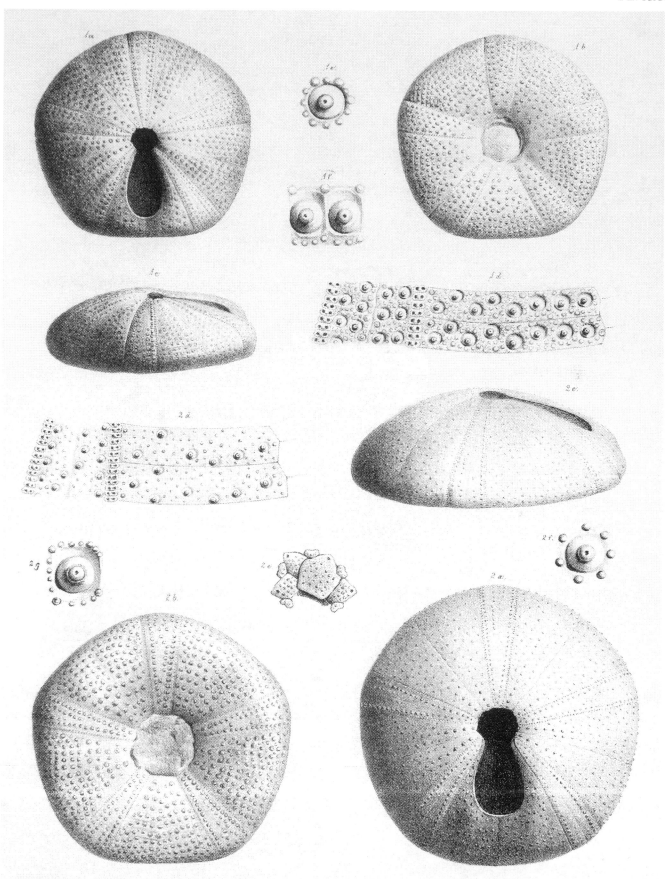

C. R. Bone, del. et lith.

Printed by Hullmandel & Walton.

PLATE XXI.

Hyboclypi from the Inferior Oolite.

Fig.

1 *a*. HYBOCLYPUS AGARICIFORMIS, *Forbes*, p. 292.   Upper surface, natural size.

   b. Under surface of the same test.

   c. Lateral view of the same.

   d. Ambulacral area, poriferous zones, and two inter-ambulacral plates, magnified two and a half times.

   e. Basal portion of an ambulacral area, showing the trigeminal pores near the peristome.

   f. Primary tubercle from the upper surface.

   g. Primary tubercle from the base, both highly magnified.

2 *a*. HYBOCLYPUS GIBBERULUS, *Agassiz*, p. 298.   Upper surface, natural size.

   b. Under surface of the same test.

   c. Lateral view of the same.

   d. Front view, showing the elevation of the gibbous crest, and the depression of the single inter-ambulacrum.

   e. The apical disc, magnified two and a half times.

   f. Ambulacral area, poriferous zones, and two inter-ambulacral plates, magnified two and a half times.

   g. Primary tubercle from the upper surface, surrounded by miliary granules, highly magnified.

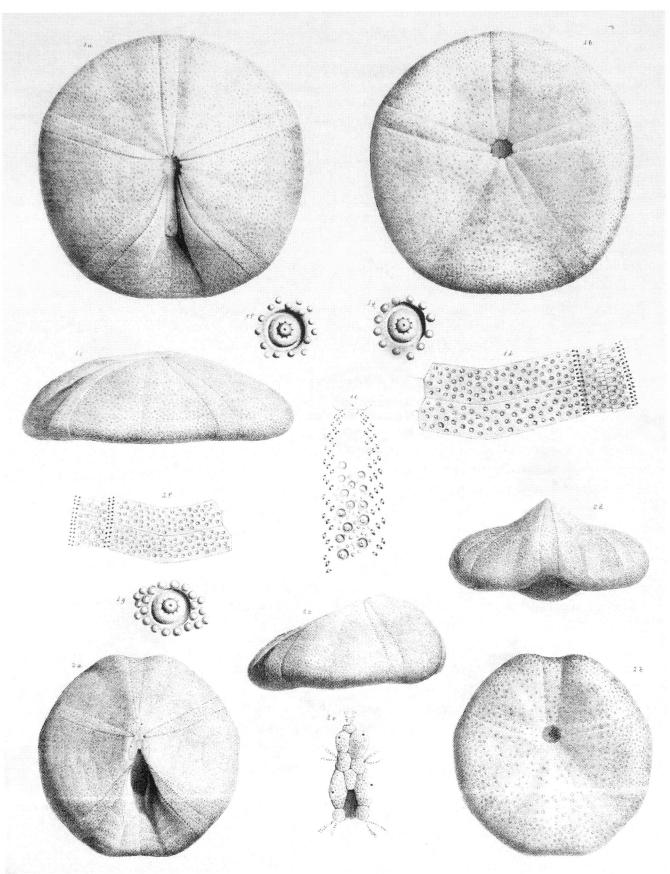

C. R. Bone, del. et lith.

Printed by Hullmandel & Walton.

PLATE XXII.

Hyboclypi from the Inferior Oolite.

Fig.

1 *a.* HYBOCLYPUS OVALIS, *Wright,* p. 301. Upper surface, natural size.

   b. Under surface of the same test.

   c. Lateral view of the same.

   d. Ambulacral area, poriferous zones, and two inter-ambulacral plates, magnified three diameters.

   e. The apical disc, magnified three diameters.

   f. Basal portion of an ambulacral area, showing the distance of the pores apart, in this region, from the size of the plates, magnified three times.

2 *a.* HYBOCLYPUS CAUDATUS, *Wright,* p. 296. Upper surface, natural size.

   b. Under surface, natural size.

   c. Lateral outline of the same.

   d. Upper surface of the common form of this urchin, with the disc.

   e. Base of the same test.

   f. Lateral view of a broad variety.

   i. Upper surface of the same test.

   k. Outline of a conical variety.

   g. Ambulacral area, poriferous zones, and inter-ambulacral plates, magnified three times.

   h. The apical disc of *d,* magnified four diameters.

Collyrites from the Inferior Oolite.

3 *a.* COLLYRITES RINGENS, *Agassiz,* p. 306. Upper surface, natural size.

   b. Under surface of the same test.

   c. Lateral view of the same.

   d. Posterior view, showing the position of the vent, and the arch formed by the posterior pair of ambulacra.

   e. Ambulacral area, poriferous zones, and inter-ambulacral plates, magnified three times.

   f. Ambulacral area, poriferous zones, and inter-ambulacral plates, magnified three times.

   g. Primary tubercles, and miliary granules, highly magnified.

   h. Oviductal holes, madreporiform body, and terminations of the three anterior ambulacra.

   i. Diagram of the structure of the test; the plates of the disc, and those above the vent not made out.

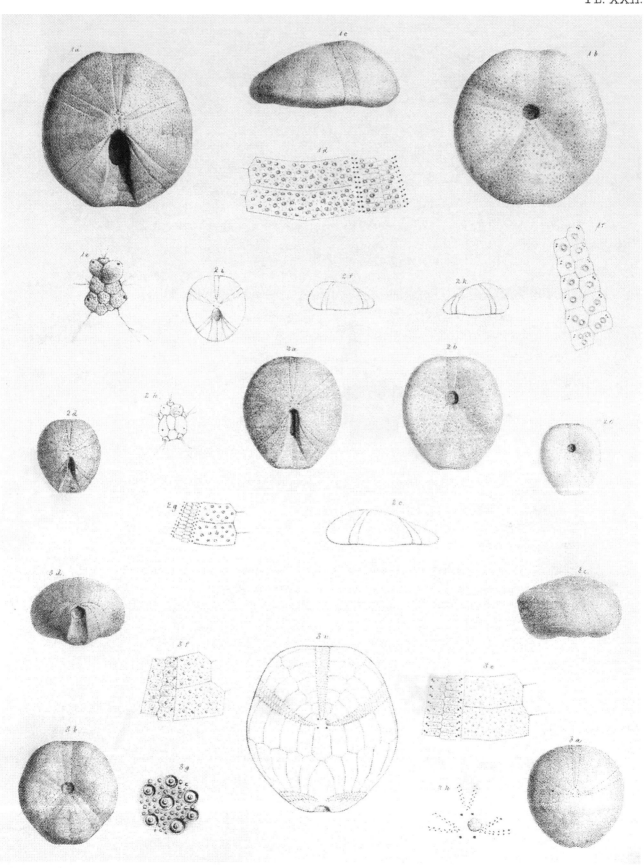

C. R. Bone, del. et lith.

Printed by Hullmandel & Walton.

PLATE XXIII.

COLLYRITES FROM THE INFERIOR OOLITE AND CORAL RAG.

Inferior Oolite Species.

1 *a.* COLLYRITES OVALIS, *Leske*, p. 314.   Upper surface, natural size.
 b. Under surface of the same test.
 c. Lateral view of ditto.
 d. Posterior view of ditto.
 e. Ambulacral areas, poriferous zones, and inter-ambulacral plates, magnified thrice.
 f. Primary tubercles and miliary granules, highly magnified.
 g. Anterior view of a smaller test.

Coral Rag Species.

2 *a.* COLLYRITES BICORDATA, *Leske*, p. 319.   Upper surface, natural size.
 b. Under surface of the same test.
 c. Lateral view of ditto.
 d. Posterior view of the same, showing the anal opening.
 e. Ambulacral area, poriferous zones, and inter-ambulacral plates, magnified thrice.
 f. Anterior pair of ovarial plates, and madreporiform body, with the single, and anterior pair of ocular plates, magnified four times.
 g. Primary tubercles, and miliary granules, of the dorsal plates, highly magnified.
 h. Primary tubercle, and hexagonal areolas from the base ;   ditto.

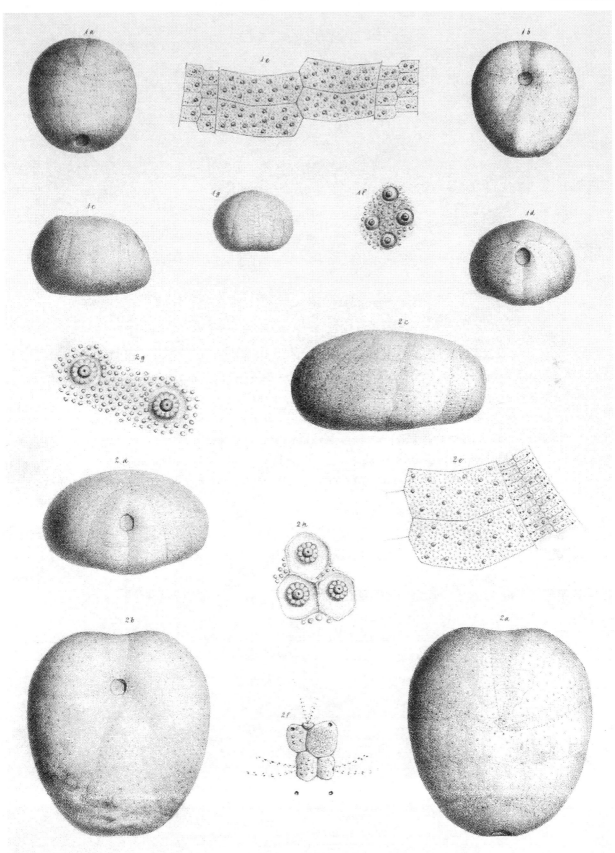

C. R. Bone, del. et lith.

Printed by Hullmandel & Walton.

PLATE XXIV.

ECHINOBRISSI FROM THE INFERIOR OOLITE, GREAT OOLITE, AND CORNBRASH.

Inferior Oolite and Cornbrash Species.

1 *a.* ECHINOBRISSUS CLUNICULARIS, *Llhwyd*, p. 332.   Upper surface, natural size.
   b. Under surface of the same test, natural size.
   c. Lateral view of ditto   ditto.
   d. Posterior view, showing the anal valley, and inter-ambulacrum.
   e. Ambulacral area, poriferous zones, and inter-ambulacral plates ; magnified thrice.
   f. Ambulacral area, and petaloid portion of the poriferous zones ; highly magnified.
   g. Apical disc, and madreporiform body ; magnified four times.
   h. Primary tubercles, and miliary granules, highly magnified.
   i. Diagram of the upper surface, showing the relative anatomy of the shell.
   k. The largest test of this species yet found ; from the Rev. A. W. Griesbach's cabinet ;
        natural size.

Great Oolite Species.

2 *a.* ECHINOBRISSUS WOODWARDI, *Wright*, p. 337.   Upper surface, natural size.
   b. Under surface of the same test, natural size.
   c. Lateral view of ditto, showing its tumid sides ; natural size.
   d. Ambulacral area, poriferous zones, and inter-ambulacral plates ; magnified four times.
   e. Upper surface of a large test.

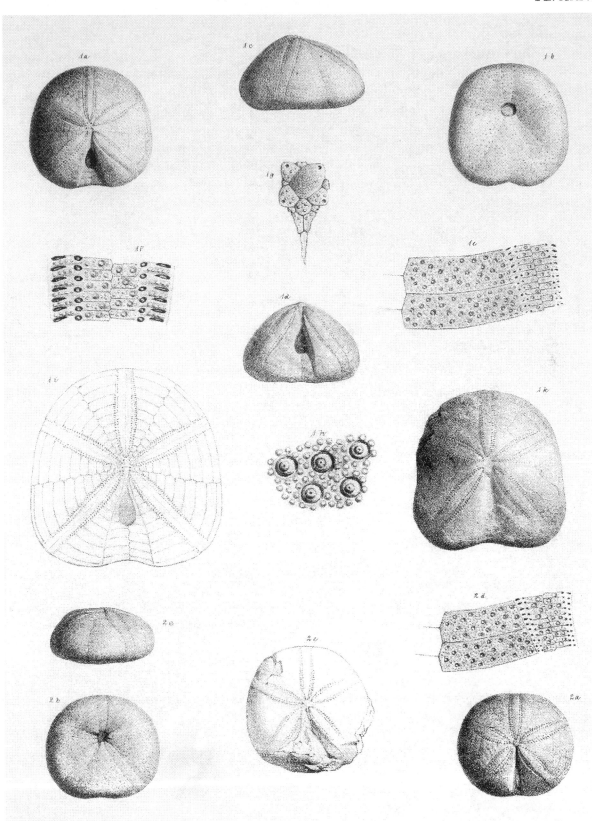

C. R. Bone, del. et lith.

Printed by Hullmandel & Walton.

PLATE XXV.

Echinobrissi from the Great Oolite, and Cornbrash.

Great Oolite Species.

1 *a*. ECHINOBRISSUS GRIESBACHII, *Wright*, p. 340.   Upper surface, natural size.
  b. Upper surface of the same, magnified twice.
  c. Under surface, natural size.
  d. Ditto, magnified twice.
  e. Lateral view of the test, natural size.
  f. Ambulacral area, poriferous zones, and inter-ambulacral plates, magnified four times.

Cornbrash Species.

2 *a*. ECHINOBRISSUS ORBICULARIS, *Phillips*, p. 341.   Upper surface, natural size.
  b. Under surface of the same test, natural size.
  c. Lateral view of ditto, showing the poriferous zones.
  d. Apical disc, with its internal complementary plates, magnified five times.
  e. Ambulacral area, poriferous zones, and inter-ambulacral plates, magnified thrice.
  f. Lateral view of a conoidal variety of this species, from Northamptonshire.
  g. Upper surface of another specimen, showing a greater eccentricity of the apical disc,
         a wider anal valley, and a larger vent.

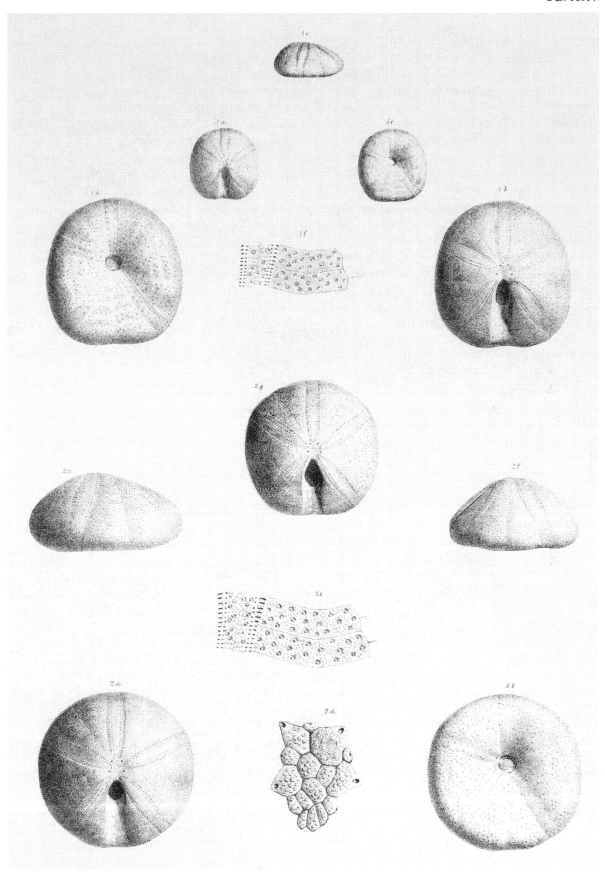

C. R. Bone, del. et lith.

Printed by Hullmandel & Walton.

PLATE XXVI.

ECHINOBRISSI FROM THE CORNBRASH, AND CORAL RAG.

Cornbrash Species.

1 *a.* ECHINOBRISSUS QUADRATUS, *Wright*, p. 344. Upper surface, natural size.
  b. Under surface of the same test, natural size.
  c. Lateral view of the same, showing the long dorsal slope of the posterior half.
  d. Ambulacral area, poriferous zones, and inter-ambulacral plates, magnified four times.

Coral Rag Species.

2 *a.* ECHINOBRISSUS SCUTATUS, *Lamarck*, p. 346. Upper surface, natural size.
  b. Under surface of the same test, natural size.
  c. Lateral view of the same test, showing the tumidity of its sides.
  d. Posterior view, showing the anal valley, and large vent.
  e. Apical disc, magnified more than four times.
  f. Ambulacral area, zones, and inter-ambulacral plates, magnified four times.

3 *a.* ECHINOBRISSUS DIMIDIATUS, *Phillips*, p. 350. Upper surface, natural size.
  b. Under surface of the same test, natural size.
  c. Lateral view of ditto, showing the great tumidity of its sides.
  d. Ambulacral area, zones, and inter-ambulacral plates, magnified four times.

4 *a.* ECHINOBRISSUS SCUTATUS, *Lamk.* A small round variety.
  b. Lateral view of this test.
  c. Ambulacral area, zones, and inter-ambulacral plates, magnified four times.

Portland Oolite Species.

ECHINOBRISSUS BRODIEI, *Wright*, p. 353. See Pl. XXXV, fig. 1, *a, b, c, d, e.*

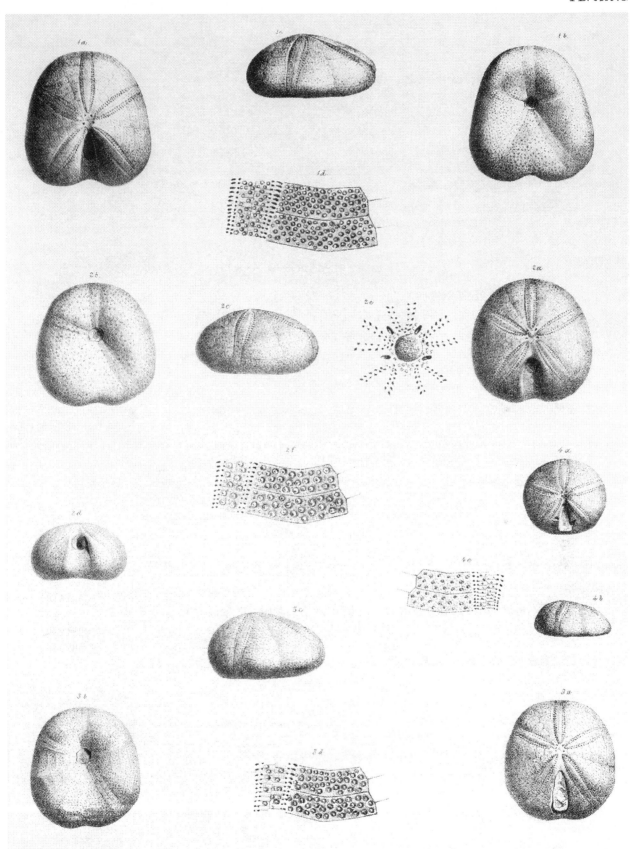

C. R. Bone, del. et lith.

Printed by Hullmandel & Walton.

PLATE XXVII.

CLYPEI FROM THE INFERIOR OOLITE.

1 *a.* CLYPEUS ALTUS, *M'Coy*, p. 366.   Upper surface, natural size.

  b. Under surface of the same test, natural size.

  c. Lateral view of ditto, showing the convexity of the dorsal surface, and the deep undulations of the border.

  d. Posterior view, showing the structure of the inter-ambulacrum ; the length, and narrowness of the anal valley, and the shape of the vent.

  e. Ambulacral area, zones, and inter-ambulacral plates, magnified four times.

  f. Apical disc, with the large madreporiform body, small ocular plates, and posterior complementary plates, magnified five times.

2 *a.* CLYPEUS MICHELINI, *Wright*, p. 369.   Upper surface, natural size.

  b. Lateral view of the same test, natural size.

  c. Ambulacral area, zones, and inter-ambulacral plates, magnified five times.

    This is a rare urchin, on a slab of Inferior Oolite limestone, from Whitwell, Yorkshire, where it was found associated with *Stomechinus germinans*, Phillips.

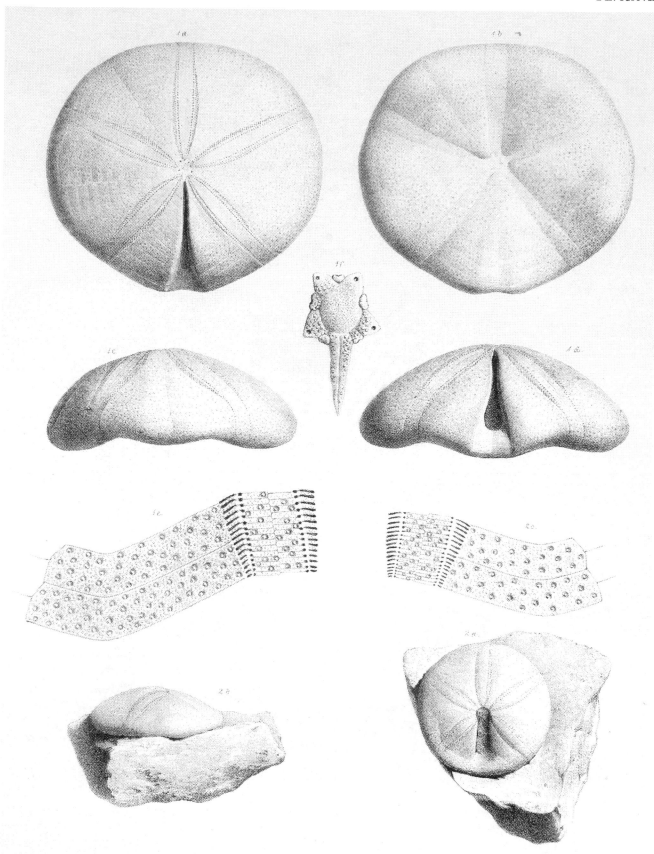

C. R. Bone, del. et lith.

Printed by Hullmandel & Walton.

PLATE XXVIII.

Clypeus Plotii, *Klein*.

From the Inferior Oolite.

1 *a*. Clypeus Plotii, *Klein*, p. 360.   Upper surface, natural size.

  b. Under surface of the same test, natural size.

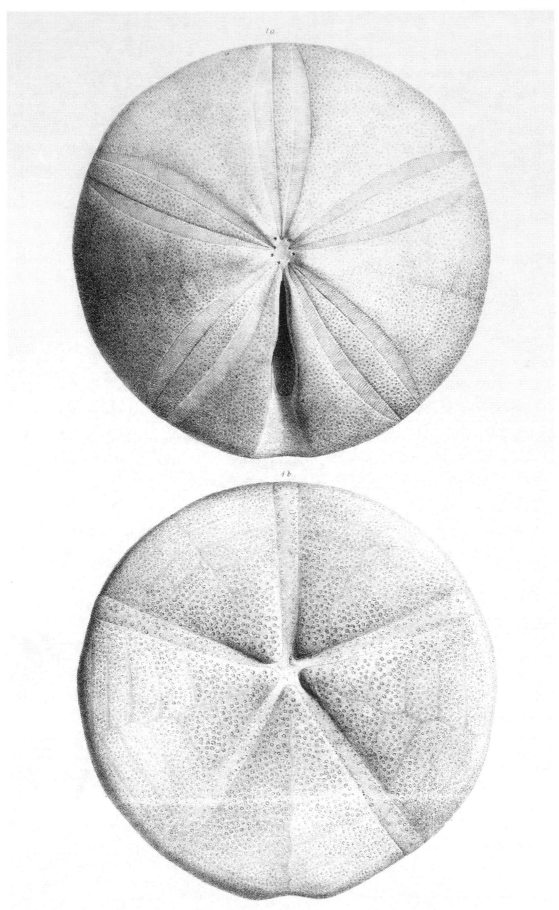

C. R. Bone, del. et lith.

Printed by Hullmandel & Walton.

PLATE XXIX.

Clypeus Plotii, *Klein*.

From the Inferior Oolite.

1 *a*. Clypeus Plotii, *Klein*, p. 360. Posterior view, showing the anal valley, oblong vent, and postero-lateral ambulacra.

  b. Lateral view of the same test.

  c. Ambulacral area, zones, and inter-ambulacral plates, magnified two-and-a-half times.

  d. Apical disc, magnified two-and-a-half times.

  e. Primary tubercles, and miliary granules, from the upper surface, highly magnified.

  f Ditto, from the base, ditto   ditto.

  g. Ambulacral area, and poriferous zones, at the base; magnified four times.

  h. Mouth-opening, peristome, and oral lobes; magnified two-and-a-half times.

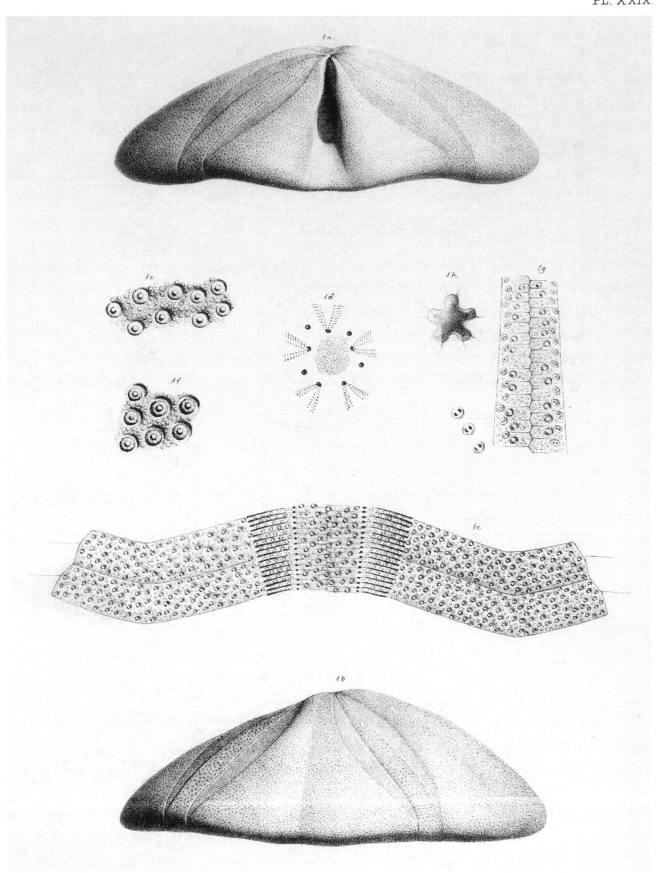

C. R. Bone, del. et lith.

Printed by Hullmandel & Walton.

PLATE XXX.

Clypei from the Inferior Oolite.

1 *a*. Clypeus Hugii. *Agassiz,* p. 375, upper surface, natural size.

  b. Under surface of the same test, natural size.

  c. Lateral view of the same test,   ditto.

  d. Posterior view of the same, showing the shortness of the anal valley, and the size of the vent.

  e. Ambulacral area, zones, and inter-ambulacral plates, magnified thrice.

  f. Apical disc, madreporiform body, and summits of the ambulacra, magnified.

2 *a*. Clypeus Michelini, *Wright,* p. 369.   Upper surface, natural size.

  b. Under surface of another test, natural size.

  c. Lateral view of *a*, showing the obliquity of the posterior half of the dorsal surface.

  d. Ambulacral area, zones, and inter-ambulacral plates, magnified thrice.

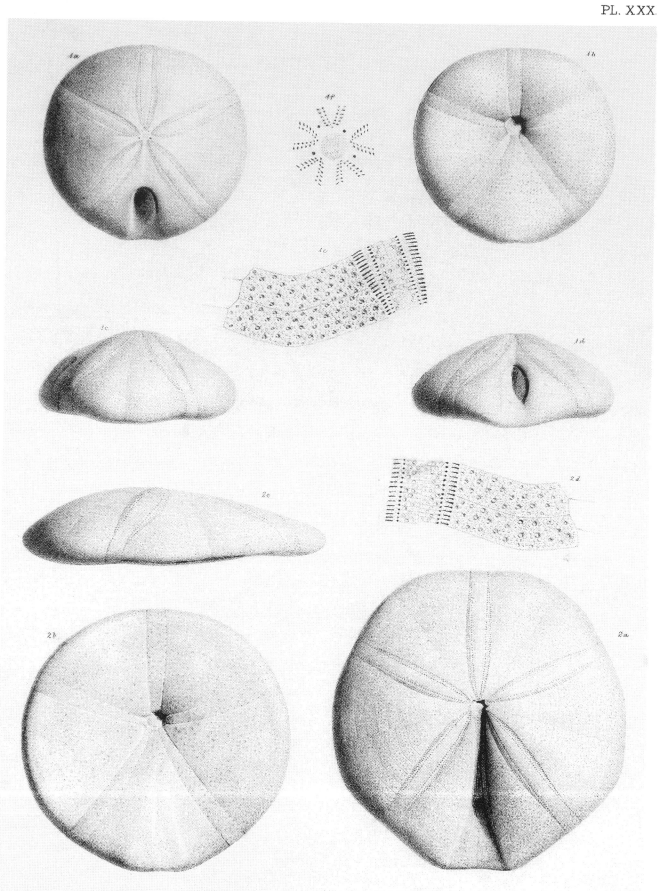

C. R. Bone, del. et lith.

Printed by Hullmandel & Walton.

PLATE XXXI.

Clypeus Agassizii, *Wright.*

From the Inferior Oolite.

1 *a*. Clypeus Agassizii, *Wright*, p. 378.   Upper surface, natural size.
   b. Under surface of the same test, natural size.

C. R. Bone, del. et lith.

Printed by Hullmandel & Walton.

PLATE XXXII.

CLYPEUS AGASSIZII, *Wright.*

From the Inferior Oolite.

1 *a.* CLYPEUS AGASSIZII, *Wright,* p. 378.   Lateral view, natural size.

   b. Posterior view, showing the shallow anal valley, and superficial oblong vent.

   c. Ambulacral area, poriferous zones, and inter-ambulacral plates, magnified twice and a
       half times.

   d. Portion of the basal ambulacral area near the mouth, showing the arrangement of
       the pores in triple oblique ranks, magnified four times.

   e. Tubercles from the base, with miliary granules around them, magnified.

   f. Tubercles from the upper surface, showing the regular disposition of the granules
       around the areolas, and on the intermediate surface.

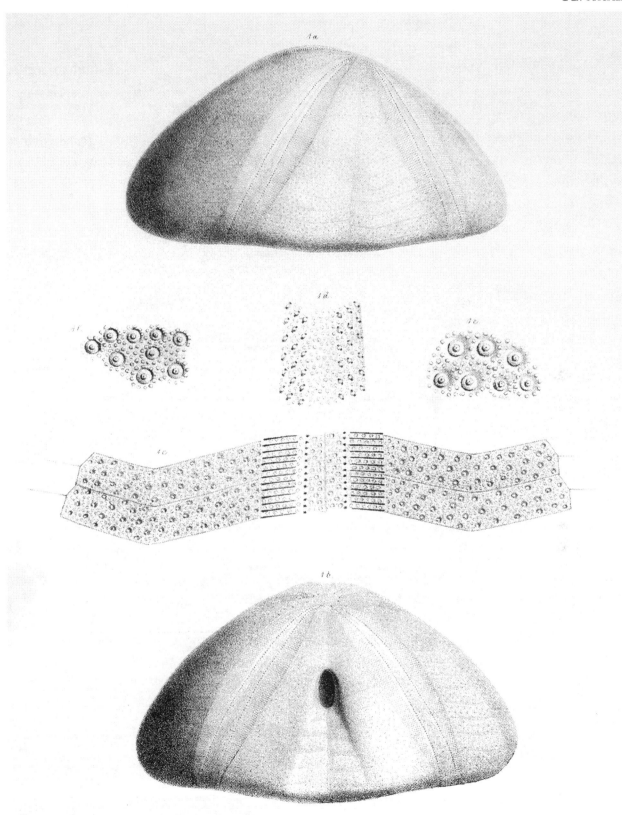

C. R. Bone, del. et lith.

Printed by Hullmandel & Walton.

PLATE XXXIII.

CLYPEUS MÜLLERI, *Wright*.

From the Great Oolite.

1. CLYPEUS MÜLLERI, *Wright*, p. 371. Upper surface of a large specimen lying on a slab of forest marble, natural size. From Mr. Frederick Bravender's Collection.

2 *a*. Upper surface of a medium sized specimen from the Great Oolite, natural size.
   b. Under surface of the same test, showing the central mouth-opening, natural size.
   c. Lateral view of the same test, showing the shelving character of the posterior half of the upper surface.

3 *a*. Common variety of the species, having the apical disc very ex-central posteriorly, natural size.
   b. Under surface of the same test, showing the mouth-opening and peristome, ditto.

4. Apical disc, showing the two pairs of large, triangular genital plates, the small oculars, the large madreporiform body, and single imperforate genital plate, magnified about four times.
5. Ambulacral area, poriferous zones, and inter-ambulacral plates, magnified four times.
6. Portion of a poriferous zone, showing the inner round holes, the external slit-like apertures, and the tubercles on the septa, highly magnified.

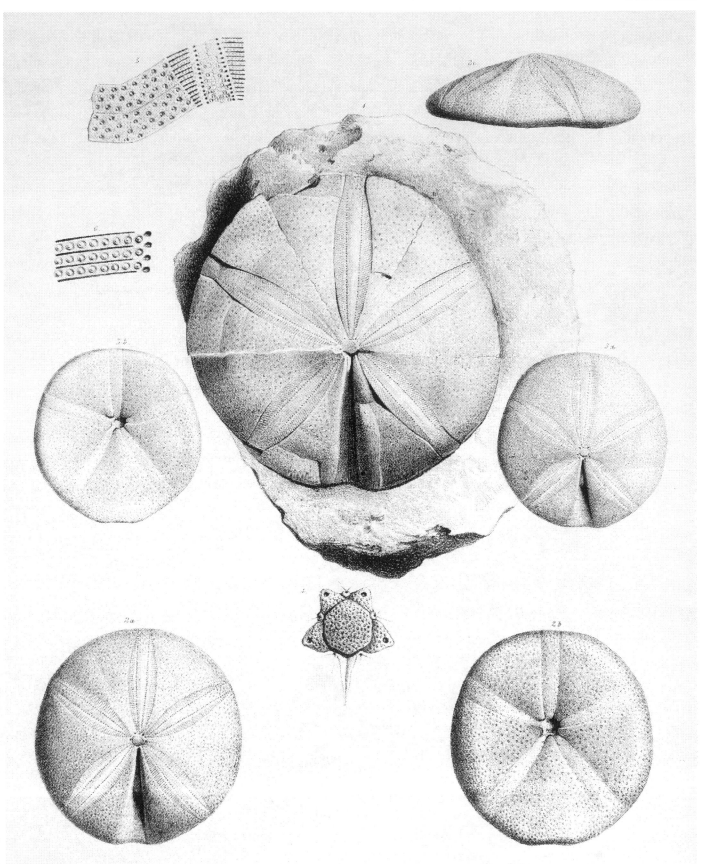

C. R. Bone, del. et lith.

Printed by Hullmandel & Walton.

PLATE XXXIV.

CLYPEUS SUBULATUS, *Young and Bird.*

From the Coralline Oolite.

1 *a.* CLYPEUS SUBULATUS, *Young and Bird*, p. 382.   Upper surface, natural size.

  b. Under surface of the test *c*, slightly enlarged.

  c. Lateral view of another test, showing the border, natural size.

  d. Posterior view of the same test, showing the short anal valley and vent.

  e. Ambulacral area, poriferous zones, and inter-ambulacral plates, magnified thrice.

  f. Tubercles and miliary granules from the upper surface, magnified.

  g. Apical disc, showing the madreporiform tubercle covering the other elements of the disc.

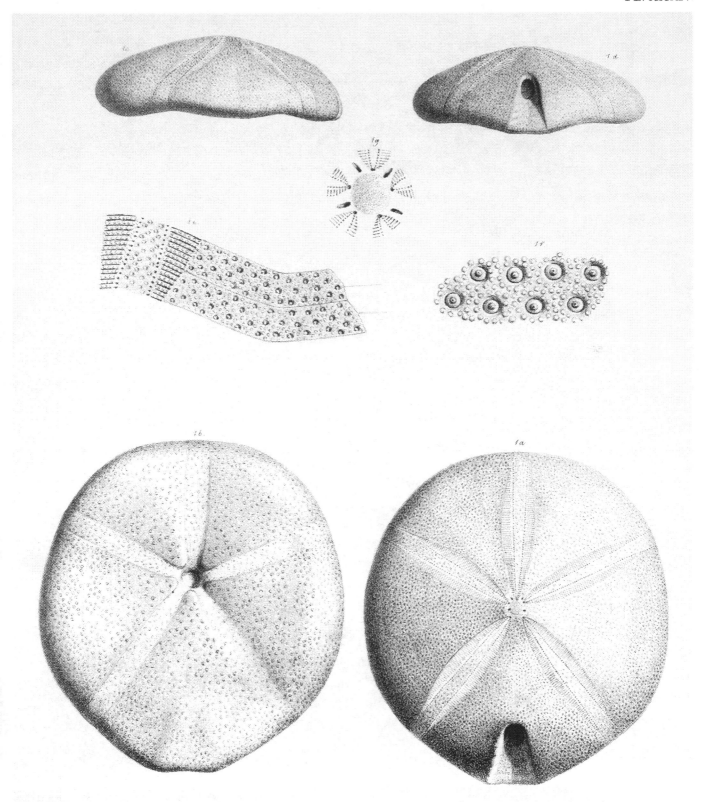

C. R. Bone, del. et lith.

Printed by Hullmandel & Walton.

PLATE XXXV.

ECHINOBRISSUS BRODIEI, *Wright.*

From the Portland Oolite.

1 *a.* ECHINOBRISSUS BRODIEI, *Wright*, p. 353.   Upper surface, natural size.
   b. Under surface of the same test, natural size.
   c. Lateral view of the same test, ditto.
   d. Posterior view of the same, showing the short anal valley, and large vent.
   e. Basal tubercles, with hexagonal areolas, highly magnified.

PYGURUS MICHELINI, *Cotteau.*

From the Cornbrash.

2 *a.* PYGURUS MICHELINI, *Cotteau*, p. 392.   Upper surface, natural size.
   b. Under surface of the same test, showing the vent and its areola, natural size.
   c. Posterior view of the same, showing the relative position of the vent, and border.
   d. Lateral view of the same, showing the excentral position of the apical disc, and the shelving character of the posterior half of the upper surface.
   e. Ambulacral area, poriferous zones, and ambulacral plates, magnified twice and a half times.
   f. Apical disc, magnified four times.

3. Apical disc of the same species, copied from M. Cotteau's work.   This figure shows the relative anatomy of the discal elements better than in our best specimen, and exhibits the magnitude of the large madreporiform body, which covers nearly all the other plates.

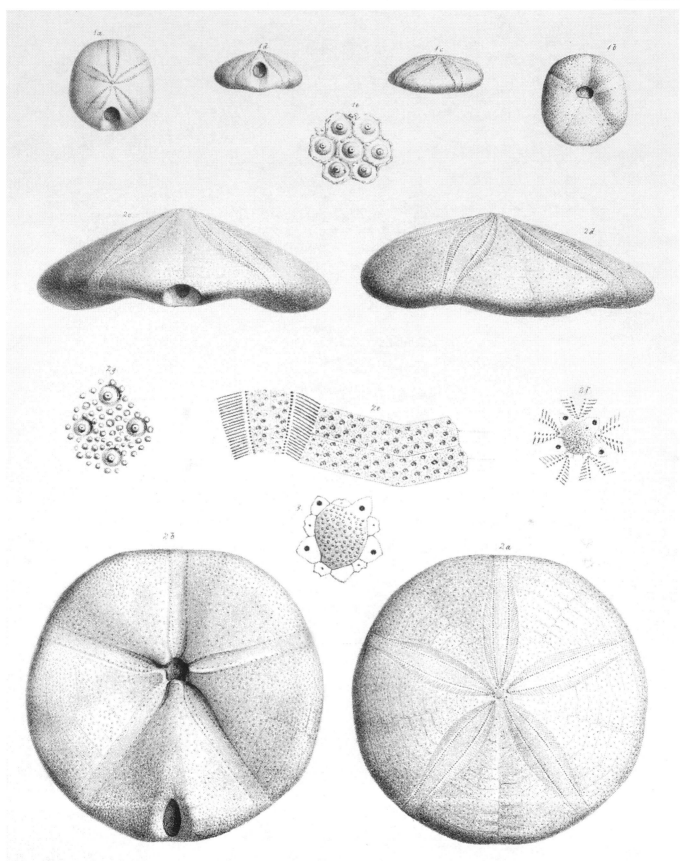

C. R. Bone, del. et lith.

Printed by Hullmandel & Walton.

PLATE XXXVI.

PYGURUS PENTAGONALIS, *Phillips.*

From the Coralline Oolite.

1 *a.* PYGURUS PENTAGONALIS, *Phillips*, p. 394.   Upper surface, natural size.

  b. Under surface of the same test, natural size.

  c. Lateral view of the same test, ditto.

  d. Apical disc with madreporiform tubercle, magnified.

2 *a.* PYGURUS PENTAGONALIS, *Phillips.*   A large specimen from the Coralline Oolite of Malton imbedded in a fragment of that rock.   This beautiful fossil belongs to the Scarborough Museum.

  b. Ambulacral area, poriferous zones, and inter-ambulacral plates, magnified twice and a half times.

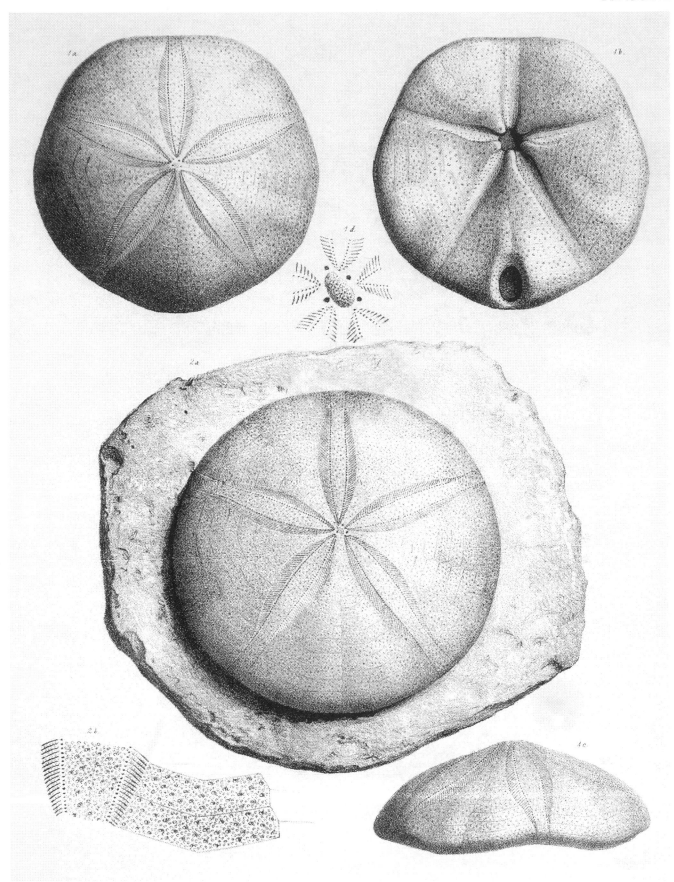

C. R. Bone, del. et lith.

Printed by Hullmandel & Walton.

PLATE XXXVII.

Pygurus Costatus, *Wright.*

From the Coralline Oolite.

1 *a*. Pygurus Costatus, *Wright*, p. 397.   Upper surface, natural size.

  b. Under surface of the same test, natural size.

  c. Posterior view, showing the flatness of the base and the extreme depression of the upper surface.

  d. Two inter-ambulacral plates, poriferous zones, and ambulacral area, magnified three times.

  e. Phylloidal expansion of the ambulacral area near the mouth-opening, showing the crowding together of the pores in this region, magnified three times.

  f. One of the lip-like processes of the mammillated, oval lobes which project over the border of the peristome, magnified twice.

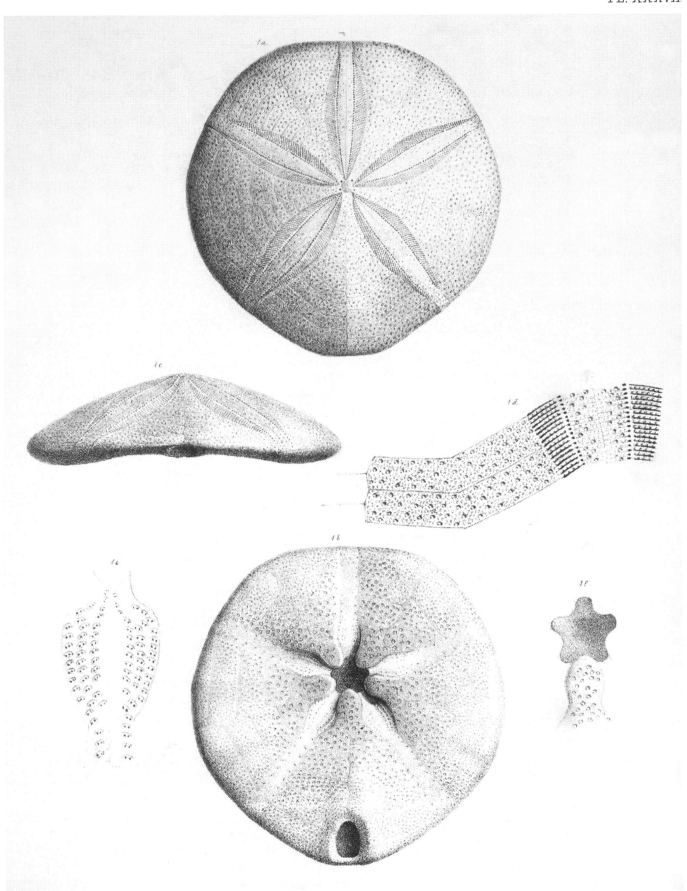

C. R. Bone, del. et lith.

Printed by Hullmandel & Walton.

PLATE XXXVIII.

Pygurus Blumenbachii, *Koch and Dunker.*

From the Coralline Oolite.

1 *a.* Pygurus Blumenbachii, *Koch and Dunker*, p. 400. Upper surface, natural size.

   b. Under surface of the same test, natural size.

   c. Posterior view of ditto, showing the undulations of the base and the elevation of the dorsal surface.

   d. Lateral view of ditto, showing the great elevation of the anterior half of the test, and the eccentricity of the vertex.

   e. Tubercles from the base, with their hexagonal areas.

2 *a.* Upper surface of another specimen in my collection.

   b. Two inter-ambulacral plates, a portion of the poriferous zones, and ambulacral area, magnified four times.

   c. Apical disc and madreporiform tubercle, magnified.

   d. Tubercles from the upper surface, greatly magnified.

3.   Apical disc from a French specimen after M. Cotteau, showing the size and arrangement of the plates.

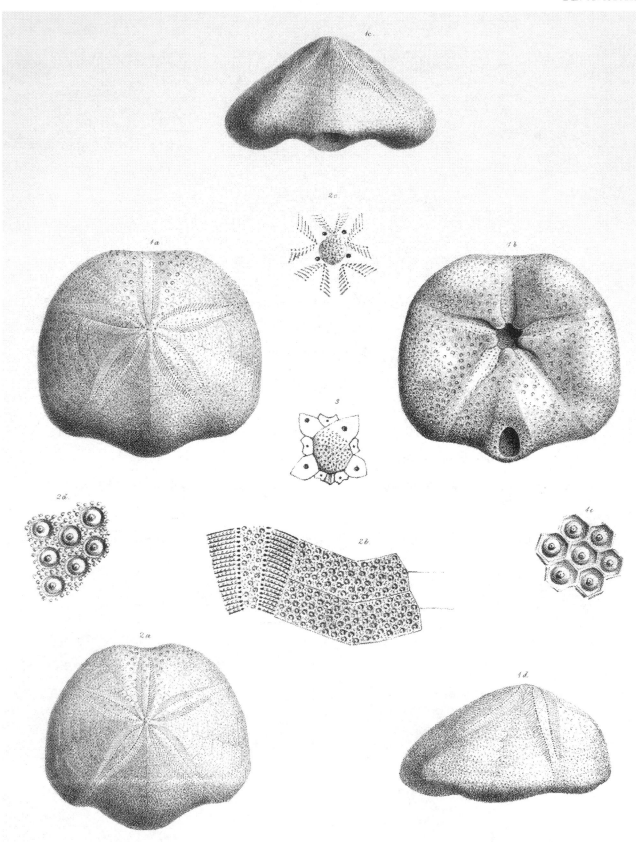

C. R. Bone, del. et lith.

Printed by Hullmandel & Walton.

PLATE XXXIX.

PYGURUS PHILLIPSII, *Wright.*

From the Coralline Oolite.

1 *a.* PYGURUS PHILLIPSII, *Wright*, p. 403.   Upper surface, natural size.

  b. Posterior view of the same test, showing the depression of the upper surface and the undulation of the base.

  c. Two inter-ambulacral plates, a portion of the poriferous zones, and ambulacral area, magnified three times.

  d. A portion of an ambulacral area, with its poriferous zones, magnified eight times.

2.   PYGURUS HAUSMANNI, *Koch and Dunker*, p. 405.   Two of the inter-ambulacral plates, a portion of the poriferous zones, and ambulacral area, magnified three times.

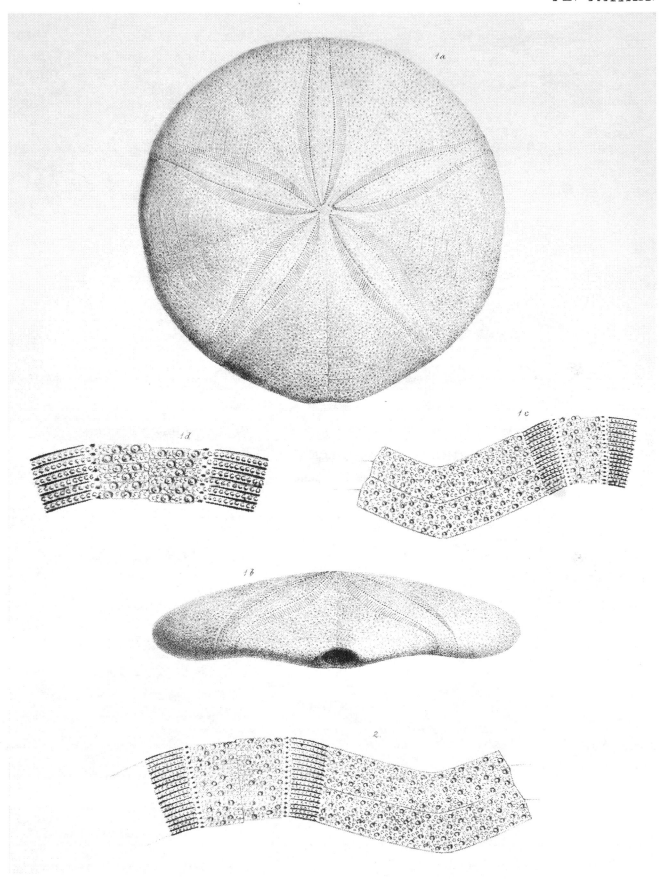

C. R. Bone, del. et lith.

Printed by Hullmandel & Walton.

PLATE XL.

Pygurus Hausmanni, *Koch and Dunker.*

From the Coralline Oolite.

Pygurus Hausmanni, *Koch and Dunker*, p. 405.   The upper surface restored, natural size.   Most of the adult specimens of this largest British Oolitic urchin are fractured and distorted, but some of the smaller individuals show the true outline of the test.

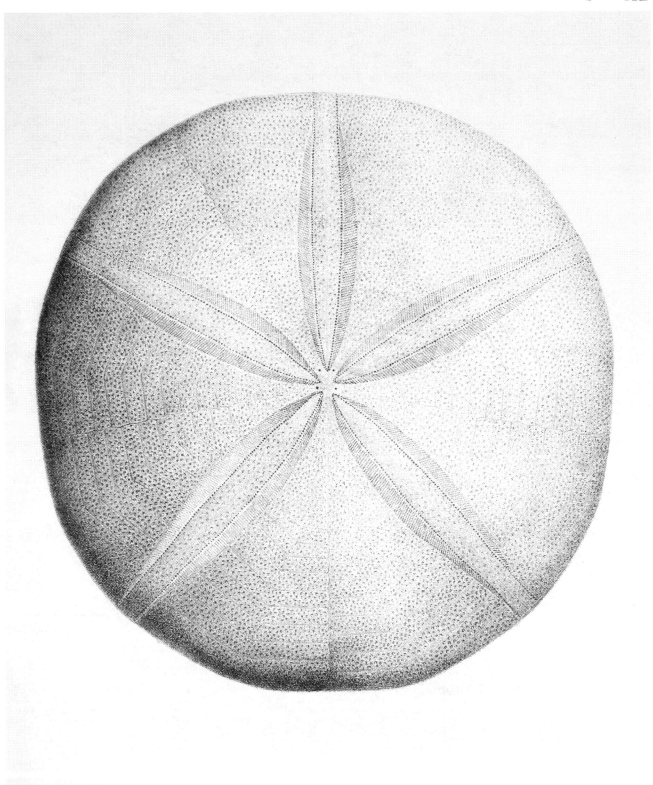

C. R. Bone, del. et lith.

Printed by Hullmandel & Walton.

PLATE XLI.

1 *a.* NUCLEOLITES RECENS, *Edwards.* The living representative of this group from Australia, showing the test and spines copied from Professor Milne-Edwards' figure in the illustrated edition of Cuvier's 'Règne Animal Zoophytes,' tab. xiv, fig. 3.

  b. The anal aperture, and its surrounding anal plates.

2 *a.* PEDINA SMITHII, *Forbes,* p. 176. Upper surface of the test, natural size.

  b. Inter-ambulacral plate, portion of the poriferous zones, and ambulacral area, magnified three times.

  c. Three pair of pores, greatly magnified.

3 *a.* PSEUDODIADEMA LOBATUM, *Wright.* Fragments of tests, with spines, on a slab of Lower Lias shale from Pinhay Bay.

  b. Spine of the same, magnified.

4 *a.* PYGASTER MEGASTOMA, *Wright,* n. sp. Upper surface, natural size.

  b. Under surface of the same test, natural size.

  c. Lateral view of ditto, ditto.

5 *a.* Two inter-ambulacral plates, poriferous zones, and ambulacral areas, of another specimen of the same species.

  b. Apical disc of a small specimen of *Pygaster megastoma,* Wr.

6. Large spine of DIPLOCIDARIS WRIGHTII, *Desor.*

7. Spine of CIDARIS PUSTULATA, *Wright,* with a portion greatly magnified.

8. Spine of CIDARIS.

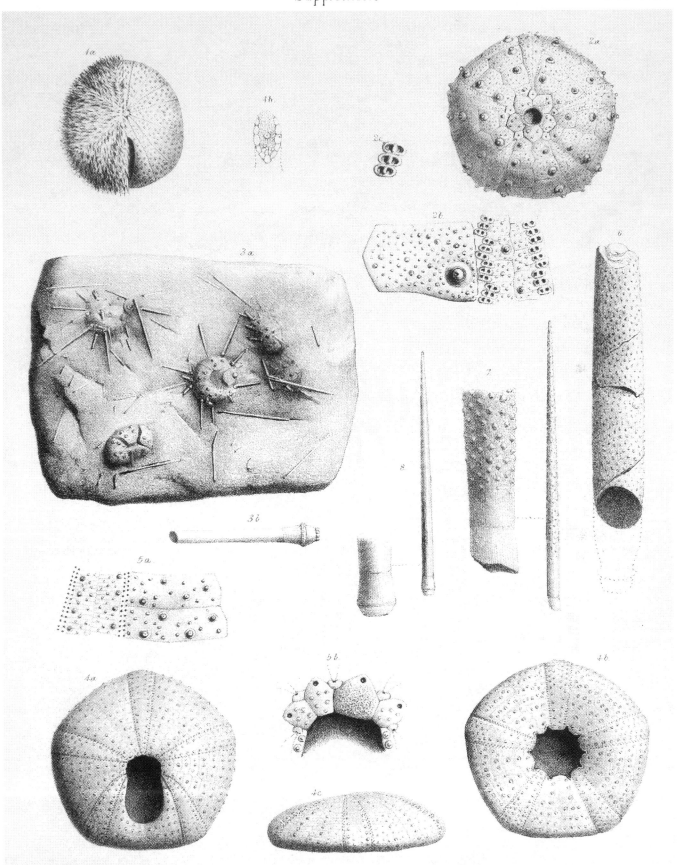

C. R. Bone, del. et lith.   Printed by Hullmandel & Walton.

SUPPLEMENT.

PLATE XLII.

1 *a.* CIDARIS FOWLERI, *Wright,* p. 32.   Test, with spines attached, of the natural size.

   b. Inter-ambulacral plate, zones, and ambulacral area, magnified.

   c. Primary tubercle, magnified.

   d. One of the jaws, magnified.

   e. One of the primary spines, magnified twice.

   f. One of the secondary spines, magnified three times.

2 *a.* HYBOCLYPUS AGARICIFORMIS, *Forbes.*   Upper surface, natural size, showing the apical disc *in situ.*

   b. Under surface of the same, natural size.

3.   The apical disc of the same urchin, magnified twice.

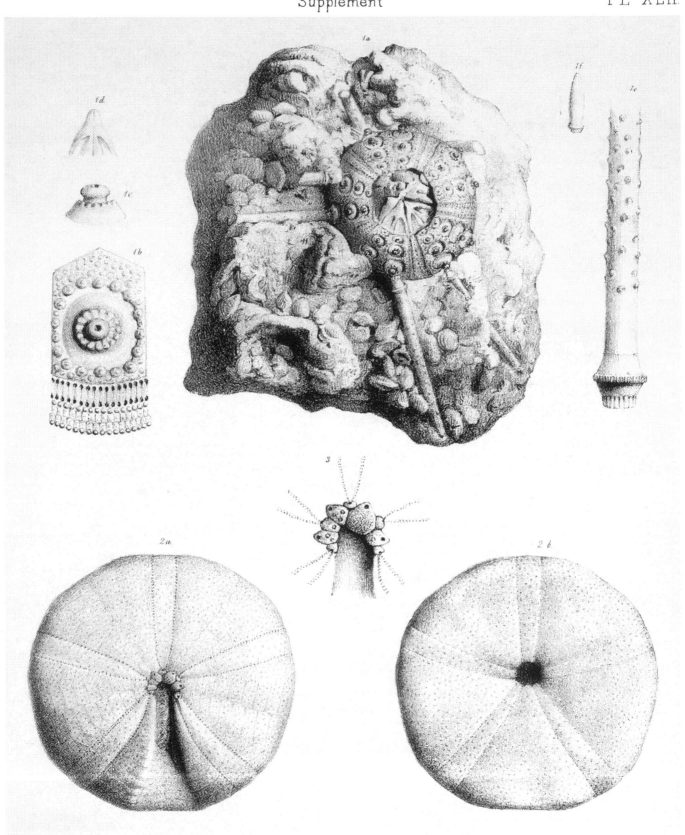

C. R. Bone, del. et lith.

Printed by Hullmandel & Walton.

PLATE XLIII.

1 *a.* PEDINA SMITHII, *Forbes*, p. 176.   Upper surface, natural size.

  b. Under surface, natural size, showing the mouth-opening.

  c. Lateral view of the same test, natural size.

  d. Base of an ambulacral area, showing the crowding of the pores in the zones, magnified four diameters.

2 *a.* HEMICIDARIS BRILLENSIS, *Wright* (Supplement).   Upper surface, natural size.

  b. Lateral view of the same test, natural size.

  c. Two inter-ambulacral plates, a portion of the ambulacral area, and two poriferous zones, magnified three and a half times.

  d. Apical disc, magnified two diameters.

3.   ECHINOBRISSUS BRODIEI, *Wright*, p. 353.   Upper surface, natural size.

4. *a.* CLYPEUS PLOTII, *Klein*, p. 361.   A weathered portion of the upper surface of a test of this species, showing perforated tubercles on all the inter-ambulacral and ambulacral plates, with the madreporiform body extending over all the pieces of the apical disc, magnified three diameters.

  b. CLYPEUS PLOTII, *Klein*, a portion of the upper surface of the test, highly magnified, to show the perforations of the tubercles.

5 *a.* HETEROCIDARIS WICKENSE, *Wright* (Supplement).   Four rows of plates, natural size.

  b. One of the plates and primary tubercles, magnified three times.

  c. A lateral view of one of the primary tubercles, highly magnified.

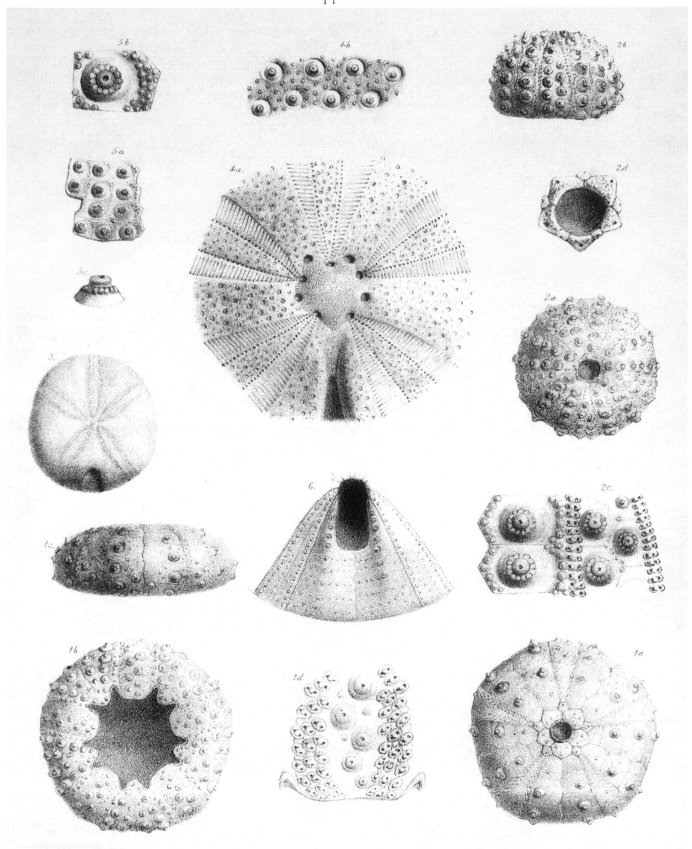

C. R. Bone, del. et lith.

Printed by Hullmandel & Walton.